BASIC TRANSPORT PHENOMENA
IN BIOMEDICAL ENGINEERING

BASIC TRANSPORT PHENOMENA IN BIOMEDICAL ENGINEERING

Second Edition

Ronald L. Fournier, PhD, PE

Professor of Bioengineering
The University of Toledo

Taylor & Francis
Taylor & Francis Group

New York • London

Vice President	Denise Schanck
Senior Editor	Robert L. Rogers
Associate Editor	Summers Scholl
Senior Publisher	Jackie Harbor
Production Editor	Karin Henderson
Copyeditor	Annette Musker
Cover Designer	Aktiv
Indexer	Ronald Fournier
Typesetter	Phoenix Photosetting, Chatham, Kent
Printer	Sheridan Books, Inc., Ann Arbor

Library of Congress Cataloging-in-Publication Data

Fournier, Ronald L.
  Basic transport phenomena in biomedical engineering / Ronald L. Fournier.—2nd ed.
      p.   ;   cm.
  Includes bibliographical references and index.
  ISBN 1–59169–026–9 (alk. paper)
 1.   Biological transport. 2.   Biomedical engineering. 3.   Biotechnology.
  [DNLM: 1.   Biological Transport. 2.   Biomedical Engineering. 3.   Blood—metabolism.
4.   Cell Membrane—metabolism.   QU 120 F778b 2006]   I. Title.
  R857.B52F68 2006
  571.6′4—dc22

                                                                                    2006004326

Published in 2007 by Taylor & Francis Group, LLC,
270 Madison Avenue, New York, NY 10016, USA and
2 Park Square, Milton Park, Abingdon, Oxon, OX14 4RN, UK.

Printed in the United States of America on acid-free paper.

10   9   8   7   6   5   4   3   2   1

Taylor & Francis Group, an informa business   Visit our web site at http://www.garlandscience.com

To my wife Lynn, and our sons Joshua Adam and Ryan Michael

CONTENTS

PREFACE

The challenge in presenting a text on biomedical engineering is the fact that it is a very broad field that encompasses a variety of different engineering disciplines and life sciences. There are many recent edited works that provide specialized treatments and summaries of current research in various areas of biomedical engineering. These works tend to be written more for the sophisticated biomedical researcher than the student entering the field perhaps for the first time. There is therefore a need for an entry level book that introduces some of the basic concepts. This book is designed to meet this goal. This work brings together fundamental engineering and life science principles to provide a focused coverage of key transport concepts in biomedical engineering.

In accomplishing this task within a finite volume, the coverage was limited to those areas that emphasize chemical and physical transport processes with applications towards the development of artificial organs, bioartificial organs, controlled drug delivery systems, and tissue engineering. With this focus, the book first covers in a new chapter introductory material to provide the student with a background and review of units and dimensions, some tips for solving engineering problems, and a discussion on material balances. This is then followed by another new chapter that reviews thermodynamic concepts with an emphasis on solutions. Following this the remaining chapters focus on such topics as the body fluids, the physical and flow properties of blood, capillary solute transport, oxygen transport, and pharmacokinetic analysis. New material has also been added to these chapters related to fluid mechanics, diffusion, and mass transfer in boundary layers. This provides a more theoretical basis for understanding mass transfer processes and mass transfer coefficients. This is followed by application of these principles to extracorporeal devices and the relatively new areas of tissue engineering and bioartificial organs. Throughout the book considerable emphasis is placed on developing a quantitative understanding of the underlying physical, chemical, and biological phenomena. Therefore, mathematical models are developed using the conceptually simple "shell balance" or compartmental approaches. Numerous examples are presented based on these mathematical models and they are compared in many cases with experimental data. In this way the student should be able to gain confidence in the development of mathematical models for relatively simple systems, and then be able to apply these concepts to a wide variety of problems of even greater complexity.

Since there are a variety of mathematical software packages available for solving engineering problems, the examples are no longer worked out with Mathcad. However, key results to the examples are shown and the student is encouraged to rework these examples using the mathematical software package of their choice. Working through the examples and the end of chapter problems using mathematical software packages will develop skill and expertise in engineering problem solving while at the same time making the mathematics less formidable. These mathematical software packages also provide the student with the opportunity to explore various aspects of the solution on their own, or apply these techniques as starting points for the solution to their own problems. In addition, a section on units conversions and the notation used in the book are also included.

I would also like to take this opportunity to thank the following students that have helped me during the preparation of the second edition of this book: Jennifer Bergstrom, Kanak Chatterjee and Carrie Fihe. Also many thanks to Linda Ruiz for her assistance in many tasks related to the preparation of this text. I also would like to thank Michael King, University of Rochester; Guillermo Aguilar, UC Riverside; Harold Knickle, University of Rhode Island; George Pins, Worcester Polytechnic Institute; H. Wally Wu, Texas A&M University; and Michael Deem, Rice University; for their constructive criticism of the first edition and suggestions for the second edition. Support for an earlier version of this work from the University of Toledo Center for Teaching Excellence is also greatly appreciated.

It is hoped that this book is timely and useful to engineers and researchers in the biomedical community, as well as to students in chemical engineering, mechanical engineering, biotechnology, bioengineering, and the life sciences.

Ronald L. Fournier

NOTATION

a	Molecular radius
a_f	Radius of a macromolecule
A	Helmholtz free energy defined as $U - TS$
A	Total amount of a drug in the body
A_E	Total amount of a drug in the extracellular fluid
A_P	Total amount of a drug in the plasma
A_R	Total amount of a drug everywhere else in the body
A_P	Total cross-sectional area of pores in a membrane
A_S	Surface area
A_{xs}	Cross-sectional area of a reactor
$\text{AUC}^{0 \to \infty}$	Area under the concentration time curve
B, C, D and B', C', D'	Virial coefficients
c	Weight concentration
C	Concentration
C	Total concentration of drug in the apparent distribution volume
C_b	Concentration in blood
C_d	Concentration in exchange fluid
C_g	Concentration in the gas
C_G	Concentration of glucose
C_{CAPD}	Concentration of a solute in the CAPD exchange fluid
C_E	Total concentration of drug in the extracellular fluid
C_{EB}	Concentration of bound drug in the extracellular fluid
$C_{\max}$	Maximum concentration
C_{plasma}	Concentration in plasma
C_{tissue}	Concentration in tissue
C_R	Total concentration of drug everywhere else in the body
C_{SS}	Steady state concentration
C_U	Unbound concentration of a drug

C'	Concentration of oxy-hemoglobin
C'_{SAT}	Concentration of oxygen saturated hemoglobin
C'	Concentration in the tissue
CL	Clearance
CL_{CAPD}	Clearance of continuous ambulatory peritoneal dialysis
CL_D	Clearance of dialysis
CL_{plasma}	Plasma clearance
CL_{renal}	Clearance due to the kidneys
C_P	Heat capacity at constant pressure
C_{total}	Total concentration of a drug, bound and unbound
C_V	Heat capacity at constant volume
d	Diameter
D	Diameter
D	Diffusivity
D	Dilution rate
D	Dose of a drug
D_0	Diffusivity in interstitial fluid
Da	Dahmkohler number
D_B	Dialysance
D_{cell}	Diffusivity in the cell
D_e	Effective diffusivity
$D_{effective}$	Effective diffusivity in blood
$D_{equivalent}$	Equivalent diameter
D_H	Hydraulic diameter
D_m	Membrane diffusivity
D_{plasma}	Diffusivity in plasma
D_T	Diffusivity in tissue
E	Enzyme concentration
E	Extraction factor
E	Extraction ratio
E	Solute extraction factor
E_K	Kinetic energy
E_P	Gravitational potential energy
f	Fraction of drug absorbable
f	Friction factor
f	Fugacity
f_i	Fugacity of pure component i
$\hat{f}_i$	Fugacity of component i in a mixture
f_U	Fraction of unbound drug
f_{UT}	Fraction of unbound drug in the tissue
F	Faraday's constant
F	Feed flow rate
F	Force
g	Acceleration of gravity
g	Membrane conductance

G	Gibbs free energy defined as H − TS
G_B	Glucose concentration in the plasma
G^E	Excess Gibbs free energy
$\bar{G}_i^E$	Partial molar excess Gibbs free energy
GFR	Glomerular filtration rate
h	Height, capillary rise
h_{friction}	Frictional effects affecting the flow of a fluid
H	Enthalpy defined as U + PV
H	Hematocrit
H_C	Hematocrit of core
H_F	Hematocrit of feed
H_T	Hematocrit of the tube
H	Henry's constant
ΔH^m	Enthalpy of fusion
ΔH^{VAP}	Heat of vaporization
i	Refers to the current flow of a particular component
I	Current
I_B	Plasma insulin concentration
I_0	Drug infusion rate
j_S	Diffusion flux
J_S	Solute diffusion rate
k_a	Absorption rate constant
k_b	Mass transfer coefficient for blood
k_{cat}	Enzyme rate constant
k_e	Mass transfer coefficient for exchange fluid
k_g	Mass transfer coefficient for a gas
k_i	Elimination rate constant
k_m	Mass transfer coefficient
$\bar{k}_m$	Length averaged mass transfer coefficient
k_{renal}	Renal elimination rate constant
k_{te}	Total elimination rate constant
K	Partition coefficient
K_a	Drug–protein affinity constant
K_{fitting}	Fitting friction factor
K_i	Distribution coefficient for component i
K_m	Constant in Michaelis–Menten model
$K_{O/W}$	Octanol–water partition coefficient
K_0	Overall mass transfer coefficient
K_S	Monod constant
L	Latent heat of a phase change
L	Length
L	Liquid flow rate
L_P	Hydraulic conductance of a membrane
m	Slope of the oxygen–hemoglobin dissociation curve
M	Molar property value

M^E	Excess property value
M_i	Molar property value of component i
$\bar{M}_i$	Partial molar property value of component i
M_{urine}	Mass of drug in the urine
MW	Molecular weight
n	Number of moles
n_i	Number of moles of component i
N	Number of equilibrium stages
N_A	Avogadro's number
N_{fiber}	Number of fibers in a hollow fiber unit
N_S	Solute diffusion rate across a permeable membrane
N_T	Number of transfer units
N_V	Volumetric gas transport rate
P	Absolute pressure
pCO_2	Partial pressure of carbon dioxide
pO_2	Partial pressure of oxygen
$<pO_2>$	Average partial pressure of oxygen
P	Power
P	Product concentration
P_{50}	Constant in the Hill equation
P_c	Permeability of the capillary wall to oxygen
P_C	Hydrodynamic pressure in the capillary
P_C	Critical pressure
Pe	Peclet number
P_i	Partial pressure of component i
P_{IF}	Hydrodynamic pressure of interstitial fluid
P_m	Membrane permeability
P^{Sat}	Saturation or vapor pressure
P_{SC}	Permeability of the stratus corneum
$\overline{\Delta P}$	Effective pressure drop
q	Filtration flux
q_b	Tissue blood perfusion rate
Q	Blood flow rate
Q	Filtration rate
Q	Heat
Q	Volumetric flow rate of a fluid
Q_b	Volumetric flow rate of blood
Q_d	Volumetric flow rate of exchange fluid
Q_g	Volumetric flow rate of a gas
$Q_{capillary}$	Volumetric flow rate in a capillary
r	Radius of a bubble or droplet
r_C	Capillary radius
r_{islet}	Islet insulin release rate
r_S	Rate of substrate consumption in an enzyme reaction
$r_{metabolic}$	Metabolic rate of drug consumption

r_T	Krogh tissue cylinder radius
R	Universal gas constant
R	Radius
$R(\bar{C})$	Volumetric reaction rate
R_0	Zero-order volumetric reaction rate
Re	Reynolds number
Re_x	Local Reynolds number at x
$R_{E/l}$	Ratio of the amount of drug binding protein in the extracellular fluid to that in the plasma
s	Parameter in the Casson equation
s	Capillary surface area per volume of tissue
S	Entropy
S	Substrate concentration
S	Surface area of a membrane
S_a	Sieving coefficient
S_c	Capillary surface area
S_0	Sieving coefficient at high filtration rates
Sc	Schmidt number
Sh	Sherwood number
SSE	Sum of the square of the errors
t	Temperature in °C or °F
t	Time
$t_{1/2}$	Half life
t_m	Membrane thickness
T	Absolute temperature in K or R
T_C	Critical temperature
T_f	Freezing temperature of a mixture
T_m	Normal melting temperature
U	Internal energy
$\bar{U}$	Reduced average velocity
v	Capillary volume fraction in tissue
v_i	Molar volume
v_x	Velocity in the x direction
v_y	Velocity in the y direction
v_z	Velocity in the z direction
V	Molar, specific volume, or total volume
V	Velocity, usually of a plate or bulk solution
V	Voltage
$V_{apparent}$	Apparent volume of distribution
$V_{average}$	Average velocity
V_{Bg}	Glucose distribution volume
V_{Bi}	Insulin distribution volume
V_C	Critical volume
V_{CAPD}	Volume of exchange fluid in continuous ambulatory peritoneal dialysis

V_{device}	Device volume
V_{IF}	Interstitial fluid distribution volume
V_{max}	Constant in Michaelis–Menten model
V_{plasma}	Plasma volume
V_{tissue}	Tissue volume
V_T	Total gas volume
V_T	Total volume of tissue
W	Width
W	Work
W	Membrane area per unit length of membrane
W_p	Work of a pump
W_S	Shaft work as in a flow process
x_i	Mole fraction of component i, usually the liquid phase
X	Cell density
y_i	Mole fraction of component i, usually the vapor or gas phase
Y	Fraction of hemoglobin that is saturated
$Y_{X/S}$, $Y_{P/S}$, $Y_{P'/S}$	Yield coefficients
z	Ion charge
z	Ratio of blood flow rate to exchange fluid flow rate
Z	Compressibility factor
Z	Elevation relative to a datum level

Superscripts

c	Core
E	Excess property value
G	Refers to the gas
$I, II, \ldots \pi$	Phase I, phase II, phase π
L	Refers to the liquid
p	Plasma
(P)	Refers to a planar surface
R	Denotes a residual thermodynamic property
S	Refers to the solid
SAT	Refers to the saturated state
STP	Refers to standard temperature and pressure
T	Refers to tissue
V	Refers to the vapor
∞	Denotes the value at infinite dilution
∞	Denotes the value after a very long time

Subscripts

0	Denotes initial value
0	Denotes interstitial fluid value

A	Denotes arterial value
AVG	Denotes average value
b	Refers to the value in a bulk solution
b	Refers to the value in blood
bs	Refers to the value at the surface of a membrane or particle
BP	Refers to the boiling point
c	Core
C	Capillary
CAPD	Refers to continuous ambulatory peritoneal dialysis
f	Filtrate side of the membrane
g	Refers to the gas
IF	Interstitial fluid
L	Local value at x = L
m	Denotes membrane value
max	Denotes a maximum of the value
p	Plasma
plasma	Refers to plasma
tube	Refers to a tube or other conduit
T	Denotes tissue
urine	Refers to the urine
V	Denotes venous value
W	Refers to the wall
x	Local value at position x

Greek letters

ρ	Mass density
δ	Marginal or plasma layer thickness
δ	Thickness of the velocity boundary layer
δ_C	Concentration boundary layer thickness
δ_{cell}	Thickness of a cellular layer
δ_{device}	Device thickness
μ	Specific growth rate
μ	Viscosity
$\mu_{apparent}$	Apparent viscosity
μ_i	Chemical potential of component i
ν	Kinematic viscosity, μ/ρ
τ	Residence time
τ	Shear stress
τ	Tortuosity of a membrane
τ_{max}	Time when concentration peaks
τ_{rz}	Shear stress, viscous transport of z momentum in the r direction
τ_w	Wall shear stress
τ_y	Yield stress

τ_{yx}	Shear stress, viscous transport of x momentum in the y direction
γ	Heat capacity ratio defined as C_P/C_V
γ	Surface energy
$\dot{\gamma}$	Shear rate
$\dot{\gamma}_w$	Shear rate at the wall
γ_i	Activity coefficient of component i
ϕ	Cell volume fraction
ϕ	Thiele modulus
ϕ_i	Fugacity coefficient of pure component i
$\hat{\phi}_i$	Fugacity coefficient of component i in a mixture
δ_i	Solubility parameter of component i
Φ_1	Volume fraction of the solvent
π	Osmotic pressure
η	Effectiveness factor
η	Overall efficiency of an equilibrium stage
η	Efficiency of a pump
Δ	Ratio of concentration and velocity boundary thicknesses
λ	Ratio of the molecular radius to the pore radius
Γ	Gamma function
$\Gamma_{\text{metabolic}}$	Metabolic oxygen consumption rate
ω_r	Hindered diffusion parameter
σ	Reflection coefficient
ε	Tissue void volume fraction
ε	Void volume
ε_{SC}	Void volume fraction of stratus corneum

Additional Notes

$^{-}$	An overbar denotes a partial molar property
$\wedge$	Circumflex denotes the value of a property in a mixture
Δ	Difference operator, final state minus initial state

ONE

INTRODUCTION

Before we can begin our study of biomedical engineering transport phenomena, we first must spend some time reviewing the basic concepts that are essential for understanding the material in this book. You may have come across some of this in other courses such as in chemistry, physics, and perhaps thermodynamics. Reviewing these concepts is still very important since they form the basis of our approach to analyzing and solving biomedical engineering problems.

1.1 REVIEW OF UNITS AND DIMENSIONS

1.1.1 Units

Careful attention must be given to *units* and *dimensions* when solving engineering problems. Otherwise, serious errors can occur in your calculations.

Units are how we describe the size or amount of a dimension. For example, the second is a common unit that is used for the dimension of time. In this book we will primarily use the *International System of Units* which is also known by its abbreviation of SI, for *Système International*. There are other systems of units that are also still in use such as the English and American engineering systems and the cgs (centimeter–gram–second) system. We will also come across some of these non-SI units in our study. The units of these other systems may be related to the SI units by appropriate conversion factors. Table 1.1 provides a convenient summary of common conversion factors that relate these other units to the SI system.

It is important to remember that in engineering calculations you must attach units to the numbers that arise in your calculations, unless they are already unitless. Furthermore, within a calculation, it is important to use a consistent system of units, and

Table 1.1 Conversion factors

Dimension	Conversion Factors
Length	1 m = 100 cm
	1 m = 3.28084 ft
	1 m = 39.37 in
Volume	$1 \text{ m}^3 = 10^6 \text{ cm}^3$
	$1 \text{ m}^3 = 35.3147 \text{ ft}^3$
	$1 \text{ m}^3 = 61{,}023.38 \text{ in}^3$
Mass	1 kg = 1000 g
	$1 \text{ kg} = 2.20462 \text{ lb}_m$
Force	$1 \text{ N} = 1 \text{ kg m sec}^{-2}$
	1 N = 100,000 dynes
	$1 \text{ N} = 0.22481 \text{ lb}_f$
Pressure	$1 \text{ bar} = 100{,}000 \text{ kg m}^{-1} \text{ sec}^{-2}$
	$1 \text{ bar} = 100{,}000 \text{ N m}^{-2}$
	$1 \text{ bar} = 106 \text{ dynes cm}^{-2}$
	1 bar = 0.986923 atm
	1 bar = 14.5038 psia
	1 bar = 750.061 mm Hg or torr
Density	$1 \text{ g cm}^{-3} = 1000 \text{ kg m}^{-3}$
	$1 \text{ g cm}^{-3} = 62.4278 \text{ lb}_m \text{ ft}^{-3}$
Energy	$1 \text{ J} = 1 \text{ kg m}^2 \text{ sec}^{-2}$
	1 J = 1 N m
	$1 \text{ J} = 1 \text{ m}^3 \text{ Pa}$
	$1 \text{ J} = 10^{-5} \text{ m}^3 \text{ bar}$
	$1 \text{ J} = 10 \text{ cm}^3 \text{ bar}$
	$1 \text{ J} = 9.86923 \text{ cm}^3 \text{ atm}$
	$1 \text{ J} = 10^7 \text{ dyne cm}$
	$1 \text{ J} = 10^7 \text{ erg}$
	1 J = 0.239 cal
	$1 \text{ J} = 5.12197 \times 10^{-3} \text{ ft}^3 \text{ psia}$
	$1 \text{ J} = 0.7376 \text{ ft lb}_f$
	$1 \text{ J} = 9.47831 \times 10^{-4} \text{ Btu}$
Power	$1 \text{ kW} = 10^3 \text{ W}$
	$1 \text{ kW} = 1000 \text{ kg m}^2 \text{ sec}^{-3}$
	$1 \text{ kW} = 1000 \text{ J sec}^{-1}$
	$1 \text{ kW} = 239.01 \text{ cal sec}^{-1}$
	$1 \text{ kW} = 737.562 \text{ ft lb}_f \text{ sec}^{-1}$
	$1 \text{ kW} = 0.947831 \text{ Btu sec}^{-1}$
	1 kW = 1.34102 HP
Viscosity	1 P = 100 cP
	$1 \text{ P} = 1 \text{ g cm}^{-1} \text{ sec}^{-1}$
	$1 \text{ P} = 1 \text{ dyne sec cm}^{-2}$
	$1 \text{ P} = 0.1 \text{ N sec m}^{-2}$

in this book we recommend that you work with the SI system. In the event a number has a non-SI unit, you will need to convert those units into the SI units using the conversion factors found in Table 1.1. Also, treat the units associated with a number as algebraic symbols. Then as long as the units are the same, you can perform operations such as addition, subtraction, multiplication, and division on them thus combining and in some cases even cancelling them out.

1.1.2 Fundamental Dimensions

The measurement of the physical properties in which we will be interested derive from the *fundamental dimensions* of length, mass, time, temperature, and mole. Table 1.2 summarizes the basic SI units for these fundamental dimensions. The SI unit for *length* is the meter (m) and that for *time* is the second (sec).

Table 1.2 SI Units for the fundamental dimensions

Fundamental dimension	Unit	Abbreviation
Length	meter	m
Mass	kilogram	kg
Time	second	sec or s
Temperature	kelvin	K
Mole	mole	mol

1.1.2.1 Mass and weight The *mass* of an object refers to the total amount of material that is in the object. The mass is a property of matter and is the same no matter where the object is located. For example, the mass of an object is the same on Earth, on Neptune, or if it is just floating along somewhere in space. In SI units we measure the mass of an object in kilograms (kg). Remember that the mass of an object is different from the weight of an object. *Weight* is the force exerted by gravity on the object. Since weight is a force, we can use Newton's second law ($F = ma$) to relate weight (F) and mass (m).

Example 1.1 Calculate the weight in SI units of an average person on the Earth.

SOLUTION On Earth, the acceleration due to gravity (a) is 9.8 m sec^{-2}. The average mass of a human is 75 kg, so on Earth the weight of the average human is 75 kg × 9.8 m sec^{-2} = 735 kg m sec^{-2}. In a few moments we will see that 1 Newton (N) is a special SI unit defined as 1 kg m sec^{-2}, so our average human also has a weight of 735 N.

In the USA with its American engineering system of units, a very confusing situation can occur when working with mass and force (weight). The following discussion should be sufficient for you to see the advantages of working in the SI system of units. However, the American engineering system of units is still widely used and it is important that you understand the following in order to handle properly these units in the event they arise in your calculations.

What is commonly referred to as a pound is really a *pound force* (weight) in the American engineering system of units. The pound force is abbreviated as "lb_f" where the subscript "f" must be present to indicate that this is a pound force. Mass is also expressed as pounds but in reality what is meant is the *pound mass* (lb_m) and once again the subscript "m" must be present to indicate that this is a pound mass. Using Newton's second law we can develop the following relationship between the pound force and the pound mass.

$$\text{(force) } lb_f = \text{(mass) } lb_m \times \text{(acceleration of gravity) ft sec}^{-2} \times 1/g_c \qquad (1.1)$$

The g_c term in the above equation is a conversion factor needed to make the units in Equation 1.1 work out properly, that is convert the lb_m to lb_f. Note that unlike the acceleration of gravity, the value of g_c does not depend on location and is a constant. In the American engineering system of units the value of the force (pound force or lb_f) and the mass (pound mass or lb_m) are defined in such a way so as to be the same at sea level at 45 degrees latitude. The acceleration of gravity (a) in the American system of units at sea level and 45 degrees latitude is 32.174 ft sec^{-2}. The value of g_c in Equation 1.1 is therefore 32.174 ft lb_m sec^{-2} lb_f^{-1}. It is important to recognize though that the acceleration of gravity depends on location. A pound force and a pound mass have the same value only at sea level where the acceleration of gravity is 32.174 ft sec^{-2} or whenever the ratio of the local acceleration of gravity to g_c is equal to one. For calculations made on the surface of the Earth, it is usually just assumed that a pound mass (lb_m) equals a pound force (lb_f) since the acceleration of gravity really does not vary that much. So, if an object with a mass of 10 lb_m is on Earth, then for practical purposes its weight is 10 lb_f. However, if this object is taken to the moon where the acceleration of gravity is 5.309 ft sec^{-2}, the mass is still 10 lb_m, however by Equation 1.1 its weight is quite a bit lower as shown in the calculation below.

$$10 \; lb_m \times 5.309 \, \frac{\text{ft}}{\text{sec}^2} \times \frac{\text{sec}^2 \; lb_f}{32.174 \text{ ft } lb_m} = 1.65 \; lb_f$$

1.1.2.2 Temperature Our *temperature* is defined as an *absolute* temperature instead of a relative temperature like the commonly used Fahrenheit and Celsius scales. The Fahrenheit scale is based on water freezing at 32°F and boiling at 212°F, whereas the Celsius scale sets these at 0°C and 100°C, respectively. Recall that the following equations may be used to relate a Celsius temperature to a Fahrenheit temperature and vice versa.

$$t \; °F = 1.8 \, t \; °C + 32 \text{ or } t \; °C = \frac{5}{9} \, (t \; °F - 32) \qquad (1.2)$$

Our SI unit for the measurement of absolute temperature is the Kelvin (K). The absolute temperature scale sets the absolute zero point, or 0 degrees Kelvin (0 K), as the lowest possible temperature at which matter can exist. The temperature unit of the absolute Kelvin temperature scale is the same as that of the Celsius scale, so 1 K = 1°C. The following equation may be used to relate the Celsius temperature (t) scale to the absolute Kelvin temperature (T) scale. In Celsius units the absolute zero point is therefore $-273.15°C$.

$$t \, °C = T \, K - 273.15 \tag{1.3}$$

The absolute temperature can also be defined by the same unit of temperature measurement as the Fahrenheit scale. This is referred to as the absolute Rankine (R) temperature scale and $1 \, R = 1°F$. Note that $0 \, K$ is the same as $0 \, R$ since the absolute Kelvin and the absolute Rankine temperature scales are based on the same reference point of absolute zero. The following equation may be used to relate the Fahrenheit temperature scale (t) to the absolute Rankine temperature (T) scale. In Fahrenheit units the absolute zero point is $-459.67°F$.

$$t \, °F = T \, R - 459.67 \tag{1.4}$$

Using the above equations, one can also show that the relationship between the Rankine and Kelvin temperatures is given by the following equation.

$$T \, R = 1.8 \, T \, K \tag{1.5}$$

1.1.2.3 Mole The *mole* is used to describe the amount of a substance that contains the same number of atoms or molecules as there are atoms in 0.012 kg (or 12 g) of carbon-12, that is 6.023×10^{23} atoms which is also called *Avogadro's number*. This is also called the gram mole which is also written as *gmole* or just *mole*. In the American engineering system the *pound mole* (lbmole or lbmol) is used and is defined as $6.023 \times 10^{23} \times 454$ atoms. Therefore, 1 pound mole is equal to 454 moles.

Recall that the *molecular weight* is defined as the mass per mole of a given substance. For example, the molecular weight of glucose is 0.180 kg/mole whereas the molecular weight of water is 0.018 kg/mole. A mole of glucose or a mole of water would both contain an Avogadro's number of molecules. However, even though one mole of glucose and water contains the same number of molecules, the masses of each are quite different because the molecules of glucose are much larger than those of water. That is, one mole of glucose has a mass of 0.180 kg and one mole of water has a mass of 0.018 kg.

1.1.3 Derived Dimensional Quantities

From these fundamental dimensions we can derive the units for a variety of other dimensional quantities of interest. Some of these quantities will occur so often that it is useful to list them as shown in Table 1.3. In addition, some of these quantities, like force and energy, have their own special SI units associated with them, whereas others, like mass density, are based on the fundamental dimensions listed in Table 1.2. Many times we will use special symbols to denote a quantity. For example the Greek symbol ρ is used to denote the mass density and μ is often used for the viscosity of a substance. The viscosity, discussed in Chapter 4, is a physical property of a fluid and is a measure of its resistance to flow. A listing of the symbols used in this book to denote a variety of quantities is given in the *Notation* section that follows the *Table of Contents*.

Table 1.3 SI Units for other common dimensional quantities

Quantity	Unit	Abbreviation	Fundamental units	Derived units
Force	newton	N	kg m sec^{-2}	J m^{-1}
Energy, work, heat	joule	J	kg m^2 sec^{-2}	N m
Pressure and stress	pascal	Pa	kg m^{-1} sec^{-2}	N m^{-2}
Power	watt	W	kg m^2 sec^{-3}	J sec^{-1}
Volume	–	–	m^3	–
Mass density	–	–	kg m^{-3}	–
Concentration	–	–	mol m^{-3}	–
Specific volume	–	–	m^3 kg^{-1}	–
Viscosity	–	–	kg m^{-1} sec^{-1}	Pa sec
Diffusivity	–	–	m^2 sec^{-1}	–
Permeability, mass transfer coefficient	–	–	m sec^{-1}	–

Often a particular unit is expressed as a multiple or a decimal fraction. For example 1/1000th of a meter is known as a millimeter where the prefix "milli-" means to multiply the base unit of meter by the factor of 10^{-3}. The prefix "kilo-" means to multiply the base unit by 1000. So a kilogram is the same as 1000 grams. Table 1.4 summarizes a variety of prefixes that are commonly used to scale a base unit.

1.1.3.1 Pressure Pressure is defined as a force per unit area. Pressure is also expressed on either a relative or an absolute scale, and its unit of measure in the SI system is the pascal (Pa) which is equal to a N m^{-2}. How the pressure is measured will affect whether or not it is a relative or an absolute pressure. *Absolute pressure* is based on reference to a perfect vacuum. A *perfect vacuum* on the absolute pressure scale defines the zero point or 0 Pa. The zero point for a relative pressure scale is usually the pressure of the air or atmosphere where the measurement is taken. This local pressure is called either the *atmospheric pressure* or the *barometric pressure*. Barometric pressure is measured by a *barometer* which is a device for measuring atmospheric pressure. It is important to remember that the pressure of the surrounding air or atmosphere is not a constant, but

Table 1.4 Common prefixes for SI units

Prefix	Multiplication factor	Symbol
femto	10^{-15}	f
pico	10^{-12}	p
nano	10^{-9}	n
micro	10^{-6}	μ
milli	10^{-3}	m
centi	10^{-2}	c
deci	10^{-1}	d
deka	10	da
hecto	10^2	h
kilo	10^3	k
mega	10^6	M
giga	10^9	G
tera	10^{12}	T

depends on location, elevation, and other factors related to the weather. That is why it is always important to log in your record book the atmospheric pressure in the laboratory when doing experiments that involve pressure measurements.

Most pressure measurement devices, or pressure gauges, measure the pressure relative to the surrounding atmospheric pressure or barometric pressure. These so-called *gauge pressures* are also relative pressures. A good example of a gauge pressure is the device used to measure the inflation pressure of your tires. If the tire pressure gauge has a reading of 250 kPa, then that means that the pressure in the tire is 250 kPa *higher* than the surrounding atmospheric pressure. A gauge pressure of 0 Pa means that the pressure is the same as the local atmospheric pressure, or you have a flat tire! A negative gauge pressure (i.e. suction or partial vacuum) means that the pressure is that amount *below* the atmospheric pressure. If the pressure measurement device measures the pressure relative to a perfect vacuum, then that pressure reading is referred to as an absolute pressure reading. The relationship between relative or gauge pressure and absolute pressure is given by the following equation.

$$\text{(gauge pressure)} + \text{(atmospheric pressure)} = \text{(absolute pressure)} \qquad (1.6)$$

The *standard atmosphere* is equivalent to the absolute pressure exerted at the bottom of a column of mercury that is 760 mm in height at a temperature of 0°C. This value is nearly the same as the atmospheric pressure that one may find on a typical day at sea level. The standard atmosphere in a variety of units is summarized in Table 1.5. Using Equation 1.6, and the pressure conversion factors shown in Table 1.5, we can say that your tire pressure is 250 kPa gauge or 351.325 kPa absolute.

1.1.3.2 Volume *Volume* refers to the amount of space that an object occupies and depends on the amount of the object or its mass. The *mass density* is the mass of an object divided by the volume of space that it occupies. *Specific volume* is the volume of an object divided by its mass. Specific volume is therefore just the inverse of the mass density. For solids and liquids, the volume of an object of a given mass is not strongly dependent on the temperature or the pressure. For solids and liquids, the volume is then the mass multiplied by the specific volume or the mass divided by the density of the object. For gases, the volume of a gas is also strongly dependent on both the temperature and the pressure. For gases we need to use an equation of state to relate the volume of the gas to the temperature, pressure, and moles (mass divided by molecular weight).

Table 1.5 The standard atmosphere in a variety of units

1.000 atmosphere (atm)
14.696 pounds force per square inch (psi)
760 millimeters of mercury (mm Hg)
33.91 feet of water (ft H_2O)
29.92 inches of mercury (in. Hg)
101,325 pascals (Pa)

1.1.3.3 Equations of state An *equation of state* is a mathematical relationship that relates the pressure, volume, and temperature of a gas, liquid, or solid. For our purposes, the ideal gas law is an equation of state that will work just fine for the types of problems involving gases that we will consider here. Recall that an ideal gas has mass, however the gas molecules themselves have no volume and these molecules do not interact with one another. The ideal gas law works well for gases like hydrogen, oxygen, and air at pressures around atmospheric. Recall that the ideal gas law is given by the following relationship.

$$P V = n R T \tag{1.7}$$

In this equation, P denotes the absolute pressure, V the volume, T the absolute temperature, and n the number of moles. R is called the *universal gas constant* and a suitable value must be used to make the units in Equation 1.7 work out properly. Table 1.6 summarizes commonly used values of the gas constant in a variety of different units.

1.2 DIMENSIONAL EQUATION

A *dimensional equation* is an equation that contains both numbers and their associated units. The dimensional equation usually arises from the use of a specific formula that is being used to solve a problem or when a particular number has non-standard units associated with it. In the latter case, a series of conversion factors are used to put the number in some other units, for example the SI system of units. To arrive at the final answer, which is usually a number with its associated units, the dimensional equation is solved by performing the arithmetical operations on the numbers and as discussed earlier, algebraic operations on the various units.

The first example below illustrates the use of the dimensional equation to convert a quantity in non-standard units to SI units. The next example illustrates the use of the dimensional equation to determine the volume of a gas at a given temperature and pressure. When solving engineering problems, make sure to use the dimensional equation approach illustrated in these examples. In this way you can easily handle the conversion of the units associated with the quantities involved in your calculation and arrive at the correct answer.

Table 1.6 The Gas Constant (R)

Gas Constant
8.314 m^3 Pa mol^{-1} K^{-1}
8.314 J mol^{-1} K^{-1}
82.06 cm^3 atm mol^{-1} K^{-1}
0.0821 liters atm mol^{-1} K^{-1}
0.7302 ft^3 atm lbmol^{-1} R^{-1}
10.73 ft^3 psi absolute lbmol^{-1} R^{-1}

Example 1.2 A particular quantity was reported to have a value of 1.5 cm³ hr⁻¹ m⁻² mm Hg⁻¹. Recall that a mm Hg is a measurement of pressure, where 760 mm of mercury (Hg) is equal to a pressure of 1 atmosphere (1 atm). Convert this dimensional quantity to SI units of m² sec kg⁻¹.

SOLUTION The dimensional equation for the conversion of this quantity to SI units is shown below.

$$1.5 \frac{cm^3}{hr\ m^2\ mm\ Hg} \times \frac{1\ hr}{3600\ sec} \times \frac{760\ mm\ Hg}{1\ atm} \times \frac{1\ atm}{101,325\ Pa} \times \frac{1\ Pa}{1\ N\ m^{-2}} \times \frac{1\ N}{kg\ m\ sec^{-2}}$$

$$\times \frac{1\ m^3}{(100\ cm)^3} = 3.125 \times 10^{-12} \frac{m^2\ sec}{kg}$$

Example 1.3 Calculate the volume occupied by 100 lb$_m$ of oxygen (molecular weight equals 0.032 kg mole⁻¹) at a pressure of 40 feet of water and a temperature of 20°C. Express the volume in m³.

SOLUTION The dimensional equation for the calculation of the volume of oxygen is based on Equation 1.7 rearranged to solve for the volume, i.e. $V = nRT/P$.

$$V = \left(100\ lb_m \times \frac{1\ kg}{2.2046\ lb_m} \times \frac{1\ mole}{0.032\ kg}\right) \times 8.314 \frac{m^3\ Pa}{mole\ K} \times (273.15 + 20)K$$

$$\times \frac{1}{\left(40\ ft\ of\ water \times \dfrac{1\ atm}{33.91\ ft\ of\ water} \times \dfrac{101,325\ Pa}{1\ atm}\right)} = 28.90\ m^3$$

1.3 TIPS FOR SOLVING ENGINEERING PROBLEMS

In this section we outline a basic strategy that can be used to solve the type of engineering problems considered in this book and those you are likely to encounter as a biomedical engineer. This basic strategy is useful since it eliminates approaching the solution of each new problem or system as if it is a unique situation. The numerous examples throughout the book as well as the problems at the end of each chapter will provide you with the opportunity to apply the strategy outlined here and at the same time develop your skills in using the engineering analysis techniques discussed in this book. One of the unique and challenging aspects of engineering is that the number of problems that you may come across during the course of your career is limitless. However, it is important to understand that the underlying engineering principles and the solution strategy are always the same.

The engineering problems we will consider derive from some type of process. A *process* is defined as any combination of physical operations and chemical reactions that act on or change the substances involved in the process. A process also occurs within what is also more generally called the *system*. Everything else that is affected by the system or interacts with the system is known as the *surroundings*. A biological example of a process or system is the kidney. Blood enters the kidney (system) and through a variety of physical and chemical operations the essential nutrients in the blood are recovered and the blood is purified of waste products. These waste products are stored as urine in the bladder. Several times a day the urine that is stored in the bladder (another system) is eliminated through the process of urination. The urine eliminated from the bladder is then added to the surroundings. The engineering analysis of all problems can therefore be considered within the context of the system and its surroundings.

There are several important steps that you should follow before attempting to make any engineering calculations on a system and its surroundings. For case of reference, these steps are also summarized in Table 1.7. One of the first things that you should do when solving a particular problem is to draw a sketch that defines your system. Drawing a sketch of the system for a given problem will allow you to identify and focus your attention on the key features of the problem. The sketch does not have to be elaborate; you can use boxes to represent the system and lines with arrows can be used to indicate the flow of streams that enter and leave the system. Figure 1.1 shows a simple sketch

Table 1.7 Steps for solving engineering problems

Problem-solving steps

1. Draw a sketch that defines your system.
2. Indicate with lines and arrows the flow of streams that enter and leave the system.
3. Indicate the flow rates and the amounts or concentration of each substance in each of the flowing streams. For those that are unknown see if you can easily calculate them using material balance equations.
4. Write down any chemical reactions that need to be considered and make sure that these are also balanced.
5. Write down the available data that you already have concerning the problem.
6. Obtain any additional information or data needed to solve the problem.
7. Identify the specific engineering principles and formulas that govern the physical and chemical operations that occur within your system.
8. Solve the problem using the above information.

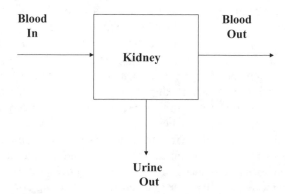

Figure 1.1   Engineering sketch of the kidney.

for the kidney example discussed earlier. If you know them or can easily calculate them, make sure you also write down the flow rates of the flowing streams and the amounts or concentration of each substance in each of the flowing streams. On your sketch you will also write down the available data that you already have concerning the problem. Examples of additional data that may be important in solving your problem include temperatures, pressures, experimental data, and even conversion factors. Next, you should obtain any additional information or data needed to solve the problem. You should also write down any chemical reactions that need to be considered and make sure that these are also balanced.

Obtaining the answer to the problem illustrated by your sketch of the system will require the use of specific engineering principles and formulas that govern the physical and chemical operations that occur within your process. Formulating the solution and solving engineering problems is a skill that one needs to develop and refine in order to be successful in engineering. That is the goal of this book: to provide you with the mathematical techniques and skills for solving a variety of biomedical engineering problems that involve chemical reactions and such transport processes as fluid flow and mass transfer. Building on the skills learned in this book you will then be able to design biomedical equipment and devices.

1.4 CONSERVATION OF MASS

One of the most fundamental relationships that should always be considered when performing an engineering analysis on a system is the *conservation of mass*. This is also called a *material balance*. In its simplest terms, conservation of mass is the statement of the fact that the sum of the masses of all substances that enter the system must equal the sum of the masses of all the substances that leave the system. However, there can be changes in the individual masses or amounts of various chemical entities because of chemical reactions that may also be occuring within the system. Remember that mass is always conserved and moles are not conserved if chemical reactions occur in the system. The only exception to conservation of mass is for nuclear reactions where energy is converted to mass and vice versa. Nuclear reactions are not included in the systems that we will be considering.

1.4.1 Law of Conservation

The following equation provides a generalized statement of the *law of conservation* for our system that we will find extremely useful for the solution of a variety of problems.

$$\text{Accumulation} = \text{In} - \text{Out} + \text{Generation} - \text{Consumption} \tag{1.8}$$

This equation may be used to account for changes in such quantities as mass, moles, energy, momentum, and even money.

The "Accumulation" term accounts for the change with time of the quantity of interest within the system volume. The accumulation term in this equation is a time derivative of the quantity of interest and defines what is called an *unsteady problem*.

The accumulation term can be either positive or negative. If the accumulation term is zero, then the quantity of interest within the volume of the system is not changing with time and we refer to this situation as a *steady-state problem*. The "In" term accounts for the entry of the quantity of interest into the system by all routes. For example, as carried in by the flow of fluid, by diffusion, or in the case of energy, by heat conduction. The "Out" term is similar to the "In" term but accounts for the loss of the quantity of interest from the system. The "Generation" term accounts for the production of the quantity of interest within the system volume whereas the "Consumption" term represents the loss of the quantity of interest within the system. The generation and consumption of the quantity of interest occurs through chemical reactions and the metabolic processes that occur within cells.

The following "non-engineering" example illustrates the use of Equation 1.8 for calculating the periodic payment on a loan. Remember this example since you can use the results to calculate payments for such items as automobiles, your house, and your credit card balance.

Example 1.4 You have just graduated from college as a bioengineer and have accepted your first job with a medical device company. You decide to buy yourself a nice sports car and in order to buy it you will need to borrow from the bank a total of \$65,000. The bank is offering you a 5-year loan at an annual interest rate of 6% compounded continuously. What is your monthly payment? What is the total amount that you will pay back to the bank? How much interest will you pay?

SOLUTION We will let A represent the amount of money at any time that you owe the bank and R will represent the annual payment that you make to the bank on the loan. The interest rate is represented by i. The amount of interest that the bank charges you for the loan is equal to the interest rate multiplied by the amount at any time that you owe them. Using Equation 1.8 we can then write the following equation that expresses how the amount you owe the bank changes with time.

$$\frac{dA}{dt} = i A - R$$

Note that the amount you owe the bank "Accumulates" according to the term dA/dt. The amount you owe is also "Generated" by the interest term represented by iA and the amount you owe is decreased (Out) by the amount of your annual payment represented by R. Since R will be greater than $i A$, then dA/dt will be < 0 and the amount that you owe the bank will decrease with time.

This is a first order differential equation and represents what is also called an initial value problem. In order to solve this equation we need to specifiy an initial condition (IC) which is the amount you owe the bank at time equal to zero which is when you take out the loan. The initial condition can be written as follows.

$$\text{IC: } t = 0 \ A = A_0 = \$65,000$$

There are several methods that can be used to solve the above differential equation to include integrating factors and Laplace transforms. Table 4.5 summarizes the Laplace transforms for a variety of functions and one of the end of chapter problems asks you to resolve this example using Laplace transforms. However, an easy way to solve the above equation (since R is a constant) is to let A be the sum of the homogeneous solution (i.e. when $R = 0$) represented by A_h and a constant C_1 yet to be determined. So we let $A = A_h + C_1$. We then substitute this equation into the differential equation above and we obtain the following two equations.

$$\frac{dA_h}{dt} = i\,A_h \quad \text{and} \quad -i\,C_1 = -R$$

We immediately see that $C_1 = R/i$ and the homogeneous differential equation can be readily integrated to give $A_h = C_2\,e^{it}$ where C_2 is another constant that we evaluate from the initial condition. Our solution for the amount that we owe the bank at any time is then given by this equation.

$$A(t) = C_2\,e^{it} + \frac{R}{i}$$

Now we use the initial condition that at $t = 0$, $A = A_0$ and we find that the constant $C_2 = A_0 - R/i$. Our result for the amount that you owe the bank at any time after the loan inception is then given by the next equation.

$$A(t) = A_0\,e^{it} + \left(\frac{R}{i}\right)(1 - e^{it})$$

Now what we want to find is the annual payment needed to pay off the loan after T years. Then since $A(T) = 0$ we can solve the above equation for R and obtain the following equation for the annual loan payment.

$$R = \frac{i\,A_0}{1 - e^{-it}}$$

We can now insert the numbers to calculate the loan payment amount.

$$R = \frac{\dfrac{0.06}{\text{yr}} \times \$65{,}000}{1 - e^{-\frac{0.06}{\text{yr}} \times 5\,\text{yrs}}} = \$15047.35 / \text{yr} = \$1253.95 / \text{month}$$

So we see that the monthly payment for the sports car is about \$1254 per month. The total amount paid to the bank over the 5 year period of the loan is \$15,047.35/ yr $\times$ 5 yrs = \$75,236.77 and the total amount of interest paid to the bank for the loan is \$75,236.77 − \$65,000 = \$10,236.77.

1.4.2 Chemical Reactions

In the case of chemical reactions or cellular metabolism occuring within the system, you must also take into account the reaction stoichiometry as given by your balanced chemical equations for those substances that are involved in the chemical reactions. In addition, it is important to remember that for biological reactions there can be an increase in the number of cells and your balanced chemical reaction must also consider the formation of biomass from the substances that take part in the reaction as shown in the example below.

Example 1.5 Consider the anaerobic fermentation of glucose to ethanol by yeast. Glucose ($C_6H_{12}O_6$) is converted into yeast, ethanol (C_2H_5OH), the byproduct glycerol ($C_3H_8O_3$), carbon dioxide, and water. An empirical chemical formula for yeast can be taken as $CH_{1.74}N_{0.2}O_{0.45}$ (Shuler and Kargi 2001). Assuming ammonia (NH_3) is the nitrogen source, we can write the following empirical chemical equation to describe the fermentation.

$$C_6H_{12}O_6 + a\ NH_3 \rightarrow b\ CH_{1.74}N_{0.2}O_{0.45}\ (\text{yeast}) + c\ C_2H_5OH +$$

$$d\ C_3H_8O_3 + e\ CO_2 + f\ H_2O$$

Suppose we found that 0.12 moles of glycerol were formed for each mole of ethanol produced and 0.08 moles of water were formed for each mole of glycerol. Determine the *stoichiometric coefficients* which are the letters (a, b, c, d, e, f) in front of the chemical formulas in the above equation. Remember that the stoichiometric coefficients need to be determined so that the total number of carbon, hydrogen, oxygen, and nitrogen atoms are the same on each side of the equation.

SOLUTION Balancing the above equation by inspection is not an easy thing to do. Hence we use the more formal approach as outlined below where we write a balance for each element, i.e. carbon, hydrogen, oxygen, and nitrogen.

Carbon balance: $6 = b + 2c + 3d + e$

Hydrogen balance: $12 + 3a = 1.74b + 6c + 8d + 2f$

Oxygen balance: $6 = 0.45b + c + 3d + 2e + f$

Nitrogen balance: $a = 0.2b$

Other constraints $d = 0.12\ c$ and $f = 0.08\ d$

The above relationships for each element can then be arranged into matrix form as shown below. We see that the solution for $a, b, c, d, e,$ and f involves the solution of six algebraic equations for these six unknowns.

$$
\begin{bmatrix}
0 & 1 & 2 & 3 & 1 & 0 \\
-3 & 1.74 & 6 & 8 & 0 & 2 \\
0 & 0.45 & 1 & 3 & 2 & 1 \\
1 & -0.2 & 0 & 0 & 0 & 0 \\
0 & 0 & -0.12 & 1 & 0 & 0 \\
0 & 0 & 0 & 0.08 & 0 & -1
\end{bmatrix}
\begin{bmatrix}
a \\ b \\ c \\ d \\ e \\ f
\end{bmatrix}
=
\begin{bmatrix}
6 \\ 12 \\ 6 \\ 0 \\ 0 \\ 0
\end{bmatrix}
$$

Solution of the above matrix gives $a = 0.048, b = 0.239, c = 1.68, d = 0.202, e = 1.796,$ and $f = 0.016$. The balanced chemical equation for the anaerobic fermentation of glucose by yeast can then be written as shown below.

$$C_6H_{12}O_6 + 0.048\ NH_3 \rightarrow 0.239\ CH_{1.74}N_{0.2}O_{0.45}\ (yeast) + 1.68\ C_2H_5OH$$
$$+ 0.202\ C_3H_8O_3 + 1.796\ CO_2 + 0.016\ H_2O$$

Example 1.6 Now determine the biomass yield coefficient ($Y_{X/S}$) and the product yield coefficients ($Y_{P/S}$, $Y_{P'/S}$) where X, S, P, and P' denote biomass, glucose, ethanol, and glycerol, respectively. Use the balanced chemical equation from the previous example which takes into account the formation of biomass. Compare these results with the theoretical maximum which is based on the conversion of all the glucose to just ethanol and carbon dioxide, i.e. $C_6H_{12}O_6 \rightarrow 2\ C_2H_5OH + 2\ CO_2$.

SOLUTION The yield coefficients based on the balanced empirical equation obtained in the previous example are defined as follows:

$$Y_{X/S} = \frac{b}{1} = \frac{0.239\ \text{moles yeast}}{1\ \text{mole}\ \text{glucose}} \times \frac{23.74\ g\,/\,\text{mole}}{180\ g\,/\,\text{mole}} = \frac{0.032\ g\ \text{yeast}}{1\ g\ \text{glucose}}$$

$$Y_{P/S} = \frac{c}{1} = \frac{1.68\ \text{moles ethanol}}{1\ \text{mole glucose}} \times \frac{46\ g\,/\,\text{mole}}{180\ g\,/\,\text{mole}} = \frac{0.4293\ g\ \text{ethanol}}{1\ g\ \text{glucose}}$$

$$Y_{P'/S} = \frac{d}{1} = \frac{0.202\ \text{moles glycerol}}{1\ \text{mole glucose}} \times \frac{92\ g\,/\,\text{mole}}{180\ g\,/\,\text{mole}} = \frac{0.1032\ g\ \text{glycerol}}{1\ g\ \text{glucose}}$$

For comparison the theoretical yield of ethanol is given as:

$$Y_{P/S\ \text{theoretical}} = \frac{2\ \text{moles ethanol}}{1\ \text{mole glucose}} \times \frac{46\ g\,/\,\text{mole}}{180\ g\,/\,\text{mole}} = \frac{0.511\ g\ \text{ethanol}}{1\ g\ \text{glucose}}$$

1.4.3 Material Balances

In applying the law of mass conservation to your system, you can write a material balance using Equation 1.8 for each component. If you have N components then this will provide a total of N equations. You can also write a total material balance giving a total of $N+1$ equations. However, these $N+1$ equations are not independent since we can also obtain the total material balance equation by just summing the N component material balances. Remember that you need as many equations as you have unknowns in order to describe your system. If you have more unknowns than equations relating these unknowns, then you have that many *degrees of freedom*, that is:

Degrees of Freedom = Number of Unknowns – Number of Equations   (1.9)

and you must specify the values of some of these unknowns. In solving engineering problems or designing equipment and devices, it is quite common to have more unknowns than equations that relate these unknowns. These degrees of freedom are also referred to as design variables or quantities that you specify as part of the design of the system.

The following two examples illustrate both unsteady and steady state material balance calculations for a bioreactor.

Example 1.7 Consider the production of ethanol by the fermentation of glucose by yeast in a batch fermentor. Let X represent the yeast cell density (g/L) in the bioreactor. We then define μ (1/hr) as the specific growth rate of the yeast in the fermentor and by definition $\mu = \dfrac{1}{X}\dfrac{dX}{dt}$. The *Monod equation* is an empirical expression that describes how the specific growth rate of cells depends on a limiting substrate, which in this case is glucose:

$$\mu = \frac{\mu_{max} S}{K_S + S}$$

where S (g/L) represents the concentration of glucose in the fermentor, K_S is a constant, and μ_{max} is the maximum specific growth rate, i.e. when $S \gg K_S$. Assuming that the glucose fermentation stoichiometry is given by the empirical chemical equation obtained in Example 1.5, calculate as a function of time the concentration in the fermentor of the yeast cells (X), glucose (S), ethanol (P), and glycerol (P'). Assume that the initial glucose concentration in the fermentor is 100 g/L and that the initial yeast cell density after inoculation of the fermentor is 0.2 g/L. The Monod constants are $K_S = 1.5$ g/L and $\mu_{max} = 0.075$ hr^{-1}.

SOLUTION After the growth medium in the fermentor is inoculated with the yeast, there is a short period of time called the lag phase (t_0) where the yeast cells are adapting to their new environment. The yeast then enters a phase where they grow very rapidly according to the Monod equation shown above. A material balance on the yeast cells in the fermentor may then be written using Equation 1.8 as shown below.

$$\text{Yeast} \quad V\frac{dX}{dt} = \mu X V = \frac{\mu_{max} S}{K_S + S} X V$$

Note there is no yeast flowing into or out of the fermentor, hence the accumulation of yeast $\left(V\dfrac{dX}{dt}\right)$ within the fermentor is due only to the production of yeast as described by the Monod equation $\left(\dfrac{\mu_{max} S}{K_S + S} XV\right)$. We can also write material balance equations as well for glucose, ethanol, and glycerol as shown below. Note that the definition of the yield coefficients from Example 1.6 allows us to relate the changes in the glucose (S), ethanol (P), and glycerol (P') concentrations to the change in the yeast concentration (X).

$$\text{Glucose} \quad V\frac{dS}{dt} = -\frac{V}{Y_{X/S}}\frac{dX}{dt}$$

$$\text{Ethanol} \quad V\frac{dP}{dt} = -Y_{P/S}V\frac{dS}{dt} = \frac{VY_{P/S}}{Y_{X/S}}\frac{dX}{dt}$$

$$\text{Glycerol} \quad V \frac{dP'}{dt} = -Y_{P'/S} V \frac{dS}{dt} = \frac{VY_{P'/S}}{Y_{X/S}} \frac{dX}{dt}$$

Since the volume of the fermentor (V) is constant, we can cancel out the volumes and rewrite the above equations as shown below.

$$\text{Yeast} \quad \frac{dX}{dt} = \mu X = \frac{\mu_{max} S}{K_s + S} X$$

$$\text{Glucose} \quad \frac{dS}{dt} = -\frac{1}{Y_{X/S}} \frac{dX}{dt}$$

$$\text{Ethanol} \quad \frac{dP}{dt} = \frac{Y_{P/S}}{Y_{X/S}} \frac{dX}{dt}$$

$$\text{Glycerol} \quad \frac{dP'}{dt} = \frac{Y_{P'/S}}{Y_{X/S}} \frac{dX}{dt}$$

We see that we now have four differential equations that can be solved to provide X, S, P, and P' as a function of time. The differential equations for glucose, ethanol, and glycerol all depend on X. If we multiply both sides of these equations by dt, then we obtain:

$$dS = -\frac{1}{Y_{X/S}} dX$$

$$dP = \frac{Y_{P/S}}{Y_{X/S}} dX$$

$$dP' = \frac{Y_{P'/S}}{Y_{X/S}} dX$$

These equations may be integrated from an initial condition where $t = t_0$ and $S = S_0$, $P = P_0$, $P' = P'_0$, and $X = X_0$, to some later time in the fermentation (t):

$$S(t) = S_0 - \frac{1}{Y_{X/S}}(X(t) - X_0)$$

$$P(t) = P_0 + \frac{Y_{P/S}}{Y_{X/S}}(X(t) - X_0)$$

$$P'(t) = P'_0 + \frac{Y_{P'/S}}{Y_{X/S}}(X(t) - X_0)$$

So now we have replaced three differential equations for glucose, ethanol, and glycerol, with three algebraic equations whose solution depends only on $X(t)$. Once we solve the remaining differential equation shown below for $X(t)$, we can then obtain the glucose, ethanol, and glycerol concentrations using the three equations above. Note that the maximum yeast cell concentration occurs when $S = 0$, hence $X_{max} = Y_{X/S} S_0 + X_0$.

$$\frac{dX}{dt} = \mu X = \frac{\mu_{max} S}{K_S + S} X$$

Note that the right-hand side of the above differential equation depends on both S and X. However, we can use the algebraic relationship we obtained above to relate S to X, that is:

$$S = S_0 - \frac{1}{Y_{X/S}} (X - X_0).$$

Hence our differential equation for X becomes:

$$\frac{dX}{dt} = \frac{\mu_{max} \left[S_0 - \dfrac{1}{Y_{X/S}} (X - X_0) \right]}{K_S + \left[S_0 - \dfrac{1}{Y_{X/S}} (X - X_0) \right]} X$$

The initial condition needed for the solution of the above equation is that at $t = t_0$, $X = X_0$ where t_0 represents the end of the lag period and the beginning of rapid cell growth. All of the terms involving X may be collected on the left-hand side of the above equation as shown below.

$$\int_{X_0}^{X} \frac{\left\{ K_S + \left[S_0 - \dfrac{1}{Y_{X/S}} (X - X_0) \right] \right\}}{\left[S_0 - \dfrac{1}{Y_{X/S}} (X - X_0) \right] X} dX = \mu_{max} \int_{t_0}^{t} dt = \mu_{max} (t - t_0)$$

The integral on the left side can then be integrated to provide the analytical solution that is shown below.

$$\mu_{max} (t - t_0) = \left(\frac{K_S Y_{X/S}}{Y_{X/S} S_0 + X_0} + 1 \right) \ln \left(\frac{X}{X_0} \right) + \left(\frac{K_S Y_{X/S}}{Y_{X/S} S_0 + X_0} \right)$$

$$\ln \left(\frac{Y_{X/S} S_0}{Y_{X/S} S_0 - (X - X_0)} \right)$$

The above equation may then be solved for the time needed to achieve a certain cell density in the fermentor. Using the values of the parameters from the problem statement for this example, we can solve this equation, as well as the algebraic equations, for the concentrations of glucose, ethanol, and glycerol. The graphs in Figure 1.2 show the solution assuming that the lag time is negligible.

Example 1.8 Suppose the fermentation considered in the previous example takes place at steady state in a continuous stirred-tank fermentor. Assume that the feed stream to the fermentor contains no yeast cells (i.e. sterile), ethanol, or glycerol. In addition, the glucose concentration in the feed stream is 100 g/L. The fermentor has a volume of 2000 L and the feed flow rate to the fermentor is 100 L/hr. Determine the exiting concentrations of the yeast cells, glucose, ethanol, and glycerol.

SOLUTION   Once again we use Equation 1.8 and write a material balance for yeast, glucose, ethanol, and glycerol as shown below.

$$\text{Yeast } V \frac{dX}{dt} = FX_{in} - FX_{out} + \frac{V \mu_{max} S}{K_S + S} X$$

$$\text{Glucose } V \frac{dS}{dt} = FS_{in} - FS_{out} - \frac{V}{Y_{X/S}} \frac{\mu_{max}}{K_S + S} X$$

$$\text{Ethanol } V \frac{dP}{dt} = FP_{in} - FP_{out} + \frac{VY_{P/S}}{Y_{X/S}} \frac{\mu_{max}}{K_S + S} X$$

$$\text{Glycerol } V \frac{dP'}{dt} = FP'_{in} - FP'_{out} + \frac{VY_{P'/S}}{Y_{X/S}} \frac{\mu_{max}}{K_S + S} X$$

X, S, P, and P' represent the concentrations of cells, glucose, ethanol, and glycerol respectively within the fermentor. Since the fermentor is considered to be well-mixed, the exiting concentrations, i.e. X_{out}, S_{out}, P_{out}, and P'_{out}, are the same as X, S, P, and P'. Since the fermentor operates at steady state we can also eliminate the

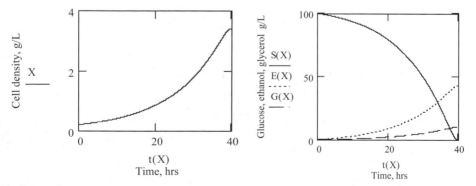

Figure 1.2   Results from the simulation of a batch fermentor.

time derivatives on the left side of the above equations. With X_{in}, P_{in}, and P'_{in} also equal to zero, the above equations may then be written as shown below.

$$\text{Yeast } FX = \frac{V\mu_{max}S}{K_c + S}X$$

$$\text{Glucose } FS = FS_{in} - \frac{V}{Y_{X/S}}\frac{\mu_{max}S}{K_S + S}X$$

$$\text{Ethanol } FP = \frac{VY_{P/S}}{Y_{X/S}}\frac{\mu_{max}S}{K_S + S}X$$

$$\text{Glycerol } FP' = \frac{VY_{P/S}}{Y_{X/S}}\frac{\mu_{max}S}{K_S + S}X$$

The dilution rate (D) is defined as F/V and has units of 1/time. For this problem:

$$D = \frac{100 \text{ L hr}^{-1}}{2000 \text{ L}} = 0.05 \text{ 1 hr}^{-1}$$

and from the yeast equation we can solve for the value of S as given by:

$$S = \frac{DK_S}{\mu_{max} - D} = \frac{0.05 \text{ 1/ hr}^{-1} \times 1.5 \text{ g/L}^{-1}}{0.075 \text{ 1/ hr}^{-1} - 0.05 \text{ 1/ hr}^{-1}} = 3 \text{ g/L}$$

giving a glucose conversion of 97%. Note that the above equation is only valid provided that $\mu_{max} > D$. If this condition is not met, then the growth rate of the cells is not sufficient to keep up with the cells leaving in the outlet stream. This then leads to washout of the cells from the fermentor. From the glucose material balance equation we can then solve for the exiting yeast concentration as shown below.

$$X = \frac{F\left(S_{in} - S\right)}{\dfrac{V}{Y_{X/S}}\dfrac{\mu_{max}S}{K_S + S}} = \frac{100 \text{ L hr}^{-1}\left(100 - 3\right)\text{g L}^{-1}}{\dfrac{2000 \text{ L}}{0.032}\dfrac{0.075 \text{ 1 hr}^{-1}}{1.5 \text{ g L}^{-1}}\dfrac{3 \text{ g L}^{-1}}{+ 3 \text{ g L}^{-1}}} = 3.104 \text{ g L}^{-1}$$

Similarly we can rearrange the material balance equations for ethanol and glycerol and obtain their exiting concentrations as follows:

$$P = \frac{VY_{P/S}}{FY_{X/S}}\frac{\mu_{max}S}{K_S + S}X = \frac{2000 \text{ L} \times 0.4293}{100 \text{ L hr}^{-1} \times 0.032}\frac{0.075 \text{ 1 hr}^{-1} \times 3 \text{ g L}^{-1}}{1.5 \text{ g L}^{-1} + 3 \text{ g L}^{-1}} \times 3.104 \text{ g L}^{-1}$$

$$P = 41.64 \text{ g L}^{-1}$$

$$P' = \frac{VY_{P/S}}{FY_{X/S}} \frac{\mu_{max} S}{K_s + S} X = \frac{2000 \text{ L} \times 0.1032}{100 \text{ L hr}^{-1} \times 0.032} \frac{0.075 \text{ 1 hr}^{-1} \times 3 \text{ g L}^{-1}}{1.5 \text{ g L}^{-1} + 3 \text{ g L}^{-1}} \times 3.104 \text{ g L}^{-1}$$

$$P' = 10.01 \text{ g L}^{-1}$$

Table 1.8 compares the results obtained in Example 1.7 for the batch fermentor to those obtained in this example for a continuous stirred-tank fermentor (CSTF). The batch reactor results shown in the table are based on the time needed to obtain 98% conversion of the glucose fed to the fermentor which is 38.8 hrs. This is also known as the residence time. In the CSTF the residence time is equal to the volume of the fermentor divided by the flowrate or:

$$\text{Residence time } \tau = \frac{V}{F} = \frac{1}{D} = \frac{2000 \text{ L}}{100 \text{ L hr}^{-1}} = 20 \text{ hrs}$$

Productivity is a measure of the amount of product produced per unit time per unit volume of fermentor. For the batch fermentor, the productivity is equal to the final product concentration divided by the batch residence time. For the CSTF the productivity is equal to $\frac{PF}{V}$. Note that the CSTF is nearly twice as productive as the batch fermentor.

Table 1.8 Comparison of results obtained for the batch and continuous fermentors

	Batch	CSTF
Glucose conversion	97%	97%
Residence time, hrs	38.58	20
Productivity, g ethanol/L/hr	1.08	2.082
Yeast, g/L	3.304	3.104
Glucose, g/L	3.0	3.0
Ethanol, g/L	41.64	41.64
Glycerol, g/L	10.01	10.01

PROBLEMS

1. $R = 8.314 \text{ J mol}^{-1} \text{ K}^{-1}$, find the equivalent in cal mol^{-1} K^{-1}.

2. Convert 98.6°F to °C and then K.

3. During cell migration, fibroblasts can generate traction forces of approximately 2×10^4 μdynes. What is the equivalent force in kN?

4. Endothelial cells in random motion were recorded to move at the *lightning speed* of 27 μm/hr, what would their speed be in miles per second?

5. A process requires 48 MW of power to convert A to B. What is the power needed in cg cm^2 hr^{-3}? What about in kJ per minute?

6. Which is moving faster: a plane moving at 400 miles per hour or a molecule moving at 6.25×10^{15} nm per minute? Show the answer in units of mm per second.

7. Resolve Example 1.4 using the Laplace transform technique.

8. Rework Example 1.5 assuming there are 0.21 moles of glycerol formed for each mole of ethanol and 0.13 moles of water formed for each mole of glycerol.

9. Draw a sketch of Example 1.7 and label all of the components added to the batch fermentor. Show for a glucose conversion of 97% that the material balance on all the components that take part in the fermentation is satisfied.

10. Draw a sketch of Example 1.8 and label all of the streams entering and exiting the fermentor. Show for a glucose conversion of 97% that the material balance on all the components that take part in the fermentation is satisfied.

11. Rework Example 1.7 assuming the specific growth rate (μ_{max}) of the yeast cells is 0.15 1/hr.

12. Rework Example 1.8 assuming the specific growth rate (μ_{max}) of the yeast cells is 0.15 1/hr.

13. Explore the productivity of the fermentor discussed in Example 1.8 as a function of the residence time $\tau = V/F$. Is there an optimal residence time that maximizes the ethanol productivity in the fermentor?

14. An artificial kidney is a device that removes water and wastes from blood. In one such device, the hollow fiber hemodialyzer, blood flows from an artery through the insides of a bundle of cellulose acetate fibers, and dialyzing fluid, which consists of water and various dissolved salts, flows on the outside of the fibers. Water and wastes – principally urea, creatinine, uric acid, and phosphate ions – pass through the fiber walls into the dialyzing fluid, and the purified blood is returned to a vein. At some time during the dialysis of a patient in kidney failure, the arterial and venous blood conditions are as follows:

	Arterial blood – In	Venous blood – Out
Flow rate, ml/min	200	195
Urea concentration, mg/ml	2.1	1.2

a. Calculate the rates at which urea and water are being removed from the blood.

b. If the dialyzing fluid enters at the rate of 1500 ml/min and the exiting dialyzing solution (dialysate) leaves at about the same rate, calculate the concentration of urea in the dialysate.

c. Suppose we want to reduce the patient's urea level from an initial value of 2.7 mg/ml to a final value of 1.1 mg/ml. If the total distribution volume is 40 liters and the average rate of urea removal is that calculated in part (a), how long must the patient be dialyzed? (Neglect the loss in total blood volume due to the removal of water in the dialyzer.)

TWO

A REVIEW OF THERMODYNAMIC CONCEPTS

Thermodynamics (power from heat) as a science began its development during the 19th century and was first used to understand the operation of work producing devices such as steam engines. In broadest terms, thermodynamics is concerned with the relationships between different types of energy. Here we will be mainly interested in a specialized area of thermodynamics related to solutions. But before we can understand solution thermodynamics we first must review some basic concepts of thermodynamics that will then lead us to the relationships we need for understanding the thermodynamics of solutions. Our goal here is to understand the mathematical basis of the field and apply the most useful results to our study of biomedical engineering transport phenonema.

2.1 THE FIRST LAW OF THERMODYNAMICS

There are three general laws of thermodynamics and these will be stated here in the forms that do not consider the effects of nuclear reactions. The *first law* is a statement of energy conservation and must be applied to both the system and the surroundings. In order to describe how the system and surroundings may exchange energy, we first need to define the type of system that is being considered. The most basic type of system is called an *isolated system*. In an isolated system there is no exchange of mass or energy between the system and the surroundings. Therefore for an isolated system its energy and mass is constant. We define a *closed system* as a system that can exchange energy with its surroundings but not mass. Hence the mass of a closed system does not change. The *open system* is defined as one that can exchange both mass and energy with its surroundings. We will discuss open systems in greater detail later during our discussion of solution thermodynamics.

2.1.1 Closed Systems

Energy exchange between a closed system and its surroundings is in the form of *heat* (Q) and *work* (W). Careful attention must be given to the sign of Q and W. The sign convention that is used is that Q and W are positive for transfer of energy from the surroundings to the system. Hence if heat is added to the system, or work done on the system, then Q and W are positive.

The total energy of a closed system consists of a summation of its *internal energy* (U) and the external energies known as the *potential energy* (E_P) and the *kinetic energy* (E_K). The potential energy depends on the position of the system in a gravitational field and the kinetic energy is a result of system motion. The internal energy of the system is that energy possessed by the molecules that make up the chemical substances within the system. These molecular energies include the kinetic energy of translation, rotation, vibration, and intermolecular forces.

With respect to the system, we are usually not really interested in the absolute values of these energies but only in their changes. Hence we can state that the change in the total energy of the system is equal to $\Delta U + \Delta E_P + \Delta E_K$ where the Greek symbol delta (Δ) signifies the final state minus the initial state.

For a closed system the change in the total energy of the system $(\Delta U + \Delta E_P + \Delta E_K)$ must be equal to the energy transferred as heat and work with the surroundings. The following equation is therefore a statement of the *first law of thermodynamics* for a closed system.

$$\Delta U + \Delta E_P + \Delta E_K = Q + W \tag{2.1}$$

If there are no changes in the kinetic and potential energies of the system, then we simply have that for a closed system

$$\Delta U = Q + W \tag{2.2}$$

Most of the time ΔE_P and ΔE_K are much smaller than ΔU, Q, and W and we can just use Equation 2.2. The exception are those cases where large changes in system position or velocity are expected or desired. For an isolated system we have $\Delta U = 0$ since both Q and W are zero.

2.1.2 Steady Flow Processes

We can also apply the first law of thermodynamics to the steady flow process shown in Figure 2.1. We now consider the energy changes that occur within a unit mass of material that enters at plane 1 and leaves the process at plane 2. We take our system to be this unit mass of material. The total energy of this unit mass of fluid can change as a result of changes in its internal energy, potential energy, and kinetic energy. Equation 2.1 still applies to this situation as well, however the kinetic energy term represents the change in kinetic energy of the unit mass of flowing fluid and the potential energy term represents the change in potential energy of the unit mass of flowing fluid due to changes in its elevation within its gravitational field. Also it is important to remember that the work term W also includes the work done on or by the unit mass of fluid as it

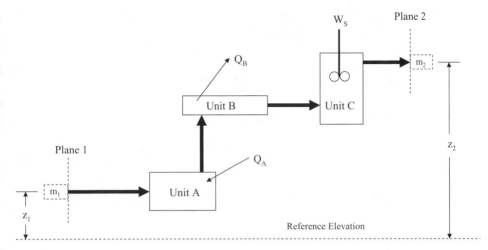

Figure 2.1   A steady state flow process.

flows through the process. This work that is done on or by the unit mass of fluid is called the *shaft work* (W_S).

Examples of shaft work include the work effect associated with reciprocating engines, pumps, turbines, and compressors. Therefore, W in Equation 2.1 represents the shaft work as well as the work done on the unit mass of fluid as it enters and exits the process. Recall that work is defined as the displacement of an external force, that is $W = - \int_{x_1}^{x_2} F_{external} dx$. Here the force would be the pressure at the entrance of the process (P_1) multiplied by the cross-sectional area (A_1) normal to the flow of the unit mass of fluid. The displacement of our unit mass of fluid would equal the specific volume of our fluid (V_1) divided by the cross-sectional area (A_1). So the work performed by the surroundings to push our unit mass of fluid into the process at the entrance is given by the following equation.

$$W_1 = P_1 A_1 \times \frac{V_1}{A_1} = P_1 V_1 \tag{2.3}$$

Note that W_1 is positive since work is done by the surroundings on the unit mass of fluid. At the exit of the process the work is done by the system on the surroundings and W_2 is given by the following expression.

$$W_2 = - P_2 V_2 \tag{2.4}$$

For our system taken as a unit mass of fluid, we can now rewrite Equation 2.1 and obtain the following equation.

$$\Delta U + \Delta E_P + \Delta E_K = Q + W_S + P_1 V_1 - P_2 V_2 \tag{2.5}$$

This equation can also be rearranged and written as follows:

$$\Delta(U + PV) + \Delta E_P + \Delta E_K = Q + W_S \tag{2.6}$$

The combination of the internal energy (U) and the quantity PV is defined as the *enthalpy* (H) and we can write Equation 2.6 as:

$$\Delta H + \Delta E_P + \Delta E_K = Q + W_S \qquad (2.7)$$

In many cases as mentioned earlier the changes in the potential and kinetic energy of the fluid are negligible in comparison to the other terms so Equation 2.7 simplifies to:

$$\Delta H = Q + W_S \qquad (2.8)$$

Equations 2.7 and 2.8 are expressions of the first law of thermodynamics for steady flow processes. Recall that for a closed non-flow system the first law is given by Equations 2.1 and 2.2.

As written, Equations 2.2 and 2.8 apply for finite changes in the energy quantities ΔU and ΔH. We will find it useful later to work with differential changes of these energy quantities and Equations 2.2 and 2.8 may then be written in their differential form as

$$dU = dQ + dW \text{ and } dH = dQ + dW_S \qquad (2.9)$$

2.2 THE SECOND LAW OF THERMODYNAMICS

The *second law of thermodyamics* states that there is a property of matter called the *entropy* (S) and that for any process, the sum of the entropy changes of the system and the surroundings (i.e. ΔS_{Total}) is always greater than or equal to zero. Therefore, entropy is not conserved like energy. No process is feasible if the total change in entropy is less than zero. This statement of the second law of thermodynamics is given by the following equation.

$$\Delta S_{\text{Total}} = \Delta S_{\text{System}} + \Delta S_{\text{Surroundings}} \geq 0 \qquad (2.10)$$

For an isolated system there is no exchange of mass or energy between the system and the surroundings, so $\Delta S_{\text{Surroundings}} = 0$ and $\Delta S_{\text{System}}(\text{isolated}) \geq 0$.

At the molecular level entropy is a measure of disorder. The more disorder, the higher the entropy. It is important to recognize that a system can become more organized ($\Delta S_{\text{System}} < 0$), however there must be an even greater increase in the disorder of the surroundings such that in an overall sense Equation 2.10 is satisfied. The *third law of thermodynamics* sets a lower limit on the entropy. The third law states that at a temperature of absolute zero the entropy for a perfectly ordered crystal of a given substance is zero.

2.2.1 Reversible Processes

Processes for which $\Delta S_{\text{Total}} > 0$ are known as *irreversible processes* whereas processes for which $\Delta S_{\text{Total}} = 0$ are known as *reversible processes*. A reversible process is one that can be reversed by an infinitesimal change in the variable that controls the process, in other words the system is only infinitesimally removed from its equilibrium state.

In thermodynamics, an equilibrium state is defined as one in which there is no tendency for change on a macroscopic scale. For example, for a mechanically reversible process imagine a frictionless piston that confines a gas within a cylinder closed at one end. At equilibrium the pressure of the gas within the cylinder exactly balances the weight of the piston and the pressure of the surroundings. A reversible expansion of this gas requires that the pressure within the cylinder be increased an infinitesimal amount. Similarly a reversible compression requires that the pressure of the surroundings be increased by an infinitesimal amount. The driving force for the reversible expansion or compression of the gas within the cylinder is the infinitesimal difference in pressure between the confined gas and the surroundings.

Similarly, thermal equilibrium requires that an object and its surroundings be at the same temperature. Reversible heat transfer occurs when the object and its surroundings differ in temperature by an infinitesimal amount. The driving force for reversible transfer of heat being the infinitesimal difference in temperature between the object and its surroundings.

A reversible process is an idealization since no real process is reversible. However, it is a useful concept since it does allow us to set limits on a real process and perform useful calculations. For example, for a mechanically reversible process like the piston and cylinder example discussed above, the work term in Equations 2.1 and 2.2 can be evaluated as follows. Recall that work is defined as the displacement of an external force, that is $W = -\int_{x_1}^{x_2} F_{external} dx$. Then for the reversible work we can replace the external force by the product of the gas pressure P and the cross-sectional area of the cylinder A, since these forces are only infinitesimally different, hence $F_{external} = PA$. The differential displacement of the piston (i.e. dx) by the expanding or contracting gas within the cylinder would be equivalent to the differential change in volume of the gas divided by the cross-sectional area of cylinder, i.e. $dx = dV/A$. Therefore, the following equation is obtained for the mechanically reversible work of the gas contained within our piston cylinder system:

$$W_{Reversible} = -\int_{V_1}^{V_2} P \, dV \qquad (2.11)$$

or in differential form we can say that $dW_{Reversible} = -P \, dV$. Provided we know how the pressure P varies with the change in the volume we can evaluate the integral in Equation 2.11 and obtain the reversible work $W_{Reversible}$.

Example 2.1 Consider the reversible expansion or compression of one mole of an ideal gas at constant temperature (isothermal) from an initial volume V_1 to a final volume V_2. Find an expression for the heat and work effects of this isothermal process.

SOLUTION For an isothermal process involving an ideal gas there is no change in the internal energy so $\Delta U = 0$ and from Equation 2.2 we have that $Q = W_{Reversible}$. For an ideal gas we can also use the ideal gas law with $P = RT/V$ and substitute this into Equation 2.11. Upon integration we obtain the following equations for the heat and reversible work per mole of gas.

$$Q = -W_{\text{Reversible}} = \int_{V_1}^{V_2} \frac{RT}{V} dV = RT \ln\left(\frac{V_2}{V_1}\right) = RT \ln\left(\frac{P_1}{P_2}\right)$$

2.3 PROPERTIES

Through the above discussion of the three laws of thermodynamics we have introduced three new properties of matter. These are the internal energy (U), the enthalpy (H), and the entropy (S). Along with the pressure (P), volume (V), and temperature (T), these are all known as properties or state variables. State variables only depend on the state of the system and not on how one arrived at that state. Q and W (or W_S) are not properties since according to Equations 2.2 and 2.8 one can envision different processes or paths involving the transfer of Q and W (or W_S) all giving the same change in the internal energy (ΔU), the enthalpy (ΔH), or the entropy (ΔS). Therefore, Q and W (or W_S) are path-dependent quantities and depend on how the process is actually carried out. For example, for a mechanically reversible process, Equation 2.11 allows calculation of the work, provided one knows how P changes with V.

For a pure substance only two of these properties (i.e. T, P, V, U, and S) need to be specified in order to define completely the thermodynamic state of the substance. Our equation of state also provides an additional relationship between P, V, and T. Usually we fix the state of a pure component by specifying the temperature and the pressure. So from the equation of state we can calculate the volume. The other properties such as internal energy, enthalpy, and entropy would also be known since they too will depend only on the temperature and pressure as shown in the following discussion.

2.3.1 Heat Capacity

Calculating the changes in the internal energy, enthalpy, and entropy is facilitated by the definition of heat capacities. The *heat capacity* relates the change in temperature of an object to the amount of heat that was added to it. However, heat or Q is not a property but is dependent on how the process of heat transfer was carried out. So it is convenient to define heat capacity in such a manner that it too is a property. Therefore, we have two types of heat capacity, one defined at constant volume (C_V) and the other defined at constant pressure (C_P) as given by the following equations.

$$C_V \equiv \left(\frac{\partial U}{\partial T}\right)_V \quad \text{Heat capacity at constant volume} \tag{2.12}$$

$$C_P \equiv \left(\frac{\partial H}{\partial T}\right)_P \quad \text{Heat capacity at constant pressure} \tag{2.13}$$

Since they only depend on properties, C_P and C_V are also properties of a substance and their values for specific substances may be found or estimated as described in such reference books as Poling *et al.* (2001). For monoatomic gases, C_p is approximately

$5/2\ R$, and for diatomic gases like nitrogen, oxygen, and air, C_p is approximately $7/2\ R$. We will also show later on in Example 2.2 that for an ideal gas $C_p = C_v + R$.

Now for a constant volume process, or for that matter, any process where the final volume is the same as the initial volume, Equation 2.12 when integrated tells us that the change in internal energy is given by the following equation.

$$\Delta U = \int_{T_1}^{T_2} C_V\, dT \tag{2.14}$$

If the process occurs reversibly and at constant volume, i.e. the volume never changes, then from Equation 2.11 $W_{\text{Reversible}} = 0$ and $Q = \Delta U$. So for a constant volume process the heat transferred is equal to the change in the internal energy.

In a similar manner, for the constant pressure process, or a process in which the initial and final pressures are the same, Equation 2.13 can be integrated to obtain the change in enthalpy.

$$\Delta H = \int_{T_1}^{T_2} C_p\, dT \tag{2.15}$$

If the process occurs reversibly and at constant pressure, i.e. the pressure never changes, then from Equations 2.2 and 2.11 we have that $W_{\text{Reversible}} = -P\,\Delta V$ and that $\Delta U = Q - P\Delta V$ or $Q = \Delta H$. Hence for a constant pressure process the heat transferred is equal to the change in the enthalpy.

2.3.2 Calculating the Change in Entropy

The entropy change of a closed system during a reversible process has been shown to be given by the following equation.

$$\Delta S = \int \frac{dQ_{\text{Reversible}}}{T} \tag{2.16}$$

In differential form we can write this as $dQ_{\text{Reversible}} = T\, dS$.

2.3.2.1 Entropy change of an ideal gas
We can use this result to calculate the entropy change for a substance. First we consider an ideal gas and assume that we have one mole of gas and write the first law of thermodynamics for a reversible process as $dU = dQ_{\text{Reversible}} - PdV = TdS - PdV$. Next we use the definition of enthalpy ($H = U + PV$) to arrive at the fact that $dH = dU + PdV + VdP$. We now solve this relationship for dU and use the result shown a few lines above for dU to obtain the following expression for dS.

$$dS = \frac{dH}{T} - \frac{V}{T}\, dP \tag{2.17}$$

For one mole of an ideal gas we know that $dH = C_p\, dT$ and from the ideal gas law $V/T = R/P$ so Equation 2.17 becomes:

$$dS = C_p \frac{dT}{T} - R\frac{dP}{P} \tag{2.18}$$

The above equation can then be integrated from an initial state of (T_1, P_1) to a final state (T_2, P_2) as given by the following equation.

$$\Delta S = \int_{T_1}^{T_2} C_P \frac{dT}{T} - R \ln\left(\frac{P_2}{P_1}\right) \tag{2.19}$$

Note that all that is needed to calculate the entropy change of an ideal gas is the heat capacity C_P. Note that at constant P, $\Delta S = \int_{T_1}^{T_2} C_P \frac{dT}{T}$.

For a constant volume process we can use the fact that from the ideal gas law $\frac{dP}{P} = \frac{R \, dT}{PV} = \frac{dT}{T}$ and from the example shown below we have that for an ideal gas $C_P = C_V + R$, so when these two relationships are used in Equation 2.18 and the result is integrated, we obtain $\Delta S = \int_{T_1}^{T_2} C_V \frac{dT}{T}$ for the constant volume process.

It is important to note that although we started the derivation of Equations 2.18 and 2.19 assuming a reversible process, the resulting equations only contain properties and are therefore independent of how the process was actually carried out between the initial state of (T_1, P_1) and the final state of (T_2, P_2). Therefore Equations 2.18 and 2.19 can be used to calculate the entropy change of an ideal gas regardless of whether or not the process is reversible.

Example 2.2 Calculate the the final temperature and the work produced for the reversible adiabatic expansion of 10 moles of an ideal gas from 10 atm and 500 K to a final pressure of 1 atm. Assume that $C_V = 5/2 \, R$ and that the heat capacity ratio $\gamma = C_P/C_V$ is also a constant and equal to 1.4.

SOLUTION Since the process is reversible we can state that $\Delta S_{Total} = \Delta S_{System} + \Delta S_{Surroundings} = 0$. An adiabatic process means that there is no heat transfer between the system (i.e. the gas) and its surroundings. Therefore $\Delta S_{Surroundings} = 0$ and we then have that $\Delta S_{System} = 0$. A process for which $\Delta S_{System} = 0$ is also called an *isentropic process*. Next we can use Equation 2.19 from above and set that equal to zero and obtain the equation shown below.

$$C_P \ln\left(\frac{T_2}{T_1}\right) = R \ln\left(\frac{P_2}{P_1}\right)$$

Next we solve the above equation for the final temperature T_2 which is given by the next equation.

$$\frac{T_2}{T_1} = \left(\frac{P_2}{P_1}\right)^{R/C_P} = \left(\frac{P_2}{P_1}\right)^{\gamma-1/\gamma}$$

Note that we have used the fact that since $H = U + PV$, then for an ideal gas $H = U + RT$ and $dH = dU + R \, dT$. Since $dH = C_P \, dT$ and $dU = C_V \, dT$ we have that for an ideal gas $C_P = C_V + R$. The heat capacity ratio is defined by $\gamma = C_P/C_V$. We can now use the above equation to calculate the final temperature of the gas following the reversible expansion.

$$T_2 = 500\,\text{K} \left(\frac{1\,\text{atm}}{10\,\text{atm}} \right)^{\frac{1.4-1}{1.4}} = 258.98\,\text{K}$$

Next we can calculate the amount of work that is done by the 10 moles of gas. For a closed system, we have from the first law of thermodynamics that $n\,\Delta U = Q + W$, where n is the number of moles of gas. Since the process is adiabatic then $Q = 0$ and $W = n\,\Delta U = n\,C_V\,(\,T_2 - T_1\,)$. Now $C_V = 5/2\,R$ and using the value of $R = 8.314$ J mol^{-1} K^{-1} from Table 1.6 gives a value of $C_V = 20.78$ J mol^{-1} K^{-1}. So the work effect of this expansion can now be calculated as shown below.

$$W = 10 \text{ moles} \times 20.78 \text{ J mol}^{-1} \text{ K}^{-1} \times (\,258.98 \text{ K} - 500 \text{ K})$$

$$= -50{,}084 \text{ J} = -50.084 \text{ kJ}$$

The work is negative since our sign convention states that if the system does work its value is negative. The change in the internal energy is also the same as the work in this case so the energy used to perform work by the system is at the expense of the internal energy of the gas. That is why the final temperature of the gas is much lower than the initial temperature of the gas.

2.3.3 The Gibbs and Helmholtz Free Energy

In addition to the internal energy, entropy, and enthalpy $H = U + PV$, there are also two additional properties that can be derived from the primary properties of P, V, T, U, and S. These are the *Gibbs free energy* (G) and the *Helmholtz free energy* (A).

2.3.3.1 Gibbs free energy The Gibbs free energy (G) is also a property and is defined in terms of the enthalpy, temperature, and entropy as $G = H - T\,S$ or $G = U + PV - TS$. The Gibbs free energy change (ΔG) at constant temperature can be calculated from the change in the enthalpy and the entropy as $\Delta G = \Delta H - T\,\Delta S$. The Gibbs free energy can be shown to equal the maximum amount of useful work that can be obtained from a reversible process at constant T and P. This can be shown as follows.

From above, we have at constant temperature that $\Delta G = \Delta H - T\,\Delta S = \Delta U + P\,\Delta V - T\,\Delta S$. But for a reversible process we also know that $\Delta U = Q_{\text{Reversible}} + W_{\text{Reversible}} = T\,\Delta S + W_{\text{Reversible}}$. Therefore $\Delta G = W_{\text{Reversible}} + P\,\Delta V$. At constant pressure, $- P\,\Delta V$ represents the work (W_{PV}) obtained as a result of any volume changes that occur in the system. So we define the useful work over and above any PV work as $W_{\text{Useful}} = W_{\text{Reversible}} - W_{PV} = \Delta G$.

The Gibbs free energy is also useful for determining whether or not a given process will occur at constant temperature and pressure. For example, from the second law of thermodynamics we know that for the system $\Delta S = Q_{\text{Reversible}}/T \geq 0$. From the first law of thermodynamics we know that for a reversible process at constant pressure $Q_{\text{Reversible}} =$

$\Delta U + P\Delta V = \Delta H$. Therefore, we have that $\Delta S \geq \Delta H / T$ or $\Delta H - T \Delta S \leq 0$. But $\Delta H - T \Delta S$ is the same as ΔG at constant temperature so $\Delta G \leq 0$. We thus obtain the following criterion for determining the feasibility of a given process at constant temperature and pressure.

Spontaneous process	$\Delta G < 0$	
Equilibrium	$\Delta G = 0$	(2.20)
No spontaneous process	$\Delta G > 0$	

2.3.3.2 Helmholtz free energy The Helmholtz free energy (A) is defined in terms of the internal energy, temperature, and entropy as $A = U - TS$. The Helmholtz free energy is useful for determining whether or not a given process will occur at constant temperature and volume. The criterion for feasibility of a process at constant temperature and volume is shown below.

Spontaneous process	$\Delta A < 0$	
Equilibrium	$\Delta A = 0$	(2.21)
No spontaneous process	$\Delta A > 0$	

2.4 THE FUNDAMENTAL PROPERTY RELATIONS

With this background on the first and second laws of thermodynamics we can now develop what are known as the fundamental property relations for one mole of a single fluid phase of constant composition. From Equation 2.9 in differential form we have that $dU = dQ + dW$. For a reversible process we also have that $dQ_{\text{Reversible}} = TdS$ and $dW_{\text{Reversible}} = -P\,dV$. Hence,

$$dU = TdS - PdV \qquad (2.22)$$

As we discussed before even though this equation was derived for the special case of a reversible process, since it only contains properties of the system, it is valid for all processes, reversible or not.

Now since $H = U + PV$, we can differentiate this to obtain $dH = dU + PdV + VdP$. We can then substitute for dU from Equation 2.22 to obtain $dH = TdS - PdV + PdV + VdP$. So we obtain that $\underline{dH = TdS + VdP}$. Next we have that $G = H - TS$, so then $dG = dH - T\,dS - S\,dT$. Using the relationship we just obtained for dH, we then have $dG = TdS + VdP - TdS - SdT$, and obtain the result that $\underline{dG = -SdT + VdP}$. Finally, since $A = U - TS$, we then have $dA = dU - TdS - SdT$. Using Equation 2.22 we get $dA = TdS - PdV - TdS - SdT$ and we have $\underline{dA = -PdV - SdT}$.

So for one mole of a single fluid phase of constant composition, we can state our fundamental property relationships as follows:

$$dU = TdS - PdV \qquad (2.23)$$

$$dH = TdS + VdP \qquad (2.24)$$

$$dG = -SdT + VdP \tag{2.25}$$

$$dA = -PdV - SdT \tag{2.26}$$

The above equations also provide functional relationships for U, H, G, and A. Therefore, we have that $U = U(S,V)$, $H = H(S,P)$, $G = G(T,P)$, and $A = A(V,T)$.

2.4.1 Exact Differentials

The fundamental property relations shown above are also exact differentials. Recall from calculus that the total differential of a function $F(x,y)$ is given by:

$$dF = \left(\frac{\partial F}{\partial x}\right)_y dx + \left(\frac{dF}{dy}\right)_x dy = M\,dx + N\,dy \tag{2.27}$$

with $M = \left(\dfrac{\partial F}{\partial x}\right)_y$ and $N = \left(\dfrac{\partial F}{\partial y}\right)_x$. We can also differentiate M and N with respect to y and x and obtain the following relationships:

$$\left(\frac{\partial M}{\partial y}\right)_x = \frac{\partial^2 F}{\partial y \partial x} \quad \text{and} \quad \left(\frac{\partial N}{\partial x}\right)_y = \frac{\partial^2 F}{\partial x \partial y} \tag{2.28}$$

Because the order of the differentiation with respect to x and y in the above equations is not important, we obtain the following result:

$$\left(\frac{\partial M}{\partial y}\right)_x = \left(\frac{\partial N}{\partial x}\right)_y \tag{2.29}$$

which is the criterion for exactness of the total differential given by Equation 2.27. We can now use Equation 2.29 on our fundamental property relations given in Equations 2.23–2.26 to obtain a set of relationships between our properties. These relationships are known as the *Maxwell equations* and several of these derived from Equations 2.23–2.26 are shown below.

$$\left(\frac{\partial T}{\partial V}\right)_S = -\left(\frac{\partial P}{\partial S}\right)_V$$

$$\left(\frac{\partial T}{\partial P}\right)_S = \left(\frac{\partial V}{\partial S}\right)_P$$

$$\left(\frac{\partial V}{\partial T}\right)_P = -\left(\frac{\partial S}{\partial P}\right)_T \tag{2.30}$$

$$\left(\frac{\partial P}{\partial T}\right)_V = \left(\frac{\partial S}{\partial V}\right)_T$$

The Maxwell equations and the fundamental property relations are useful for deriving thermodynamic relationships between the various properties. The following example illustrates this.

Example 2.3 Obtain expressions for the T and P dependence of the enthalpy and the entropy, i.e. $H(T,P)$ and $S(T,P)$.

SOLUTION The approach is to use the fundamental property relations and the Maxwell equations to obtain final expressions that only depend on P, V, and T and heat capacity. P, V, and T can then be related by an equation of state, the simplest being the ideal gas law for which $PV = RT$, or experimental PVT data. With this as our strategy, we first take the total differential of H and S and obtain:

$$dH = \left(\frac{\partial H}{\partial T}\right)_P dT + \left(\frac{\partial H}{\partial P}\right)_T dP \quad \text{and} \quad dS = \left(\frac{\partial S}{\partial T}\right)_P dT + \left(\frac{\partial S}{\partial P}\right)_T dP$$

Now since $\left(\dfrac{\partial H}{\partial T}\right)_P = C_P$ and from Equation 2.24 we can also obtain that

$\left(\dfrac{\partial H}{\partial T}\right)_P = T\left(\dfrac{\partial S}{\partial T}\right)_P$ and $\left(\dfrac{\partial S}{\partial T}\right)_P = \dfrac{1}{T}\left(\dfrac{\partial H}{\partial T}\right)_P = \dfrac{C_P}{T}$. From Equation 2.24 we

can also write that $\left(\dfrac{\partial H}{\partial P}\right)_T = T\left(\dfrac{\partial S}{\partial P}\right)_T + V$. Using the third Maxwell equation in

Equation 2.30, replace $\left(\dfrac{\partial S}{\partial P}\right)_T$ with $-\left(\dfrac{\partial V}{\partial T}\right)_P$ and we then have that

$\left(\dfrac{\partial H}{\partial P}\right)_T = V - T\left(\dfrac{\partial V}{\partial T}\right)_P$. Our expressions for dH and dS can then be written as follows.

$$dH = C_P\, dT + \left[V - T\left(\frac{\partial V}{\partial T}\right)_P\right] dP \quad \text{and} \quad dS = C_P\frac{dT}{T} - \left(\frac{\partial V}{\partial T}\right)_P dP$$

These equations can be integrated to obtain the change in the enthalpy and entropy between two states (T_1, P_1) and (T_2, P_2).

$$\Delta H = \int_{T_1}^{T_2} C_P\, dT + \int_{P_1}^{P_2}\left[V - T\left(\frac{\partial V}{\partial T}\right)_P\right] dP \quad \text{and} \quad \Delta S = \int_{T_1}^{T_2} C_P\frac{dT}{T} - \int_{P_1}^{P_2}\left(\frac{\partial V}{\partial T}\right)_P dP$$

The above equations provide the temperature and pressure dependence for the enthalpy and the entropy for one mole of a single phase fluid of constant composition. The heat capacity and experimental PVT data, or an equation of state, is needed to solve these equations. For the special case of an ideal gas we can use the fact that $PV = RT$ and that for an ideal gas $\left(\dfrac{\partial V}{\partial T}\right)_P = \dfrac{R}{P}$. Upon substituting this expression into the above equations we obtain the following results for an ideal gas.

$$\Delta H = \int_{T_1}^{T_2} C_P \, dT \quad \text{and} \quad \Delta S = \int_{T_1}^{T_2} C_P \frac{dT}{T} - R \ln\left(\frac{P_2}{P_1}\right)$$

Note that we obtain the result that for an ideal gas the enthalpy and, for that matter, the internal energy, only depends on the temperature. However, the entropy of an ideal gas depends on both the temperature and the pressure. Also note that this result for ΔS for the special case of the ideal gas is the same result we obtained earlier as Equation 2.19.

2.5 SINGLE PHASE OPEN SYSTEMS

Now suppose that we have a single phase open system. For an open system there can be an exchange of matter between the system and the surroundings. Therefore the composition or the number of moles of each substance present in the system can change. Let n represent the total moles of these N chemical substances that are present. So for a total of n moles and any property M, we can state that $nM = nM$ (T, P, n_1, n_2, n_3,..., n_N) and n_i represents the moles of each substance that is present in the system. This tells us that the property M depends on the temperature, the pressure, and the number of moles of each species present in the solution. The total differential of nM is then given by:

$$d(nM) = \left[\frac{\partial(nM)}{\partial T}\right]_{P,n} dT + \left[\frac{\partial(nM)}{\partial P}\right]_{T,n} dP + \sum_i \left[\frac{\partial(nM)}{\partial n_i}\right]_{T,P,n_j} dn_i \quad (2.31)$$

In the last term of Equation 2.31 the summation is over all substances that are present and the partial derivative within the bracketed term in the summation is taken with respect to substance i at constant temperature, pressure, and the moles of all other substances held constant.

2.5.1 Partial Molar Properties

In solution thermodynamics, the partial derivative of a property (M) with respect to the number of moles of a given substance (n_i) is called a partial molar property ($\overline{M}_i$) and in general is defined by the following equation.

$$\overline{M}_i = \left[\frac{\partial(nM)}{\partial n_i}\right]_{T,P,n_j} \quad (2.32)$$

Of particular interest in subsequent discussions is the partial molar Gibbs free energy ($\overline{G}_i$) which is also known as the *chemical potential* (μ_i) of component i in the mixture. The chemical potential is therefore defined as follows:

$$\mu_i = \bar{G}_i = \left[\frac{\partial (nG)}{\partial n_i} \right]_{T,P,n_j} \tag{2.33}$$

We can use Equation 2.32 and rewrite Equation 2.31 as follows:

$$d(nM) = \left[\frac{\partial (nM)}{\partial T} \right]_{P,n} dT + \left[\frac{\partial (nM)}{\partial P} \right]_{T,n} dP + \sum_i \bar{M}_i \, dn_i \tag{2.34}$$

Next we can define the mole fraction of substance i as $x_i = n_i/n$ so $n_i = x_i n$ and $\sum_i x_i = 1^1$. We also have that $dn_i = x_i \, dn + n \, dx_i$ and that $d(nM) = ndM + Mdn$. Using these relations, and recognizing that n in the first two bracketed terms of Equation 2.34 is constant, Equation 2.34 can be rearranged into the following form:

$$\left[dM - \left(\frac{\partial M}{\partial T} \right)_{P,x} dT - \left(\frac{\partial M}{\partial P} \right)_{T,x} dP - \sum_i \bar{M}_i dx_i \right] n + \left[M - \sum_i x_i \bar{M}_i \right] dn = 0 \tag{2.35}$$

Since n and dn are arbitrary, the only way the above equation can be satisfied is for the bracketed terms to both equal zero. Hence we obtain the following equations:

$$dM = \left(\frac{\partial M}{\partial T} \right)_{P,x} dT + \left(\frac{\partial M}{\partial P} \right)_{T,x} dP + \sum_i \bar{M}_i dx_i \tag{2.36}$$

$$M = \sum_i x_i \bar{M}_i \quad \text{or} \quad nM = \sum_i n_i \bar{M}_i \tag{2.37}$$

Now Equation 2.36 is nothing more than Equation 2.34 written on a mole fraction basis. Equation 2.37 shows how the mixture property M depends on the composition of the solution. Note that the mixture property is a mole fraction or mole weighted average of the component partial molar property $\bar{M}_i$. The partial molar property for component i represents the property value for that component as it exists in the solution which can be quite different from the pure component value of that property, M_i.

From Equation 2.37 we can also write that $dM = \sum_i x_i d\bar{M}_i + \sum_i \bar{M}_i dx_i$. Using this relationship along with Equation 2.36 results in the following equation.

$$\left(\frac{\partial M}{\partial T} \right)_{P,x} dT + \left(\frac{\partial M}{\partial P} \right)_{T,x} dP - \sum_i x_i d\bar{M}_i = 0 \tag{2.38}$$

Equation 2.38 places another restriction on the property changes for a single phase solution. This equation is known as the *Gibbs–Duhem equation* and can be used to test the thermodynamic consistency of experimental mixture property data since this

---

[1]Note that most of the time x is used to denote liquid phase mole fractions and y is used for vapor or gas phase mole fractions.

equation must be satisfied. For the special case of constant temperature and pressure, we have:

$$\sum_i x_i d\bar{M}_i = 0 \qquad (2.39)$$

2.5.1.1 Binary systems To illustrate the application of the above equations, let us apply them to a binary solution at constant temperature and pressure. For a binary solution we can write from Equation 2.37 that:

$$M = x_1 \bar{M}_1 + x_2 \bar{M}_2 \quad \text{and} \quad dM = x_1 d\bar{M}_1 + \bar{M}_1 dx_1 + x_2 d\bar{M}_2 + \bar{M}_2 dx_2 \qquad (2.40)$$

The Gibbs–Duhem equation for a binary can be written as follows:

$$x_1 d\bar{M}_1 + x_2 d\bar{M}_2 = 0 \qquad (2.41)$$

Since $x_1 + x_2 = 1$ and $dx_1 = -dx_2$, we can combine Equations 2.40 and 2.41 and obtain:

$$dM = \bar{M}_1 dx_1 - \bar{M}_2 dx_1 \quad \text{or} \quad \frac{dM}{dx_1} = \bar{M}_1 - \bar{M}_2 \qquad (2.42)$$

Now we can solve for $\bar{M}_1$ and $\bar{M}_2$ from Equation 2.40 and substitute these results into Equation 2.42 to obtain the following expressions for $\bar{M}_1$ and $\bar{M}_2$.

$$\bar{M}_1 = M + (1 - x_1)\frac{dM}{dx_1} \quad \text{and} \quad \bar{M}_2 = M - x_1 \frac{dM}{dx_1} \qquad (2.43)$$

The above equations are very important since they allow for the calculation at constant temperature and pressure of the partial molar properties $\bar{M}_1$ and $\bar{M}_2$ from the composition dependence of the mixture property M. Note that as $x_2 \rightarrow 0$, then component 2 is becoming infinitely dilute and from Equation 2.43, $\bar{M}_2 \rightarrow \bar{M}_2^\infty = M_1 - \frac{dM}{dx_1}\Big|_{x_1 \rightarrow 1}$ where $\bar{M}_1 \rightarrow M_1$ or the pure component property value M_1 and the superscript ∞ on $\bar{M}_2$ denotes that component 2 is infinitely dilute. Likewise as $x_1 \rightarrow 0$, then component 1 is becoming infinitely dilute and $\bar{M}_2 \rightarrow M_2$ or the pure component property value M_2 and $\bar{M}_1 \rightarrow \bar{M}_1^\infty = M_2 + \frac{dM}{dx_1}\Big|_{x_2 \rightarrow 1}$. Figure 2.2 shows the composition dependence of these property values at constant T and P. Note that $\frac{dM}{dx_1}\Big|_{x_1 \rightarrow 1}$ and $\frac{dM}{dx_1}\Big|_{x_2 \rightarrow 1}$ are the respective tangents of the M curve as $x_2 \rightarrow 0$ and $x_1 \rightarrow 0$.

2.5.1.2 Property changes of mixing The difference between the mixture property value (M) and the mole fraction weighted sum of the pure component property values $\left(\sum_i y_i M_i\right)$ evaluated at the same T and P is known as the *property change of mixing* (ΔM^{mix}) and is given by the following equation.

$$\Delta M^{\text{mix}} = \sum_i y_i \bar{M}_i - \sum_i y_i M_i = M - \sum_i y_i M_i \qquad (2.44)$$

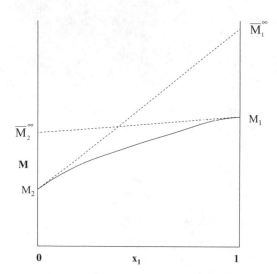

Figure 2.2   Partial molar properties in a binary system at constant T and P.

2.5.1.3 Ideal gas

The above relationships are valid for describing the solution behavior of real solutions whether they are solids, liquids, or gases. Now let us see what these relationships tell us for the special case of an ideal gas. Recall that the ideal gas model assumes that the molecules have negligible volume and that the molecules do not interact with one another. In other words, each species in the ideal gas acts as if no other species are present and all of the gas molecules can move independently within the whole volume (V_T) of the container.

Consider an ideal gas consisting of a total of n moles and N components at a given temperature T and in a total volume V_T. We know from the ideal gas law that $P = nRT/V_T$. For the same value of the temperature, each of the n_i moles of component i considered separately will also occupy the same total volume V_T. Therefore, each component i exerts a contribution to the total pressure P called the *partial pressure* (P_i) that is given by $P_i = n_iRT/V_T$. The ratio of P_i to P is given by $P_i/P = n_i/n$ which defines the gas phase mole fraction of component i in the mixture, that is y_i. Hence we obtain the fact that the *partial pressure* of component i in an ideal gas mixture is given by the following relationship.

$$P_i = y_i P \text{ with } y_i = n_i/n = P_i/P \tag{2.45}$$

Since the $\sum_i y_i = 1$ then $\sum_i P_i = P$, or the sum of the partial pressures of all components equals the total pressure. In addition, we can deduce that for an ideal gas the partial molar volume for component i is $\overline{V}_i^{\text{ideal gas}} = V_i^{\text{ideal gas}} = \dfrac{RT}{P}$. Since $V_T =$

$\sum_i n_i \overline{V}_i^{\text{ideal gas}}$, then we have that $V_T = \sum_i n_i \overline{V}_i^{\text{ideal gas}} = \sum_i n_i \dfrac{RT}{P} = \dfrac{nRT}{P}$, and we see that this definition of the ideal gas partial molar volume satisfies the ideal gas law for the mixture. Therefore, for an ideal gas the partial molar volume is the same for all components and is equal to RT/P.

With the exception of the volume, for any other property, the partial molar value of the property for component i in an ideal gas mixture at T and P is the same as the corresponding pure component molar property at the mixture temperature and at a pressure that is equal to that component's partial pressure in the mixture, that is:

$$\overline{M}_i^{\text{ideal gas}}(T,P) = M_i^{\text{ideal gas}}(T,P_i) \tag{2.46}$$

In Example 2.3 we showed that for an ideal gas the enthapy and the internal energy do not depend on the pressure. Furthermore, since there are no molecular interactions in an ideal gas, the enthalpy or internal energy of component i in the mixture for a given T and P is the same as its pure component value at T and P_i. Therefore, for an ideal gas we can state that $\overline{H}_i(T,P) = H_i^{\text{ideal gas}}(T,P_i)$ and $\overline{U}_i(T,P) = U_i^{\text{ideal gas}}(T,P_i)$, where $H_i^{\text{ideal gas}}$ and $U_i^{\text{ideal gas}}$ represent the pure component ideal gas values of the enthalpy and internal energy at the mixture T and at component i's partial pressure, P_i. Hence from Equation 2.37, then for an ideal gas mixture we have that:

$$H^{\text{ideal gas}} = \sum_i y_i H_i^{\text{ideal gas}} \quad \text{and} \quad U^{\text{ideal gas}} = \sum_i y_i U_i^{\text{ideal gas}} \tag{2.47}$$

and this can be generalized using Equation 2.46 for any ideal gas property M as:

$$M^{\text{ideal gas}}(T,P) = \sum_i y_i M_i^{\text{ideal gas}}(T,P_i) \tag{2.48}$$

We also saw in Example 2.3 that the entropy of an ideal gas depends on the pressure and the temperature. Consider pure component i at T and P_i. We can ask, what is the entropy change for component i when it is placed in an ideal gas mixture at T and P? Since the T is constant and the pressure changes from P_i to P, we can use the result for the entropy change of an ideal gas that was obtained in Example 2.3, that is:

$$\Delta S_i^{\text{ideal gas}} = S_i^{\text{ideal gas}}(T,P) - S_i^{\text{ideal gas}}(T,P_i) = -R\ln\left(\dfrac{P}{P_i}\right) = -R\ln\left(\dfrac{P}{y_i P}\right) = R\ln y_i \tag{2.49}$$

and after rearranging we obtain the result that:

$$S_i^{\text{ideal gas}}(T,P_i) = S_i^{\text{ideal gas}}(T,P) - R\ln y_i \tag{2.50}$$

Using Equation 2.48 we then obtain the following expression for the entropy of an ideal gas mixture

$$S^{\text{ideal gas}} = \sum_i y_i S_i^{\text{ideal gas}}(T,P) - R\sum_i y_i \ln y_i \tag{2.51}$$

Example 2.4 Determine an expression for the enthalpy, internal energy, and entropy change that occurs when N pure components at T and P are mixed to form one mole of a solution at the same T and P. Assume that the pure components and the resulting mixture are ideal gases.

SOLUTION Equation 2.44 above defines the property change of mixing as:

$$\Delta M^{\text{mix}} = M - \sum_i y_i M_i$$

Comparing the above equation with Equation 2.47 shows that the enthalpy and internal energy change of mixing is zero for the formation of an ideal gas mixture from its pure components. However, using Equation 2.51, the entropy change of mixing for this case is given by the following expression.

$$\Delta S^{\text{mix}} = S^{\text{ideal gas}} - \sum_i y_i S_i^{\text{ideal gas}}(T,P) = -R\sum_i y_i \ln y_i$$

The $-R\sum_i y_i \ln y_i$ term is always positive indicating that the process of mixing results in an increase in the entropy of the system. Since the enthalpy of mixing is also zero there is no change in the entropy of the surroundings, so the total entropy change is also positive. Therefore, mixing these pure ideal gas components is an irreversible process.

2.5.1.4 Gibbs free energy of an ideal gas mixture With the entropy of an ideal gas mixture given by Equation 2.51, we can now calculate the Gibbs free energy of an ideal gas mixture, that is $G^{\text{ideal gas}} = H^{\text{ideal gas}} - T\,S^{\text{ideal gas}}$. Using the above relationships that were derived for $H^{\text{ideal gas}}$ and $S^{\text{ideal gas}}$, we can obtain the following expression for the Gibbs free energy of an ideal gas mixture:

$$G^{\text{ideal gas}} = \sum_i y_i H_i^{\text{ideal gas}} - T\left(\sum_i y_i S_i^{\text{ideal gas}} - R\sum_i y_i \ln y_i\right) \tag{2.52}$$

which can also be written as follows:

$$G^{\text{ideal gas}} = \sum_i y_i G_i^{\text{ideal gas}} + RT\sum_i y_i \ln y_i \tag{2.53}$$

Comparing Equation 2.53 with Equations 2.46 and 2.48 we see that the partial molar Gibbs free energy, or the chemical potential, of component i in an ideal gas mixture is then given by:

$$\mu_i^{\text{ideal gas}} = \bar{G}_i^{\text{ideal gas}} = G_i^{\text{ideal gas}} + RT \ln y_i \tag{2.54}$$

We also know from our fundamental property relations (see Equation 2.25) for pure component i at constant T, that $dG_i^{\text{ideal gas}} = V_i^{\text{ideal gas}}\,dP = \dfrac{RT}{P}dP = RT\,d\ln(P)$. This result can be integrated from an arbitrary pressure P_0 to the pressure P to give the following result for the pure component Gibbs free energy of an ideal gas:

$$G_i^{\text{ideal gas}} = (G_i^{0,\,\text{ideal gas}} - RT \ln P_0) + RT \ln P = G_i^{0,\text{ideal gas}} + R\,T \ln \frac{P}{P_0} = \mu_i^{0,\,\text{ideal gas}}$$

$$+RT \ln \frac{P}{P_0} \tag{2.55}$$

where $G_i^{0,\text{ideal gas}}$ or $\mu_i^{0,\text{ideal gas}}$ depends only on the temperature and is the Gibbs free energy or chemical potential per mole of pure component i at a pressure equal to P_0. Combining the above result with Equation 2.54 provides the following alternative expression for the chemical potential of component i in an ideal gas mixture.

$$\mu_i^{\text{ideal gas}} = \mu_i^{0,\text{ideal gas}} + RT \ln\left(\frac{y_i P}{P_0}\right) \tag{2.56}$$

2.5.2 Pure Component Fugacity

For a pure component (i.e. $y_i = 1$) in the ideal gas state, Equation 2.56 tells us that the pure component Gibbs free energy of an ideal gas is given by the following expression: $G_i^{\text{ideal gas}} = (\mu_i^{0,\text{ideal gas}} - RT \ln P_0) + RT \ln P$. This result can be generalized to the real gas by defining a new property called the *pure component fugacity* (f_i) to replace the pressure. The fugacity may be thought of as a "corrected" pressure that provides the value of the pure component Gibbs free energy of a real gas. Therefore the following expression defines the pure component fugacity for a real gas:

$$G_i = \mu_i^{0,\text{ideal gas}} - RT \ln P_0 + RT \ln f_i \tag{2.57}$$

Now if we subtract Equation 2.55 from Equation 2.57 we obtain the following result.

$$G_i - G_i^{\text{ideal gas}} = RT \ln \frac{f_i}{P} = RT \ln \phi_i \tag{2.58}$$

The pure component fugacity coefficient, ϕ_i, is defined as the ratio of the pure component fugacity, f_i, to the pressure, P. Therefore, $\phi_i = f_i/P$. Clearly we see that for an ideal gas the pure component fugacity, f_i, is equal to the pressure, or $f_i^{\text{ideal gas}} = P$, and $\phi_i^{\text{ideal gas}} = 1$.

The difference in the Gibbs free energy in Equation 2.58, i.e. $G_i - G_i^{\text{ideal gas}}$, defines what is also called a *residual property*, or in this case, the residual Gibbs free energy, $G_i^R = G_i - G_i^{\text{ideal gas}}$, where both G_i and $G_i^{\text{ideal gas}}$ are evaluated at the same T and P. This leads to the following general definition of a residual property as the difference between the actual value of a property and its value in the ideal gas state.

$$M^R = M - M^{\text{ideal gas}} \tag{2.59}$$

Equation 2.59 applies to any of our properties, i.e. U, H, S, G, A, V.

2.5.2.1 Calculating the pure component fugacity
As shown in the following development, the pure component fugacity can be calculated from experimental PVT data or an appropriate equation of state. First it is convenient to define the *compressibility factor* (Z) which describes the deviation of a real gas from the ideal gas state. The compressibility factor is a dimensionless quantity and is defined by the following equation.

$$Z = \frac{P V}{R T} \tag{2.60}$$

Note that for an ideal gas $Z = 1$. A variety of equations of state have been developed for calculating the compressibility factor for real gases and liquids at high pressures (Poling *et al.* 2001).

We can also use the following mathematical relationship for $d\left(\dfrac{G}{RT}\right)$ that is

$$d\left(\frac{G}{RT}\right) = \frac{RTdG - GRdT}{(RT)^2} = \frac{1}{RT}dG - \frac{G}{RT^2}dT \qquad (2.61)$$

and then substituting for dG from Equation 2.25 and recognizing that $G = H - TS$, we have that:

$$d\left(\frac{G}{RT}\right) = \frac{V}{RT}dP - \frac{H}{RT^2}dT \qquad (2.62)$$

From this equation we can also write the following additional relationships at constant pressure and constant temperature.

$$\frac{H}{RT} = -T\left[\frac{\partial\left(\dfrac{G}{RT}\right)}{\partial T}\right]_P \quad \text{and} \quad \frac{V}{RT} = \left[\frac{\partial\left(\dfrac{G}{RT}\right)}{\partial P}\right]_T \qquad (2.63)$$

The above relationships are also valid for the residual properties G^R, H^R, and V^R by simply adding a superscript to the property.

Now the residual volume, V^R, is defined as $V - V^{\text{ideal gas}}$. So we can write the residual volume as follows using the definition from above for the compressibility factor.

$$V^R = \frac{RT}{P}(Z - 1) \qquad (2.64)$$

Using the above relationships, we can now obtain the following expressions for the residual properties. First we can take the second expression in Equation 2.63 at constant temperature and write it for the residual property. Then we integrate at constant temperature from zero pressure (G^R is equal to zero since this is an ideal gas state) to any pressure P.

$$\frac{G^R}{RT} = \int_0^P \frac{V^R}{RT}dP = \int_0^P (Z - 1)\frac{dP}{P} \qquad (\text{constant } T) \qquad (2.65)$$

To obtain the residual enthalpy we can use the first expression in Equation 2.63. Next

we evaluate $\left[\dfrac{\partial\left(\dfrac{G^R}{RT}\right)}{\partial T}\right]_P$ using Equation 2.65 and then we obtain:

$$\frac{H^R}{RT} = -T \int_0^P \left(\frac{\partial Z}{\partial T}\right)_P \frac{dP}{P} \qquad \text{(constant } T) \qquad (2.66)$$

Since $TS^R = H^R - G^R$, we can also write that $\dfrac{S^R}{R} = \dfrac{H^R}{RT} - \dfrac{G^R}{RT}$ and we can use the above two relationships for H^R and G^R to obtain:

$$\frac{S^R}{R} = -T \int_0^P \left(\frac{\partial Z}{\partial T}\right)_P \frac{dP}{P} - \int_0^P (Z-1)\frac{dP}{P} \qquad \text{(constant } T) \qquad (2.67)$$

Equations 2.65–2.67 are very important relationships since they allow us to calculate the real property value, M, from the residual property value, M^R, and the ideal gas property value $M^{\text{ideal gas}}$. Provided we have experimental PVT data, or an appropriate equation of state, we can use these equations to calculate the values of the Gibbs free energy, the enthalpy, and the entropy for real gases and even liquids at high pressures. In addition, we can also use Equation 2.65 to calculate the fugacity of a pure component at any T and P. Comparing Equation 2.65 with Equation 2.58, we see that Equation 2.65 is also equal to the natural logarithm of the fugacity coefficient (ϕ_i) or f_i/P, that is:

$$\ln \phi_i = \ln\left(\frac{f_i}{P}\right) = \int_0^P (Z-1)\frac{dP}{P} \qquad (2.68)$$

Equation 2.68 therefore allows us to determine the fugacity of a pure component at any T and P from knowledge of its PVT behavior.

Example 2.5 Obtain an expression for the fugacity coefficient using the pressure explicit form of the virial equation of state truncated after the second coefficient. If the second virial coefficient ($\bar{B}$) of a particular gas is $- 0.01$/atm, calculate the compressibility factor and the fugacity of this gas at a temperature of 500 K and a pressure of 10 atm.

SOLUTION The simplest equation of state for a real gas is the virial equation of state given by the following volume and pressure explicit forms

$$Z = 1 + \frac{B}{V} + \frac{C}{V^2} + \frac{D}{V^3} + \ldots$$

$$Z = 1 + \bar{B}P + \bar{C}P^2 + \bar{D}P^3 + \ldots$$

where B, C, D and $\bar{B}$, $\bar{C}$, $\bar{D}$ are known as the second, third, and fourth virial coefficients and are only a function of temperature. If we truncate the above series after the second coefficient then we have for the pressure explicit form that $Z = 1 + \bar{B}P$ or $Z - 1 = \bar{B}P$. Using Equation 2.68 we then obtain:

$$\ln \phi_i = \bar{B} \int_0^P dP = \bar{B}P$$

Hence the value of $Z = 1 - 0.01/\text{atm} \times 10 \text{ atm} = 0.90$ and the fugacity coefficient and fugacity is calculated as follows:

$$\ln \phi_i = -0.01/\text{atm} \times 10 \text{ atm} = -0.10$$

$\phi_i = 0.905$ and since $\dfrac{f_i}{P} = \phi_i$, then $f_i = 0.905 \times 10 \text{ atm} = 9.05 \text{ atm}$

and we see that the fugacity of this gas at these conditions is 9.05 atm compared with 10 atm if the gas were an ideal gas.

2.5.3 Fugacity of a Component in a Mixture

For a component in a mixture of real gases or a liquid solution, we can generalize Equation 2.57 for an ideal gas mixture to provide the fugacity of component i as it exists in the real mixture

$$\mu_i = (\mu_i^{0, \text{ ideal gas}} - RT \ln P_0) + RT \ln \hat{f}_i \tag{2.69}$$

where $\hat{f}_i$ is defined as the fugacity of component i as it exists in the mixture. We can also solve Equation 2.57 for the value of $(\mu_i^{0,\text{ideal gas}} - RT \ln P_0)$ in terms of G_i and f_i to obtain the following result.

$$\mu_i = \bar{G}_i = G_i + RT \ln \left(\frac{\hat{f}_i}{f_i} \right) \tag{2.70}$$

Equation 2.70 shows the relationship between the chemical potential of component i, its pure component Gibbs free energy and fugacity (G_i and f_i), and its fugacity as it exists in the mixture ($\hat{f}_i$).

Now if we subtract Equation 2.56 from Equation 2.69 for the same T and P we obtain:

$$\mu_i - \mu_i^{\text{ideal gas}} = \bar{G}_i^R = RT \ln \left(\frac{\hat{f}_i}{y_i P} \right) = RT \ln \hat{\phi}_i \tag{2.71}$$

where $\hat{\phi}_i = \dfrac{\hat{f}_i}{y_i P}$ is defined as the component fugacity coefficient. For component i in an ideal gas $\hat{\phi}_i = 1$ since $\mu_i = \mu_i^{\text{ideal gas}}$ and we have:

$$\hat{f}_i = \hat{f}_i^{\text{ideal gas}} = y_i P \tag{2.72}$$

We can calculate the values of $\hat{\phi}_i$ provided we have mixture PVT data or an equation of state that describes the mixture (Poling et al. 2001). To see how this is done, we can rewrite Equation 2.65 for n moles of our mixture as follows:

$$\frac{nG^R}{RT} = \int_0^P (nZ - n) \frac{dP}{P} \tag{2.73}$$

Then from Equation 2.71 we have that $\ln \hat{\phi}_i = \bar{G}_i^R / RT = \left[\dfrac{\partial \left(n\, G^R / RT \right)}{\partial n_i} \right]_{T, P, n_j}$ which allows us to rewrite Equation 2.73 as:

$$\ln \hat{\phi}_i = \int_0^P \left[\frac{\partial \left(nZ - n \right)}{\partial n_i} \right]_{T, P, n_j} \frac{dP}{P} = \int_0^P \left(\bar{Z}_i - 1 \right) \frac{dP}{P} \tag{2.74}$$

with the partial molar compressibility factor $\bar{Z}_i = \left[\dfrac{\partial \left(nZ \right)}{\partial n_i} \right]_{T, P, n_j}$ and $\dfrac{\partial n}{\partial n_i} = 1.$

Provided we have mixture PVT data or a mixture equation of state, we can use Equation 2.74 to calculate the component fugacity coefficient (Poling *et al.* 2001).

2.5.4 The Ideal Solution

The ideal solution model includes ideal gas mixtures as well as liquids and solids. The ideal solution model is useful for describing mixtures of substances whose molecules do not differ much in their size and chemical nature. For example, a liquid solution of ethanol and propanol would be expected to form an ideal solution whereas a solution of ethanol and water would be expected to form a non-ideal or real solution. The ideal solution model, therefore, serves as a useful basis of comparison to real solution behavior.

Recall that for an ideal gas the chemical potential of component i was given by Equation 2.54. For an ideal solution we generalize this result by replacing the Gibbs free energy of component i in the ideal gas state, $G_i^{\text{ideal gas}}$, with G_i, the pure component Gibbs free energy at the same T, P, and physical state (i.e. solid, liquid, or gas) as the mixture. Therefore, an ideal solution is defined by the following equation for the partial molar Gibbs free energy or chemical potential of component i:

$$\mu_i^{\text{ideal solution}} = \bar{G}_i^{\text{ideal solution}} = G_i + RT \ln x_i \tag{2.75}$$

where x_i is defined as the mole fraction of component i in the mixture. We can also write Equation 2.70 for an ideal solution as $\mu_i^{\text{ideal solution}} = G_i + RT \ln \left(\dfrac{\hat{f}_i^{\text{ideal solution}}}{f_i} \right)$ and combining this result with Equation 2.75 above, we obtain the important result that:

$$\hat{f}_i^{\text{ideal solution}} = x_i f_i \tag{2.76}$$

Equation 2.76 is also known as the *Lewis–Randall rule* and says that the fugacity of a component in an ideal solution is proportional to its mole fraction. Furthermore, the proportionality constant is the pure component fugacity evaluated at the same T and P as the solution being considered. Since $\hat{\phi}_i = \dfrac{\hat{f}_i}{x_i P}$, we then see using Equation 2.76 that for an ideal solution $\hat{\phi}_i^{\text{ideal solution}} = \dfrac{f_i}{P} = \phi_i.$

Example 2.6 Show that with the above definition of the partial molar Gibbs free energy of component i in an ideal solution, the enthalpy of mixing, the internal energy of mixing, and the volume of mixing are equal to zero for an ideal solution.

SOLUTION Based on Equation 2.31 we can write the mixture Gibbs free energy as follows:

$$d(nG) = \left[\frac{\partial(nG)}{\partial T}\right]_{P,n} dT + \left[\frac{\partial(nG)}{\partial P}\right]_{T,n} dP + \sum_i \left[\frac{\partial(nG)}{\partial n_i}\right]_{T,P,n_j} dn_i$$

For a closed system containing n moles we can also write from Equation 2.25 that $d(nG) = -(nS)\, dT + (nV)\, dP$. Comparing this result to the first two terms of the above equation for constant n we see that the first two partial derivatives can be replaced by $-(nS)$ and (nV) respectively. In addition, we know that $\left[\dfrac{\partial(nG)}{\partial n_i}\right]_{T,P,n_j}$

$= \overline{G}_i = \mu_i$, the chemical potential of component i. Therefore, the above equation may be written as:

$$d(nG) = -(nS)dT + (nV)dP + \sum_i \mu_i \, dn_i$$

We can now use the criterion of exactness (Equation 2.29) on the above equation and obtain the following relationships.

$$\left(\frac{\partial \mu_i}{\partial T}\right)_{P,n} = -\left(\frac{\partial(nS)}{\partial n_i}\right)_{P,T,n_j} = -\overline{S}_i \quad \text{and} \quad \left(\frac{\partial \mu_i}{\partial P}\right)_{T,n} = \left(\frac{\partial(nV)}{\partial n_i}\right)_{P,T,n_j} = \overline{V}_i$$

Using the above relationships along with Equation 2.75, we then have that for an ideal solution:

$$\left(\frac{\partial \mu_i^{\text{ideal solution}}}{\partial T}\right)_{P,n} = -\overline{S}_i^{\text{ideal solution}} = \left(\frac{\partial \overline{G}_i}{\partial T}\right)_P + R \ln x_i$$

$$\left(\frac{\partial \mu_i^{\text{ideal solution}}}{\partial P}\right)_{T,n} = \overline{V}_i^{\text{ideal solution}} = \left(\frac{\partial \overline{G}_i}{\partial P}\right)_T$$

Next we can use our fundamental property relations for component i (Equation 2.25) to show that $\left(\dfrac{\partial \overline{G}_i}{\partial T}\right)_P = -\overline{S}_i$ and $\left(\dfrac{\partial \overline{G}_i}{\partial P}\right)_T = \overline{V}_i$. Hence we obtain the following equations for the partial molar entropy and partial molar volume for an ideal solution.

$$\overline{S}_i^{\text{ideal solution}} = \overline{S}_i - R \ln x_i \quad \text{and} \quad \overline{V}_i^{\text{ideal solution}} = \overline{V}_i$$

By Equation 2.44 we easily can see that the volume change of mixing for an ideal solution is zero and that $V^{\text{ideal solution}} = \sum_i x_i V_i$. For the enthalpy change of mixing we can use the fact that $\overline{H}_i^{\text{ideal solution}} = \overline{G}_i^{\text{ideal solution}} + T \overline{S}_i^{\text{ideal solution}}$ and using the above relationships we obtain that $\overline{H}_i^{\text{ideal solution}} = G_i + T S_i = H_i$. Therefore, the enthalpy change of mixing for an ideal solution is also zero and $H^{\text{ideal solution}} = \sum_i x_i H_i$. In a similar fashion, we can use the fact that $\overline{U}_i^{\text{ideal solution}} = \overline{H}_i^{\text{ideal solution}} - P\overline{V}_i^{\text{ideal solution}} = H_i - PV_i = U_i$ and show that the internal energy change of mixing is also zero.

2.6 PHASE EQUILIBRIUM

With this background on solution thermodynamics, we can now address the criterion for equilibrium between two phases (I and II) contained within a closed system. Equilibrium between the two phases also implies that the two phases are at the same temperature. Also, the two phases must have the same pressure, unless they are separated by a semipermeable rigid barrier or membrane. We will consider the situation where the pressures may be different later in this chapter when we discuss osmotic equilibrium.

Each phase is also an open system and is free to exchange mass with the other phase. Since we are considering the closed system containing the two phases to be at constant T and P, we can use the result from Example 2.6 that $d(nG) = -(nS) \, dT + (nV) \, dP + \sum_i \mu_i \, dn_i$ and write this equation for each phase as follows:

$$d(nG)^{\text{I}} = -(nS)^{\text{I}} \, dT + (nV)^{\text{I}} \, dP + \sum_i \mu_i^{\text{I}} \, dn_i^{\text{I}}$$

$$(2.77)$$

$$d(nG)^{\text{II}} = -(nS)^{\text{II}} \, dT + (nV)^{\text{II}} \, dP + \sum_i \mu_i^{\text{II}} \, dn_i^{\text{II}}$$

Using the above equations, then for the closed system we can also write that:

$$d(nG) = -(nS) \, dT + (nV) \, dP = d(nG)^{\text{I}} + d(nG)^{\text{II}} =$$

$$-\left[(nS)^{\text{I}} + (nS)^{\text{II}}\right] dT + \left[(nV)^{\text{I}} + (nV)^{\text{II}}\right] dP + \sum_i \mu_i^{\text{I}} dn_i^{\text{I}} + \sum_i \mu_i^{\text{II}} dn_i^{\text{II}} \quad (2.78)$$

Since $(nS)^{\text{I}} + (nS)^{\text{II}} = (nS)$ and $(nV)^{\text{I}} + (nV)^{\text{II}} = (nV)$, the above equation simplifies to:

$$\sum_i \mu_i^{\text{I}} dn_i^{\text{I}} + \sum_i \mu_i^{\text{II}} dn_i^{\text{II}} = 0 \quad (2.79)$$

The dn_i^{I} and dn_i^{II} represent the differential changes in the moles of component i as a result of mass transfer between phases I and II. Clearly, the law of mass conservation requires that $dn_i^{\text{I}} = -dn_i^{\text{II}}$. With this result the above equation may then be written as follows:

$$\sum_i \left(\mu_i^{\text{I}} - \mu_i^{\text{II}}\right) dn_i^{\text{I}} = 0 \quad (2.80)$$

The values of dn_i^{I} and dn_i^{II} are arbitrary so the only way the above equation can be satisfied is for each term in parentheses (i.e. $\mu_i^{\mathrm{I}} - \mu_i^{\mathrm{II}}$) to be equal to zero. So our criterion for phase equilibrium in a two phase system is expressed by the following equation:

$$\mu_i^{\mathrm{I}} = \mu_i^{\mathrm{II}} \qquad (2.81)$$

which simply states that the chemical potential for each component i is the same in phase I and phase II.

This result can be easily generalized to provide the criterion for phase equilibrium in π phases at the same T and P as:

$$\mu_i^{\mathrm{I}} = \mu_i^{\mathrm{II}} = \ldots = \mu_i^{\pi} \qquad (2.82)$$

If we only have a pure component in each phase, Equation 2.81 or 2.82 still applies and since $\mu_i = \bar{G}_i = G_i$, then for a pure component, phase equilibrium requires that:

$$G_i^{\mathrm{I}} = G_i^{\mathrm{II}} = \ldots = G_i^{\pi} \quad \text{(pure component)} \qquad (2.83)$$

Using Equation 2.82 along with $\mu_i = \bar{G}_i = G_i + RT \ln\left(\dfrac{\hat{f}_i}{f_i}\right)$ from Equation 2.70, we can write an alternative expression for the criterion for phase equilibrium in multicomponent systems in terms of the component mixture fugacities as:

$$\hat{f}_i^{\mathrm{I}} = \hat{f}_i^{\mathrm{II}} = \ldots = \hat{f}_i^{\pi} \qquad (2.84)$$

2.6.1 Pure Component Phase Equilibrium

Figure 2.3 shows a pressure-volume or PV curve for a pure component. Shown in the figure are several lines of constant temperature or isotherms. The area within the dome shaped region defines conditions where two phases, for example vapor and liquid, are at equilibrium for a given T and P. At the top of the dome is the *critical point* (C) where the liquid and vapor phases become identical and thus have the same physical properties. The isotherm that passes through the critical point is known as the critical isotherm, T_C. The critical temperature, pressure, and volume (T_C, P_C, and V_C) are also pure component physical properties and play an important role in the estimation of a variety of physical properties (Poling *et al.* 2001).

Below this critical isotherm we see that the other isotherms (i.e. subcritical isotherms) have three separate segments. Within the dome, we see that the isotherm is horizontal and represents the constant temperature and pressure phase change between the saturated liquid and saturated vapor states. The left most point on this horizontal segment denotes the saturated liquid and the right most point on the horizontal segment denotes the saturated vapor. Saturated means that for the pure component, the vapor and liquid phases are at equilibrium at $T = T^{\mathrm{Sat}}$ and $P = P^{\mathrm{Sat}}$, where P^{Sat} is called the saturation pressure or the vapor pressure of the pure component. P^{Sat} for a pure component only depends on the saturation temperature, i.e. $P^{\mathrm{Sat}} = P^{\mathrm{Sat}}(T^{\mathrm{Sat}})$. The normal boiling point (T_{BP}) of a pure component occurs when the value of P^{Sat} is the same as the

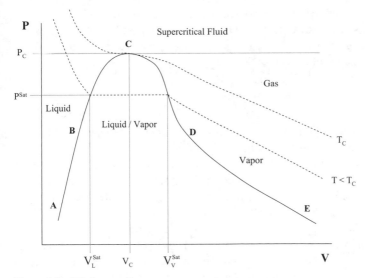

Figure 2.3   PV diagram for a pure component.

atmospheric pressure or $P^{\text{Sat}} = 1$ atmosphere $= P^{\text{Sat}}(T^{\text{Sat}}) = P^{\text{Sat}}(T_{\text{BP}})$. Points along the curve ABC define saturated liquids and the points along the curve CDE define saturated vapor. Points along the horizontal isotherm within the dome represent saturated mixtures of liquid and vapor. The actual molar volume of the vapor/liquid mixture will depend on the relative amounts of each phase present, i.e. $V = (1 - x)\, V^{\text{Lsat}} + x\, V^{\text{Vsat}}$, where x is the vapor fraction of the vapor/liquid mixture.

The region to the left of the curve ABC denotes the liquid state, and the isotherms here rise very quickly because liquid volume changes are very small as the pressure is increased, and for the most part, liquids at modest pressures are therefore considered to be incompressible. The region to the right of curve CDE denotes the vapor state provided $T < T_C$ and $P < P_C$, and a gas for $T > T_C$ and $P < P_C$. The conditions where $T > T_C$ and $P > P_C$ is known as the supercritical fluid region.

Now let us consider the vapor (V) and liquid (L) equilibrium of a pure component i at T and $P = P^{\text{Sat}}$. From Equation 2.83 we can write that $G^V = G^L$ and it follows that $dG^V = dG^L$. Next we can use Equation 2.25 and express the previous result as:

$$dG^V = -S^V\, dT + V^V\, dP^{\text{Sat}} = -S^L\, dT + V^L\, dP^{\text{Sat}} = dG^L \tag{2.85}$$

This equation may be rearranged as follows:

$$\frac{dP^{\text{Sat}}}{dT} = \frac{S^V - S^L}{V^V - V^L} = \frac{\Delta S^{LV}}{\Delta V^{LV}} \tag{2.86}$$

Now ΔS^{LV} and ΔV^{LV} represent the changes in the entropy and volume of the pure component when it is transferred from the liquid phase to the vapor phase at the equilibrium temperature (T) and pressure (P^{Sat}). If we integrate Equation 2.24 for the change in phase from liquid to vapor at constant T and P we obtain $\Delta H^{LV} = T\Delta S^{LV}$ or $\Delta S^{LV} = \Delta H^{LV}/T$. ΔH^{LV} is also known as the *heat of vaporization* (ΔH^{Vap}) and represents the difference in enthalpy between the saturated vapor and the saturated liquid. These results can be substituted into Equation 2.86 to give:

$$\frac{dP^{Sat}}{dT} = \frac{\Delta H^{Vap}}{T\Delta V^{LV}} \qquad (2.87)$$

The above equation is also known as the *Clausius–Clapeyron* equation and is an equilibrium relationship between the saturated liquid and the saturated vapor. For a given change in the temperature (dT), Equation 2.87 provides the change in the pressure (dP^{Sat}) required to maintain equilibrium between the saturated liquid and its saturated vapor. For solid–liquid or solid–vapor equilibrium, one may substitute the *enthalpy of melting* or the *enthalpy of sublimation* for the enthalpy of vaporization in Equation 2.87.

For liquid–vapor and solid–vapor phase changes at saturation pressures near atmospheric, one can neglect the molar volume of the solid or liquid phase in comparison to the volume of the vapor and assume that the vapor phase behaves as an ideal gas, i.e. $\Delta V^{LV} = V^{V} = RT/P^{Sat}$. Equation 2.87 may then be written as follows:

$$\frac{d \ln P^{Sat}}{d\dfrac{1}{T}} = -\frac{L}{R} \qquad (2.88)$$

where L now represents either ΔH^{LV} or ΔH^{SV}. Equation 2.88 provides a relationship between the saturation pressure and the temperature in terms of the enthalpy change associated with the phase change. Over narrow ranges of temperature, L is pretty much constant, and the above equation can be integrated from an arbitrary temperature (T_0) to provide an approximate relationship between the saturation pressure and the temperature, i.e.

$$\ln P^{Sat} = \left(\ln P_0^{Sat} + \frac{L}{RT_0} \right) - \left(\frac{L}{R} \right)\frac{1}{T} = A - \frac{B}{T} \qquad (2.89)$$

where A and B represent constants that can be fitted to vapor pressure data. The above equation shows that over a narrow range of temperatures there is a linear relationship between the natural logarithm of the saturation pressure and the inverse of the temperature. Additional empirical equations for determining the saturation or vapor pressure of a given component as a function of temperature can be found in Poling *et al.* (2001).

Since we have a pure component i, Equation 2.84 can be written in terms of the pure component fugacities as:

$$f_i^{L} = f_i^{V} = f_i^{Sat} \qquad (2.90)$$

where f_i^{Sat} denotes the fugacity of either the saturated liquid or the saturated vapor. Since $f_i^{Sat} = \phi_i^{Sat} P_i^{Sat}$, we then have that:

$$f_i^{L} = f_i^{V} = f_i^{Sat} = \phi_i^{Sat} P_i^{Sat} \qquad (2.91)$$

where ϕ_i^{Sat} would be given by Equation 2.68 with $P = P^{\text{Sat}}$ using an appropriate set of PVT data or an equation of state to perform the integration. Therefore,

$$f_i^L = f_i^V = f_i^{\text{Sat}} = P_i^{\text{Sat}} \exp\left(\int_0^{P_i^{\text{Sat}}} (Z-1)\frac{dP}{P} \right) \tag{2.92}$$

2.6.1.1 Fugacity of a pure component as a compressed liquid

For a given temperature (T), if the pressure (P) is greater than the value of $P^{\text{Sat}}(T)$ for a pure component, then the liquid is considered to be subcooled or a compressed liquid and is not in equilibrium with its vapor. Calculation of the fugacity of a pure component i as a compressed liquid starts with our fundamental property relation, Equation 2.25, i.e. $dG_i = -S_i\,dT + V_i\,dP$. For constant temperature we can write this as $dG_i = V_i\,dP$ and integrate this equation from P_i^{Sat} to P as follows:

$$G_i - G_i^{\text{Sat}} = \int_{P_i^{\text{Sat}}}^{P} V_i\,dP \tag{2.93}$$

Using Equation 2.57 we can rewrite the above equation as:

$$G_i - G_i^{\text{Sat}} = RT \ln \frac{f_i}{f_i^{\text{Sat}}} = \int_{P_i^{\text{Sat}}}^{P} V_i\,dP \tag{2.94}$$

As mentioned earlier liquid molar volumes do not depend that strongly on P, so Equation 2.94 can be written as follows after setting the subcooled molar liquid volume (V_i) equal to the saturated liquid volume (V_i^{Sat}) at the same T:

$$f_i = f_i^{\text{Sat}} \exp\left(\frac{V_i^{\text{Sat}}\left(P - P_i^{\text{Sat}}\right)}{RT} \right) = \phi_i^{\text{Sat}} P_i^{\text{Sat}} \exp\left(\frac{V_i^{\text{Sat}}\left(P - P_i^{\text{Sat}}\right)}{RT} \right) \tag{2.95}$$

Comparing Equation 2.95 with Equation 2.91, we see that the exponential term is a correction of the saturated fugacity to account for the fact that the pressure for the subcooled liquid is greater than its saturation pressure for a given T. The exponential term is also known as the *Poynting factor*.

2.6.2 Excess Properties

Recall that we previously defined a residual property as the difference between the real property value and its corresponding value as an ideal gas $(M^R = M - M^{\text{ideal gas}})$. We showed in Section 2.5.2.1 that the residual properties can be determined from PVT data or an equation of state. We also found that the definition of the residual properties provided a convenient method for calculating the pure component fugacity coefficient and the fugacity coefficient of a component in a mixture, i.e. Equations 2.68 and 2.74. Accordingly, residual properties and fugacity coefficients are usually used for describing the behavior of real gases since the residual properties and the fugacity coefficients express the deviation of the real gas from ideal gas behavior.

For liquids it is usually easier to compare real liquid solution behavior with an ideal solution. An excess property (M^E) is then defined as the difference between the property value in a real solution and the value it would have in an ideal solution at the same T, P, and composition. Therefore we have:

$$M^E = M - M^{\text{ideal solution}} \qquad (2.96)$$

In terms of partial molar properties the above equation becomes:

$$\bar{M}_i^E = \bar{M}_i - \bar{M}_i^{\text{ideal solution}} \qquad (2.97)$$

For phase equilibrium calculations we are primarily interested in the Gibbs free energy, so the partial molar excess Gibbs free energy is given by the following equation.

$$\bar{G}_i^E = \bar{G}_i - \bar{G}_i^{\text{ideal solution}} \qquad (2.98)$$

Using Equations 2.70 and 2.75 for $\bar{G}_i$ and $\bar{G}_i^{\text{ideal solution}}$, we obtain using Equation 2.98 that:

$$\bar{G}_i^E = \bar{G}_i - \bar{G}_i^{\text{ideal solution}} = RT \ln \frac{\hat{f}_i}{x_i f_i} \qquad (2.99)$$

The term $\dfrac{\hat{f}_i}{x_i f_i}$ is dimensionless and is also known as the *activity coefficient*, γ_i. Note that for an ideal solution the γ_i all equal unity and $\hat{f}_i^{\text{ideal solution}} = x_i f_i$. We can then write the partial molar excess Gibbs free energy in terms of the activity coefficient as follows:

$$\bar{G}_i^E = \left[\frac{\partial \left(nG^E \right)}{\partial n_i} \right]_{T,P,n_j} = RT \ln \gamma_i \qquad (2.100)$$

Using Equation 2.37 and the Gibbs–Duhem relationship, Equation 2.39, at constant T and P, we also have that:

$$G^E = RT \sum_i x_i \ln \gamma_i \qquad (2.101)$$

$$\sum_i x_i d \ln \gamma_i = 0 \qquad (2.102)$$

Activity coefficients for a component in a mixture are obtained from experimental data, and the results are correlated with a model of how the excess Gibbs free energy depends on the composition of all the components within the mixture, i.e. $\dfrac{G^E}{RT} = f(x_1, x_2,\dots,x_N)$ at constant T, since for the most part, we can ignore the effect of pressure on the activity coefficients.

Equation 2.102 serves as a thermodynamic consistency check of experimentally determined activity coefficients or empirical equations that are used to calculate activity coefficients. For a binary system at constant T and P, Equation 2.102 can also be written as follows:

$$x_1\left(\frac{\partial \ln \gamma_1}{\partial x_1}\right)_{T,P} + x_2\left(\frac{\partial \ln \gamma_2}{\partial x_1}\right)_{T,P} = 0 \tag{2.103}$$

For a binary system we can also write Equation 2.101 as:

$$\frac{G^E}{RT} = x_1 \ln \gamma_1 + x_2 \ln \gamma_2 \tag{2.104}$$

The above equation can then be differentiated with respect to x_1 at constant T and P to give:

$$\frac{d\left(G^E/RT\right)}{dx_1} = x_1 \frac{\partial \ln \gamma_1}{\partial x_1} + \ln \gamma_1 + x_2 \frac{\partial \ln \gamma_2}{\partial x_1} + \ln \gamma_2 \frac{dx_2}{dx_1} \tag{2.105}$$

Since $dx_2/dx_1 = -1$ and using the Gibbs–Duhem equation from 2.103 above, we then have that:

$$\frac{d\left(G^E/RT\right)}{dx_1} = \ln\left(\frac{\gamma_1}{\gamma_2}\right) \tag{2.106}$$

This equation can then be integrated over x_1 as shown in the next equation.

$$\int_0^1 \frac{d\left(G^E/RT\right)}{dx_1} dx_1 = \left(G^E/RT\right)\Big|_{x_1=1} - \left(G^E/RT\right)\Big|_{x_1=0} = \int_0^1 \ln\left(\frac{\gamma_1}{\gamma_2}\right) dx_1 \tag{2.107}$$

Now the excess Gibbs free energy at $x_1 = 0$ and $x_1 = 1$ by definition is zero, hence we obtain the following as a condition on the activity coefficients:

$$\int_0^1 \ln\left(\frac{\gamma_1}{\gamma_2}\right) dx_1 = 0 \tag{2.108}$$

Equation 2.108 is also called the *area test* of activity coefficients. It provides a convenient method for testing the thermodynamic consistency of activity coefficient data. One can simply plot the values of $\ln\frac{\gamma_1}{\gamma_2}$ versus x_1 and, if the area under the resulting curve is equal to zero, then the test for thermodynamic consistency is satisfied.

A variety of models of varying complexity for describing the excess Gibbs free energy of liquid solutions have been developed (Poling *et al.* 2001). Most of the modern activity coefficient models easily handle multicomponent solutions. In addition, the so-called UNIFAC model allows the activity coefficients to be predicted from the molecular structure of the species that are in the solution. The following example illustrates the calculation of an expression for the activity coefficients for a binary system from a model of the excess Gibbs free energy.

Example 2.7 The simplest model to describe the excess Gibbs free energy of a binary liquid solution is given by the following power series.

$$\frac{G^E}{x_1 x_2 RT} = A_0 + A_1 \left(x_1 - x_2\right) + A_2 \left(x_1 - x_2\right)^2 + \dots$$

Retaining only the lead constant A_0, find expressions for γ_1 and γ_2.

SOLUTION We therefore have that $\dfrac{G^E}{RT} = A_0\, x_1\, x_2 = A_0 \left(\dfrac{n_1}{n}\right)\left(\dfrac{n_2}{n}\right)$ where n_1 and n_2 are the moles of components 1 and 2 and $n = n_1 + n_2$. Multiplying both sides of this expression by n and using Equation 2.100 above we have that:

$$\ln\,\gamma_1 = \left[\frac{\partial\left(nG^E/RT\right)}{\partial n_1}\right]_{T,\,P,\,n_2} = \frac{\left(n_1 + n_2\right)A_0\, n_2 - A_0\, n_1\, n_2}{\left(n_1 + n_2\right)^2} = A_0\,\frac{n_2}{n} - A_0\,\frac{n_1\, n_2}{n^2}$$

$$= A_0 \left(x_2 - x_1\, x_2\right) = A_0\, x_2 \left(1 - x_1\right) = A_0\, x_2^2$$

So we find that the activity coefficient for component 1 based on this model for G^E is given by $\ln\,\gamma_1 = A_0\, x_2^2$. In a similar manner we can show that the activity coefficient for component 2 is given by $\ln\,\gamma_2 = A_0\, x_1^2$. Expressions for the activity coefficients based on more complex models of G^E for both binary and multicomponent systems may be found in Poling *et al.* (2001).

Example 2.8 From the example above determine the expressions for the activity coefficients of components 1 and 2 at infinite dilution.

SOLUTION At infinite dilution for component 1, we have that $x_1 = 0$ and $x_2 = 1$. Therefore, from the results in the above example we have that $\ln\,\gamma^\infty_1 = A_0$ and $\ln\,\gamma_2 = 0$ or $\gamma_2 = 1$. At infinite dilution for component 2 we have that $x_1 = 1$ and $x_2 = 0$. Hence $\ln\,\gamma_1 = 0$ or $\gamma_1 = 1$ and $\ln\,\gamma^\infty_2 = A_0$. In this example, for either component, the natural logarithm of the activity coefficient at infinite dilution is equal to A_0.

Example 2.9 Show that the activity coefficient expressions found in Example 2.7 satisfy the Gibbs–Duhem relationship, i.e. Equation 2.103.

SOLUTION For a binary system Equation 2.103 states that:

$$x_1 \left(\frac{\partial \ln \gamma_1}{\partial x_1}\right)_{T,P} + x_2 \left(\frac{\partial \ln \gamma_2}{\partial x_1}\right)_{T,P} = 0$$

Substituting in the fact that $\ln\,\gamma_1 = A_0\, x_2^2$ and $\ln\,\gamma_2 = A_0\, x_1^2$ we then have that:

$$2\, A_0\, x_1 \left(1 - x_1\right)\left(-1\right) + 2\, A_0\, x_1 \left(1 - x_1\right) = 0$$

or

$$0 = 0$$

which shows that the expression used in Example 2.7 for the excess Gibbs free energy satisfies the Gibbs–Duhem consistency test.

2.6.3 Phase Equilibrium in Mixtures

With the above phase equilibrium relationships developed, we can now address specific topics in solution thermodynamics such as the solubility of a solid in a liquid solvent, freezing point depression, solid–gas equilibrium, gas solubility, osmotic pressure, the distribution of a solute between two liquid phases, and vapor–liquid equilibrium. The discussion below forms the foundation for our understanding in later chapters of solute transport in biological systems. Solute transport occurs across the interface between phases, and at the interface, we assume that the solute is in phase equilibrium.

2.6.3.1 Solubility of a solid in a liquid solvent   We will consider a binary system where the solvent is denoted by subscript 1 and the solute by subscript 2. We assume that the solvent has negligible solubility in the solid, hence the solid solute will exist as a pure phase. We can then write Equation 2.84 as follows to express the phase equilibrium between the solid and liquid phases:

$$f_2^S = \gamma_2\, x_2\, f_2^L \tag{2.109}$$

where f_2^S refers to the fugacity of the pure solid solute, x_2 represents the equilibrium solubility of the solid in the solvent phase expressed as a mole fraction, and f_2^L represents the fugacity of pure liquid solute at the equilibrium temperature and pressure of the solution. Since this temperature must be less than the melting temperature of the solid solute, it follows that the solute in the solution exists as a subcooled liquid. Since the solute does not ordinarily exist as a liquid at these conditions, we will have to estimate the value of f_2^L as shown in the discussion below.

With these assumptions we can rearrange Equation 2.109 to solve for the solubility (x_2) as follows:

$$x_2 = \frac{f_2^S}{\gamma_2\, f_2^L} \tag{2.110}$$

Equation 2.110 shows that the solubility is directly proportional to the ratio of the pure component fugacities of the solid and its subcooled liquid and inversely proportional to the value of the activity coefficient of the solid in the solvent solution.

To determine an expression for the solid solubility, we first must calculate the ratio of these fugacities for the pure solid. This can be accomplished by recognizing the fact that the *triple point* (T_{tp}) for a pure substance (see Figure 2.4) defines that unique equilibrium temperature and pressure where all three phases (solid, liquid, and vapor) coexist in equilibrium. In addition, as shown in Figure 2.4, the melting temperature (T_m) of a pure substance does not depend strongly on the pressure and is therefore very close to the triple point temperature. Now from Equation 2.57 we can write that at constant temperature $dG = RT \ln f$. Combining this result with Equation 2.62 we have that:

$$d\left(\frac{G}{RT}\right) = d\,\ln f = -\frac{H}{RT^2}\,dT + \frac{V}{RT}\,dP \tag{2.111}$$

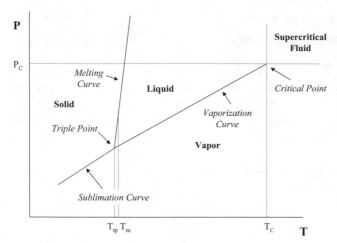

Figure 2.4 PT diagram for a pure component.

This equation may then be written for the solid and subcooled liquid phases, with the difference between the solid and subcooled liquid fugacities given by:

$$d \ln \frac{f_2^S}{f_2^L} = \frac{H^L - H^S}{RT^2} dT - \frac{V^L - V^S}{RT} dP \qquad (2.112)$$

The term $(H^L - H^S)$ is the *enthalpy of fusion* and is dependent on the temperature. Using the enthalpy of fusion at the triple point temperature (T_{tp}) as a reference, we can write that:

$$H^L - H^S = (H^L \big|_{tp} + \int_{T_{tp}}^{T} C_{PL} dT) - (H^S \big|_{tp} + \int_{T_{tp}}^{T} C_{PS} dT) \qquad (2.113)$$

which can also be written as follows assuming that the heat capacities do not vary much with the temperature between T_{tp} and T:

$$H^L - H^S = \Delta H_{tp} + \Delta C_P (T - T_{tp}) \qquad (2.114)$$

ΔH_{tp} is the enthalpy of fusion of the solid at the triple point temperature and ΔC_P represents the difference in the heat capacities.

We can now substitute Equation 2.114 into Equation 2.112 and integrate between (T_{tp}, P_{tp}) and the conditions of interest, i.e. (T,P). In performing this integration, we also make use of the fact that for the most part, the difference in the specific volumes $(\Delta V = V^L - V^S)$ is independent of the pressure. In addition, we recognize that at the triple point $f^L{}_2 = f^S{}_2$.

$$\ln \frac{f_2^S}{f_2^L} = \frac{\Delta H_{tp}}{R} \left(\frac{1}{T_{tp}} - \frac{1}{T} \right) - \frac{\Delta C_P}{R} \left(\ln \frac{T_{tp}}{T} - \frac{T_{tp}}{T} + 1 \right) - \frac{\Delta V}{RT} (P - P_{tp}) \qquad (2.115)$$

We can now substitute the above equation into Equation 2.110 and obtain the following general equation for the solubility of a solid in a liquid solvent at T and P.

$$x_2 = \frac{1}{\gamma_2} \exp\left[\frac{\Delta H_{tp}}{R}\left(\frac{1}{T_{tp}} - \frac{1}{T} \right) - \frac{\Delta C_P}{R}\left(\ln\frac{T_{tp}}{T} - \frac{T_{tp}}{T} + 1 \right) - \frac{\Delta V}{RT}\left(P - P_{tp} \right) \right] \quad (2.116)$$

We can now make some reasonable approximations that simplify the use of the above equation. In most cases, the pressure correction in the last term of Equation 2.116 is negligible. In addition, the heat capacity difference also provides only a minor contribution and can be ignored. Also, as mentioned earlier, the triple point temperature is close to the melting point temperature (T_m) of the solid at atmospheric pressure and we can then replace the enthalpy of fusion at T_{tp} with the enthalpy of fusion at the normal melting temperature, i.e. ΔH^m. For a variety of organic molecules, the ratio of the enthalpy of fusion (ΔH^m, calories/gmol) to the normal atmospheric melting point (T_m, K) is approximately in the range from 9 to 11. With these simplifications, Equation 2.116 may be written as follows:

$$x_2 = \frac{1}{\gamma_2} \exp\left[\frac{\Delta H^m}{R}\left(\frac{1}{T_m} - \frac{1}{T} \right) \right] \quad (2.117)$$

If the solid solute and the solvent are chemically similar, then we would expect them to form an ideal solution. For an ideal solution we can set γ_2 in the above equation equal to unity. For an ideal solution ($\gamma_2 = 1$), the solubility of the solid solute can be predicted from only the enthalpy of fusion and the melting temperature. Note that the ideal solubility is based only on the pure component properties of the solute and is the same regardless of the solvent. For non-ideal solutions, one must use an appropriate solution model for the activity coefficient (γ_2) (Poling *et al.* 2001) in order to calculate the solubility. In the limit of negligible solubility in the solvent, one may use the infinite dilution activity coefficient (γ^{∞}_2).

In the case of non-polar solutes and solvents, it is worth mentioning at this point that the *Scatchard–Hildebrand equation* allows estimation of the solute activity coefficient from the pure component properties of the solute and the solvent. Using the Scatchard–Hildebrand equation, the solute activity coefficient is given by the following relation:

$$\ln\gamma_2 = \frac{V^L_2 \left(\delta_1 - \delta_2 \right)^2 \Phi^2_1}{RT} \quad (2.118)$$

In this equation, V^L_2 represents the molar volume of the solute as a subcooled liquid, which in practice, is usually taken to be the same as the molar volume of the solute as a liquid at the melting temperature, the δ's are the *solubility parameters* for the solute and the solvent, and Φ_1 is the volume fraction of the solvent defined by the following equation.

$$\Phi_1 = \frac{x_1 V^L_1}{x_1 V^L_1 + x_2 V^L_2} \quad (2.119)$$

The square of the solubility parameter (δ_i) is defined as the ratio of the change in the internal energy for complete vaporization and the molar liquid volume. The internal energy change of vaporization is the same as the enthalpy of vaporization minus RT assuming an ideal gas. The solubility parameter for component i is then given by the following equation.

$$\delta_i = \left(\frac{\Delta H_i^{\text{Vap}} - RT}{V_i^L} \right)^{1/2} \tag{2.120}$$

One usually ignores the temperature dependence of $\Delta H^{\text{Vap}}{}_i$ and $V^L{}_i$ and simply uses the values at the normal melting point for the solute and at 25°C for the solvent. For a solid solute, the heat of vaporization would also be equal to the difference between the heat of sublimation and the heat of fusion, i.e. $\Delta H^{\text{Vap}} = \Delta H^{\text{Sub}} - \Delta H^m$. If an expression for the vapor pressure of the solid is known, for example from an equation like 2.89, then the heat of sublimation can be calculated using the Clausius–Clapeyron equation, i.e. Equation 2.88. The more chemically similar the solute and solvent are, the closer the values of their respective solubility parameters and from Equation 2.118 we see that the resulting solution then approaches ideality. Hence comparing the solubility parameters provides a quick method for determining chemical similarity and the degree of non-ideality between the solute and the solvent.

> **Example 2.10** A drug has a molecular weight of 230 and a melting temperature of 155°C. Estimate the solubility of this drug in benzene and in n-hexane at 25°C assuming they form an ideal solution. Also determine the solubility based on the Scatchard–Hildebrand equation. The following data is also provided:
>
> | | |
> |---|---|
> | Heat of fusion of the drug | 4300 cal gmol^{-1} |
> | Density of the drug | 1.04 g cm^{-3} at 25°C |
> | Vapor pressure of the solid drug | ln P^{Sat} (mm Hg) $= 27.3 - 8926/T$ (K) |
> | Molar volume of benzene | 89.4 cm^3 gmol^{-1} |
> | Solubility parameter for benzene | 9.2 (cal cm^{-3})$^{1/2}$ |
> | Molar volume of n-hexane | 131.6 cm^3 gmol^{-1} |
> | Solubility parameter for n-hexane | 7.3 (cal cm^{-3})$^{1/2}$ |
>
> SOLUTION To calculate the ideal solubility of the drug we use Equation 2.117 with $\gamma_2 = 1$.
>
> $$x_2 = \exp\left[\frac{4300 \text{ cal gmol}^{-1}}{1.987 \text{ cal gmol}^{-1}\text{K}^{-1}} \left(\frac{1}{273.15+155} - \frac{1}{298.15} \right) \text{K}^{-1} \right] = 0.110$$
>
> This is the same value whether the solvent is benzene or n-hexane. To calculate the solubility based on the Scatchard–Hildebrand equation, we need to estimate the solubility parameter for the drug. First, the heat of sublimation can be found as follows from the vapor pressure of the solid drug using Equation 2.88:
>
> $$\Delta H^{\text{Sub}} = -R \frac{d \ln P^{\text{Sat}}}{d\frac{1}{T}} = 1.987 \text{ cal gmol}^{-1} \text{ K}^{-1} \times 8926 \text{ K} = 17736.9 \text{ cal gmol}^{-1}$$

Then the solubility parameter for the drug is found from Equation 2.120 assuming that the molar volume of the drug as a liquid and as a solid are similar:

$$
\delta_2 = \left(\frac{17736.9\,\text{cal gmol}^{-1} - 4300\,\text{cal gmol}^{-1} - 1.987\ \text{cal gmol}^{-1}\,\text{K}^{-1}\,298\,\text{K}}{\dfrac{1}{1.04}\ \text{cm}^3\ \text{g}^{-1} \times 230\ \text{g gmol}^{-1}} \right)^{1/2}
$$

$$
= 7.62 \left(\text{cal gmol}^{-1} \right)^{1/2}
$$

Next we can combine Equations 2.117, 2.118, and 2.119 to give:

$$
x_2 = \frac{\exp\left[\dfrac{\Delta H^m}{R} \left(\dfrac{1}{T_m} - \dfrac{1}{T} \right) \right]}{\exp\left\{ \dfrac{V_2^L \left(\delta_1 - \delta_2 \right)^2}{RT} \left[\dfrac{\left(1 - x_2\right) V_1^L}{\left(1 - x_2\right) V_1^L + x_2\, V_2^L} \right]^2 \right\}}
$$

The above equation is an implicit algebraic equation in the solubility x_2. Numerical techniques need to be used to solve for the value of x_2. One approach that can be used is *direct iteration*. In this technique one assumes a value for x_2, substitutes this value on the right-hand side of the above equation, and calculates a new value of x_2 called $\hat{x}_2$. One then lets $x_2 = \hat{x}_2$, and this new value of x_2 is then substituted into the right-hand side of the above equation. The process is repeated until there is no longer any significant difference in the values of x_2 and $\hat{x}_2$, and convergence is then obtained. Although direct iteration is simple to implement, there is no guarantee for convergence to the solution. In addition, numerous iterations are sometimes required. The second approach is based on *Newton's method*. Newton's method has more reliable convergence and requires fewer iterations. To implement Newton's method one first forms the following difference equation:

$$
f(x_2) = x_2 - \frac{\exp\left[\dfrac{\Delta H^m}{R} \left(\dfrac{1}{T_m} - \dfrac{1}{T} \right) \right]}{\exp\left\{ \dfrac{V_2^L \left(\delta_1 - \delta_2 \right)^2}{RT} \left[\dfrac{\left(1 - x_2\right) V_1^L}{\left(1 - x_2\right) V_1^L + x_2\, V_2^L} \right]^2 \right\}}
$$

The goal is to find the value of x_2 that makes $f(x_2) = 0$. This is accomplished through the following iterative equation which provides an update for the value of x_2 based on the evaluation of $f(x_2)$ and $df(x_2)/dx_2$ at the previous value of x_2.

$$
\hat{x}_2 = x_2 - \frac{f\left(x_2\right)}{\dfrac{df\left(x_2\right)}{dx_2}}
$$

One then assumes $x_2 = \hat{x}_2$ and the process is repeated until convergence. In terms of choosing an initial starting value of x_2 for either approach, one could use the ideal solubility or simply assume that $x_2 = 0$. After substituting in the values of the known quantities into the above equation for $f(x_2)$, and performing the iterative calculations, it is found that the solubility based on the Scatchard–Hildebrand model for the solute activity coefficient is 0.054 when the solvent is benzene and 0.107 when the solvent is n-hexane. The drug and n-hexane form an ideal solution since their solubility parameters are nearly identical. On the other hand, for benzene as the solvent, the drug solubility is considerably lower than the value based on an ideal solution.

2.6.3.2 Depression of the freezing point of a solvent by a solute Consider the dissolution of a small amount of a solute in a solvent. As the temperature of this solution is decreased, a temperature (T_f) is reached where the pure solvent just starts to separate out of the solution as a solid phase. T_f is the freezing temperature of this mixture. This temperature is lower than the freezing point of the pure solvent (T_m). One is usually interested in determining the freezing point depression which is then defined as $\Delta T = T_m - T_f$. This problem is very similar to the one we just addressed concerning the solubility of a solute in a solvent. However, now the focus is on the solvent which is component 1.

To solve this problem, we assume that equilibrium exists between the first amount of solvent that freezes and the rest of the solution. We also assume that the solvent that freezes exists as pure solvent and we assume that the minute amount of solid solvent formed has no effect on the composition of the solution. We can then write an equilibrium equation like Equation 2.109 for the solvent:

$$f_1^S = \gamma_1 \, x_1 \, f_1^L \tag{2.121}$$

which can be rearranged to give:

$$\ln \gamma_1 \, x_1 = \ln \frac{f_1^S}{f_1^L} \tag{2.122}$$

Following a similar development we used for the solute in the previous section we can write that:

$$d \ln \frac{f_1^S}{f_1^L} = \frac{H^L - H^S}{RT^2} \, dT - \frac{V^L - V^S}{RT} \, dP \tag{2.123}$$

and we have that:

$$H^L - H^S = \Delta H^m + \Delta C_P \, (T_f - T_m) \tag{2.124}$$

where for the solvent we use the normal melting temperature of the pure solvent as the reference temperature, and T_f is the mixture freezing temperature. We also assume that the heat capacities do not vary much as the temperature changes from T_m to T_f. We can now substitute Equation 2.124 into Equation 2.123 and integrate between (T_m, P_m) and the conditions of interest, i.e. (T_f, P_f). In performing this integration we also make use of the fact that for the most part the difference in the specific volumes ($\Delta V = V^L - V^S$) is independent of the pressure. In addition, we recognize that at the melting point of the pure solvent, $f^L_1 = f^S_1$.

$$\ln\frac{f_1^S}{f_1^L}=\frac{\Delta H^m}{R}\left(\frac{1}{T_m}-\frac{1}{T_f}\right)-\frac{\Delta C_P}{R}\left(\ln\frac{T_m}{T_f}-\frac{T_m}{T_f}+1\right)-\frac{\Delta V}{RT}\left(P_f-P_m\right) \quad (2.125)$$

Combining the above equation with Equation 2.122 we obtain:

$$\ln\gamma_1 x_1=\frac{\Delta H^m}{R}\left(\frac{1}{T_m}-\frac{1}{T_f}\right)-\frac{\Delta C_P}{R}\left(\ln\frac{T_m}{T_f}-\frac{T_m}{T_f}+1\right)-\frac{\Delta V}{RT}\left(P_f-P_m\right) \quad (2.126)$$

Once again we can ignore the effect of the pressure term and since $T_f \approx T_m$ we obtain:

$$\ln\gamma_1 x_1=\frac{\Delta H^m}{R}\left(\frac{1}{T_m}-\frac{1}{T_f}\right)=-\frac{\Delta H^m}{RT_m^2}\left(T_m-T_f\right) \quad (2.127)$$

The freezing point depression is then given by:

$$\Delta T=T_m-T_f=-\frac{RT_m^2}{\Delta H^m}\ln\gamma_1 x_1 \quad (2.128)$$

Note that if the freezing depression is measured, then the above equation provides a convenient means for determining the activity coefficient of the solvent for a solution of given composition.

For ideal solutions we have that $\gamma_1 = 1$. For solutions that are very dilute in the solute, then $\gamma_1 \approx 1$ and $\ln x_1 = \ln(1-x_2) \approx -x_2$. So for dilute solutions we can write that:

$$\Delta T=T_m-T_f=\frac{RT_m^2}{\Delta H^m}x_2 \quad (2.129)$$

where x_2 is the mole fraction of the solute in the mixture.

Example 2.11 Estimate the freezing point depression of a benzene solution containing the drug considered earlier in Example 2.10. Assume that the drug concentration is 0.04 g/cm^3. The normal melting point for benzene is 278.7 K.

SOLUTION We expect the mole fraction of the drug at this concentration to be very small. We can therefore assume that the density of the liquid solution is the same as that of pure benzene which is 0.885 g/cm^3. As a basis for the calculation of the mole fraction, assume that we have 1 cm^3 of the liquid mixture. Then for the drug mole fraction in benzene we can write that:

$$x_2=\frac{\dfrac{0.04\,\text{g}}{\text{cm}^3}\times 1\,\text{cm}^3\,\text{solution}\times\dfrac{1\,\text{mole of drug}}{230\,\text{g}}}{\left(\dfrac{0.04\,\text{g}}{\text{cm}^3}\times\dfrac{1\,\text{mole of drug}}{230\,\text{g}}+\dfrac{0.885-0.04\,\text{g}}{\text{cm}^3}\times\dfrac{1\,\text{mole benzene}}{78\,\text{g}}\right)\times 1\,\text{cm}^3\,\text{solution}}$$

and we get that $x_2 = 0.0158$. Now using Equation 2.129:

$$\Delta T=T_m-T_f=\frac{1.987\ \text{cal gmol}^{-1}\,\text{K}^{-1}\ 278.7^2\ \text{K}^2}{2350\ \text{cal gmol}^{-1}}\times 0.0158=1.038\,\text{K}$$

2.6.3.3 Equilibrium between a solid and a gas phase Now we consider a pure solid phase that is in equilibrium with a surrounding gas phase. Here the question that can be addressed is for a given T and P, what is the equilibrium composition of the gas phase? We let component 1 represent the gas and component 2 is the material that makes up the pure solid phase. We also assume that the gas has negligible solubility in the solid phase so the solid phase is pure. Equilibrium requires that the solid phase fugacity of component 2 equal the fugacity of component 2 in the gas phase. We use Equation 2.71 to describe the fugacity of component 2 in the gas phase and obtain the following statement for equilibrium between the pure solid phase and the gas.

$$f_2^S = \hat{\phi}_2 \, y_2 \, P \tag{2.130}$$

The mole fraction of component 2 in the gas phase is then given by:

$$y_2 = \frac{f_2^S}{\hat{\phi}_2 \, P} \tag{2.131}$$

If the pressure is significantly above atmospheric, then a mixture equation of state is needed to calculate the fugacity coefficient of component 2 in the gas phase as described by Equation 2.74 (Poling *et al.* 2001). However, for most applications the pressure will be near atmospheric and we can treat the gas phase as an ideal gas for which $\hat{\phi}_2 = 1$. The fugacity of the pure solid would be given by Equation 2.95 recognizing that we are now considering a subcooled or compressed solid rather than that of a liquid for which we derived Equation 2.95.

$$f_2^S = f_2^{\text{Sat}} \exp\left(\frac{V_2^{S,\,\text{Sat}}\left(P - P_2^{\text{Sat}}\right)}{RT}\right) = \phi_2^{\text{Sat}} \, P_2^{\text{Sat}} \exp\left(\frac{V_2^{S,\,\text{Sat}}\left(P - P_2^{\text{Sat}}\right)}{RT}\right) \tag{2.132}$$

P_2^{Sat} would be the sublimation pressure (or vapor pressure) wherein the solid will vaporize directly to a vapor without first forming a liquid. Since the sublimation pressure is generally very small at ambient temperatures, the saturated fugacity coefficient (ϕ_2^{Sat}) is equal to one. If the pressure is low and on the order of P_2^{Sat}, then the exponential term is also equal to unity. Therefore, for this case, we obtain that the pure component solid phase fugacity of component 2 is simply equal to its sublimation or vapor pressure. Hence we can write that:

$$y_2 = \frac{P_2^{\text{Sat}}}{P} \tag{2.133}$$

Example 2.12 Estimate the gas phase equilibrium mole fraction of the drug considered in Example 2.10 at a temperature of 35°C. The pressure of the gas is 1 atm.

SOLUTION Since the pressure is 1 atm we assume that the gas phase is ideal and set $\phi_2^{\text{Sat}} = 1$. We then calculate the fugacity of the drug as a solid using Equation 2.132.

$$f_2^S = P_2^{\text{Sat}} \exp\left(\frac{V_2^S\left(P - P_2^{\text{Sat}}\right)}{RT}\right)$$

We also assume that the molar volume of the solid at the saturation T and P is the same as the molar volume of the solid at ambient conditions. For P^{Sat} we were given an expression in Example 2.10. For the given T, we calculate that $P^{Sat} = 0.189$ mm Hg. Using the above equation we then calculate the fugacity of the pure solid as follows:

$$f_2^S = 0.189 \, \text{mm Hg}$$

$$\times \exp\left(\frac{1 \text{cm}^3 \, 1.04^{-1} \, \text{g}^{-1} \times 1 \text{L} \, 1000^{-1} \, \text{cm}^{-3} \times 230 \, \text{g gmol}^{-1} \times (760 - 0.189) \text{mm Hg} \times 1 \, \text{atm}/760^{-1} \, \text{mm Hg}^{-1}}{0.082 \, \dfrac{\text{atm L}}{\text{K g mol}} \times 308.15 \, \text{K}} \right)$$

$$f_2^S = 0.189 \text{ mm Hg} \times 1.0088 = 0.191 \text{ mm Hg}$$

As expected, because of the low pressure we see in the above calculation that the exponential correction to the solid fugacity is negligible and that the fugacity of the solid is nearly the same as its vapor pressure. Using Equation 2.133 we then find that the gas phase mole fraction of the drug is $y_2 = 0.191$ mm Hg 760^{-1} mm Hg^{-1} = 2.513×10^{-4}.

2.6.3.4 Solubility of a gas in a liquid

We now consider the calculation of the equilibrium solubility of a sparingly soluble gas (taken as component 2) in a liquid, i.e. $x_2 \to 0$. Our phase equilibrium relationship, i.e. Equation 2.84, requires for the solute gas that:

$$\hat{f}_2^L = \hat{f}_2^G \tag{2.134}$$

The definition of the activity coefficient, where $\gamma_2 = \dfrac{\hat{f}_2^L}{x_2 \, f_2^L}$, provides an approach for calculating the fugacity of component 2 within the liquid phase:

$$\hat{f}_2^L = \gamma_2 \, x_2 \, f_2^L \tag{2.135}$$

where f_2^L is the pure component liquid fugacity at the mixture T and P and is given by Equation 2.95. Recall from Equation 2.71 we also defined the component mixture fugacity coefficient as $\hat{\phi}_2 = \dfrac{\hat{f}_2^G}{y_2 \, P}$ which also says that the fugacity of component 2 within a vapor or gas mixture is given by:

$$\hat{f}_2^G = y_2 \, \hat{\phi}_2 \, P \tag{2.136}$$

By Equation 2.84 the above two equations are equal so:

$$y_2 \, \hat{\phi}_2 \, P = \gamma_2 \, x_2 \, f_2^L \tag{2.137}$$

We then assume that the solvent has a negligible vapor pressure hence $y_2 = 1$ and $\hat{\phi}_2 = \phi_2$. The gas solubility in the solvent expressed as a mole fraction is then given by the next equation.

$$x_2 = \frac{\phi_2 \, P}{\gamma_2 \, f_2^L} = \frac{f_2^G}{\gamma_2 \, f_2^L} \tag{2.138}$$

Usually the temperature of interest will be much higher than the critical temperature of the soluble gas being considered, so we have that $T > T_{C2}$. For example, the critical temperature and other properties for several common gases are summarized in Table 2.1.

Although Equation 2.138 provides a convenient means of calculating the gas solubility, it is encumbered by the fact that the soluble gas does not really exist as a liquid since $T > T_{C2}$. If by chance the soluble gas is below its critical temperature, then this problem of the gas existing as a hypothetical liquid is not an issue.

One approach to solve this problem involves using Equation 2.138 above with the Scatchard–Hildebrand equation for the activity coefficient of the solute. When Equation 2.118 is substituted into Equation 2.138 one obtains the following equation:

$$x_2 = \frac{f_2^G}{f_2^L} \exp\left[-\frac{V_2^L \left(\delta_1 - \delta_2\right)^2 \Phi_1^2}{RT} \right] \tag{2.139}$$

Prausnitz and Shair (1961) used the above equation to correlate known solubility data for a number of gases in a variety of solvents and thus obtained the three parameters in the above equation that describe the soluble gas as a hypothetical liquid, i.e. the pure liquid fugacity (f_2^L), the liquid volume (V_2^L), and the gas solubility parameter (δ_2). Table 2.1 summarizes for several gases the values of (V_2^L) and (δ_2) that were found and the equation below provides the fugacity of the soluble gas as a hypothetical pure liquid at a pressure of 1.013 bar.

$$\ln\left(\frac{f_2^L}{P_{C2}}\right) = 7.81 - \frac{8.06}{T/T_{C2}} - 2.94 \ln\left(\frac{T}{T_{C2}}\right) \tag{2.140}$$

If the pressure is greater than 1.013 bar, then the liquid phase fugacity (f_2^L) calculated from the above equation needs to be corrected for the effect of the pressure as shown in Equation 2.95 after changing the reference pressure from P^{Sat} to atmospheric pressure of 1.013 bar.

$$f_2^L(P) = f_2^L(1.013\,\text{bar}) \exp\left[\frac{V_2^L(P - 1.013\,\text{bar})}{RT} \right] \tag{2.141}$$

Example 2.13 Estimate the solubility of oxygen and carbon dioxide in toluene at a gas partial pressure of 1 atm and 25°C. The solubility parameter for toluene is 8.91 $(\text{cal/cm}^3)^{1/2}$.

Table 2.1 Physical properties of some common gases

Gas	Critical properties T_C (K) & P_C (MPa)		Liquid volume V^L (cm³ gmol⁻¹)	Solubility parameter δ (cal cm⁻³)¹/²
Oxygen	154.6	5.046	33.0	4.0
Nitrogen	126.2	3.394	32.4	2.58
Carbon Dioxide	304.2	7.376	55	6.0

SOLUTION Since the pressure is atmospheric we can neglect the pressure correction to the liquid fugacity. From Equation 2.140 we then calculate the fugacity of oxygen and carbon dioxide in the liquid state at 25°C. For carbon dioxide, we have that:

$$\ln\left(\frac{f_2^L}{P_{C2}}\right) = 7.81 - \frac{8.06}{298.15/304.2} - 2.94\ln\left(\frac{298.15}{304.2}\right) = -0.3545$$

$$f_2^L = 7.376\,\text{MPa} \times \exp(-0.3545) \times \frac{10^6\,\text{Pa}}{1\,\text{MPa}} \times \frac{1\,\text{atm}}{101,325\,\text{Pa}} = 51.07\,\text{atm}$$

and for oxygen we have that:

$$\ln\left(\frac{f_2^L}{P_{C2}}\right) = 7.81 - \frac{8.06}{298.15/154.6} - 2.94\ln\left(\frac{298.15}{154.6}\right) = 1.70$$

$$f_2^L = 5.046\,\text{MPa} \times \exp(1.70) \times \frac{10^6\,\text{Pa}}{1\,\text{MPa}} \times \frac{1\,\text{atm}}{101,325\,\text{Pa}} = 272.6\,\text{atm}$$

Since the gas partial pressure is only 1 atm we can treat the gas phase as an ideal gas and set the gas phase fugacity equal to the partial pressure. From Equation 2.139 we can then calculate the mole fraction of carbon dioxide and oxygen in toluene. Note that we have set $\Phi_1 = 1$ since we expect the solubility of these gases in toluene to be very small. For carbon dioxide we then obtain that:

$$x_2 = \frac{1\,\text{atm}}{51.07\,\text{atm}}\exp\left[-\frac{55\,\text{cm}^3\,\text{gmol}^{-1}\left(6-8.91\right)^2\,\text{cal cm}^{-3}\times 1}{1.987\,\text{cal gmol}^{-1}\,\text{K}^{-1}\times 298.15\,\text{K}}\right] = 0.0089$$

and for oxygen we obtain:

$$x_2 = \frac{1\,\text{atm}}{272.6\,\text{atm}}\exp\left[-\frac{33\,\text{cm}^3\,\text{gmol}^{-1}\left(4-8.91\right)^2\,\text{cal cm}^{-3}\times 1}{1.987\,\text{cal gmol}^{-1}\,\text{K}^{-1}\times 298.15\,\text{K}}\right] = 0.0010$$

The reported solubility of carbon dioxide in toluene at 25°C and at a partial pressure of 1 atm is 0.010 and that for oxygen in toluene is 0.0009. Calculating the % error, we see that the error for prediction of the carbon dioxide and oxygen solubility using the Scatchard–Hildebrand model is 11% for both gases.

The use of Equation 2.139 for predicting the solubility of a gas in a liquid is limited by the assumptions made in the development of the Scatchard–Hildebrand model for estimating activity coefficients. Generally, the prediction will be better for non-polar gases and liquids. Also the accuracy will rapidly diminish as the absolute magnitude of the difference between the solubility parameter of the gas and the solvent increases. For example, consider the solubility of oxygen in ethanol ($\delta_{\text{ethanol}} = 12.8$ (cal/cm^3)$^{1/2}$). The difference in the solubility parameters of oxygen and ethanol is quite large and we would therefore expect greater error in the prediction of the solubility. We can use the results from the above example and calculate the predicted oxygen solubility in ethanol as 4.91×10^{-5} whereas the reported value is about 6×10^{-4}. So in this case we have underestimated the solubility of oxygen by over a factor of ten.

Because of these limitations on predicting gas solubility, a more empirical approach is often used. At a given temperature and pressure, it has been found that for the most part the gas solubility expressed as a mole fraction (x_2) is directly proportional to the gas phase fugacity, i.e. $\hat{f}_2^G = y_2 \hat{\phi}_2 P = H x_2$. The proportionality constant H is called Henry's constant. If the gas phase is ideal, then $\hat{\phi}_2 = 1$, and $y_2 P$ is also just the partial pressure of the solute gas, i.e. P_2 or pX where X is the name of the solute gas. So we can write that $P_2 = pX = H\, x_2$. It is important to remember though that the Henry constant for a specific gas will depend on the solvent, the temperature, and to some extent the pressure.

In some cases Henry's constant is based on the solute concentration of the solute in the solvent. This is common for example in describing the solubility of oxygen in blood where we write that the $pO_2 = H\, C_{oxygen}$. pO_2 represents the partial pressure of oxygen in the gas phase and C_{oxygen} is the molar concentration of dissolved oxygen in the blood. The value of H for describing the solubility of oxygen in blood is 0.74 mm Hg/μM.

The thermodynamic basis for Henry's constant can be obtained from Equation 2.135 where we have the following expression for the liquid phase fugacity, i.e. $\hat{f}_2^L = \gamma_2 x_2 f_2^L$. The product of γ_2 and f_2^L as $x_2 \to 0$ is the Henry constant. Substituting H for γ_2 and f_2^L results in Henry's law for describing gas solubility, i.e.

$$\hat{f}_2^L = \gamma_2 x_2 f_2^L = H x_2 \qquad (2.142)$$

For sparingly soluble gases, where $x_2 \to 0$, the activity coefficient approaches in the limit of infinite dilution a value that becomes independent of the solute mole fraction, i.e. $\gamma_2 \to \gamma_2^\infty$. Hence from Equation 2.142 we would expect Henry's constant to be nearly independent of the solute mole fraction. Since H is also related to the pure component liquid fugacity, we would also expect H to be somewhat dependent on the temperature and only weakly dependent on the pressure.

Example 2.14 Calculate the concentration of oxygen dissolved in blood for a gas phase partial pressure of oxygen equal to 95 mm Hg.

SOLUTION Using Henry's law and the value of H from above for oxygen and blood we can write that:

$$C_{oxygen} = \frac{pO_2}{H} = \frac{95 \, \text{mm Hg}}{0.74 \, \text{mm Hg}/\mu M} = 128.4 \; \mu M$$

2.6.3.5 Osmotic pressure Consider the situation illustrated in Figure 2.5 where a rigid membrane separates region A from region B. Region B contains only pure solvent whereas region A contains the solvent and a solute. The membrane that separates region

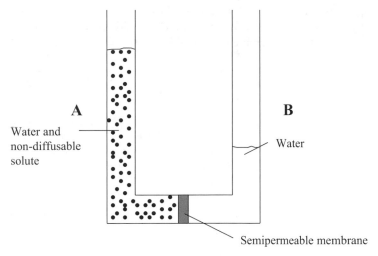

Figure 2.5   Concept of osmotic pressure.

A from region B is only permeable to the solvent, hence the solute is completely retained in region A. Solvent will diffuse from region B into region A in an attempt to satisfy the requirement that at equilibrium (see Equations 2.82 and 2.84) the chemical potential or fugacity of the solvent on each side of the membrane must be equal. This movement of solvent from region B to region A is called *osmosis* and will tend to increase the level of region A until its increase in hydrostatic pressure stops the flow of solvent from region B and equilibrium is attained. The pressure difference between regions A and B at this point of equilibrium is the osmotic pressure of region A.

A quantitative understanding of osmosis is needed for a variety of calculations and concepts involving solutions and mass transport across semipermeable membranes. We can easily derive an expression for the osmotic pressure of a solution by starting with the fundamental requirement that at equilibrium the temperature and fugacity of the solvent in regions A and B must be equal.

$$\hat{f}^A_{solvent}\left(T,\,P^A\right)=f^B_{solvent}\left(T,\,P^B\right) \tag{2.143}$$

Note that at equilibrium the temperatures of regions A and B will be the same, however the pressures will not be the same and the difference $(P^A - P^B)$ will be the osmotic pressure of region A. The above equation can be rewritten as follows using an activity coefficient model to describe the fugacity of the solvent in region A:

$$\gamma^A_{solvent}\,x^A_{solvent}\,f^A_{solvent}\left(T,\,P^A\right)=f^B_{solvent}\left(T,\,P^B\right) \tag{2.144}$$

The pure component solvent fugacity in region A at T and P^A can be related to that in region B at T and P^B through the use of Equation 2.95 after changing the reference pressure from P^{Sat} to P^B. Hence we can write that:

$$f^A_{solvent}\left(T,\,P^A\right)=f^A_{solvent}\left(T,\,P^B\right)\exp\left(\frac{V^L_{solvent}\left(P^A-P^B\right)}{R\,T}\right) \tag{2.145}$$

Substituting Equation 2.145 into Equation 2.144 results in the following equation for the osmotic pressure ($P^A - P^B$). This osmotic pressure difference is given the symbol π and represents the osmotic pressure of region A.

$$\pi = \left(P^A - P^B \right) = \frac{-RT}{V^L_{solvent}} \ln \left(\gamma^A_{solvent} x^A_{solvent} \right)$$ (2.146)

Note that measurement of the osmotic pressure of a solution provides a convenient method through the use of Equation 2.146 to determine the activity coefficient of the solvent. For non-ideal solutions a variety of theoretical and empirical models exist for estimating the activity coefficient in Equation 2.146 (Poling *et al.* 2001, also see problems 28 and 29 at the end of this chapter).

The above equation may be simplified for the special case where the solvent and solute form an ideal solution. In this case $\gamma^A_{solvent} = 1$ and since $x^A_{solvent}$ is close to unity, we can rewrite Equation 2.146 as follows:

$$\pi = \frac{RT}{V^L_{solvent}} x^A_{solute} = RTC_{solute}$$ (2.147)

where x^A_{solute} represents the mole fraction of the solute in region A. Since the solute mole fraction is generally quite small, we may approximate this as $x^A_{solute} = V^L_{solvent} C_{solute}$, where C_{solute} is the concentration of the solute in gmoles/liter of solution and $V^L_{solvent}$ is the molar volume of the solvent. This ideal dilute solution osmotic pressure, described by Equation 2.147, is known as *van't Hoff's law*. If the solution contains N ideal solutes, then the total osmotic pressure of the solution would be the summation of the osmotic pressure generated by each solute according to Equation 2.147.

$$\pi = RT \sum_{i=1}^{N} C_{solute_i}$$ (2.148)

Osmotic pressure is not determined on the basis of the mass of the solute in the solution, but rather on the number of particles that are formed by a given solute. Each non-diffusing particle in the solution contributes the same amount to the osmotic pressure regardless of the size of the particle. Thus if we take 1/1000th of an Avogadro's number of glucose molecules or 1/1000th of an Avogadro's number of albumin molecules, and form a 1 liter solution of each with water, we have a 1 mM (1 millimole/liter) solution of each solute. According to Equation 2.147, the osmotic pressure of these two solutions would be the same. However, the mass of glucose added to the solution would be 180 mg, whereas the mass of albumin added to the solution would be nearly 70 g, demonstrating that it is not the mass of solute in the solution that is important in determining the osmotic pressure.

Example 2.15 Consider the situation shown in Figure 2.5 where on side A of the membrane we have a solution that has a water mole fraction of 0.99. On side B of the membrane we have pure water. If the temperature is 25°C and side B is at 1 atmosphere pressure, estimate the pressure needed on side A to stop the osmosis of water from region B. What would happen if the pressure on side A was increased to a value above this pressure?

SOLUTION Assuming the solution in region A is an ideal solution we can use Equation 2.147 to calculate the osmotic pressure of region A, i.e.

$$\pi = \left(P_A - P_B\right) = \frac{8.314 \ \frac{Pa\ m^3}{gmol\ K} \times \frac{1\,atm}{101,325\ Pa} \times 298\,K}{18 \times 10^{-6} \ \frac{m^3}{gmol}} \times 0.01 = 13.6\,atm$$

Therefore the pressure in region A would have to be 14.6 atm to prevent the osmosis of water from region B into region A. If the pressure in region A is greater than 14.6 atm, then water will move from region A into region B and this process is called *reverse osmosis*.

Since small pressures are easy to measure, the osmotic pressure is also useful for determining the molecular weight of macromolecules. For example, if we let m_{solute} represent the mass concentration (g liter^{-1}) of the solute in the solution, then from Equation 2.147 we can write the molecular weight of the solute (MW_{solute}) in terms of the osmotic pressure of the solution as follows:

$$MW_{solute} = \frac{RT\,m_{solute}}{\pi} \qquad (2.149)$$

Example 2.16 The osmotic pressure of a solution containing a macromolecule is equivalent to the pressure exerted by 8 cm of water. The mass concentration of the protein in the solution is 15 g liter^{-1}. Estimate the molecular weight of this macromolecule.

SOLUTION We can use Equation 2.149 to estimate the molecular weight as follows.

$$MW_{solute} = \frac{8.314 \ \frac{m^3\,Pa}{mol\,K} \times 298\,K \times \frac{15\,g}{liter} \times \frac{1\,liter}{1000\,cm^3} \times \frac{\left(100\,cm\right)^3}{m^3} \times \frac{1\,atm}{101,325\,Pa}}{8\,cm\,H_2O \times \frac{1\,in}{2.54\,cm} \times \frac{1\,ft}{12\,in} \times \frac{1\,atm}{33.91\,ft\,H_2O}}$$

$MW_{solute} = 47400$ g/mol

2.6.3.6 Distribution of a solute between two liquid phases

Many purification processes are based on the unequal distribution of a solute between two partially miscible liquid phases. For example, through the process of liquid extraction, a drug produced by fermentation can be extracted into a suitable solvent and purified from the aqueous fermentation broth. Factors to consider in the selection of the solvent include its toxicity, cost, degree of miscibility with the fermentation broth, and selectivity for the solute.

Consider the resulting equilibrium when N_1 moles of solute is mixed with N_2 moles of solvent 2, and N_3 moles of solvent 3. The equilibrium distribution of these three components between the two resulting liquid phases (I and II) may be written for component i as follows:

$$\gamma_i^I \, x_i^I \, f_i^I \left(T,P\right) = \gamma_i^{II} \, x_i^{II} \, f_i^{II} \left(T,P\right) \qquad (2.150)$$

Now the pure component fugacities of component i are the same in phase I and II since component i exists in the same state at the same temperature and pressure in each of the two phases. Therefore, Equation 2.150 becomes:

$$\gamma_i^I \, x_i^I = \gamma_i^{II} \, x_i^{II} \qquad (2.151)$$

The ratio of the mole fractions of component i in the two phases is called the *distribution coefficient* (K_i) and is defined as follows after rearranging the above equation:

$$K_i = \frac{x_i^I}{x_i^{II}} = \frac{\gamma_i^{II}}{\gamma_i^I} \qquad (2.152)$$

It is important to remember that in liquid–liquid equilibrium problems, the solvents that form the two partially miscible liquid phases form highly non-ideal solutions. Therefore, the activity coefficients of each component in each phase tend to be strong functions of the composition and need to be determined by multicomponent activity coefficient models that have the capability to describe liquid–liquid equilibrium problems as discussed in Poling *et al.* (2001).

In this type of problem, it is desired to determine the values of the mole fractions of the three components in the two liquid phases and the amounts of each phase, that is N^I and N^{II}. For the three components and two phases, we therefore have eight unknowns, i.e. the six mole fractions and N^I and N^{II}. So we need a total of eight relationships between these variables in order to obtain a solution. The equilibrium relationships provide us with three equations and we can also write a mole balance for each component as follows:

$$N_i = x_i^I \, N^I + x_i^{II} \, N^{II} \qquad (2.153)$$

providing us with an additional three equations. In addition, we can use the fact that the mole fractions in each phase must sum to unity, that is $\sum_{i=1}^{3} x_i^I = 1$ and $\sum_{i=1}^{3} x_i^{II} = 1$. Solution of these eight equations provides the desired solution. However, it should be pointed out that this is a somewhat challenging problem to solve because of the strong non-linear dependence of the activity coefficients on the composition of each phase.

The above problem can be simplified by assuming that the two solvents are immiscible. Usually solvents are selected to miminize their mutual solubility so in a practical sense this is a pretty good assumption. As given by Equation 2.152 the distribution coefficient for the solute taken as component 1 will depend on the composition of each of the two phases. In many biological applications the solute concentration is so low that the activity coefficients approach their infinite dilution values and the distribution coefficient is a constant. Also, since the solute is present at such a small amount the value of N_2 and N_3 are assumed to be constant and equal to N^I and N^{II} respectively. A mole balance on the solute can then be written as follows:

$$N_1 = x_1^I N_2 + x_1^{II} N_3 \qquad (2.154)$$

and with $x_1^I = K x_1^{II}$ we can solve for the mole fraction of the solute in phase II as follows:

$$x_1^{II} = \frac{N_1}{K N_2 + N_3} \qquad (2.155)$$

Example 2.17 The octanol–water partition coefficient ($K_{o/w}$) is frequently used to describe the lipophilicity and to estimate the distribution of a drug between the aqueous and lipid regions of the body. $K_{o/w}$ is the same as the distribution coefficient described above with component 2 and phase I being the octanol phase and component 3 and phase II being the aqueous phase. Suppose we have 0.01 moles (N_1) of drug dissolved in 100 moles of water (N_3). The mole fraction of the drug in the aqueous phase is therefore equal to $0.01/100 = 10^{-4}$. We then add to this phase 100 moles (N_2) of octanol. Estimate the mole fractions of the drug in the two phases once equilibrium has been attained. Also, determine the % extraction of the drug from the aqueous phase. The octanol–water partition coefficient for the drug in this example is 89.

SOLUTION Using Equation 2.155 we can calculate the mole fraction of the drug in the aqueous phase once equilibrium has been reached.

$$x_1^{II} = \frac{0.01}{89 \times 100 + 100} = 1.11 \times 10^{-6}$$

The corresponding mole fraction of the drug in the octanol phase would be given by Equation 2.152.

$$x_1^I = 89 \times 1.11 \times 10^{-6} = 9.89 \times 10^{-5}$$

The % extraction of the drug from the aqueous phase is calculated as shown below.

$$\% \text{ extraction} = \frac{N_1 - x_1^{II} N_3}{N_1} = \frac{0.01 - 1.11 \times 10^{-6} \times 100}{0.01} \times 100 = 98.89\%$$

Now consider the situation shown in Figure 2.6 where a pure flowing solvent stream at molar flow rate L^I is contacted in an extractor with an aqueous stream flowing at L^{II} with a solute of mole fraction x_{in}^{II}. Also, the solvent and water are immiscible and we assume that there is no change in either L^I or L^{II}. Our mole balance on the solute can then be written as follows:

$$L^{II} x_{in}^{II} = L^I x_{out}^I + L^{II} x_{out}^{II} \qquad (2.156)$$

The streams exiting the extractor are at equilibrium so from Equation 2.152 we can write that $x_{out}^I = K x_{out}^{II}$. Using this relationship in Equation 2.156 we can solve for x_{out}^{II} as shown below:

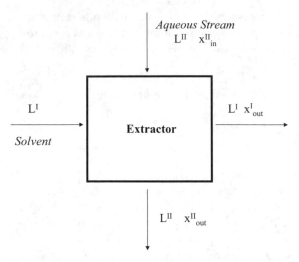

Figure 2.6   A single stage liquid–liquid extractor.

$$\frac{x_{out}^{II}}{x_{in}^{II}} = \frac{1}{1+E} \tag{2.157}$$

where E is defined as the extraction factor, $\dfrac{L^I K}{L^{II}}$. The % extraction of the solute from phase II is given by the next equation:

$$\% \text{ extraction} = \left(1 - \frac{x_{out}^{II}}{x_{in}^{II}}\right) \times 100 \tag{2.158}$$

Example 2.18   Suppose a drug is in an aqueous stream flowing at 100 moles min^{-1} at a drug mole fraction of 0.01. The aqueous stream is then contacted within an extractor with a solvent that has a distribution coefficient of 6 for this particular drug and is flowing at 200 moles min^{-1}. What is the equilibrium mole fraction of the drug in the streams exiting the extractor? What is the % extraction of the drug from the aqueous stream?

SOLUTION   From the above data we find that the extraction factor is 12. Then from Equation 2.157 the mole fraction of the drug in the aqueous stream that leaves the extractor is:

$$x_{out}^{II} = 0.01 \times \frac{1}{1+12} = 7.69 \times 10^{-4}$$

The mole fraction of the drug in the exiting solvent stream is:

$$x_{out}^{I} = 6 \times 7.69 \times 10^{-4} = 4.62 \times 10^{-3}$$

and the % extraction of the solute works out to be 92.3%.

A % extraction in the above example of 92.3% results in the potential for loss of valuable product. Therefore, extraction is frequently carried out in a countercurrent multistage extractor. Figure 2.7 illustrates this for N equilibrium stages of extraction. Since the phases are immisicible and the solute concentration is low, we will assume that L^I and L^{II} are constant. In addition, the distribution coefficient K is also assumed to be constant. A solute mole balance around the dashed box enclosing stages 1 through n of the extractor provides the following equation for the mole fraction of the solute entering stage n in phase II.

$$x_{n+1}^{II} = \frac{L^I}{L^{II}} x_n^I + \frac{L^{II} x_1^{II} - L^I x_0^I}{L^{II}} \tag{2.159}$$

The streams exiting stage n will be in equilibrium so from Equation 2.152 we can write that $x_n^I = K\, x_n^{II}$ where K is the distribution coefficient. Substituting this result into the above equation results in:

$$x_{n+1}^{II} = E\, x_n^{II} + x_1^{II} - \left(\frac{E}{K}\right) x_0^I \tag{2.160}$$

We can now use the above equation in a stepwise manner to calculate successive values of x_{n+1}^{II} starting with stage 1. Note that below we have also made use of the fact that we can replace x_0^I in Equation 2.160 with $K\, x_e^{II}$, where x_e^{II} is the solute concentration in phase II that would be in equilibrium with the solute concentration that enters in phase I. So for stage 1 we have:

$$n = 1: \quad x_2^{II} = E\, x_1^{II} + x_1^{II} - E\, x_e^{II} = (E+1) x_1^{II} - E\, x_e^{II} \tag{2.161}$$

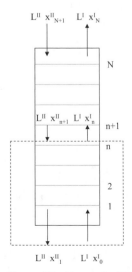

Figure 2.7   N stage liquid–liquid extractor.

For stage 2 we can eliminate x_2^{II} by using the above equation which then gives:

$$n = 2: \quad x_3^{II} = E\,x_2^{II} + x_1^{II} - E\,x_e^{II} = x_1^{II}\left(1 + E + E^2\right) - x_e^{II}\left(E + E^2\right)$$

and for stage 3 we obtain the following result after using the above equation to eliminate x_3^{II}:

$$n = 3: \quad x_4^{II} = x_1^{II}\left(1 + E + E^2 + E^3\right) - x_e^{II}\left(E + E^2 + E^3\right)$$

The above equation can then be generalized for the nth stage with the following result:

$$x_{n+1}^{II} = x_1^{II}\left(1 + E + E^2 + \ldots + E^n\right) - x_e^{II}\left(E + E^2 + \ldots + E^n\right) \tag{2.162}$$

Using the above equation we can then write for stage N that:

$$x_{N+1}^{II} = x_1^{II}\left(1 + E + E^2 + \ldots + E^N\right) - x_e^{II}\left(E + E^2 + \ldots + E^N\right) \tag{2.163}$$

The summations in Equation 2.163 are geometric progressions wherein the ratio of each successive term to the previous term is a constant, in this case E. The first series is then given by $\dfrac{1 - E^{N+1}}{1 - E}$ and the second series is given by $E\dfrac{1 - E^N}{1 - E}$. Using these relationships for the series, Equation 2.163 can then be written as shown below:

$$x_{N+1}^{II} = x_1^{II}\left(\frac{1 - E^{N+1}}{1 - E}\right) - x_e^{II}\,E\left(\frac{1 - E^N}{1 - E}\right) \tag{2.164}$$

In the above equation for example, phase II represents the aqueous phase and we would know the incoming mole fraction of the solute, i.e. x_{N+1}^{II}. Recall that $x_e^{II} = \dfrac{x_0^I}{K}$ and we would also know the amount of solute in the solvent feed stream, i.e. x_0^I. Therefore, for a given number of equilibrium stages (N), the above equation can be solved to provide the amount of solute present in phase II as it exits the extractor, i.e. x_1^{II}. If there is no solute in the entering solvent phase, then $x_0^I = 0$ and we find that:

$$\frac{x_1^{II}}{x_{N+1}^{II}} = \frac{E - 1}{E^{N+1} - 1} \tag{2.165}$$

The % extraction of the solute from phase I is then given by:

$$\% \text{ extraction} = \left(1 - \frac{x_1^{II}}{x_{N+1}^{II}}\right) \times 100 \tag{2.166}$$

Equation 2.164 can also be rearranged and solved to give the number of equilibrium stages (N) for specified solute mole fractions.

$$N = \frac{\ln\left(\dfrac{x_{N+1}^{II} - x_N^{II}}{x_1^{II} - x_e^{II}}\right)}{\ln E} \tag{2.167}$$

with $x_N^{II} = \dfrac{x_N^I}{K}$ and $x_e^{II} = \dfrac{x_0^I}{K}$. The number of actual stages will be greater than that calculated from the above equation since equilibrium is seldom achieved in each stage of the extractor. To find the actual number of stages, we can use the overall efficiency (η) of an equilibrium stage defined as follows:

$$\eta = \frac{N}{N_{\text{actual}}} \qquad (2.168)$$

Equilibrium stage efficiencies may be obtained from manufacturer's information on their extractor or from experimental data.

Example 2.19 Rework the previous example and determine the number of equilibrium and actual stages needed to achieve a 99.99% extraction of the drug from the aqueous phase. Assume that the efficiency of an equilibrium stage is 60%. Let phase II represent the aqueous phase and phase I the solvent. No solute is present in the solvent stream that enters the extractor.

SOLUTION From Equation 2.166 we can calculate the mole fraction of the solute in the aqueous phase that exits the extractor.

$$x_1^{II} = \left(1 - \% \text{ extraction} / 100\right) x_{N+1}^{II} = \left(1 - 0.9999\right)0.01 = 1 \times 10^{-6}$$

Next we can write an overall solute mole balance around the extractor that can be used to solve for the mole fraction of the solute in the exiting solvent stream.

$$x_N^I = x_0^I + \left(\frac{L^{II}}{L^I}\right)\left(x_{N+1}^{II} - x_1^{II}\right) = 0 + \left(\frac{100}{200}\right)\left(0.01 - 1\times 10^{-6}\right) = 4.9995 \times 10^{-3}$$

Next we use Equation 2.167 to find the number of equilibrium stages.

$$N = \frac{\ln\left(\dfrac{x_{N+1}^{II} - \dfrac{x_N^I}{K}}{x_1^{II} - \dfrac{x_0^I}{K}}\right)}{\ln E} = \frac{\ln\left(\dfrac{0.01 - \dfrac{4.9995\times 10^{-3}}{6}}{10^{-6}}\right)}{\ln 12} = 3.67 \approx 4$$

So we need four equilibrium stages or $4/0.6 \approx 7$ actual stages.

2.6.3.7 Vapor–liquid equilibrium

Vapor–liquid equilibrium plays an important role in understanding and designing separation processes that involve distillation, gas absorption, and the removal of solutes from liquids by contact with another gas stream, also known as stripping. These separation processes are widely used in the chemical process industry and the design of these processes falls within the realm of chemical engineering. Here we will illustrate vapor–liquid equilibrium calculations by determining the bubble point (boiling point) and dew point (condensation) temperatures of mixtures. We will also use these concepts to develop an understanding of the flammability or explosive limits of combustible materials.

Consider now the equilibrium distribution of N components at a given T and P between a vapor and a liquid. Our phase equilibrium relationship, i.e. Equation 2.84, requires for component i that:

$$\hat{f}_i^L = \hat{f}_i^V \tag{2.169}$$

For non-ideal liquid solutions, an activity coefficient model is once again required to describe the behavior of the components of the liquid phase. In addition, at pressures considerably greater than atmospheric pressure, the vapor phase will deviate from that of an ideal gas and a mixture equation of state is also needed to calculate the mixture fugacity coefficient of each species in the vapor phase (Poling *et al.* 2001). For a non-ideal liquid solution, we can then use the activity coefficient, where $\gamma_i = \dfrac{\hat{f}_i^L}{x_i f_i^L}$, to calculate the fugacity of component i within the liquid phase:

$$\hat{f}_i^L = \gamma_i \, x_i \, f_i^L \tag{2.170}$$

where f_i^L is the pure component liquid fugacity at the mixture T and P and is given by Equation 2.95. Recall from Equation 2.71, we also defined the component mixture fugacity coefficient as $\hat{\phi}_i = \dfrac{\hat{f}_i^V}{y_i P}$ which says that the fugacity of component i within a non-ideal vapor mixture is given by:

$$\hat{f}_i^V = y_i \, \hat{\phi}_i \, P \tag{2.171}$$

By Equation 2.84 the above two equations are equal so we can write the following equation that expresses the equilibrium condition for each component i in the vapor–liquid mixture:

$$y_i \, \hat{\phi}_i \, P = \gamma_i \, x_i \, f_i^L \tag{2.172}$$

If the total pressure (P) and the component vapor pressures are on the order of atmospheric pressure, then we can usually treat the vapor phase as an ideal gas and set $\hat{\phi}_i = 1$ and $\phi_i^{Sat} = 1$. Then from Equation 2.95, we have for these conditions that the $f_i^L \approx P_i^{Sat}$. Therefore, we can write Equation 2.172 as follows:

$$y_i \, P = \gamma_i \, x_i \, P_i^{Sat} = P_i \tag{2.173}$$

with P_i equal to the partial pressure of component i. Once again, we have the constraint on the mole fractions that $\sum_i x_i = \sum_i y_i = 1$.

For an ideal liquid solution we also have $\gamma_i = 1$, and then we have that:

$$y_i \, P = x_i \, P_i^{Sat} = P_i \tag{2.174}$$

The above equation is also known as *Raoult's law* which shows that the partial pressure of a component in an ideal liquid solution is equal to the product of its mole fraction and its vapor pressure.

Now consider a liquid solution consisting of N components and a mole fraction x_i for each component i. For a given pressure P, we can ask at what temperature would this liquid solution just start to boil, that is, when does it first form a vapor bubble? This temperature is known as the *bubble point temperature* and can be calculated as follows. When the liquid mixture is at its bubble point temperature, it is also referred to as a saturated liquid solution. Below the bubble point temperature, the liquid is a subcooled liquid.

First, we solve Equation 2.172 for the equilibrium mole fraction of component i in the vapor bubble that is formed when the solution reaches the bubble point temperature.

$$y_i = \frac{\gamma_i x_i f_i^L}{\hat{\phi}_i P} \tag{2.175}$$

Since the $\sum_i y_i = 1$, we then have the following equation that can be solved to find the bubble point temperature:

$$\sum_i \frac{\gamma_i x_i f_i^L}{\hat{\phi}_i} = P \tag{2.176}$$

It is important to recognize that Equation 2.176 depends on temperature through the terms that include the component liquid phase fugacities and the vapor phase fugacity coefficient. The activity coefficient may also have a temperature dependence. One therefore solves the above non-linear algebraic equation for the temperature that makes the left-hand side of Equation 2.176 equal to the pressure. This temperature is then known as the bubble point temperature or the boiling temperature of the liquid mixture. The composition of the first vapor bubble that is formed is then given by Equation 2.175. If the liquid continues to boil then the composition of the liquid and vapor phases will change as the more volatile components escape from the liquid to the vapor phase. This is the essence of distillation. The design of distillation processes falls within the study of staged processes and will not be considered here.

If the pressure is low we can once again replace the liquid phase fugacity of each component with its vapor pressure and set the component mixture fugacity coefficient to one. If the liquid also forms an ideal solution, then the activity coefficients can also be set to unity. In this latter case, which is Raoult's law, we have that:

$$\sum_i x_i P_i^{\text{Sat}}(T) = P \tag{2.177}$$

The only temperature dependence in the above equation that remains is in the component vapor pressures. The above equation can then be solved for the bubble point temperature.

A similar analysis leads to the calculation of the dew point or condensation temperature of a vapor mixture of composition y_i for component i. When the vapor mixture is at its *dew point temperature* it is also referred to as a saturated vapor. Above the dew point temperature, the vapor is referred to as superheated.

First, we solve Equation 2.172 for the equilibrium mole fraction of component i in the first drop of liquid that is formed when the vapor is cooled to its dew point temperature:

$$x_i = \frac{\hat{\phi}_i y_i P}{\gamma_i f_i^L} \tag{2.178}$$

Since the $\sum_i x_i = 1$, we then have the following equation that can be solved to find the dew point temperature:

$$\sum_i \frac{\hat{\phi}_i \, y_i}{\gamma_i \, f_i^L} = \frac{1}{P} \tag{2.179}$$

where once again $\hat{\phi}_i$, γ_i, and f_i^L will depend on the temperature. The temperature that satisfies Equation 2.179 is then the dew point temperature of the vapor mixture. The composition of the liquid drop that is just formed is given by Equation 2.178.

If the pressure is low, we can once again replace the liquid phase fugacity of each component with its vapor pressure and set the component mixture fugacity coefficient to one. If the liquid also forms an ideal solution, then the activity coefficients can also be set to unity. In this latter case, which is Raoult's law, we have:

$$\sum_i \frac{y_i}{P_i^{Sat}(T)} = \frac{1}{P} \tag{2.180}$$

The temperature dependence is only in the component vapor pressures and the above equation can then be solved for the dew point temperature.

Example 2.20 Estimate the normal (i.e. $P = 1$ atmosphere) bubble point or boiling temperature of a liquid solvent solution consisting of 25 mole % benzene (C_6H_6, MW = 78.11, $T_B = 353.3$ K), 45 mole % toluene (C_7H_8, MW = 92.14, $T_B = 383.8$ K), and 30 mole % ethylbenzene (C_8H_{10}, MW = 106.17, $T_B = 409.3$ K), where the chemical formula, molecular weight, and normal boiling point is given in parenthesis after each of the components. What is the composition of the vapor bubble that is formed at the bubble point temperature? The vapor pressures of these components are given by the Antoine vapor pressure equation (Reid *et al.* 1977) with the vapor pressure in mm Hg and the temperature in K.

$$\ln P_{benzene}^{Sat} = 15.9008 - \frac{2788.51}{T - 52.36}$$

$$\ln P_{toluene}^{Sat} = 16.0137 - \frac{3096.52}{T - 53.67}$$

$$\ln P_{ethylbenzene}^{Sat} = 16.0195 - \frac{3279.47}{T - 59.95}$$

SOLUTION Since these molecules are all chemically similar we will assume that they form an ideal solution. Equation 2.177 may therefore be used to calculate the bubble point temperature. The approach for solving this problem is as follows:

1. Assume a value of the bubble point temperature, T
2. Calculate the vapor pressures for each component at T
3. Calculate the $\sum_i x_i P_i^{Sat} = \hat{P}$
4. If $\hat{P} = P$, which in this problem is 1 atm, then T is the bubble point temperature

5. If $\hat{P} > P$, then the assumed T is too high, decrease T and go back to step 2
6. If $\hat{P} < P$, then the assumed T is too low, increase T and go back to step 2

After following the above procedure we obtain the following results:

Bubble point temperature	= 377.81 K
$y_{benzene}$	= 0.503
$y_{toluene}$	= 0.379
$y_{ethylbenzene}$	= 0.118

Note the change in the distribution of the components between the liquid and the vapor bubble that is formed. Benzene being the most volatile component has a near two-fold increase in its composition whereas ethylbenzene being considerably less volatile than benzene, has its mole fraction in the vapor decreased by nearly a factor of three as compared with its composition in the liquid phase.

Example 2.21 Repeat the above calculation but this time find the dew point temperature and the composition of the first drop of liquid that is formed.

SOLUTION Equation 2.180 may be used to calculate the dew point temperature. The approach for solving this problem is as follows:

1. Assume a value of the dew point temperature, T
2. Calculate the vapor pressures for each component at T
3. Calculate the $\sum_i \dfrac{y_i}{P_i^{Sat}} = \dfrac{1}{\hat{P}}$
4. If $\hat{P} = P$ which in this problem is 1 atm, then T is the dew point temperature
5. If $\hat{P} > P$, then the assumed T is too high, decrease T and go back to step 2
6. If $\hat{P} < P$, then the assumed T is too low, increase T and go back to step 2

After following the above procedure we obtain the following results:

Dew point temperature	= 389.6 K
$y_{benzene}$	= 0.092
$y_{toluene}$	= 0.382
$y_{ethylbenzene}$	= 0.526

Note that the dew point temperature is greater than the bubble point temperature. This makes sense since if we first consider that we have a subcooled liquid of the given composition, then as we heat this liquid mixture, we will reach the bubble point temperature that we found in the previous example. Then as we continue to add heat at constant pressure, we will vaporize more and more of the components in the liquid phase until, when we reach the dew point temperature, we now have vaporized all of the components and we have a saturated vapor of the same composition as the original liquid mixture. As we continue heating, the temperature rises above the dew point temperature and we have a superheated vapor.

2.6.3.8 Flammability limits Most manufacturing processes employ a variety of flammable substances. It is important that these materials be handled and processed in a safe manner in order to avoid fires and explosions that can have devastating consequences. The ignition, combustion, or explosion of flammable substances in air only occurs within a narrow range of composition called the *flammability limits*. Below a certain concentration of the flammable substance in air, the mixture is too "lean" to burn, while above a certain concentration the mixture is too "rich" to burn. The flammable or explosive range lies within these limits. This is important in the design and operation of combustion engines, furnaces, and incinerators. It is also important from a safety viewpoint to prevent fires and explosions in manufacturing and other processes that involve the use of flammable compounds.

A fire or explosion requires three elements: the fuel, a source of oxygen, and an ignition source. This is referred to as the *fire triangle*. The fuel can be the flammable substances in your manufacturing process. Oxygen is in the air and air certainly exists outside of your manufacturing equipment and may also be within your manufacturing process. Also, you can count on Mother Nature to provide the ignition source, although other likely ignition sources in a manufacturing facility can include hot surfaces, and flames, sparks, unstable chemicals, and static electricity. Flammability limits for a large number of organic chemicals are available in the literature.

These limits are provided as the lower and upper flammability (or explosion) limits. The upper (UFL) and lower flammability limit (LFL) is stated as the volume % (which is the same as the mole %) of the flammable chemical in air. If the flammable compound's volume % or mole % in air at a given T and P lies within the LFL to UFL range, then that mixture is ignitable or explosive. If the volume % or mole % of the flammable chemical in air is less than the LFL, or greater than the UFL, then that mixture is not ignitable or explosive.

Table 2.2 summarizes from a variety of sources the LFL and UFL for some common chemicals in air. Note that different sources may give slightly different numbers than those reported below.

Table 2.2 Flammability limits of some common substances in air at atmospheric pressure

Chemical	Lower flammability limit (Volume % or Mole %)	Upper flammability limit (Volume % or Mole %)
Acetylene	2.5	100
Acetic acid	5.4	16.0
Acetone	2.6	13.0
Ammonia	15.0	28.0
Benzene	1.4	8.0
Ethanol	3.3	19.0
n-Butane	1.8	8.4
Methanol	6.7	36.0
Diethyl ether	1.9	36.0
Propane	2.1	9.5
Octane	1.0	6.5
Gasoline	1.3	7.6
Toluene	1.3	7.0
Hydrogen	4.0	75.0

For a flammable substance, it is important to remember that it is not the liquid that actually burns, but the vapors that are produced from that liquid. In many cases we can assume that the vapor and liquid phases are in equilibrium. We can then use our vapor liquid equilibrium calculations as discussed above to determine the equilibrium composition of the vapor phase. This vapor phase composition can then be compared with the flammability limits to determine whether or not there may be an issue regarding a fire or explosion. This type of calculation is illustrated in the next example.

Example 2.22 Consider the storage at ambient conditions (25°C and 1 atm) of a solvent such as benzene in a large storage tank in a manufacturing facility. Determine whether or not the vapor phase composition in this tank lies within the flammability limits for benzene.

SOLUTION In a closed and partially full storage tank the vapor phase may be assumed to be in equilibrium with the liquid benzene provided there have been no recent transfers of benzene into or out of the storage tank. At 25°C the vapor pressure of benzene is 94.5 mm Hg using the vapor pressure equation previously used in Example 2.20. From Raoult's law:

$$Py_{benzene} = P_{benzene} = P_{benzene}^{Vap} x_{benzene} = P_{benzene}^{Vap}$$

the partial pressure of benzene in the vapor phase is 94.5 mm Hg. Since the total pressure in the vapor space must be atmospheric this means that the partial pressure of air in the vapor space of the tank is 760 mm Hg – 94.5 mm Hg = 655.5 mm Hg. From Raoult's law we can then calculate the mole fraction of benzene in the vapor space as:

$$y_{benzene} = \frac{P_{benzene}^{Vap}}{P} = \frac{94.5\,mm\,Hg}{760\,mm\,Hg} = 0.1243$$

Therefore the mole % or volume % of benzene in the vapor space is 12.4%. Comparing this value to the LFL and UFL values for benzene in Table 2.2 indicates that the vapor space in the benzene storage tank at ambient conditions is not flammable or explosive assuming that the liquid and vapor phases are in equilibrium. However, non-equilibrium events, such as the removal of benzene from the tank, will result in the addition of fresh air that will temporarily decrease the concentration of benzene in the vapor space from the above value until equilibrium is reestablished. It is possible during these transfer periods for the benzene concentration to fall within the flammability limits. In addition, for outside storage tanks, seasonal variations of temperature should also be considered. For example, at 40°F or 4.4°C, the vapor pressure of benzene decreases to 33.5 mm Hg, giving at equilibrium, a vapor phase benzene concentration of 4.4 mole % or 4.4 volume %. Now even at equilibrium the vapor phase lies well within the flammability range. If an ignition source became present, then there is the possibility that this vapor mixture will be ignited and cause a fire or explosion.

The above example illustrates that the storage of flammable substances can result in vapor mixtures that lie within the flammability range. In large storage tanks under these conditions, nitrogen gas is frequently supplied as a replacement rather than simply using ambient air. The use of nitrogen gas eliminates the formation of flammable mixtures within a storage tank. For smaller tanks and portable containers for which this is not possible, it is important to make sure that these tanks and containers are properly grounded to eliminate static discharges that can serve as an ignition source. In addition, all vent openings into the tank or container can be fitted with flame arrestors to prevent the propagation of a fire from outside of the tank to inside the tank. Commercial flame arrestors are relatively simple and inexpensive devices made of screens or mesh-like materials with small openings on the order of a millimeter or so. The design and construction of flame arrestors is based on the fact that flame propagation can be suppressed and eliminated if the flame has to pass through a large number of narrow openings. There exists a critical opening size below which the flame is quenched and this is called the quenching distance. Quenching distances for a variety of flammable substances may be found in the combustion literature.

2.6.3.9 Thermodynamics of surfaces So far we have discussed the thermodynamic properties of solids, liquids, and gases. However, when two phases are in contact there is also a surface that is formed at the interface. The surface that is formed may have properties that are quite different from the bulk phases. Understanding the thermodynamics of surfaces is important in many biomedical applications that involve the wetting of surfaces as well as for understanding such processes as capillary action.

Consider the differential increase in surface area (dA_S) of a fluid of arbitrary shape. The Gibbs free energy change for this differential increase in the surface area may be written as follows:

$$dG = -SdT + V\,dP + \gamma\,dA_S \tag{2.181}$$

where the additional term $\gamma\,dA_S$ represents the increase in the Gibbs free energy due to the change in the surface area dA_S. The *surface energy* is denoted by γ and is defined by:

$$\gamma = \left(\frac{\partial G}{\partial A_S}\right)_{T,P} \tag{2.182}$$

For a pure component γ is the same as the *surface tension*.

The surface tension acts as a restoring force that resists the addition of new surface area. At constant T and P, the work (dW) required to change the surface area an amount dA_S is given by $dW = \gamma\,dA_S$. Typical units for the surface tension are dynes/cm and mN/m. The surface tension of pure water in air at 25°C is about 72 mN/m.

In order to develop some useful relationships let us first consider a soap bubble. A soap bubble consists of a very thin spherical film of liquid. The film is relatively thick on a molecular level so the innermost portion of the film acts like the bulk liquid. This film has two surfaces that are exposed to the outside atmosphere and the air trapped within the soap bubble. According to Equation 2.181 the soap bubble can decrease its free energy by decreasing its surface area. Hence the bubble will decrease in size placing more of the liquid film molecules in the innermost portion of the film. However, as the bubble shrinks in size the internal pressure (P^B) will increase until a point is reached where the soap bubble can shrink no further. An equilibrium state then occurs where the free energy decrease of the liquid film is equal to the work done compressing the air within the soap bubble. At equilibrium, the tension in the surface of the bubble is balanced by the difference in pressure between the inside of the bubble (P^B) and the outside of the bubble (P^A). Hence we can write that:

$$\gamma \, dA_S = (P^B - P^A) \, A_S \, dr \tag{2.183}$$

The left side of the above equation represents the free energy gained by shrinking the soap bubble an amount dA_S. The right-hand side represents the net force acting on the surface A_S, and this force multiplied by the displacement of the interface dr, is the work done against the pressure difference as a result of the change in area dA_S.

Recall that for a sphere of radius r the surface area is $4 \pi r^2$. For a soap bubble there are two interfaces, the inside and outside surfaces of the liquid film, so dA_S in this case is equal to $16 \pi r \, dr$. For a droplet or a gas bubble, where we have either a drop of liquid suspended in a gas or a gas bubble suspended in a liquid, there is only one interface and $dA_S = 8 \pi r \, dr$. Substituting in for dA_S and A_S in Equation 2.183 we can obtain the following equations for the pressure difference $P^B - P^A$:

$$2 \text{ interfaces} \quad P^B - P^A = \frac{4\gamma}{r} \tag{2.184}$$

$$1 \text{ interface} \quad P^B - P^A = \frac{2\gamma}{r} \tag{2.185}$$

The above equations are known as the *Laplace–Young* equation and show that the excess pressure $(P^B - P^A)$ is inversely proportional to the radius of the bubble or the droplet. The radius r in the above equations is always considered to have its center of curvature in the phase in which P^B is measured. Hence for a gas bubble in a liquid P_B is the pressure within the bubble which is then greater than the pressure (P_A) in the liquid phase.

Example 2.23 Consider a droplet of water that is 1 micron in diameter that is suspended in air at 1 atmosphere pressure and 25°C. What is the pressure of the water inside the droplet?

SOLUTION The center of curvature for the droplet of water lies within the droplet. Therefore, P^B represents the pressure within the droplet. Using Equation 2.185 for the droplet we have:

$$P^B - 1 \, atm = \frac{2 \times 72 \times 10^{-3} \, N \, m^{-1}}{0.5 \times 10^{-6} \, m} = 288,000 \, Pa = 2.84 \, atm$$

$$P^B = 3.84 \, atm$$

Now consider a liquid droplet suspended in its own vapor. The liquid and vapor phases are therefore in equilibrium. The Gibbs free energy for the droplet phase may be written as follows assuming we have only a single component i:

$$dG = -SdT + V\,dP + \gamma\,dA_S + \mu_i^{(P)}\,dn_i \qquad (2.186)$$

The chemical potential in this case is defined as $\mu_i^{(P)} = \left(\dfrac{\partial G}{\partial n_i}\right)\bigg|_{T,P,A_S,n_j}$. The superscript P on the chemical potential denotes the fact that the transfer of dn_i moles does not change the surface area of the droplet since in the above definition of the chemical potential A_S is constant. The only way that this can occur is if $\mu_i^{(P)}$ refers to the chemical potential of component i as a planar surface.

For a spherical droplet, the surface area of the droplet will depend on how much material is in the droplet. With v_i the molar volume of liquid i, we can then write that:

$$dV = v_i\,dn_i = d\left(\frac{4}{3}\pi r^3\right) = 4\pi r^2\,dr \qquad (2.187)$$

Similarly we can write for the surface area of the droplet that:

$$dA_S = d\left(4\pi r^2\right) = 8\pi r\,dr = \frac{2\,dV}{r} = \frac{2\,v_i}{r}\,dn_i \qquad (2.188)$$

Using Equation 2.188 we can then rewrite Equation 2.186 as follows:

$$dG = -S\,dT + V\,dP + \left(\mu_i^{(P)} + \frac{2\,v_i\,\gamma}{r}\right)dn_i \qquad (2.189)$$

The parenthetical term represents the chemical potential of component i as a droplet and takes into account the free energy change associated with changes in the droplet surface area.

Since the droplet is at equilibrium with its vapor at constant T and P, then we must have that:

$$\mu_i^V = \mu_i^L = \mu_i^{(P)} + \frac{2\,v_i\,\gamma}{r} \qquad (2.190)$$

From Equation 2.56 we can write that:

$$\mu_i^{L\,\text{ideal gas}} = C_i(T) + RT\ln\left(P_i^{VAP}\right) \text{ and } \mu_i^{(P)\,\text{ideal gas}} = C_i(T) + RT\ln\left(P_i^{(P)VAP}\right) \quad (2.191)$$

where the $y_i\,P$ for a pure component would be the same as the vapor pressure by Raoult's law. Substituting the relationships in Equation 2.191 into Equation 2.190 provides the following result:

$$\ln\left(\frac{P_i^{VAP}}{P_i^{(P)VAP}}\right) = \frac{2\,v_i\,\gamma}{r\,RT} \qquad (2.192)$$

The above equation, also known as the *Kelvin equation*, shows that the vapor pressure of a small droplet of liquid of radius r is higher than the corresponding value as a planar layer of liquid. If the surface is concave towards the vapor side, for example a gas or vapor bubble in a liquid phase, then use a $-$ sign on the right-hand side of Equation 2.192.

The above equation shows that a swarm of liquid droplets of uniform radius at equilibrium with its vapor is unstable in terms of its radius. For example, if one droplet has a slight increase in its radius, then by Equation 2.192 the vapor pressure of the droplet will decrease and there will be a corresponding transfer of mass from the vapor state to the liquid state thus increasing the size of the droplet and so on. On the other hand if a droplet has a slight decrease in radius, then its vapor pressure will increase and there will be transfer of mass from the liquid state to the vapor state thus decreasing the size of the droplet and so on. This means that in a mixture of droplets of different sizes the larger droplets will grow at the expense of the smaller droplets, and this is the phenomenon that leads to the formation of clouds and rain.

Example 2.24 Consider a droplet of water that is 1 micron in diameter that is suspended in air at 1 atmosphere pressure and 25°C. What is the vapor pressure of the water droplet? At 25°C the vapor pressure of planar water is 3.166 kPa and the molar volume of liquid water is 18.05 cm³ mol⁻¹.

SOLUTION We can use Equation 2.192 to calculate the vapor pressure of the droplet of water as shown below:

$$\ln\left(\frac{P_w^{VAP}}{P_w^{(P)VAP}}\right) = \frac{2 \times 18.05\,\text{cm}^3\,\text{mol}^{-1} \times 1\,\text{m}^3 \left(100\,\text{cm}\right)^{-3} \times 72 \times 10^{-3}\,\text{N}\,\text{m}^{-1}}{0.5 \times 10^{-6}\,\text{m} \times 8.314\,\text{m}^3\,\text{N}\,\text{m}^{-2}\,\text{mol}^{-1}\,\text{K}^{-1}\,298\,\text{K}} = 0.0021$$

$$P_w^{VAP} = e^{0.0021} \times 3.166\,\text{kPa} = 3.173\,\text{kPa}$$

PROBLEMS

1. Assuming that air is an ideal gas, calculate the change in entropy and enthalpy when 100 moles of air at 25°C and 1 atm is heated and compressed to 300°C and 10 atm. Use the following equation for C_p

$$\frac{C_p}{R} = A + BT + CT^{-2}$$

where $A = 3.355$, $B = 0.575 \times 10^{-3}$, and $C = -0.016 \times 10^5$, with T in K.

2. Calculate the work required to compress adiabatically 60 moles of carbon dioxide initially at 25°C and 1 atm to 350°C and 10 atm. Assume that $C_p = 37.1$ J/K/mol. If the process is carried out reversibly, how much work is required? Assume that carbon dioxide is an ideal gas.

3. A toy rocket consists of a Styrofoam rocket snuggly fitted into a long tube of volume equal to 0.5 liter. The pressure in the tube is increased to a final pressure of 3 atm and a temperature of 30°C at which time the rocket takes off. If the rocket weighs 50 g, estimate the maximum initial velocity of the rocket and the maximum expected height of the rocket. Assume air is an ideal gas and has a $C_v = 5/2R$. Explain why the rocket would not be expected to achieve this elevation.

4. Air at 400 kPa and 400 K passes through a turbine. The turbine is well-insulated. The air leaves the turbine at 125 kPa. Find the maximum amount of work that can be obtained from the turbine and the exit temperature of the air leaving the turbine. Assume air is an ideal gas with a $C_p = 7/2$ R.

5. Consider the slow adiabatic expansion of a closed volume of gas for which $C_p = 7/2$ R. If the initial gas temperature is 825 K, and the ratio of the final pressure to the initial pressure is 1/3, what is the change in enthalpy of the gas, the change in internal energy, the heat transferred Q, and the work W? Assume a basis of 1 mole of gas.

6. Air at 125 kPa and 350 K passes through a compressor. The compressor is well-insulated. The air leaves the compressor at 600 kPa and 650 K. Find the work required to do this compression. Is this change of state of the air possible? Assume air is an ideal gas with a $C_p = 7/2$ R.

7. Consider the slow adiabatic compression of a closed volume of gas for which $C_p = 7/2$ R. If the initial gas temperature is 320 K, and the ratio of the final pressure to the initial pressure is 4, what is the change in enthalpy of the gas, the change in internal energy, the heat transferred Q, and the work W? Assume a basis of 1 mole of gas.

8. One mole of an ideal gas, initially at 30°C and 1 atm, undergoes the following reversible changes. It is first compressed isothermally to a point such that when it is heated at constant volume to 150°C, its final pressure is 12 atm. Calculate Q, W, ΔU, ΔH for the process. Take $C_P = 7/2$ R and $C_V = 5/2$ R.

9. A rigid vessel of 75 L volume contains an ideal gas at 400 K and 1 atm. If heat in the amount of 15,000 J is transferred to the gas, determine the entropy change of the gas. Assume $C_V = 5/2$ R and $C_P = 7/2$ R.

10. An osmotic pump (Figure 7.7) is a device used to deliver a drug at a constant rate within the body. An excess of NaCl is used to form a continuously saturated solution that is contained within a compartment, called the osmotic engine, located between a semipermeable membrane which is exposed to the interstitial fluid and a piston which acts on a reservoir containing the drug. In operation, water from the surrounding tissue is brought into the osmotic engine by osmosis. The resulting expansion of the osmotic engine drives the piston that forces drug from the drug reservoir out of the osmotic pump at a constant rate. The drug delivery rate (R) can be shown to be proportional to the osmotic pressure of the fluid within the osmotic engine and the concentration (C) of the drug in the reservoir. The delivery rate of the drug is therefore given by the following equation:

$$R = k \, \pi \, C$$

At 37°C, the saturation concentration of NaCl in water is equal to 5.4 M. For a particular osmotic pump, it was found that the drug release rate (R) was 125 μg/day. The drug concentration (C) was 370 mg/ml. What is the value of the proportionality constant (k, in ml/day/atm) for this osmotic pump?

11. The Navy is considering an osmotic device in their submarines in order to desalinate water when the submarine is submerged. When the submarine is at the appropriate depth, the osmotic device converts seawater into pure water that would be available for use on the submarine. The density of seawater is 1024 kg/m^3 and the composition of the seawater is equivalent to a 0.5 M NaCl solution. At what cruising depth would this proposed desalination process work?

12. One mole of an ideal gas initially at 25°C and 1 atm pressure is heated and allowed to expand against an external pressure of 1 atm until the final temperature is 400°C. For this gas, $C_v = 20.8$ J/mol^{-1} K^{-1}. Calculate the following:

a. Calculate the work done by the gas during the expansion.
b. Calculate the change in internal energy and the change in enthalpy of the gas during the expansion.
c. How much heat is absorbed by the gas during the expansion?
d. What is the change in entropy of the gas?

13. At 50°C the vapor pressure of A and B as pure liquids are 268.0 and 236.2 mm Hg, respectively. At this temperature, calculate the total pressure and the composition of the vapor which is in equilibrium with the liquid containing a mole fraction of A of 0.25.

14. When 0.013 moles of urea are dissolved in 1000 grams of water the freezing point of water decreases by 0.024 K. How does this measured freezing point depression compare with that based on the ideal solution theory?

15. Gas in a cylinder expands slowly by pushing on a frictionless piston. The data below shows what happens to the gas in the cylinder. How much work (W) is done by the gas? Assuming C_V = 5/2 R, what is the change in the internal energy (ΔU) of the gas? How much heat was transferred during the process (Q)?

Gas pressure, bar	Volume of gas, liters	Temperature of gas, °C
10	15	70
7	17	23
5	20	10
3	24	50
6	28	150
8	35	450

16. Cubic equations of state are commonly used for PVT calculations. Most of the modern cubic equations of state derive from that first developed by van der Waals in 1873. The van der Waals equation of state is given by the following equation:

$$\left(P + \frac{a}{V^2}\right)(V - b) = RT$$

where "a" is called the attraction parameter and "b" is referred to as the repulsion parameter or the effective molecular volume. Note that with "a" and "b" equal to zero this equation simplifies to the ideal gas law. These constants are found by either fitting the above equation to actual PVT data, or as we shall see in a few moments, we can use the critical point to define them in terms of the component critical properties.

a. Show that the above equation can also be written in the following cubic forms in terms of volume or the compressibility factor:

$$V^3 - \left(b + \frac{RT}{P}\right)V^2 + \frac{a}{P}V - \frac{ab}{P} = 0$$

$$z^3 - \left(\frac{bP}{RT} + 1\right)z^2 + \frac{aP}{(RT)^2}z - \frac{abP^2}{(RT)^3} = 0$$

For isotherms (T) below the critical isotherm and $P = P^{Sat}(T)$ (also refer to Figure 2.3), the solution to either of the above equations will produce three roots for either V or z. The smallest of these roots will correspond to the saturated liquid value, i.e. point B in Figure 2.3, and the largest of these roots will correspond to the saturated vapor value, i.e. point D in Figure 2.3. The middle root has no physical meaning. At the critical point, point C in Figure 2.3, these three roots become equal, i.e. $V_L^{Sat} = V_V^{Sat} = V_C = \frac{RT_C}{P_C}$, where V_C is the critical volume. Hence at the critical point we have that $(V - V_C)^3 = 0$, or when this is expanded, we have that $V^3 - 3V_C V^2 + 3V_C^2 V - V_C^3 = 0$.

b. Comparing the like powers of the expanded volume equation at the critical point to those in the volume explicit form of the van der Waals equation, show that the van der Waals parameters are given by the following equations in terms of the critical properties.

$$a = \frac{27 R^2 T_C^2}{64 P_C} \qquad b = \frac{R T_C}{8 P_C} \qquad R = \frac{8 P_C V_C}{3 T_C} \qquad z_C = 0.375$$

Although the value for the gas constant R is found to depend on the component critical properties, the usual practice is to use the universal value of the gas constant (shown in Table 1.6) in PVT calculations. In this way the van der Waals equation of state becomes the ideal gas law at low pressures.

c. For water at 250°C the vapor pressure is 3977.6 kPa and the specific volumes of the saturated vapor and liquid are 50.04 cm³/g and 1.251 cm³/g. Compare these specific volumes to those predicted by the van der Waals equation of state using the following critical properties for water:

$$T_C = 647.1 \text{ K } P_C = 220.55 \text{ bar} \quad (1 \text{ bar} = 100 \text{ kPa})$$

17. A new drug has a molecular weight of 625 and a melting temperature of 310°C. Estimate the solubility of this drug in hexane at 25°C. Use the Scatchard–Hildebrand equation to estimate the activity coefficient of the drug. Additional information needed to solve this problem is shown below and in Example 2.10.

Heat of fusion of the drug   5850 cal gmol⁻¹
Density of the drug   1.028 g cm⁻³ at 25°C
Vapor pressure of the solid drug $\ln P^{Sat}$ (mm Hg) $= 26.3 - 14780/T$ (K)

18. Estimate the gas phase equilibrium mole fraction of the drug considered in problem 17 at a temperature of 45°C. The pressure of the gas is 1 atm.

19. A protein solution containing 0.75 g of protein per 100 ml of solution has an osmotic pressure of 22 mm H_2O at 25°C. What is the molecular weight of the protein? (1mm Hg = 13.65 mm H_2O and 1 atm = 760 mm Hg, also R=0.082 L atm/K/mol).

20. Consider the situation where a semipermeable membrane separates a bulk fluid (region A) with an osmotic pressure of 1750 mm Hg from another enclosed fluid (region B) with an osmotic pressure of 10,000 mm Hg. The hydrodynamic pressure of region A is 760 mm Hg. The solvent in both regions A and B is water. Assuming the container and membrane enclosing region B is rigid, what is the equilibrium hydrodynamic pressure in region B, that is the situation wherein there is no net flow of water across the membrane?

21. When 0.5 moles of sucrose is dissolved in 1000 grams of water the osmotic pressure at 20°C is found to be 12.75 atm. Calculate the activity coefficient of water.

22. It is desired to extract from water a drug with a mole fraction of 0.02. A single equilibrium stage extractor is to be used. The flowrate of the aqueous stream to the extractor is 50 mole/min and the flowrate of the solvent to the extractor is 150 moles/min. If 90% of the drug is to be removed from the aqueous stream what should the distribution coefficient of the solvent for this drug be?

23. Using the results from problem 22, what would the percent extraction of the drug be if four equilibrium extraction stages were used?

24. The activity coefficients of ethanol (1) and water (2) can be described by the van Laar equation, where

$$\ln \gamma_1 = A_{12} \left(\frac{A_{21} x_2}{A_{12} x_1 + A_{21} x_2} \right)^2$$

$$\ln \gamma_2 = A_{21} \left(\frac{A_{12} x_1}{A_{12} x_1 + A_{21} x_2} \right)^2$$

The infinite dilution activity coefficient (γ_i^∞) of ethanol in water is 4.66 and that for water in ethanol (γ_2^∞) is 2.64. At 25°C the vapor pressure of water is 3.166 kPa and the vapor pressure of ethanol is 7.82 kPa. Assuming a liquid phase contains 50 weight % ethanol, estimate the composition of the vapor phase that is in equilibrium with the liquid phase. Would this vapor phase composition be explosive? The density of pure water at 25°C is 0.9971 g/cm^3 and the density of pure ethanol is 0.7851 g/cm^3. The density of a 50 weight % solution of ethanol in water is 0.9099 g/cm^3.

25. Explain why a water bug can walk on water.

26. A droplet of water at 25°C has a diameter of 0.1 microns. Calculate the pressure within the droplet of water.

27. What is the vapor pressure of the water within a droplet of water at 25°C that has a diameter of 10 nm?

28. Consider the situation of a liquid mixture where the molecules of the solute are much larger than those of the solvent. An example of such a mixture would be a polymer in a solvent or a protein solution. In the solution model developed by Flory and Huggins (Sandler 1989), the entropy change as a result of mixing a solvent (1) with a much larger solute (2) is given by the following expression:

$$\Delta S^{mix} = -R \left(x_1 \ln \phi_1 + x_2 \ln \phi_2 \right)$$

where x_1 and x_2 are the mole fractions and ϕ_1 and ϕ_2 are the volume fractions of the solvent and the solute respectively. The volume fractions are given by:

$$\phi_1 = \frac{x_1}{x_1 + r x_2} \quad \text{and} \quad \phi_2 = \frac{r x_2}{x_1 + r x_2}$$

with $r = V_2/V_1$ where V_1 and V_2 are the molar volumes of each species. Using Equations 2.44 and 2.96 which respectively define the property change of mixing and the excess property, along with the result from Example 2.6 that $S_i^{\text{ideal solution}} = S_i - R \ln x_i$, show that the excess entropy of the solution is given by:

$$S^E = -R \left(x_1 \ln \frac{\phi_1}{x_1} + x_2 \ln \frac{\phi_2}{x_2} \right)$$

Note that if the volumes of the solvent and solute are comparable in size then from the above expression $S^E = 0$ and the solution is then an ideal solution. So we see that it is the difference in the size of the solvent and the solute that leads to the non-ideal solution behavior in the model proposed by Flory and Huggins. The excess enthalpy (H^E) was then expressed by the following equation:

$$H^E = \chi RT (x_1 + r x_2) \phi_1 \phi_2$$

where χ is an adjustable parameter known as the Flory–Huggins interaction parameter. The Flory–Huggins interaction parameter for non-polar systems can be shown to be related to the solubility parameters of the solvent and the solute by the following equation (Prausnitz *et al.* 1986):

$$\chi = \frac{V_1}{RT} \left(\delta_1 - \delta_2 \right)^2$$

A good solvent for the polymer or macromolecule is one for which χ is very small or $\delta_1 \sim \delta_2$. In addition, to ensure that the solvent and polymer are completely miscible, the largest value of χ is equal to 1/2. If $\chi > 1/2$, the solvent and solute are only partially miscible.

Since $G^E = H^E - TS^E$, show that the above two equations for S^E and H^E can be combined to give the Flory–Huggins model for the excess Gibbs free energy of the solution:

$$\frac{G^E}{RT} = \left(x_1 \ln \frac{\phi_1}{x_1} + x_2 \ln \frac{\phi_2}{x_2} \right) + \chi \left(x_1 + r x_2 \right) \phi_1 \phi_2$$

Using Equation 2.100, show (if you can) that the Flory–Huggins activity coefficients for the solvent (1) and the solute (2) are given by the following expressions:

$$\ln \gamma_1 = \ln \frac{\phi_1}{x_1} + \left(1 - \frac{1}{r} \right) \phi_2 + \chi \phi_2^2$$

$$\ln \gamma_2 = \ln \frac{\phi_2}{x_2} + \left(r - 1 \right) \phi_1 + \chi \phi_1^2$$

Recall from Equation 2.146 that the osmotic pressure of the solution containing our solvent (1) and solute (2) is given by:

$$\pi = -\frac{RT}{V_1} \ln \left(\gamma_1 x_1 \right)$$

Show that if the activity coefficient of the solvent (1) is described by the Flory–Huggins activity coefficient model that the osmotic pressure is given by the following expression. Note that for a dilute solute solution we can expand the solvent and solute volume fraction in an infinite series where:

$$\ln \phi_1 = \ln \left(1 - \phi_2 \right) \approx -\phi_2 - \frac{\phi_2^2}{2} - \frac{\phi_2^3}{3} - \cdots$$

$$\pi = \frac{RT}{V_1} \left[\frac{\phi_2}{r} + \left(\frac{1}{2} - \chi \right) \phi_2^2 + \frac{1}{3} \phi_2^3 + \cdots \right]$$

From the definition of the solute volume fraction from above we can then write for a dilute solution that $\phi_2 = \frac{x_2 V_2}{V_1} = V_2 C_2$, where C_2 is the molar concentration of the solute. Substituting this result for ϕ_2 into the above equation for π, show that in a general sense the osmotic pressure for a non-ideal solution can be described by a power series in the molar concentration of the solute (or virial series) as shown below:

$$\pi = R T C_2 \left(1 + \overline{B} C_2 + \overline{C} C_2^2 + \cdots \right)$$

where $\overline{B}$ and $\overline{C}$ are known as the second and third virial coefficients. Note that for an ideal solution $\overline{B}$ and $\overline{C}$ are equal to zero and $\pi = RTC_2$ which is also the result shown in Equation 2.147. Usually with polymers and macromolecules the solute concentration is expressed in weight concentration, for example grams/liter. Letting c_2 represent the weight concentration of the solute, then $C_2 = c_2/MW_2$ and then $\phi_2 = V_2 \frac{c_2}{MW_2}$. Substituting this result for ϕ_2 into the above equation for π and neglecting terms higher than the second order in ϕ_2, show that the following result is obtained for the osmotic pressure:

$$\pi = \frac{RT c_2}{MW_2}\left[1 + \frac{MW_2}{V_1 \rho_2^2}\left(\frac{1}{2} - \chi\right)c_2\right]$$

Also show that the second virial coefficient is equal to $\dfrac{MW_2}{V_1 \rho_2^2}\left(\dfrac{1}{2} - \chi\right)$,

where ρ_2 is the solute density.

29. The following data for the osmotic pressure of hemoglobin (MW = 68,000 daltons) in an aqueous solution at 0°C is presented by Freeman (1995).

Hemoglobin concentration (g 100^{-1} cm^{-3} solution)	Osmotic pressure (mm Hg)
2.5	8
5	15
8	25
10	37
12	42
15	65
19.5	100
23.4	150
24.5	167
27.5	229
28.6	254

Make a graph of the above data and compare the data to regression fits of the following virial expression derived in problem 28 for the osmotic pressure, i.e. $\pi = \dfrac{RT c_2}{MW_2}\left[1 + \bar{B}c_2 + \bar{C}c_2^2\right]$. Do three regressions where the first regression assumes an ideal solution (i.e. $\bar{B}$ and $\bar{C}$ equal zero), the second regression ($\bar{C} = 0$) truncates the osmotic pressure equation after the second virial coefficient, and the third regression includes $\bar{B}$ and $\bar{C}$.

THREE

PHYSICAL PROPERTIES OF THE BODY FLUIDS AND THE CELL MEMBRANE

3.1 BODY FLUIDS

To begin our study of transport phenomena in biomedical engineering, we first must examine the physical properties of the fluids that are within the human body. Many of our engineering calculations or the development and design of new procedures, devices, or treatments will either involve or affect the fluids that are within the human body. Therefore, we will focus our initial attention on the types and characteristics of the fluids that reside within the body.

The body has three types of fluids. These are the *extracellular*, *intracellular*, and *transcellular* fluids. As shown in Table 3.1, nearly 60% of the body weight for an average 70 kg male is composed of these body fluids resulting in a total fluid volume of about 40 liters.

The largest fraction of the fluid volume, or about 36% of the body weight, consists of *intracellular fluid*, that is the fluid contained within the body's cells, for example the fluid found within red blood cells, muscle cells, liver cells, etc. *Extracellular fluid* consists of the *interstitial fluid* that comprises about 17% of the body weight and the *blood plasma* that comprises around 4% of the body weight. Interstitial fluid circulates within the spaces (*interstitium*) between cells. The interstitial fluid space represents about one-sixth of the body volume. The interstitial fluid is formed as a filtrate from the plasma within the *capillaries*. The capillaries are the smallest element of the cardiovascular system and represent the site where the exchange of vital substances occurs between the blood and the tissue surrounding the capillary. We shall see that the interstitial fluid composition is very similar to that of plasma.

The blood volume of a 70 kg male is 5 liters, with 3 liters of the blood volume

Table 3.1 Body fluids

Fluid	Fluid volume	% of Body weight
Intracellular	25 liters	36 wt%
Extracellular	15 liters	21 wt%
Interstitial	12 liters	17 wt%
Plasma	3 liters	4 wt%
Transcellular	—	—
Total	40 liters	57 wt%

Source: Data from A.C. Guyton, *Textbook of Medical Physiology*, 8th ed. (Philadelphia: W.B. Saunders Co., 1991), 275

consisting of plasma and the remaining 2 liters representing the volume of the cells that are in blood, primarily the red blood cells. The 2 liters of cells found in the blood are filled with intracellular fluid. The fraction of the blood volume due to the red blood cells is called the *hematocrit*. The red blood cell volume fraction is found by centrifuging a given volume of blood in order to find the "packed" red cell volume. Since a small amount of plasma is trapped between the packed red blood cells, the true hematocrit (H) is about 96% of this measured hematocrit (Hct). The true hematocrit is about 40% for a male and about 36% for a female. The *transcellular fluids* are those fluids found only within specialized compartments and include the cerebrospinal, intraocular, pleural, pericardial, synovial, sweat, and digestive fluids. Some of these compartments have membrane surfaces that are in proximity to one another and have a thin layer of lubricating fluid between them.

The measurement of these fluid volumes can be achieved by using "tracer materials" that have the unique property of remaining in specific fluid compartments. The fluid volume of a specific compartment can then be found by adding a known mass of a tracer to a specific compartment, and after an appropriate period of time for dispersal, measuring the concentration (mass/volume) of the tracer in the fluid compartment. The compartment volume is then given by the ratio of the tracer mass that was added and the measured tracer concentration, i.e. (mass/mass/volume = volume).

Examples of tracers used to measure fluid volumes include radioactive water for measuring *total body water* and radioactive sodium, radioactive chloride, or inulin for measuring *extracellular fluid volume*. Tracers that bind strongly with plasma proteins may be used for measuring the *plasma volume*. Interstitial fluid volume may then be found by subtracting the plasma volume from the extracellular fluid volume. Subtraction of the extracellular fluid volume from the total body water provides the *intracellular fluid volume*.

3.2 FLUID COMPOSITIONS

The compositions of the body fluids are presented in Table 3.2 in terms of concentration in units of *milliosmolar* (mOsmole/liter of solution). We will provide a definition of milliosmolar in Section 3.4.1. For now, the definition of a milliosmolar is not important

Table 3.2 Osmolar solutes found in the extracellular and intracellular fluids

Solute	Plasma	Interstitial	Intracellular
	(mOsm/L)	(mOsm/L)	(mOsm/L)
Na^+	143	140	14
K^+	4.2	4.0	140
Ca^{++}	1.3	1.2	0
Mg^{++}	0.8	0.7	20
Cl^-	108	108	10
HCO_3^-	24	28.3	10
HPO_4^- , H2PO$_4^-$	2	2	11
SO_4^-	0.5	0.5	1
Phosphocreatine			45
Carnosine			14
Amino acids	2	2	8
Creatine	0.2	0.2	9
Lactate	1.2	1.2	1.5
Adenosine Triphosphate			5
Hexose Monophosphate			3.7
Glucose	5.6	5.6	
Protein	1.2	0.2	4
Urea	4	4	4
Others	4.8	3.9	11
Total (mOsmoles/liter)	302.8	301.8	302.2
Corrected Osmolar Activity (mOsmoles/liter)	282.5	281.3	281.3
Total Osmotic Pressure at 37°C (mm Hg)	5450	5430	5430

Source: Data from A.C. Guyton, *Textbook of Medical Physiology*, 8th ed. (Philadelphia: W.B. Saunders Co., 1991), 277.

for the following discussion. Of particular interest is the fact that nearly 80% of the total osmolarity of the interstitial fluid and plasma is produced by sodium and chloride ions. As we discussed earlier, the interstitial fluid arises from filtration of plasma through the capillaries. We would therefore expect the composition of these two fluids to be very similar. This is shown in Table 3.2. We find this to be true with the exception that the protein concentration in the interstitial fluid is significantly smaller in comparison to its value in the plasma.

3.3 CAPILLARY PLASMA PROTEIN RETENTION

The retention of proteins by the walls of the capillary during filtration of the plasma is readily explained by comparing the molecular sizes of typical plasma protein molecules to the size of the pores within the capillary wall. Figure 3.1 illustrates the relative size of various solutes as a function of their molecular weight. The wall of a capillary, illustrated in Figure 3.2, consists of a single layer of *endothelial cells* that are surrounded on their outside by a *basement membrane*. The basement membrane is a mat-like cellular support structure, or extracellular matrix, that consists primarily of a protein called *type IV collagen*, and is joined to the cells by the glycoprotein called

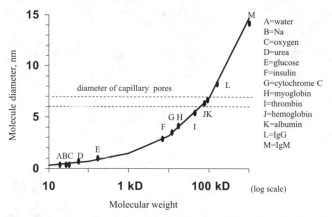

Figure 3.1 Approximate diameter of molecules as a function of their molecular weight (kD = 1000 daltons).

laminin. The basement membrane is about 50–100 nm thick. The total thickness of the capillary wall is about 0.5 microns.

As shown in Figure 3.2, there are several mechanisms that allow for the transport of solutes across the capillary wall. These include the *intercellular cleft* and *pinocytotic vesicles* and *channels*. The intercellular cleft is a thin slit or slit-pore that is formed at the interface between adjacent endothelial cells. The size of the openings or pores in this slit is about 6 to 7 nanometers, just sufficient to retain albumin and other larger proteins. The collective surface area of these openings represents less than 1/1000th of the total capillary surface area.

The plasma proteins are generally larger than the capillary slit-pores. Although ellipsoidally shaped proteins, for example the clotting protein fibrinogen, may have a minor axis that is smaller than that of the capillary slit-pore, the streaming effect caused by the fluid motion within the capillary orients the major axis of the proteins parallel to the flow axis and prevents their entry into the slit-pore. This streaming effect is shown

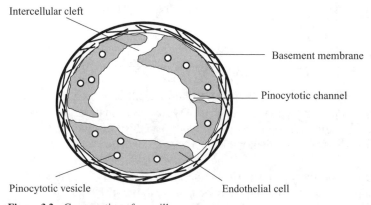

Figure 3.2 Cross section of a capillary.

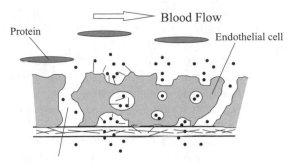

Figure 3.3   Orientation of ellipsoidal shaped proteins by streaming prevents their entry into the capillary pores.

in Figure 3.3. Therefore, the smaller substances found in the plasma, such as ions, glucose, and metabolic waste products, will readily pass through the slit-pores of the capillary wall, whereas the plasma proteins will be retained in the lumen of the capillary. The composition of the plasma and interstitial fluid should, therefore, only differ in their protein content. This is shown in Table 3.2.

3.4 OSMOTIC PRESSURE

Retention of proteins in the plasma in comparison to the interstitial fluid creates an *osmotic pressure* between the plasma and the interstitial fluid. We discussed the thermodynamics of osmosis in Section 2.6.3.5 in Chapter 2. Since the plasma proteins are the only constituent of the plasma that do not readily pass through the capillary wall, it is the plasma proteins that are responsible for the formation of the osmotic pressure between the interstitial and plasma fluids. The osmotic pressure created by these proteins is given the special name of *colloid osmotic pressure*, or *oncotic pressure*. For human plasma, the colloid osmotic pressure is about 28 mm Hg; 19 mm Hg caused by the plasma proteins and 9 mm Hg caused by the cations within the plasma that are also retained through electrostatic interaction with the negative surface charges of these proteins.

The colloid osmotic pressure is actually quite small in comparison to the osmotic pressure that is developed when a cell is placed in pure water. In this case, it is assumed that all of the species present within the intracellular fluid are retained by the cell membrane. As shown in Table 3.2, the total osmotic pressure of the intracellular fluid in this case would be 5430 mm Hg at 37°C.

3.4.1 Osmolarity

Recognizing that it is the number of non-diffusing solute molecules that contributes to the osmotic pressure, we now must make the distinction between a non-diffusing substance that dissociates, and one that does not dissociate. The term non-diffusing is used to describe solutes that would be retained by a membrane that separates the solutions of interest. For example, compounds such as NaCl are strong electrolytes. In water they completely dissociate to form two ions (i.e. Na^+ and Cl^-). Each ion or particle formed will exert its own osmotic pressure. It is important to note that the charge of the ion has no effect on the osmotic pressure, so a sodium ion with a charge of +1 is equivalent to a calcium ion of charge +2. Substances such as glucose do not dissociate and their osmotic pressure is based on their non-dissociated concentration only.

The term *osmole* has been introduced to account for the effect of a dissociating solute. One osmole is therefore defined as one mole of a non-diffusing and non-dissociating substance. Therefore, one mole of a dissociating substance such as NaCl is equivalent to two osmoles or a 1 M solution of NaCl is equivalent to a 2 Osmolar (OsM) solution. However, one mole of glucose is the same as one osmole since glucose does not dissociate in solution. *Osmolarity* simply defines the number of osmoles per liter of solution.

If a cell is placed within a solution that has a lower concentration of solutes or osmolarity, then the cell is in a *hypotonic solution*, and establishment of osmotic equilibrium requires the osmosis of water into the cell. This influx of water into the cell results in swelling of the cell and a subsequent decrease in its osmolarity. On the other hand, if the cell is placed in a solution with a higher concentration of solutes or osmolarity, i.e. *hypertonic*, then osmotic equilibrium requires osmosis of water out of the cell, concentrating the intracellular solution and resulting in shrinkage of the cell. An *isotonic* solution is a fluid that has the same osmolarity of the cell. When cells are placed in an isotonic solution there is neither swelling nor shrinkage of the cell. A 0.9 weight percent solution of sodium chloride or a 5 weight percent solution of glucose is just about isotonic with respect to a cell.

3.4.2 Calculating the Osmotic Pressure

Recall from our discussion in Chapter 2 on osmosis that for the special case where the solvent and solute form an ideal solution, the osmotic pressure given by Equation 2.147 may be written as shown below.

$$\pi = \frac{RT}{V_{solvent}^L} x_{solute}^A = RTC_{solute} \qquad (3.1)$$

Here x_{solute}^A represents the mole fraction of the solute in the solution which is region A in Figure 2.5. Recall that since the solute mole fraction is generally quite small, we may approximate this as $x_{solute}^A = V_{solvent}^L C_{solute}$, where C_{solute} is the concentration of the solute in gmoles/liter of solution and $V_{solvent}^L$ is the molar volume of the solvent. This ideal dilute solution osmotic pressure, described by the above equation, is also known as *van't Hoff's law*.

If the solution contains N ideal solutes, then the total osmotic pressure of the solution will be the summation of the osmotic pressure generated by each solute according to Equation 3.1.

$$\pi = RT \sum_{i=1}^{N} C_{solute_i}$$ (3.2)

As we discussed in Chapter 2 remember that osmotic pressure is not determined on the basis of the mass of the solute in the solution, but rather on the number of particles that are formed by a given solute. Each non-diffusing particle in the solution contributes the same amount to the osmotic pressure regardless of the size of the particle.

For physiological solutions, it is convenient to work in terms of milliosmoles (mOsm) or milliosmolar (mOsM). At a physiological temperature of 37°C, Equation 3.2 may be written as follows to give the osmotic pressure in mm Hg when the solute concentration of each non-diffusing species is expressed in mOsM:

$$\pi = 19.33 \sum_{i=1}^{N} C_{Si}$$ (3.3)

3.4.3 Other Factors that May Affect the Osmotic Pressure

The previous discussion assumed the mixture of solutes formed an ideal solution. However, the osmotic pressure should take into account the various secondary solute interactions that occur within the solution due to their charge, size, shape, and other effects. These effects will either increase or decrease the osmotic activity of a particular solute and results in the corrected osmolar activity shown in Table 3.2. As shown in Table 3.2, we see that the ratio of the corrected osmolar concentration to the total osmolar concentration is about 0.93 for each of the body fluids. This value of 0.93 represents the overall average activity coefficient. In most cases, it is conventional practice to disregard the calculation of solute activity coefficients and calculate the osmotic pressure on the basis of solute concentration only as just discussed. Since we will primarily be concerned with differences in the osmotic pressure and the generation of fluid flow across a membrane due to a difference in osmotic pressure, this constant of 0.93 will become absorbed in other constants that generally need to be determined by experiment. Thus, for our purposes, little error is introduced by simply calculating using Equation 3.2 the osmotic pressure on the basis of solute concentration only.

3.5 FORMATION OF THE INTERSTITIAL FLUID

The flow of fluid across the capillary wall, or for that matter, any porous or semipermeable membrane, is driven by a difference in pressure across the capillary wall or membrane. This pressure difference arises not only from hydrodynamic effects, but also from the difference in osmotic pressure between the fluids separated by the membrane.

Figure 3.4 shows the hydrodynamic and osmotic pressures in the capillary and the surrounding interstitial fluid. The arrows indicate for each pressure the corresponding direction of fluid flow induced by that pressure. We may write that the volumetric fluid

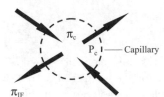

Figure 3.4 Forces acting to cause a flow of fluid across the capillary wall. Arrows indicate the direction of flow induced by each pressure.

transfer rate (Q) across the capillary membrane is directly proportional to the effective pressure drop $(\overline{\Delta P})$ across the capillary membrane as given by the following equation:

$$Q = L_P S\left[\left(P_C - P_{IF}\right) - \left(\pi_C - \pi_{IF}\right)\right] = L_P S \overline{\Delta P} \qquad (3.4)$$

In this equation, L_P represents a proportionality constant that is inversely related to the flow resistance of the capillary membrane and is called the *hydraulic conductance*. The hydraulic conductance is usually best determined by experiment. For example, Renkin (1977) reports values for L_P as low as 3×10^{-14} m² sec/kg for the tight junctions between the endothelial cells found in the capillaries of the rabbit brain, 5×10^{-12} m² sec/kg for non-fenestrated or continuous capillaries, and as high as 1.5×10^{-9} m² sec/kg for the capillaries in the kidney's glomeruli. S is simply the total circumferential surface area of the capillary membrane or the total membrane area.

The first difference term in the brackets of Equation 3.4 represents the difference between the hydrodynamic pressure of the capillary (P_C) and the interstitial fluid (P_{IF}). When this term is positive, fluid will leave the capillary. When this term is negative, fluid will flow from the interstitial fluid into the capillary. The second difference term within the brackets represents the difference in the osmotic pressure between the capillary (π_C) and the interstitial fluid (π_{IF}). When this term is positive, there will be an osmotic flow of water from the interstitial fluid into the capillary whereas if this difference is less than zero there will be an osmotic flow of water out of the capillary. If the entire term in brackets in the above equation is positive, then there is a net flow of fluid from the capillary into the interstitium. If the bracketed term is negative, then there will be a net flow of fluid from the interstitium into the capillary.

Equation 3.4 represents the fluid flow across the capillary membrane. It can also be used to describe the flow across any semipermeable membrane regardless of whether or not it is of biological or synthetic origin. We will find, for example, that this equation applies to the dialysis membranes used in the artificial kidney.

We can also develop a model for describing how the hydraulic conductance (L_P) depends on the pore geometry and the physical properties of the filtrate fluid. First we model the porous structure of the capillary wall as a series of parallel cylindrical pores. We will also show in Chapter 4 that Poiseuille's equation (Equation 4.11) provides a relationship between flow (Q) and the pressure drop $(P_0 - P_L)$ in a cylindrical duct.

$$Q = \frac{\pi R^4 \left(P_0 - P_L\right)}{8 \mu L} \qquad (3.5)$$

where μ is the viscosity (a measure of the flow resistance) of the fluid, R is the radius of the cylindrical duct or pore, and L is the length of the duct. The above equation provides the flowrate across a single pore in the capillary wall. If we have a total of N capillary pores, then we can combine the above equation for N pores with Equation 3.4 to obtain the result shown below:

$$Q = L_p S \,\overline{\Delta P} = N \left(\frac{\pi R^4}{8 \mu \tau t_m} \right) \overline{\Delta P} \tag{3.6}$$

where we have replaced $(P_0 - P_L)$ in Poiseuille's equation with the effective pressure drop $\overline{\Delta P}$. The actual pore length L is usually longer than the thickness of the capillary wall (t_m) since the pores are not straight but tortuous. Hence we replaced L with the product of the tortuosity (τ) and the capillary wall thickness t_m. The tortuosity is a correction factor and for the capillary wall it is about two. The total cross-sectional area of the pores, i.e. A_P, is equal to $N \pi R^2$. Substituting this into the above equation allows us to solve for the hydraulic conductance as shown below.

$$L_P = \left(\frac{A_P}{S} \right) \frac{R^2}{8 \mu \tau t_m} \tag{3.7}$$

The above equation provides a model that allows one to understand the factors that affect the hydraulic conductance. We see that the hydraulic conductance is directly proportional to the porosity of the capillary wall, i.e. $\left(\dfrac{A_P}{S} \right)$, and inversely proportional to the flow resistance of the fluid, i.e. its viscosity (μ), and the thickness of the capillary wall. In addition, the hydraulic conductance is directly proportional to the square of the pore radius.

3.6 NET CAPILLARY FILTRATION RATE

We can use Equation 3.4 to estimate the *net capillary filtration rate* for the human body. To perform this calculation, we will first need to define nominal values for the pressures in Equation 3.4. (Note: these are gauge pressures or relative to atmospheric pressure of 760 mm Hg.) At the arterial end of the capillary, the capillary pressure is about 30 mm Hg, and at the venous end of the capillary, the pressure is about 10 mm Hg. The mean capillary pressure is considered to be about 17.3 mm Hg, and its bias to the lower end is based on the larger volume of the venous side of the capillaries in comparison to the arterial side of the capillaries. The interstitial fluid pressure has, surprisingly, been found to be subatmospheric and the accepted value is –3 mm Hg. As mentioned before, the colloid osmotic pressure for human plasma is 28 mm Hg, and the value for the interstitial fluid is about 8 mm Hg.

At the arterial end of the capillary, we can calculate the flowrate of fluid across the capillary wall by using Equation 3.4. Therefore, $Q/L_p S = [(30 - -3) - (28 - 8)] = 13$ mm Hg which is positive, indicating that there is a net flow of fluid from the capillary into the interstitium. At the venous end of the capillary, $Q/L_p S = [(10 - -3) - (28 - 8)] = -7$ mm Hg, which is less than zero, indicating a net reabsorption of fluid from the interstitium back into the capillary.

This flow of fluid out of the capillary at the arterial end, and its reabsorption at the venous end, is called *Starling flow* after E.H. Starling who first described it over a hundred years ago. Equation 3.4 is therefore also known as the *Starling equation*. If we base the filtration rate on the mean capillary pressure, then $Q/L_pS = [(17.3--3) - (28 - 8)] = 0.3$ mm Hg. Hence, on average, there is a slight imbalance in pressure resulting in more filtration of fluid out of the capillary than is reabsorbed.

Approximately 90% of the fluid that leaves at the arterial end of the capillary is reabsorbed at the venous end. However, the 10% that is not reabsorbed by the capillary collects within the interstitium and enters the lymphatic system. The amount of this net filtration of fluid from the circulation for the human body can be estimated as illustrated in the following example.

Example 3.1 Calculate the normal rate of net filtration for the human body. Assume the capillaries have a total surface area of 500 m² and that the slit-pore surface area is 1/1000th of the total capillary surface area.

SOLUTION We model the porous structure of the capillary wall as a series of parallel cylindrical pores with a diameter of 7 nm. Plasma filtrate may be considered to be a Newtonian fluid with a viscosity of 1.2 centipoise (cP). The mean net filtration pressure for the capillary was just calculated above to be 0.3 mm Hg. Using Equation 3.7, we can calculate the hydraulic conductance and the net filtration rate.

$$L_p = \left(\frac{0.5\,\text{m}^2}{500\,\text{m}^2}\right) \frac{\left(0.5 \times 7 \times 10^{-9}\,\text{m}\right)^2}{8 \times 1.2\,\text{cP} \times \dfrac{0.01\,\text{P}}{\text{cP}} \times \dfrac{1\,\text{g}}{\text{cm sec P}} \times \dfrac{100\,\text{cm}}{\text{m}} \times \dfrac{1\,\text{kg}}{1000\,\text{g}} \times 2 \times .5 \times 10^{-6}\,\text{m}}$$

$$= 1.28 \times 10^{-12}\, \frac{\text{m}^2\,\text{sec}}{\text{kg}}$$

Note that this prediction of L_p is consistent with the values reported by Renkin (1977) for the hydraulic conductance of continuous capillaries.

$$Q = 1.28 \times 10^{-12}\, \frac{\text{m}^2\,\text{sec}}{\text{kg}} \times 500\,\text{m}^2 \times 0.3\,\text{mm Hg} \times \frac{1\,\text{atm}}{760\,\text{mm Hg}} \times 101.33\frac{\text{kPa}}{\text{atm}} \times \frac{1\text{N}/\text{m}^2}{\text{Pa}}$$

$$\times \frac{1000\,\text{Pa}}{\text{kPa}} \times \frac{\text{kgm}/\text{sec}^2}{\text{N}} \times \frac{60\,\text{sec}}{\text{min}} \times \frac{(100\,\text{cm})^3}{\text{m}^3}$$

$$Q = 1.54\,\text{cm}^3/\text{min}$$

We find that the *total net filtration rate* due to the pressure imbalance at the capillaries for the human body is on the order of 2 ml/min.

We can define the *filtration coefficient* as the ratio of Q and $\overline{\Delta P}$ which is equal to $L_p\,S$. Using the results from the above example, for the capillary wall we obtain:

$$\frac{Q}{\Delta P} = L_p S = \frac{5.1 \, \text{ml}}{\text{min mm Hg}}$$ (3.8)

For the capillary wall or a particular membrane, the filtration coefficient is a constant and represents the properties of the filtrate fluid, the membrane porosity, and the dimensions of the pores. The net filtration rate for any average pressure imbalance at the capillaries is given by simply multiplying the filtration coefficient by the pressure imbalance. For example, if the pressure imbalance increased from 0.3 to 1.5 mm Hg, then the net filtration rate would increase by a factor of 5 to about 10 ml/min.

3.7 LYMPHATIC SYSTEM

We must now address where the net filtration of fluid from the capillaries, about 120 ml/hr for a human, ultimately goes. It clearly cannot continue to collect within the interstitium since this would lead to *edema* or excess fluid (swelling) within the tissues of the body.

Two types of edema can occur. The first type involves excess extracellular fluid and is caused either by too much filtration of fluid from the capillaries as just discussed, a failure to drain this excess fluid from the interstitium, or retention of salt and water as a result of impaired kidney function. Edema can also be caused by intracellular accumulation of fluid as a result of cellular metabolic problems or inflammation. In these cases either the sodium ion pumps are impaired (see Section 3.10) or the cell membrane permeability to sodium is increased. In either case, the excess sodium in the cell causes osmosis of water into the cell.

The *lymphatic system* is an accessory flow or circulatory system in the body that drains excess fluid from the interstitial spaces and returns it to the blood. The lymphatics consist of a system of lymphatic capillaries and ducts that empties into the venous system at the junctures of the left and right internal jugular and subclavian veins. This system is also responsible for the removal from the interstitial space of large proteins and other matter that cannot be reabsorbed into the capillary. Like the vascular system capillaries, the lymphatic capillaries are also formed by endothelial cells. However, the lymphatic endothelial cells have much larger intercellular junctions and the cells overlap in such a manner to form a valve-like structure. Interstitial fluid can force the valve open to flow into the lymphatic capillary, however backflow out of lymphatic fluid from the lymphatic capillary is prevented.

3.8 SOLUTE TRANSPORT ACROSS THE CAPILLARY ENDOTHELIUM

Lipid soluble substances such as oxygen and carbon dioxide can diffuse directly through the endothelial cells that line the capillary wall without the use of the slit-pores. Accordingly, their rate of transfer across the capillary wall is significantly higher than water soluble, but lipid insoluble substances, such as sodium ions, chloride ions, and glucose, for which the cell membrane of the endothelial cell is essentially impermeable.

The transport of these latter substances across the capillary wall occurs through the use of the capillary slit-pores.

In addition to the slit-pores, there are two other pathways that can provide an additional route for the transport of large lipid insoluble solutes such as proteins across the endothelium of the capillary wall. These pathways are called *pinocytosis* and *receptor-mediated transcytosis* (Lauffenburger and Linderman 1993).

Pinocytosis is not solute specific, and as shown in Figure 3.5, involves the ingestion by the cell of the surrounding extracellular fluid and its associated solutes. A small portion of the cell's plasma membrane forms a pocket containing the extracellular fluid. This pocket grows in size and finally pinches off to form the intracellular pinocytotic vesicles. This ingestion of extracellular material by the cell is also known as *endocytosis*. The pinocytotic vesicles are either processed and their contents used internally, or they migrate through the cell and reattach to the cell membrane on the opposite side where they release their contents to the surrounding milieu. This release of material from the pinocytotic vesicle is known as *exocytosis*. Because of its non-specific nature, pinocytosis is usually not a significant solute transport mechanism.

However, receptor-mediated transcytosis can provide significant transport of specific solutes across the capillary endothelium. This process is also shown in Figure 3.5. The solute, referred to as a *ligand*, first binds with complementary receptors that are located on the surface of the cell membrane. Unlike the non-specific process of pinocytosis, receptor-mediated transcytosis can concentrate a particular ligand by many orders of magnitude. This solute concentrating mechanism by the cell-surface receptors is responsible for the significant transport rates that can be achieved by this process. The ligand-receptor complexes are then endocytosed forming transcytotic vesicles. These vesicles are either processed internally or they can move through the cell and reattach themselves to the opposite side of the cell. The transcytotic vesicle then releases its contents by exocytosis.

The importance of receptor-mediated transcytosis of large lipid insoluble solutes is provided by the following example. In vitro studies of insulin transport across vascular endothelial cells have shown that 80% of the insulin was transported by receptor-mediated processes. The remaining 20% was transported either through the slit-pores between the endothelial cells or by non-specific pinocytosis (Hachiya *et al.* 1988).

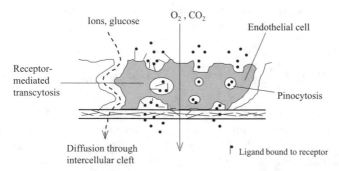

Figure 3.5   Mechanisms for solute transport across the capillary endothelium.

3.9 THE CELL MEMBRANE

A question that needs to be addressed is why smaller molecules simply do not diffuse across the endothelial cells that line the capillaries. Water can diffuse through the endothelial cell, however much more can simply pass through the slit-pores. For other substances, we must take a closer look at the chemical nature of the *cell membrane*.

The cell membrane, illustrated in Figure 3.6, is composed mainly of a *lipid bilayer*. Lipid molecules are insoluble in water but readily dissolve in organic solvents such as benzene. Three classes of lipids are found in cellular membranes. These are *phospholipids*, *cholesterol*, and *glycolipids*. The lipid bilayer results because the lipid molecule has a head and tail configuration as shown in Figure 3.6. The head of the lipid molecule is polar and thus hydrophilic, whereas the tail of the lipid molecule is non-polar and hydrophobic (typically derived from a fatty acid). Such molecules are also called *amphipathic* because the molecule has both hydrophilic and hydrophobic properties. The term lipid bilayer indicates their tendency to form bimolecular sheets when surrounded on all sides by an aqueous environment, as is the case for a cell membrane. Thus, the hydrophilic heads of the lipid molecules face into the aqueous environment, and the hydrophobic tails are sandwiched between the heads of the lipid molecules.

The lipid bilayer forms the basic structure of the cell membrane. However, other molecules such as proteins are scattered throughout the lipid bilayer of the cell membrane and serve many important functions. For example, special proteins allow for the transport of specific molecules across the cell membrane. Other proteins have catalytic activity and mediate chemical reactions that occur within the cell membrane. These proteins are known as enzymes. Still other proteins provide structural support to the cell or provide connections to surrounding cells or other extracellular materials. Some proteins found in the cell membrane act as receptors to extracellular substances or chemical signals, and through transduction of these signals, control intracellular events. Other proteins present foreign materials to the immune system or identify the cell as self.

Protein molecules associated with the cell membrane can be classified into two broad categories. The *transmembrane proteins* are also amphipathic and extend through

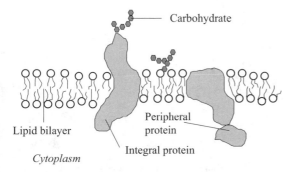

Figure 3.6   Cell membrane structure.

the lipid bilayer. They typically have hydrophobic regions that may travel across the membrane several times, and hydrophilic ends that are exposed to water on either side of the membrane. Integral membrane proteins are transmembrane proteins that are held tightly within the cell membrane through chemical linkages with other components of the cell membrane. Integral proteins have major functions related to the transport of water soluble but lipid insoluble substances across the cell membrane. The *peripheral membrane proteins* are not located within the plasma membrane but associate on either side of the membrane with transmembrane or integral proteins. Peripheral proteins mostly function as enzymes.

For the most part, the cell membrane is impermeable to polar or other water soluble molecules. Hence, large neutral polar molecules, like glucose have very low cell membrane permeabilities. Charged molecules and ions such as H^+, Na^+, K^+, and Cl^- also have very low permeabilities. Hydrophobic molecules, such as oxygen and nitrogen, readily dissolve in the lipid bilayer and show very high permeabilities. Smaller neutral polar molecules such as CO_2, urea, and water, are able to permeate the lipid bilayer because of their much smaller size and neutral charge.

The transport of essential water soluble molecules across the cell membrane is achieved through the use of special transmembrane proteins that have a high specificity for a certain type or class of molecules. These membrane transport proteins come in two basic types, *carrier proteins* and *channel proteins*. The carrier proteins bind to the solute and then undergo a change in shape or conformation ("ding" to "dong") that allows the solute to traverse the cell membrane. The carrier protein therefore changes between two shapes, alternately presenting the solute binding site to either side of the membrane. Channel proteins actually form water-filled pores that penetrate across the cell membrane. Solutes that cross the cell membrane by either carrier or channel proteins are said to be passively transported.

Figure 3.7 illustrates the passive transport of solutes through the cell membrane by either carrier proteins or channel proteins. Carrier proteins can transport either a single solute across the membrane, a process called uniport, or two different solutes, called coupled transport. The passive transport of glucose into a cell by glucose transporters is an example of a uniport. In coupled transport, the transfer of one solute occurs in combination with the transport of another solute. This coupled transport of the two solutes can occur with both solutes transported in the same direction, *symport*, or in

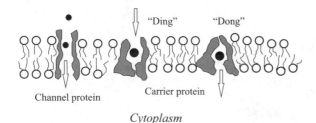

"Ding" "Dong"

Channel protein

Carrier protein

Cytoplasm

Figure 3.7 Membrane transport proteins. Carrier protein exists in two conformational states, "Ding" and "Dong".

opposite directions, *antiport*. An example of coupled transport is the sodium ion gradient driven symport of glucose and sodium ions.

In general, the driving force for the passive transport of these solutes is due to the combined effect of their concentration gradient and the electrical potential difference that exists across the membrane. Neutral molecules diffuse from regions of high concentration to regions of low concentration. However, if the molecule carries an electrical charge, then both the concentration gradient and the electrical potential difference, or voltage gradient, across the cell membrane will affect the transport of the molecule. The *electrochemical gradient* is the term used to describe the combined effect of charge and solute concentration on the transport of a molecule. The voltage gradient for a cell membrane is such that the inside of the cell membrane is negative in comparison to the outside. This membrane potential (V_M) for cells at rest is about –90 mV.

The flow of charged molecules through channels in the cell membrane is responsible for the creation of the membrane potential. For example, a higher concentration of potassium ions within the cell relative to the surroundings will tend to cause a leakage of these ions out of the cell through the potassium ion leak channels. The loss of these positive ions will make the interior of the cell negative in charge. This creates an electric field that is called the membrane potential. The growth of this membrane potential with continued loss of potassium ions will reach a point where the negative charge created within the cell begins to retard the loss of the positively charged potassium ions due to the difference in potassium concentration. When these two forces balance each other, there is no net flow of the ion. This balance, or equilibrium of the concentration and voltage gradients for an ion, is known as the *equilibrium membrane potential* for the cell.

We can obtain an equation that relates the equilibrium membrane potential (V), i.e. the difference in potential between the inside (V_{inside}) and the outside ($V_{outside}$) of the cell, to the equilibrium concentrations of the ion on either side of the cell membrane. The presence of an electrical field across the cell membrane will produce a force on the ions as a result of their charge. Hence the ions will move and their motion causes a current to flow. Recall that voltage (V) times current (I) is equal to the power and power is the rate of doing work. Electrical work is therefore done on the ion and is given by:

$$W_{electrical} = -V I t \tag{3.9}$$

If we are only infinitesimally removed from equilibrium, then $W_{electrical}$ becomes $W_{reversible\ electrical}$, and from our earlier discussion in Section 2.3.3.1, the reversible electrical work (maximum work at constant T and P) is equal to the change in the Gibbs free energy. Also, the product of the current and the time in the above equation is equal to the total charge that is transferred. Faraday's constant (F) is defined as the charge carried by one gram mole of ions of unit positive valency and is equal to 9.649×10^4 coulombs/mole or 2.306×10^4 cal V^{-1} mole^{-1}. We can now rewrite Equation 3.9 for a gram mole of ions as shown below:

$$W_{reversible\ electrical} = \Delta G_{electrical} = -z F V \tag{3.10}$$

where z is the charge on the ion being considered. For example, for a sodium ion (Na^+) $z = +1$, for a calcium ion (Ca^{++}) $z = +2$, and for a chloride ion (Cl^-) $z = -1$.

In addition to the movement of these ions by the electrical field, there is also movement of the ions as a result of the concentration gradient. The change in the Gibbs free energy due to the transport of a solute from the inside of the cell to the outside of the cells.

$$-\Delta G_{transport} = \overline{G}_{inside} - \overline{G}_{outside} = \mu_{inside} - \mu_{outside} \tag{3.11}$$

Recall from Equation 2.75 that for an ideal solution we can write that:

$$\mu_{inside}^{ideal\ solution} = G'_{inside} + RT \ln C_{inside} \quad \text{and} \quad \mu_{outside}^{ideal\ solution} = G'_{outside} + RT \ln C_{outside} \tag{3.12}$$

where we have replaced the mole fraction with the concentration and the G'''s are pure component constants. Also, since the state of the ion is the same in the two regions, then $G'_{inside} = G'_{outside}$, and we can write that:

$$\Delta G_{transport} = RT \ln \frac{C_{outside}}{C_{inside}} \tag{3.13}$$

The total change in the Gibbs free energy for movement of the ion by transport and the electrical field is given by the sum of Equations 3.10 and 3.13. At equilibrium the total change in the Gibbs free energy is zero, hence:

$$\Delta G_{total} = RT \ln \frac{C_{outside}}{C_{inside}} - z\,FV = 0 \tag{3.14}$$

Solving for V, we obtain the following equation which is known as the *Nernst equation*. This equation can be used to calculate the equilibrium membrane potential for a particular ion.

$$V = \frac{RT}{zF} \ln \frac{C_{outside}}{C_{inside}} \tag{3.15}$$

In this equation, R represents the gas constant (1.987 cal/gmole K), T is temperature in K, z is the charge on the ion, and F is Faraday's constant (2.3×10^4 cal/V/gmol). At 37°C for a univalent ion, RT/zF is equal to 26.79 mV. Considering potassium ions, the intracellular concentration from Table 3.2 is 140 mOsM and the interstitial concentration is 4 mOsM. Therefore, the equilibrium membrane potential for this ion would be equal to about –95 mV. Since this value is less than zero, this indicates that there are more negative charges within the cell than outside the cell.

The actual resting membrane potential is somewhat different than this value because of two opposing effects. First, there is a very slight leakage of sodium ions into the cell. Leakage of sodium ions into the cell will add a positive charge of about +9 mV. Secondly, the sodium–potassium pump (Section 3.10) has the effect of removing positive charge equivalent to about –4 mV. The net resting membrane potential is, therefore, about –90 mV.

We shall soon see that the cell must expend cellular energy in order to maintain these potassium and sodium ion gradients and thus maintain its resting membrane potential. Although the chloride ion concentration is very high outside of the cell, these ions are repelled by the negative charges within the cell. This is shown by the fact that the equilibrium membrane potential for chloride ions is the same as the resting membrane potential, hence there is no net driving force for the transport of chloride ions into the cell.

In nerve and muscle cells, the resting membrane potential can change very rapidly. This rapid change in the membrane potential is called an *action potential*, and provides for the conduction of a nerve signal from neuron to neuron or the contraction of a muscle fiber. Any stimulus to the cell that raises the membrane potential above a threshold value will lead to the generation of a self-propagating action potential. It is also important to note that the action potential is an all or nothing response. The development of an action potential is dependent on the presence of voltage-gated sodium and potassium channels. The voltage-gated sodium channels only open or become active when the membrane potential is less negative than during the resting state. They typically begin to open when the membrane potential is about –65 mV. These sodium gates remain open for only a few tenths of a millisecond after which time they close or become inactive. The voltage-gated sodium channels remain in this inactive or closed state until the membrane potential has returned to near its resting value. The voltage-gated potassium channels also open when the membrane potential becomes less negative than during the resting state; however, unlike the sodium channels, they open more slowly and become fully opened only after the sodium channels have closed. The potassium channels then remain open until the membrane potential has returned to near its resting potential. Figure 3.8 illustrates the events during the generation of an action potential.

The first stage of an action potential is a rapid *depolarization* of the cell membrane. The cell membrane becomes very permeable to sodium ions because of the opening of the voltage-gated sodium channels. This rapid influx of sodium ions, positive charge, increases the membrane potential in the postive direction, and in some cases, can result

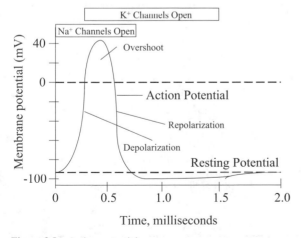

Figure 3.8   Action potential.

in a positive membrane potential (overshoot) for a brief period of time. This depolarization phase may last only a few tenths of a millisecond. Following the depolarization of the membrane, the sodium channels close and the potassium channels, which are now fully-opened, allow for the rapid loss of positively charged potassium ions from the cell, thus re-establishing within milliseconds, the normal negative resting potential of the cell membrane (repolarization). However, for a brief period of time following an action potential, the sodium channels remain inactive and they cannot open again regardless of external stimulation for several milliseconds. This is known as the refractory period.

The net driving force for the transport of an ion due to the combined effect of the concentration gradient and the membrane potential, i.e. the electrochemical gradient, is proportional to the difference between the membrane potential, V_M, and the ion's equilibrium potential, V. If the quantity $(V_M - V)$ is greater than zero, then the transport of the ion will be out of the cell through either the channel protein pores or by membrane carrier proteins. However, if this quantity is less than zero, then the ion will be transported into the cell. The proportionality constant for the transport of an ion due to its electrochemical gradient is referred to as the *membrane conductance* (g), which is the inverse of the membrane resistance.

The transport of ions generates a current (i), that is given by the product, $g \times (V_M - V)$. This current, or flow of charge, can be related to the flow of the ions themselves using the definitions summarized in Table 3.3. The following example illustrates the calculation of the flow of ions through a membrane channel.

Example 3.2 Calculate the flow of sodium ions through the voltage-gated sodium channels in a cell membrane during depolarization. Assuming the cell membrane

Table 3.3 Physical constants used to describe the electrical properties of the cell membrane

Basic electrical properties	Units
Charge	Coulomb (C), charge carried by 6.2×10^{18} univalent ions
Electric Potential	Volt (V), potential caused by separation of charges
Current	Ampere (A), flow of charge, C/sec
Capacitance	Farad (F), amount of charge needed on either side of a membrane to produce a given potential, C/V
Conductance	Siemens (S), ability of a membrane to conduct a flow of charge, A/V
Electrical Properties of Cells	Typical Values
Membrane Potential	-20 to -200 mV
Membrane Capacitance	~ 0.01 pF μm^{-2} of cell membrane surface area
Conductance of a Single Ion Channel	1 to 150 pS
Number of Specific Ion Channels	~ 75 μm^{-2} of cell membrane surface area
Other Relationships	
milli (m) = 10^{-3} , micro (μ) = 10^{-6}	
nano (n) = 10^{-9} , pico (p) = 10^{-12}	

Source: Data from Alberts *et al.*, *The Cell*, 2nd ed. (New York: Garland Publishing, Inc., 1989), 1067.

has a surface area of 1 μm^2, how long would it take to change the membrane potential by 100 mV? Assume the equilibrium membrane potential for sodium ions is 62 mV, that the threshold membrane potential for the sodium channels is –65 mV, and that at the peak of the action potential the membrane potential is 35 mV. In addition the membrane conductance for the Na ions is 4×10^{-12} S channel^{-1}.

SOLUTION Since the membrane potential rapidly changes during depolarization from the threshold value of –65 mV to the peak value of 35 mV, assume for calculation of the sodium ion flow that the "average" membrane potential during this phase is –15 mV. The flow, or current, of sodium ions may then be calculated from the following equation.

$$i_{Na} = g_{Na}\left(V_M - V_{Na}\right)$$

$$i_{Na} = \frac{4 \times 10^{-12}\,S}{channel} \times \frac{1\,A/V}{S} \times \frac{1\,C/sec\,/V}{A} \times \frac{6.2 \times 10^{18}\,\oplus}{C} \times \frac{1\,V}{1000\,mV} \times \frac{1\,Na}{\oplus} \times \frac{75\,channels}{\mu m^2}$$

$$\times (-15 - 62)\,mV = -1.43 \times 10^8\,Na\,sec^{-1}\,\mu m^{-2}$$

This flow of sodium ions into the cell is also equivalent to a current of –23 pA μm^{-2}. The flow of sodium ions transfers charge across the membrane and will, therefore, change the membrane potential. We can relate this change in charge and membrane potential to the membrane capacitance by the following relationship, $C_{membrane} = i_{Na}\,t/\Delta V_M$. Recall from Table 3.3 that the membrane capacitance is about 0.01 pF/μm^2. For the calculated sodium ion current and the 100 mV change in the membrane potential, we can then solve for the time required to achieve this change in membrane potential.

$$t = \frac{0.01\,pF\,\mu m^{-2} \times 1C\,V^{-1}F^{-1} \times 1\,V\,1000^{-1}mV^{-1} \times 100\,mV}{23\,pA/\mu m^2 \times 1C\,sec^{-1}\,A^{-1} \times 1\,sec\,1000^{-1}m\,sec^{-1}} = 0.043\,milliseconds$$

This example shows that the membrane potential is rapidly depolarized during the initial phase of the action potential.

3.10 ION PUMPS

Cells also have the ability to "pump" certain solutes against their electrochemical gradient. This process is known as active transport and involves the use of special carrier proteins. Since this is an "uphill" process, active transport requires the expenditure of cellular energy.

The large differences in the concentrations of sodium and potassium ions across the cell membrane will result in some leakage of potassium ions out of the cell and leakage of sodium ions into the cell. However, as we have discussed, proper functioning of the cell requires that these concentration differences be maintained in order to preserve the cell's resting membrane potential. Substances cannot diffuse against their own electrochemical gradient without the expenditure of energy. Therefore to compensate for the loss of potassium ions by leakage through the cell membrane, the cell must have a "pump" mechanism to shuttle potassium ions from the external environment into the cell. On the other hand, sodium ions leak into the cell and similarly the cell needs a "pump" to remove these ions. This process of shuttling substances across the cell membrane against their electrochemical gradient and at the expense of cellular energy is called *active transport*.

The energy for active transport is provided by the cellular energy storage molecule called *adenosine triphosphate* (ATP). ATP is a nucleotide and consists of the following three components: a base called adenine, a ribose sugar, and a triphosphate group. Through the action of the enzyme *ATPase*, a molecule of ATP can be converted into a molecule of ADP (adenosine diphosphate) and a free high energy phosphate bond that can cause conformational changes in special cell membrane carrier proteins.

The best example of active transport is the *sodium–potassium pump* present in all cells. The Na–K pump transports sodium ions out of the cell and, at the same time, transfers potassium ions into the cell. Figure 3.9 illustrates the essential features of the Na–K pump. The carrier protein protrudes through both sides of the cell membrane. Within the cell, the carrier protein has three receptor sites for binding sodium ions and also has ATPase activity. On the outside of the cell membrane, the carrier protein has two receptor sites for binding potassium ions. When these ions are bound to their respective receptor sites, the ATPase then becomes activated liberating the high energy phosphate bond from ATP. The energy in the phosphate bond causes shape or conformational changes in the carrier protein that allow for passage of the sodium and potassium ions. The Na–K ATPase pump is electrogenic, that is removing a net positive charge from the cell equivalent to about –4 mV. Similar active transport carrier proteins are also available to transfer other ions such as Cl^-, Ca^{+2}, Mg^{+2}, and HCO_3^-. The only difference in the carrier proteins would be their preference for a specific ion.

Active transport of a solute can also be driven by ion gradients, a process referred to as secondary active transport. For example, the higher concentration of sodium ions outside of the cell can lead to a conformational change in a carrier protein that favors the symport of another solute. The other solute is, therefore, pumped into the cell against its own electrochemical gradient. The sodium ion gradient that drives this pump

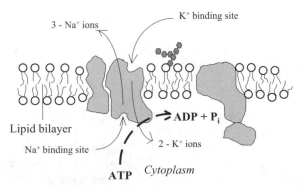

Figure 3.9   The sodium–potassium pump.

is maintained by the Na–K ATPase pump discussed earlier. Sodium ion antiports are also used to control intracellular pH. In this case, the removal of excess hydrogen ions generated by acid forming reactions within the cell is coupled with the influx of sodium ions.

PROBLEMS

1. Derive Equation 3.3.

2. A membrane has pure water on one side of it and a protein solution on the other side. The membrane has pores equivalent in size to a spherical molecule with a molecular weight of 100,000 daltons. The protein solution consists of albumin (40 g/L @ MW = 69,000), globulins (70 g/L @ MW = 150,000), and fibrinogen (60 g/L @ MW = 340,000). Assuming the proteins form an ideal solution, what is the osmotic pressure of the protein solution? What would be the osmotic pressure of the protein solution if the membrane was completely impermeable to the proteins?

3. Explain why osmotic pressure is based on the molar concentration of the solute and not the solute mass concentration.

4. Explain the difference between a mole and an osmole. Cite some examples.

5. You are concentrating a solution containing a polypeptide of modest molecular weight by pressure filtration through a membrane. The solute concentration on the feed side of the membrane is 0.10 M and the temperature is 25°C. The applied pressure on the feed side of the membrane is 6 atm, and the pressure on the opposite side of the membrane is atmospheric. If the polypeptide solute is completely rejected by the membrane, what is the "effective" pressure drop across the membrane? If the hydraulic conductance of the membrane is 3 ml/hr/m^2/mm Hg, what is the filtration rate of the solvent? Assume the membrane has a total surface area of 1 m^2. R = 0.082 liters atm mol^{-1} K^{-1}.

6. The following clean water flow rates were reported for a particular series of ultrafiltration membranes.

Nominal Molecular Weight Cutoff (NMWCO)	Clean Water Flow Flux (ml/min/cm^2@50 psi)
10,000	0.90
30,000	3.00
100,000	8.00

From this data, calculate the hydraulic conductance for each case (L_p). Assuming the membranes have similar porosity and thickness, is there a relationship between the value of L_p and the NMWCO? Hint: base your analysis on Equation 3.7.

7. Perform a literature search and write a short paper on the biomedical applications of liposomes.

8. Look up in a biochemistry text (e.g. Stryer 1988) the chemical structures of collagen, laminin, cholesterol, and the various lipids found in the cell membrane.

9. Search the literature and write a short paper on the biomedical applications of osmotic pumps.

10. How many potassium ions must a cell lose in order to produce a membrane potential of −95 mV for potassium? How does this compare to the number of potassium ions within the cell?

11. Consider a membrane that is permeable to Ca^{2+} ions. On one side of the membrane, the Ca^{2+} concentration is 100 mM, and on the other side of the membrane, the Ca^{2+} concentration is 1 mM. The electrical potential of the high concentration side of the membrane relative to the low concentration side of the membrane is +10 mV. How much reversible work is required to move each mole of Ca^{2+} from the low concentration side of the membrane to the high concentration side of the membrane? What is the equilibrium membrane potential?

12. The cell membrane is permeable to many different ions. Therefore, the equilibrium membrane potential for the case of multiple ions will depend not only on the concentrations of the ions within and outside the cell, but also on the permeability (P) of the cell membrane to each ion. The Goldman equation may be used to calculate the equilibrium membrane potential for the case of multiple ions. For example, considering Na^+ and K^+ ions, the Goldman equation may be written:

$$V = 26.71 \times \ln\left(\frac{P_{Na}C_{Na,0} + P_K C_{K,0}}{P_{Na}C_{Na,i} + P_K C_{K,i}}\right)$$

For a cell at rest, P_{Na} is much smaller than P_K, typically P_{Na} is about $0.01 \times P_K$. Calculate the equilibrium membrane potential for a cell under these conditions. During the depolarization phase of an action potential, the sodium ion permeability increases dramatically due to the opening of the voltage-gated sodium ion channels. Under these conditions, P_K is about $0.5 \times P_{Na}$. Under these conditions, recalculate the value of the equilibrium membrane potential.

FOUR

THE PHYSICAL AND FLOW PROPERTIES OF BLOOD AND OTHER FLUIDS

4.1 PHYSICAL PROPERTIES OF BLOOD

Blood is a viscous fluid mixture consisting of plasma and cells. Table 4.1 summarizes the most important physical properties of blood. Recall that the chemical composition of the plasma was previously shown in Table 3.2. Proteins represent about 7–8 wt % of the plasma. The major proteins found in the plasma are albumin (MW = 69,000; 4.5 gm/100 ml), the globulins (MW = 35,000–1,000,000; 2.5 gm/100 ml), and fibrinogen (MW = 400,000; 0.3 gm/100 ml). *Albumin* has a major role in regulating the pH and the colloid osmotic pressure. The so-called *alpha* and *beta globulins* are involved in solute transport , whereas the *gamma globulins* are the antibodies that fight infection and form the basis of the humoral component of the immune system. *Fibrinogen,* through its conversion to long strands of fibrin, has a major role in the process of blood clotting. *Serum* is simply the fluid remaining after blood is allowed to clot. For the most part, the composition of serum is the same as that of plasma, with the exception that the clotting proteins, primarily fibrinogen, has been removed.

4.2 CELLULAR COMPONENTS

The cellular component of blood consists of three main cell types. The most abundant cells are the red blood cells (RBCs) or *erythrocytes* comprising about 95% of the cellular component of blood. Their major role is the transport of oxygen by the hemoglobin contained within the red blood cell. Note from Table 4.1 that the density of a red blood cell is higher than that of plasma. Therefore, in a quiescent fluid, the red blood cells will tend to settle. The red blood cell volume fraction is called the

Table 4.1 Physical properties of adult human blood

Property	Value
Whole Blood	
pH	7.35–7.40
viscosity (37°C)	3.0 cP (at high shear rates)
specific gravity (25/4°C)	1.056
venous hematocrit – male	0.47
female	0.42
whole blood volume	~78 ml/kg body weight
Plasma or Serum	
colloid osmotic pressure	~ 330 mm H_2O
pH	7.3–7.5
viscosity (37°C)	1.2 cP
specific gravity (25/4°C)	1.0239
Formed Elements	
erythrocytes (RBCs)	
pH	7.396
specific gravity (25/4°C)	1.098
count – male	5.4×10^9 ml^{-1} whole blood
female	4.8×10^9 ml^{-1} whole blood
average life span	120 days
production rate	4.5×10^7 ml^{-1} whole blood / day
hemoglobin concentration	0.335 g ml^{-1} of erythrocyte
leukocytes	
count	~7.4×10^6 ml^{-1} whole blood
diameter	7–20 microns
platelets	
count	~2.8×10^8 ml^{-1} whole blood
diameter	~2–5 microns

Source: Data from D.O. Cooney, *Biomedical Engineering Principles* (New York: Marcel Dekker, 1976), 39.

hematocrit and typically varies between 40–50%. The true hematocrit (H) is about 96% of the measured hematocrit (Hct).

The red blood cell has a unique shape described as a *biconcave discoid*. Figure 4.1 illustrates the shape of the red blood cell and Table 4.2 summarizes the typical dimensions. Red blood cells can form stacked coin-like structures called *rouleaux*. Rouleaux can also clump together to form larger RBC structures called aggregates. Both rouleaux and aggregates break apart under conditions of increased blood flow or higher shear rates.

The next most abundant cell type are the *platelets,* comprising about 4.9% of the cell volume. The platelets are major players in blood coagulation and *hemostasis*, which is the prevention of blood loss. The remaining 0.1% of the cellular component of blood consists of the white blood cells (WBCs) or *leukocytes* which form the basis of the cellular component of the immune system. Since the white blood cells and platelets only comprise about 5% of the cellular component of blood, their effect on the macroscopic flow characteristics of blood is negligible.

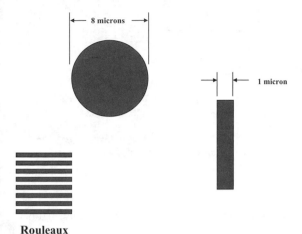

Rouleaux

Figure 4.1   The shape and dimensions of the red blood cell.

Table 4.2 Dimensions of the normal red blood cell

Property	Value
Diameter	8.1 ± 0.43 microns
Greatest thickness	2.7 ± 0.15 microns
Least thickness	1.0 ± 0.3 microns
Surface area	138 ± 17 microns^2
Volume	95 ± 17 microns^3

Source: Data from A.C. Burton, *Physiology and Biophysics of the Circulation*, 2 ed. (Year Book Medical Publishers, Inc., 1972).

4.3 RHEOLOGY

The field of rheology concerns the deformation and flow behavior of fluids. The prefix *rheo-* is Greek and refers to something that flows. Due to the particulate nature of blood, we would expect the rheological behavior of blood to be somewhat more complex than a simple fluid such as water.

To understand the flow behavior of blood, we first must define the relationship between *shear stress* (τ) and the *shear rate* ($\dot{\gamma}$). To develop this relationship, consider the situation shown in Figure 4.2. A fluid is contained between two large parallel plates both of area A. The plates are separated by a small distance equal to h. Initially the system is at rest. At time $t = 0$, the lower plate is set into motion in the x direction at a constant velocity, V. As time proceeds, momentum is transferred in the y direction to successive layers of fluid from the lower plate that is in motion in the x direction. Momentum therefore "flows or diffuses" from a region of high velocity to a region of

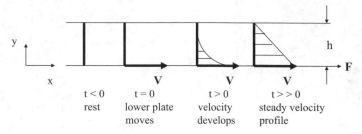

Figure 4.2   Velocity profile development for flow of a fluid between two parallel plates.

low velocity. After a sufficient length of time, a steady-state velocity profile is obtained that is a linear function of y, i.e. $v_x(y) = \dfrac{V}{h}(h - y)$. Recall that velocity is a vector and that v_x represents the component of the fluid velocity in the x direction. For the situation shown in Figure 4.2, the other components of the velocity vector, i.e. v_y and v_z are equal to zero.

At steady state, a constant force (F) must be applied to maintain the motion of the lower plate. For the situation shown in Figure 4.2, the *shear stress* on the lower plate is defined as F/A and is given the symbol τ_{yx}. The subscript yx on τ denotes the viscous flux of x momentum in the y direction (Bird *et al.* 2002). Note that the momentum flux (momentum/area/time) of the fluid adjacent to the plate that is in motion would be given by $\rho V V$, where ρ is the fluid density. The *shear rate* at any position y in the fluid is defined as $-\dfrac{dv_x(y)}{dy} = \dfrac{V}{h} = \dot{\gamma}$. Shear rate is given the symbol $\dot{\gamma}$. Notice that shear rate has units of reciprocal time.

Newton's law of viscosity states that for laminar (non-turbulent) flow the shear stress is proportional to the shear rate. The proportionality constant is called the *viscosity*, μ, which is a property of the fluid and is a measure of the flow resistance of the fluid. For the situation shown in Figure 4.2, this may be stated as follows.

$$\frac{F}{A} = \tau_{yx} = \mu \frac{V}{h} = \mu\gamma \tag{4.1}$$

Most simple homogeneous liquids and gases obey this law and are called *Newtonian fluids*.

For more complicated geometries, the steady state velocity profile is not linear. However, Newton's law of viscosity may be stated at any point in the laminar flow field as:

$$\tau_{yx} = -\mu \frac{dv_x}{dy} \tag{4.2}$$

where v_x is the velocity in the x direction at position y. This equation tells us that the momentum flows in the direction of decreasing velocity. The velocity gradient is, therefore, the driving force for momentum transport, much like the temperature gradient is the driving force for heat transfer, and the concentration gradient is the driving force for mass transport.

Fluids whose shear stress–shear rate relationship do not follow Equation 4.2 are known as non-Newtonian fluids. Figure 4.3 illustrates the type of shear stress–shear rate relationships that are typically observed for *non-Newtonian fluids*. The Newtonian fluid is shown for comparison. Note that the shear stress–shear rate relationship for a Newtonian fluid is linear with the slope equal to the viscosity. For the non-Newtonian fluids, the slope of the line at a given value of the shear rate is the *apparent viscosity*. A *dilatant* fluid thickens, or has an increase in apparent viscosity, as the shear rate increases. A *pseudoplastic* fluid on the other hand tends to thin out, or its apparent viscosity decreases, with an increase in shear rate.

Heterogeneous fluids that contain a particulate phase that forms aggregates at low rates of shear exhibit a yield stress, τ_y. Two types of fluids that exhibit this behavior are the *Bingham plastic* and the *Casson fluid*. The yield stress must be exceeded in order to get the material to flow. In the case of the Bingham plastic, once the yield stress is exceeded, the fluid behaves as if it were Newtonian. For the Casson fluid, we see that as the shear rate increases, the apparent viscosity decreases, indicating that the particulate aggregates are getting smaller and smaller and, at some point, the fluid behaves as a Newtonian fluid. Blood is a heterogeneous fluid with the particulates consisting primarily of the red blood cells. As mentioned earlier and shown in Figure 4.1, the red blood cells form rouleaux and aggregates at low shear rates. We will see in the discussion to follow that blood follows the curve shown for the Casson fluid.

The apparent viscosity of blood as a function of shear rate is shown in Figure 4.4 at a temperature of 37°C. At low shear rates, the apparent viscosity of blood is quite high due to the presence of rouleaux and aggregates. However, at shear rates above about 100/sec, only individual cells exist, and blood behaves as if it were a Newtonian fluid. We then approach the asymptotic limit for the apparent viscosity of blood, which is about 3 cP [1 P (poise) = 100 cP (centipoise) = 1 g/cm sec = 1 dyne sec/cm^2 = 0.1 Newton sec/m^2].

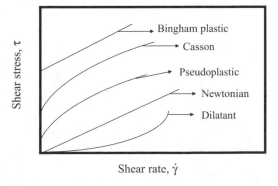

Figure 4.3   Types of shear stress–shear rate relationship.

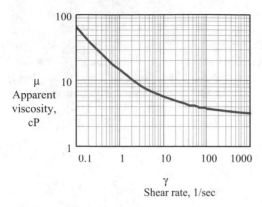

μ
Apparent
viscosity,
cP

γ
Shear rate, 1/sec

Figure 4.4   Apparent viscosity of blood at 37°C.

4.4 RELATIONSHIP BETWEEN SHEAR STRESS AND SHEAR RATE

The simplest approach for examining the shear stress–shear rate behavior of blood or other fluids is through the use of the capillary viscometer shown in Figure 4.5. The diameter of the tube is typically on the order of 500 microns. To eliminate entrance effects, the ratio of the capillary length to its radius should be greater than 100 (Rosen 1993). For a given flowrate of blood, Q, the pressure drop across the viscometer, ΔP, is measured. From this information it is possible to deduce an analytical expression for the shear stress–shear rate relationship, i.e. $\dot{\gamma} = \dot{\gamma}(\tau_{rz})$.

Analysis of the capillary flow illustrated in Figure 4.5 requires the use of cylindrical coordinates (r, θ, and z) and the following simplifying assumptions: the length of the tube (L) is much greater than the tube radius (R) (i.e. $L/R \gg 100$) to eliminate entrance effects; steady incompressible (i.e. constant density) and isothermal flow; no external forces acting on the fluid; no holes in the tube so that there is no radial velocity component v_r; axisymmetric flow or no swirls so that the tangential velocity, v_θ, is also zero; and the no-slip condition at the wall ($r = R$) requiring $v_z = 0$. Continuity or conservation of mass for an incompressible fluid with these assumptions requires that only an axial velocity component exists and that it will be a function of the tube radius only, therefore $v_z = v_z(r)$.

Now consider in Figure 4.5 a cylindrical volume of fluid of radius r and length Δz. For steady flow, the viscous force acting to retard the fluid motion, i.e. $\tau_{rz} 2\pi r \Delta z$, must

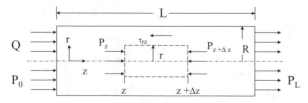

Figure 4.5   Forces acting on a cylindrical fluid element within a capillary viscometer.

be balanced by the force developed by the pressure drop acting on the volume of fluid of length Δz, i.e. $-\pi r^2 (P|_{z+\Delta z} - P|_z)$. Equating these forces, dividing by Δz, and taking the limit as $\Delta z \to 0$, the following equation is obtained for the shear stress distribution for the fluid flowing within the capillary tube.

$$\tau_{rz}(r) = -\frac{r}{2}\frac{dP}{dz} = \frac{(P_0 - P_L)r}{2L} \tag{4.3}$$

We note that the shear stress vanishes at the centerline of the capillary and achieves its maximum value, τ_w, at the wall. The *wall shear stress* is important since it is easily calculated by the following equation in terms of the measured pressure drop $(P_0 - P_L)$ and the tube dimensions (R and L).

$$\tau_w = -\frac{R}{2}\frac{dP}{dz} = \frac{(P_0 - P_L)R}{2L} \tag{4.4}$$

It should also be pointed out that Equations 4.3 and 4.4 hold whether the fluid is Newtonian or non-Newtonian. Nothing has been said so far about the relationship between a particular fluid's shear stress and shear rate. Using Equation 4.4, one may rewrite Equation 4.3 to give the shear stress in terms of the wall shear stress and the fractional distance from the centerline of the capillary tube r/R.

$$\tau_{rz}(r) = \tau_w \frac{r}{R} \tag{4.5}$$

What we want to find through the use of data obtained from the capillary viscometer is a general relationship between the shear rate and some function of the shear stress in terms of the measurable quantities Q, $P_0 - P_L$, L, and R, i.e.

$$\dot{\gamma} = -\frac{dv_z}{dr} = \dot{\gamma}(\tau_{rz}) \tag{4.6}$$

For the special case of a Newtonian fluid, we have:

$$\dot{\gamma}(\tau_{rz}) = \frac{\tau_{rz}}{\mu} \tag{4.7}$$

A general relationship between $\dot{\gamma}$ and τ_{rz}, and the measurable quantities of the capillary viscometer can be obtained by the following development. First we write the total flowrate, Q, in terms of the axial velocity profile as follows.

$$Q = 2\pi \int_0^R v_z(r)r\,dr \tag{4.8}$$

Next, we integrate this equation by parts to obtain this equation.

$$Q = -\pi \int_0^R r^2 \left(\frac{dv_z(r)}{dr} \right) dr \tag{4.9}$$

Using Equations 4.5 and 4.6, one can show that the following equation is obtained. This is called the *Rabinowitsch equation*.

$$Q = \frac{\pi R^3}{\tau_w^{\ 3}} \int_0^{\tau_w} \dot{\gamma}(\tau_{rz}) \tau_{rz}^2 d\tau_{rz} \tag{4.10}$$

For data obtained from a given capillary viscometer, the experiments would provide Q as a function of the wall shear stress, τ_w, which is related by Equation 4.4 to the observed pressure drop, i.e. $P_0 - P_L$, and the capillary dimensions, R and L. The appropriate shear rate and shear stress relationship, Equation 4.6, would be the one that best fits the data according to Equation 4.10. This will be illustrated for blood in the discussion to follow.

4.5 THE HAGAN-POISEUILLE EQUATION

Let us first consider the case of a Newtonian fluid. Equation 4.7 provides the relationship for $\dot{\gamma}(\tau_{rz})$. Substituting this equation into Equation 4.10, one readily obtains the following result for the volumetric flowrate.

$$Q = \frac{\pi R^4 (P_0 - P_L)}{8\mu L} \tag{4.11}$$

This is the *Hagan-Poiseuille law* for laminar flow of a Newtonian fluid in a cylindrical tube. It provides a simple relationship between the volumetric flowrate in the tube given by Q, and the pressure drop, $P_0 - P_L$, for the given dimensions of the tube radius, R, and tube length, L, and the fluid viscosity, μ. Using the analogy that flow is proportional to a driving force divided by a resistance, we see that the driving force is the pressure drop, $P_0 - P_L$, and the resistance is given by, $\frac{8\mu L}{\pi R^4}$. We see that for laminar flow within blood vessels, the radius of the vessel is the determining factor that controls the blood flowrate for a given pressure drop. The arterioles, the smallest elements of the arterial system with diameters less than 100 microns, consist of an endothelial cell lining that is surrounded by a layer of vascular smooth muscle cells. Contraction of the smooth muscle layer provides a reactive method for controlling arteriole diameter and hence the blood flow rate within organs and tissues.

Using Equations 4.3, 4.6, and 4.7 we can also write for a Newtonian fluid in cylindrical coordinates, i.e. tube flow, that:

$$\tau_{rz} = -\mu \frac{dv_z}{dr} = \frac{(P_0 - P_L)r}{2L} \tag{4.12}$$

This equation can be easily integrated with the boundary conditions that at $r = R$, $v_z = 0$ and that at $r = 0$, $\frac{dv_z}{dr} = 0$. The result of this integration provides an equation that describes the velocity profile for laminar tube flow of a Newtonian fluid. The following equation predicts that the velocity profile will have a parabolic shape.

$$v_z(r) = \frac{(P_0 - P_L)R^2}{4\mu L}\left[1 - \left(\frac{r}{R}\right)^2\right] \tag{4.13}$$

Note that the maximum velocity occurs at the centerline of the capillary, $r = 0$.

4.6 OTHER USEFUL FLOW RELATIONSHIPS

Some other useful relationships for Newtonian flow may be obtained by combining Equations 4.4 and 4.11. First we obtain the following expression that relates the wall shear stress to the volumetric flowrate:

$$\tau_w = \left(\frac{4\mu}{\pi}\right)\frac{Q}{R^3} \tag{4.14}$$

Also, by writing Equation 4.7 at the tube wall and using the above result, it is found that the wall or apparent shear rate is given by:

$$|\dot{\gamma}_w| = \frac{4Q}{\pi R^3} = \frac{4V_{average}}{R} \tag{4.15}$$

where $V_{average}$ is the average velocity in the tube.

The reduced average velocity $\bar{U}$ is related to the wall shear rate and is defined as the ratio of the average velocity and the tube diameter. It is given by the following equation:

$$\bar{U} = \frac{V_{average}}{D} = \frac{4Q}{\pi D^3} \tag{4.16}$$

Example 4.1 A large artery may have a diameter of about 0.5 cm whereas a large vein may have a diameter of about 0.8 cm. The average blood velocity in these arteries and veins is respectively on the order of 40 cm/sec and 20 cm/sec. Calculate for these blood flows the wall shear rate, $|\dot{\gamma}_w|$. What would the shear rate be at the center line of these blood vessels?

SOLUTION We may use Equation 4.15 to calculate the wall shear rates.

For the large artery:

$$|\dot{\gamma}_w| = \frac{4 \times 40\frac{cm}{sec}}{0.25\,cm} = 640\frac{1}{sec}$$

and for the large vein:

$$|\dot{\gamma}_w| = \frac{4 \times 20\frac{cm}{sec}}{0.40\,cm} = 200\frac{1}{sec}$$

From Equation 4.12, the shear rate in both cases is zero at the centerline $r = 0$.

4.7 RHEOLOGY OF BLOOD

As shown in Figure 4.4, blood is a non-Newtonian fluid except at shear rates above about 100/sec where the viscosity is seen to approach its asymptotic limit of about 3 cP. Therefore, as shown in Example 4.1, blood flow in arteries and veins, as well as smaller vessels, is Newtonian near the vessel wall, since the wall shear rate is significantly higher than 100/sec. However, as one gets closer to the centerline of the vessel, the shear rate approaches zero, and blood exhibits non-Newtonian behavior. Therefore, blood cannot, in general, be expected to follow the relationships summarized in Table 4.3 for Newtonian fluids.

Figure 4.6 provides the shear stress–shear rate relationship for blood under a variety of conditions as obtained by Replogle *et al.* (1967) using a capillary viscometer at ambient temperatures. This log–log plot clearly shows that below a shear stress of about 100/sec, the flow of blood (open circle) becomes non-Newtonian. Note in this figure the rapid rise in the viscosity of blood as the shear rate decreases. Above a shear rate of 100/sec, the fluid is Newtonian with a constant viscosity. If the clotting protein fibrinogen (typically about 0.3 wt % in plasma) is removed, keeping the hematocrit constant, the resulting red blood cell suspension (upsidedown triangle) behaves nearly like a Newtonian fluid for shear rates down as low as 0.01/sec. Fibrinogen and its effect of making the red blood cells stick together is, therefore, primarily responsible for the non-Newtonian flow behavior of blood at low shear rates. Thus the other plasma proteins, such as albumin and the globulins, do not contribute to the non-Newtonian behavior of blood, although their concentration will affect the viscosity of the plasma. It should also be pointed out that serum and plasma alone are Newtonian fluids.

Table 4.3 Summary of key flow equations

Relationship	Non-Newtonian	Newtonian
Shear Stress $\tau_{rz} =$	$\dfrac{(P_0 - P_L)r}{2L} = \tau_w \dfrac{r}{R}$	$\dfrac{(P_0 - P_L)r}{2L} = \tau_w \dfrac{r}{R}$
Wall Shear Stress $\tau_w =$	$\dfrac{(P_0 - P_L)R}{2L}$	$\dfrac{(P_0 - P_L)R}{2L}$
Shear Rate $\dot{\gamma} =$	$-\dfrac{dv_z}{dr} = \dot{\gamma}(\tau_{rz})$	$-\dfrac{dv_z}{dr} = \dfrac{\tau_{rz}}{\mu}$
Volumetric Flowrate $Q =$	$\pi \dfrac{R^3}{\tau_w^3} \displaystyle\int_o^{\tau_w} \dot{\gamma}(\tau_{rz}) \tau_{rz}^2 d\tau_{rz}$	$\dfrac{\pi R^4 \Delta P}{8\mu L}$
Wall Shear Stress $\tau_w =$	—	$\dfrac{4\mu Q}{\pi R^3}$
Wall Shear Rate $|\dot{\gamma}w| =$	—	$\dfrac{4Q}{\pi R^3}$
Reduced Average Velocity $\overline{U} =$	$\dfrac{4Q}{\pi D^3}$	$\dfrac{4Q}{\pi D^3}$

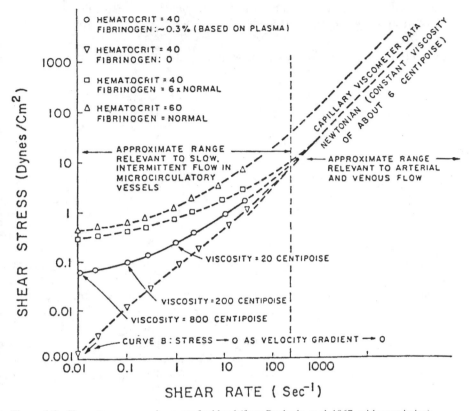

Figure 4.6   Shear stress versus shear rate for blood (from Replogle *et al.* 1967, with permission).

Increasing the hematocrit has the effect of increasing the flow resistance of blood (compare open circle $H = 40\%$ with the open triangle $H = 60\%$). The same effect is also seen when the fibrinogen concentration is increased (compare open circle with open square). Therefore, the rheology of blood can be significantly affected by the hematocrit and the plasma protein concentrations.

4.8 THE CASSON EQUATION

The shear stress–shear rate relationship for blood may be described by the following empirical equation known as the *Casson equation*.

$$\tau^{1/2} = \tau_y^{1/2} + s\gamma^{1/2} \tag{4.17}$$

In this equation, τ_y is the yield stress and s is a constant, both of which must be determined from viscometer data. The yield stress represents the fact that a minimum force must be applied to stagnant blood before it will flow. This was illustrated earlier in Figure 4.3. The yield stress for normal blood at 37°C is about 0.04 dynes/cm². It is important to point out however, that the effect of the yield stress on the flow of blood is small as the following example will show.

Example 4.2 Estimate the pressure drop in a small blood vessel that is needed to just overcome the yield stress.

SOLUTION One may use Equation 4.4 to solve for the pressure drop needed to overcome the yield stress ($\tau_y = \tau_w$).

$$(P_0 - P_L)_{min} = \frac{2L\tau_y}{R}$$

Using the yield stress of 0.04 dynes/cm^2 and an L/R of 200 for a blood vessel, the pressure drop required to just initiate flow is on the order of 0.01 mm Hg. This is considerably less than the mean blood pressure which is on the order of 100 mm Hg.

At large values of the shear rate, the Casson equation should approach the asymptotic viscosity of the Newtonian fluid as shown in Figure 4.4. Equation 4.17 at large shear rates thus provides the interpretation that the parameter (s) is the square root of this asymptotic Newtonian viscosity. Table 4.1 provides the asymptotic viscosity of blood to be 3 cP. Therefore, the parameter (s) is 1.732 (cP)$^{1/2}$ or 0.173 (dyne sec/cm^2)$^{1/2}$. It should be stressed that the values of the yield stress and the parameter (s) will also depend on plasma protein concentrations, hematocrit, and temperature. Therefore, one must be careful to calibrate the Casson equation to the actual blood and the conditions used in the viscometer or the situation under study.

4.9 USING THE CASSON EQUATION

We are now ready to use the Casson equation to represent the shear stress–shear rate relationship for blood. Consider the data shown in Figure 4.7 as obtained by Merrill *et al.* (1965) for the flow of blood in various diameter tubes. The figure provides data for the pressure drop as a function of the flowrate for five tube diameters ranging from 288 microns to 850 microns. The hematocrit is 39.3 and the temperature is about 20°C.

If the flow were Newtonian, we could use the Hagan-Poiseuille law, i.e. Equation 4.11, to express the relationship between the pressure drop and the volumetric flowrate. In a log–log system of coordinates the Hagan-Poiseuille equation becomes:

$$\log \Delta P = \log Q + \log\left(\frac{8\mu L}{\pi R^4}\right) \qquad (4.18)$$

The data shown in the log–log plot of ΔP versus Q in Figure 4.7 should, therefore, have a slope of unity. However, at the lower flowrates we see significant deviation from this requirement as we would expect since, for a given tube radius, Equation 4.15 predicts a decrease in the wall shear rate with decreasing volumetric flow. Thus, there is a transition from Newtonian to non-Newtonian flow at the lower flowrates.

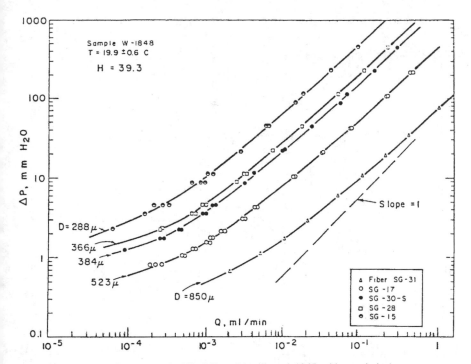

Figure 4.7   Pressure-flow curves for blood (from Merrill *et al.* 1965, with permission).

To facilitate analysis of the data shown in Figure 4.7, it is convenient to express the pressure drop in terms of the wall shear stress according to Equation 4.4, and to express the volumetric flowrate in terms of the reduced average velocity using Equation 4.16.

Figure 4.8 provides a plot of the data shown in Figure 4.7 as a plot of τ_w versus $\bar{U}$. We observe that all of the data shown in Figure 4.7 now falls nearly along a single line. This result is expected when the Rabinowitsch equation (4.10) is recast as $\bar{U}$ versus τ_w.

$$\bar{U} = \frac{4Q}{\pi D^3} = \frac{1}{2\tau_w^{\,3}} \int_0^{\tau_w} \dot{\gamma}(\tau_{rz})\tau_{rz}^{\,2}d\tau_{rz} \qquad (4.19)$$

This equation predicts that the reduced average velocity should only be a function of the wall shear stress, and this result is generally supported by the data presented in Figure 4.8. This result also applies whether or not the fluid is Newtonian or non-Newtonian.

The shear rate, or $\dot{\gamma}(\tau_{rz})$, may be written as follows for the Casson equation (4.17):

$$\dot{\gamma}(\tau_{rz}) = \frac{\left[\tau_{rz}^{1/2} - \tau_y^{1/2}\right]^2}{s^2} \qquad (4.20)$$

This equation may be substituted into Equation 4.19, and the resulting equation integrated, to obtain the following fundamental equation that describes tube flow of the

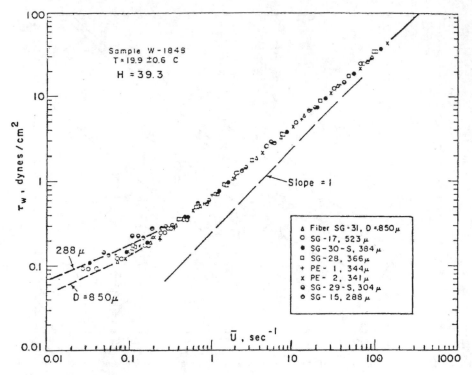

Figure 4.8 Wall shear stress versus the reduced average velocity (from Merrill *et al.* 1965, with permission).

Casson fluid. This equation depends on only two parameters, τ_y and s, that may be determined from experimental data.

$$\bar{U} = \frac{1}{2s^2}\left[\frac{\tau_w}{4} - \frac{4}{7}\sqrt{\tau_y}\sqrt{\tau_w} - \frac{1}{84}\frac{\tau_y^4}{\tau_w^3} + \frac{\tau_y}{3}\right] \tag{4.21}$$

Merrill *et al.* (1965) provides the following values of these parameters at about 20°C: $\tau_y = 0.0289$ dynes cm^{-2} and $s = 0.229$ (dynes sec cm$^{-2})^{1/2}$. It is left as an exercise to show that the Casson equation with these values of the parameters provides an excellent fit to the data shown in Figure 4.8.

4.10 THE VELOCITY PROFILE FOR TUBE FLOW OF A CASSON FLUID

We can also use the Casson equation to obtain an expression for the axial velocity profile for flow of blood in a cylindrical tube or vessel. The maximum shear stress, τ_w, is at the wall. If the yield stress, τ_y, is greater than τ_w, then there will be no flow. On the other hand, if τ_w is greater than τ_y, there will be flow, however, there will be a critical radius ($r_{critical}$) at which the the local shear stress will equal τ_y. From the tube centerline to this critical radius, this core fluid will have a flat velocity profile, i.e. $v_z(r) = v_{core}$, and will move as if it were a solid body or in what is known as "plug flow". For the region from the critical radius to the tube wall ($r_{critical} \leq r \leq R$), Equation 4.3 describes once again the shear stress distribution as a function of radial position. We can set this equation equal to the shear stress relation provided by the Casson equation (4.17), and rearrange to obtain the following equation for the shear rate.

$$\dot{\gamma} = -\frac{dv_z(r)}{dr} = \left\{ \frac{P_0 - P_L}{2L}r - 2\tau_y^{1/2} \left[\frac{(P_0 - P_L)}{2L}r \right]^{1/2} + \tau_y \right\} s^{1/2} \quad (4.22)$$

This equation may be integrated using the boundary condition that at $r = R$, $v_z(R)$ = 0, thus obtaining the following result.

$$v_z(r) = \frac{R\tau_w}{2s^2} \left\{ \left[1 - \left(\frac{r}{R} \right)^2 \right] - \frac{8}{3}\sqrt{\frac{\tau_y}{\tau_w}} \left[1 - \left(\frac{r}{R} \right)^{3/2} \right] + 2 \left(\frac{\tau_y}{\tau_w} \right) \left(1 - \frac{r}{R} \right) \right\} \quad (4.23)$$

This equation applies for the region defined by $r_{critical} \leq r \leq R$. At this point, however, we need to determine the value of $r_{critical}$. Since the shear stress at $r_{critical}$ must equal the yield stress, it is easy to show from Equation 4.5 that $r_{critical} = R\left(\dfrac{\tau_y}{\tau_w} \right)$. For locations from the centerline to the critical radius, the velocity of the core would be given by Equation 4.23 after setting $\dfrac{r}{R} = \dfrac{r_{critical}}{R} = \dfrac{\tau_y}{\tau_w}$.

4.11 TUBE FLOW OF BLOOD AT LOW SHEAR RATES

Note in Figure 4.8 the scatter in the data at lower shear rates, i.e. $\bar{U}$ less than about 1/sec. The data starts to show significant deviations from Equation 4.19 which predicts that τ_w versus $\bar{U}$ should be independent of the tube diameter. Clearly at these lower shear rates, RBC aggregates are starting to form and their characteristic size is becoming comparable to that of the tube diameter. The assumption of blood being a homogeneous fluid is no longer correct. However, for practical purposes, blood flow in the body and through medical devices will be at significantly higher shear rates. Therefore, we really need not concern ourselves with the limiting case of blood flow at these very low shear rates.

4.12 THE EFFECT OF THE DIAMETER AT HIGH SHEAR RATES

Tube flow of blood at high shear rates (> 100/sec) shows two anomolous effects that involve the tube diameter. These are the *Fahraeus effect* and the *Fahraeus-Lindquist effect* (Barbee and Cokelet 1971a,b; Gaehtgens 1980). To understand the effect of tube diameter on blood flow, consider the conceptual model illustrated in Figure 4.9. In this figure, blood flows from a larger vessel, such as an artery, into progressively smaller

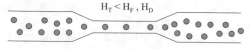

Arterial end, H_F Smaller vessel, H_T Venous end, H_D **Figure 4.9**   The Fahraeus effect.

vessels ultimately reaching the size of a capillary, after which the vessels progress in size to a vein. We make the distinction between the hematocrit found in the feed to a smaller vessel represented by H_F, the hematocrit in the vessel of interest H_T, and the discharge hematocrit H_D. In tubes with diameters from about 15 to 500 microns, it is found that the tube hematocrit (H_T) is actually less than that of the discharge hematocrit (at steady state, $H_D = H_F$, to satisfy the mass balance on the red blood cells). This is called the Fahraeus effect and is shown in Figure 4.10 where we see H_T / H_D plotted as a function of tube diameter, D, in microns. We observe a pronounced decrease in the tube hematocrit in comparison to the discharge hematocrit (or feed hematocrit). For tubes less than about 15 microns in diameter the ratio of H_T to H_D starts to increase as shown in Figure 4.10. This is because H_D is now less than H_F. One reason for this is plasma skimming. When the tube size is smaller than about 15 microns, it becomes physically very difficult for the red blood cells even to enter the smaller vessel. Hence the orientation of the branch as well as that of the red blood cells will affect the tube hematocrit as well as the resulting discharge hematocrit.

To explain the Fahraeus effect, it has been shown both by in vivo and in vitro experiments that the cells do not distribute themselves evenly across the tube cross section. Instead, the RBCs tend to accumulate along the tube axis forming, in a

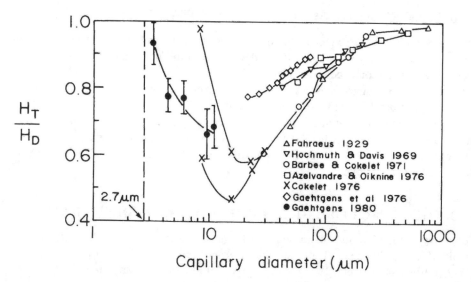

Figure 4.10   The Fahraeus effect (from Gaehtgens 1980, with permission, data source identity may be found in this reference).

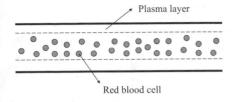

Plasma layer

Red blood cell

Figure 4.11   Axial accumulation of red blood cells.

statistical sense, a thin cell-free layer along the tube wall. This is illustrated in Figure 4.11. Recall from Equation 4.13 that the fluid velocity is maximal along the tube axis. The axial accumulation of RBCs, in combination with the higher fluid velocity, maintains the RBC balance ($H_F = H_D$) even though the hematocrit in the tube is reduced. The thin cell-free layer along the tube wall is called the *plasma layer*. The thickness of the plasma layer depends on the tube diameter and the hematocrit and is typically on the order of several microns.

Because of the Fahraeus effect, it is found that as the tube diameter decreases below about 500 microns, the viscosity of blood also decreases. The decreased viscosity is a direct result of the decrease in the tube hematocrit. The reduction in the viscosity of the blood is known as the Fahraeus-Lindquist effect.

Since the rheology of blood can also depend on the tube diameter, the simple relationship (Equation 4.11) developed to describe tube flow of blood at high shear rates no longer applies. However, since for tube flow of blood we can still measure both Q and $\Delta P/L$, we can use these measurements to calculate, for a given tube, the apparent viscosity of blood. This apparent viscosity can be written in terms of the observed values of Q and $\Delta P/L$ by rearranging Equation 4.11 (assuming the flow of blood is laminar).

$$\mu_{\text{apparent}} = \frac{\pi R^4 \Delta P}{8LQ} \tag{4.24}$$

Figure 4.12 illustrates the Fahraeus-Lindquist effect where we see the ratio of the apparent viscosity to its large tube value (μ_F) plotted as a function of the tube diameter. We observe a significant decrease in the apparent viscosity of blood between tube diameters from about 500 microns to 4–6 microns. The decrease in the apparent viscosity in the smaller tubes is explained by the fact that the viscosity in the marginal plasma layer near the tube wall is less than that of the central core that contains the cells.

As the tube diameter becomes less than about 4–6 microns, the apparent viscosity as seen in Figure 4.12 increases dramatically (Fung 1993). At this point it becomes very difficult for an RBC to enter the tube. The RBC is about 8 microns in diameter and is deformable to some extent and can enter tubes somewhat smaller than this characteristic dimension. A tube diameter of about 2.7 microns is about the smallest size that an RBC can enter. However, by Equation 4.11, in order to maintain the flowrate under these conditions, the pressure drop must increase to overcome the resistance presented by the RBC, that is the increase in the apparent viscosity.

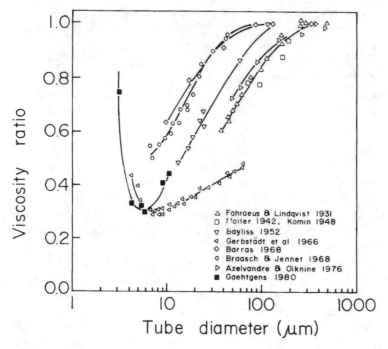

Figure 4.12   The Fahraeus–Lindquist effect (from Gaehtgens 1980, with permission, data source identity may be found in this reference).

4.13 MARGINAL ZONE THEORY

The marginal zone theory proposed by Haynes (1960) may be used to characterize the Fahraeus-Lindquist effect in the range from 4–6 microns to 500 microns. Using this theory it is possible to obtain an expression for the apparent viscosity in terms of the plasma layer thickness, tube diameter, and the hematocrit. Development of the marginal zone theory makes use of the relationships for a Newtonian fluid that we developed earlier. These are summarized in Table 4.3.

As shown in Figure 4.11, the blood flow within a tube or vessel is divided into two regions: a central core that contains the cells with a viscosity μ_c, and the cell-free marginal or plasma layer that consists only of plasma with a thickness of δ, and a viscosity equal to that of the plasma given by the symbol μ_p. In each region the flow is considered to be Newtonian and Equation 4.12 applies to each. For the core region, we may then write:

$$\tau_{rz} = \frac{\left(P_0 - P_L\right)r}{2L} = -\mu_c \frac{dv_z^c}{dr}$$

$$\text{BC1}: r = 0, \frac{dv_z^c}{dr} = 0$$

$$\text{BC2}: r = R - \delta, \tau_{rz}\big|_c = \tau_{rz}\big|_p$$

(4.25)

The first boundary condition (BC1) expresses the fact that the axial velocity is a maximum at the centerline of the tube. The second boundary condition (BC2) derives from the fact that the transport of momentum must be continuous across the interface between the core and the plasma layer. For the plasma layer the following equations apply.

$$\tau_{rz} = \frac{(P_0 - P_L)r}{2L} = -\mu_p \frac{dv_z^p}{dr}$$

$$BC3 : r = R - \delta, \; v_z^c = v_z^p \tag{4.26}$$

$$BC4 : r = R, \; v_z^p = 0$$

Boundary condition three (BC3) states the requirement that the velocity in each region must be the same at the interface. The last boundary condition (BC4) simply requires that the axial velocity is zero at the tube wall.

Equations 4.25 and 4.26 may be readily integrated to give the following expressions for the axial velocity profiles in the core and plasma regions.

$$v_z^p(r) = \frac{(P_0 - P_L)R^2}{4\mu_p L}\left[1 - \left(\frac{r}{R}\right)^2\right] \quad \text{for } R - \delta \le r \le R \tag{4.27}$$

$$v_z^c(r) = \frac{(P_0 - P_L)R^2}{4\mu_p L}\left\{1 - \left(\frac{R - \delta}{R}\right)^2 - \frac{\mu_p}{\mu_c}\left(\frac{r}{R}\right)^2 + \frac{\mu_p}{\mu_c}\left(\frac{R - \delta}{R}\right)^2\right\} \quad \text{for } 0 \le r \le R - \delta \tag{4.28}$$

The core and plasma volumetric flowrates are given by the following equations.

$$Q_p = 2\pi \int_{R-\delta}^{R} v_z^p(r)r \, dr$$

$$Q_c = 2\pi \int_{0}^{R-\delta} v_z^c(r)r \, dr \tag{4.29}$$

Integration of the above equations with the values of v_z^p and v_z^c from Equations 4.27 and 4.28 provides the following result for the volumetric flowrates of the plasma layer and the core:

$$Q_p = \frac{\pi(P_0 - P_L)}{8\mu_p L}\left[R^2 - (R - \delta)^2\right]^2 \tag{4.30}$$

$$Q_c = \frac{\pi(P_0 - P_L)R^2}{4\mu_p L}\left[(R - \delta)^2 - \left(1 - \frac{\mu_p}{\mu_c}\right)\frac{(R - \delta)^4}{R^2} - \frac{\mu_p}{\mu_c}\frac{(R - \delta)^4}{2R^2}\right] \tag{4.31}$$

The total flowrate of blood within the tube would equal the sum of the flowrates in the core and plasma regions. After adding the above two equations, we obtain the following expression for the total flowrate:

$$Q = \frac{\pi R^4 (P_0 - P_L)}{8\mu_p L} \left[1 - \left(1 - \frac{\delta}{R} \right)^4 \left(1 - \frac{\mu_p}{\mu_c} \right) \right]$$

(4.32)

Comparison of this equation with Equation 4.24 allows one to develop a relationship for the apparent viscosity based on the marginal zone theory. We then arrive at the following expression for the apparent viscosity in terms of δ, R, μ_c, and μ_p.

$$\mu_{apparent} = \frac{\mu_p}{1 - \left(1 - \dfrac{\delta}{R} \right)^4 \left(1 - \dfrac{\mu_p}{\mu_c} \right)}$$

(4.33)

As $\delta/R \to 0$, then $\mu_{apparent} \to \mu_c \to \mu$ which is the bulk viscosity of blood in a large tube at high shear rates, i.e. a Newtonian fluid, as one would expect. We can use Equations 4.32 and 4.33 for blood flow calculations in tubes less than 500 microns in diameter provided that we know the thickness of the plasma layer, δ, and the viscosity of the core, μ_c. Alternatively, one could simply estimate the apparent viscosity from the data shown in Figure 4.12. However, oftentimes for design calculations, it is necessary and more convenient to have an equation to describe the flow behavior of blood.

4.14 USING THE MARGINAL ZONE THEORY

We can now use Equation 4.32 based on the marginal zone theory to fit the apparent viscosity data shown in Figure 4.12. This will allow us to obtain values of the plasma layer thickness and the core hematocrit as a function of the tube diameter. In order to do this, we must develop the relationship for how the core hematocrit (H_C) depends on the feed hematocrit (H_F) and the thickness of the plasma layer.

We will also need an equation to describe the dependence of the blood viscosity on the hematocrit since the value of H_C will be larger than H_F because of the axial accumulation of the RBCs. This relative increase in the core hematocrit will make the blood in the core have a higher viscosity than the blood in the feed.

The following equation may be used to express the dependence of the viscosity of blood at high shear rates on the hematocrit and temperature (Charm and Kurland 1974).

$$\mu = \mu_p \frac{1}{1 - \alpha H}, \quad \text{for } H \le 0.6$$

$$\text{where } \alpha = 0.070 \exp \left[2.49 H + \frac{1107}{T} \exp(-1.69 H) \right]$$

(4.34)

In the above equations, the temperature (T) is in K. These equations may be used to a hematocrit of 0.60, the stated accuracy is within 10%. If a subscript occurs on the viscosity, then the corresponding values of H and α in the above equations will carry the same subscript. For example, if we are talking about the core region, then $\mu = \mu_c$, $H = H_C$, and $\alpha = \alpha_C$.

Going back to Equation 4.32, we can rewrite the total volumetric flowrate (Q) using Equation 4.34 in place of the viscosity ratio $\left(\dfrac{\mu_p}{\mu_c}\right)$ to obtain the following equation:

$$Q = \frac{\pi R^4 (P_0 - P_L)}{8L\mu_p}\left[1 - \left(1 - \frac{\delta}{R}\right)^4 \alpha_c H_c\right] \tag{4.35}$$

Continuity or conservation of the cells allows us to write the following expression that relates the feed hematocrit to the core hematocrit: $Q_F H_F = Q_C H_C$. Recall that the core volumetric flowrate Q_C can also be written as $Q_F - Q_P$, where Q_P is the volumetric flowrate of the marginal plasma layer. We can now use Equations 4.30, 4.31, and 4.34 to obtain an expression for the hematocrit of the core, $\dfrac{H_C}{H_F} = \dfrac{Q_F}{Q_C} = 1 + \dfrac{Q_P}{Q_C}$, with $\sigma \equiv 1 - \dfrac{\delta}{R}$.

$$\frac{H_C}{H_F} = 1 + \frac{(1 - \sigma^2)^2}{\sigma^2\left[2(1 - \sigma^2) + \sigma^2\,\dfrac{\mu_p}{\mu_c}\right]} \tag{4.36}$$

The tube hematocrit (averaged over the entire tube cross section) would also be given by the expression $H_T = \sigma^2 H_C$.

Equation 4.36 is non-linear in H_c [since $\mu_c = \mu_c(H_c)$ by Equation 4.34] and also depends for a given tube diameter, D, on the thickness of the plasma layer, $\delta = (1 - \sigma) R$, which, at this point, is not known. Therefore, we have two unknowns, H_c and δ.

The viscosity ratio for a given tube diameter provides the other relationship that may be used, along with Equation 4.36, to determine H_c and δ. The viscosity ratio, which is also the ordinate of Figure 4.12, may be determined by dividing both sides of Equation 4.33 by the feed or large tube viscosity, μ_F, and using Equation 4.34 to eliminate μ_p / μ_F in favor of the term $(1 - \alpha_f H_F)$.

$$\frac{\mu_{\text{apparent}}}{\mu_F} = \frac{1 - \alpha_F H_F}{\left[1 - \sigma^4 \alpha_c H_c\right]} \tag{4.37}$$

For a given tube diameter and corresponding value of $\dfrac{\mu_{\text{apparent}}}{\mu_F}$, Equations 4.36 and 4.37 may be solved to find δ and H_C. This is illustrated in the following example for some of the data shown in Figure 4.12.

Example 4.3 Find values of δ and H_c as a function of tube diameter for the data of Bayliss and Gaehtgens shown in Figure 4.12.

SOLUTION To obtain the solution we need to solve simultaneously Equations 4.36 and 4.37. These two algebraic equations are non-linear meaning that they cannot be explicitly solved for δ (or $\sigma = 1 - \delta/R$) and H_c. Numerical techniques based on the Newton-Raphson method can be used to solve these equations in an interative manner. For example, to solve these equations by the Newton-Raphson method, one must first rearrange Equations 4.36 and 4.37 as shown below.

$$f\left(\sigma, H_c\right) = \frac{H_C}{H_F} - 1 - \frac{\left(1 - \sigma^2\right)^2}{\sigma^2 \left[2(1 - \sigma^2) + \sigma^2 \dfrac{\mu_p}{\mu_c}\right]}$$

$$g\left(\sigma, H_c\right) = \frac{\mu_{apparent}}{\mu_F} - \frac{1 - \alpha_F H_F}{\left[1 - \sigma^4 \alpha_c H_C\right]}$$

When the problem is recast in terms of $f(\sigma, H_c)$ and $g(\sigma, H_c)$ the goal is to find the values of σ and H_c that make these functions equal to zero. Clearly if these values were the solution, then $f(\sigma, H_c)$ and $g(\sigma, H_c)$ would both equal zero, and we would have the solution. However, a Taylor series written around the i-th iteration values of σ_i and H_{ci} and truncated after the second term allows us to predict the next best values of σ and H_c based on the current values of σ_i and H_{ci} , i.e.

$$f\left(\sigma_{i+1}, H_{ci+1}\right) \approx f\left(\sigma_i, H_{ci}\right) + \frac{\partial f}{\partial \sigma}\bigg|_{\sigma_i, H_{ci}} \left(\sigma_{i+1} - \sigma_i\right) + \frac{\partial f}{\partial H_c}\bigg|_{\sigma_i, H_{ci}} \left(H_{ci+1} - H_{ci}\right)$$

$$g\left(\sigma_{i+1}, H_{ci+1}\right) \approx g\left(\sigma_i, H_{ci}\right) + \frac{\partial g}{\partial \sigma}\bigg|_{\sigma_i, H_{ci}} \left(\sigma_{i+1} - \sigma_i\right) + \frac{\partial g}{\partial H_c}\bigg|_{\sigma_i, H_{ci}} \left(H_{ci+1} - H_{ci}\right)$$

We want σ_{i+1} and H_{ci+1} to be such that $f(\sigma_{i+1}, H_{ci+1})$ and $g(\sigma_{i+1}, H_{ci+1})$ both equal zero. Letting the left-hand sides of the above equations equal zero allows for the solution of $\varepsilon_i = \sigma_{i+1} - \sigma_i$ and $\tau_i = H_{ci+1} - H_{ci}$ as shown below.

$$\begin{bmatrix} \varepsilon_i \\ \tau_i \end{bmatrix} = - \begin{bmatrix} f_i \\ g_i \end{bmatrix} \begin{bmatrix} \dfrac{\partial f}{\partial \sigma}\bigg|_i & \dfrac{\partial f}{\partial H_c}\bigg|_i \\ \dfrac{\partial g}{\partial \sigma}\bigg|_i & \dfrac{\partial g}{\partial H_c}\bigg|_i \end{bmatrix}^{-1}$$

Solution of the above matrix equation then provides the values of ε_i and δ_i from which $\sigma_{i+1} = \sigma_i + \varepsilon_i$ and $H_{ci+1} = H_{ci} + \tau_i$. If σ_{i+1} and H_{ci+1} differ significantly from the previous values, i.e. σ_i and H_{ci} , then convergence has not been obtained and one repeats the calculation by letting $\sigma_i = \sigma_{i+1}$ and $H_{ci} = H_{ci+1}$. Table 4.4 summarizes the results of these calculations for tube diameters from 130 to 25 microns using the Haynes marginal zone layer theory. Tubes smaller than 25 microns in diameter give a core hematocrit greater than 0.60 which is beyond the range allowed for by Equation 4.34. Knowing how δ and H_c vary with tube diameter for a particular data set allows one to then calculate the apparent viscosity from Equation 4.37 and the flowrate, for a given pressure drop, from Equation 4.24.

Table 4.4 Calculation of H_c and δ from the data of Bayliss and Gaehtgens (refer to Example 4.3)

d, microns	μ_{app}/μ_f	δ, microns	H_C	H_T	δ/d
130	1.0	0	0.40	0.40	0
100	0.93	1.15	0.40	0.38	0.012
80	0.90	1.4	0.40	0.38	0.018
60	0.82	2.27	0.41	0.35	0.038
40	0.77	2.22	0.43	0.34	0.055
30	0.69	2.92	0.48	0.31	0.093
25	0.63	3.87	0.59	0.28	0.155

4.15 BOUNDARY LAYER THEORY

Describing the flow of a fluid near a surface is extremely important in a wide variety of engineering problems. Generally, the effect of the fluid viscosity is such that the fluid velocity changes from zero at the surface to the free stream value over a narrow region near the surface that is referred to as the *boundary layer*. It is the presence of this boundary layer that affects the rates of mass transfer and heat transfer between the surface and the fluid.

Analysis of these types of problems using boundary layer theory for relatively simple cases can provide a great deal of insight on how the flow of the fluid affects the transport of mass and energy and leads to the rational development of correlations to describe mass and energy transfer in more complex geometries and flow systems. We will use these correlations in later chapters in problems that involve the transport of mass across a bounding surface into a flowing fluid.

4.15.1 The Flow Near a Wall that is Set in Motion

Consider the situation shown in Figure 4.13. A semi-infinite quantity of a viscous fluid is contacted from below by a flat and horizontal plate. For $t < 0$, the plate and the fluid are not moving. At $t = 0$, the plate is set in motion with a constant velocity to the right as shown in the figure. There are no pressure gradients or gravitational forces acting on the fluid so the motion of the fluid is solely due to the momentum transferred from the plate to the fluid. The flow is also laminar meaning there is no mixing of fluid elements in the y direction.

As time progresses the velocity profile penetrates further into the fluid. We can arbitrarily define the boundary layer thickness (δ) for this problem as that distance perpendicular to the plate surface where the fluid velocity has decreased to a value equal to 1% of V. Our goal here is to determine the velocity profile $v_x(y,t)$ and the boundary layer thickness, $\delta(t)$. v_y is zero since we have no holes in the plate and the flow is laminar. v_z is zero since the fluid and plate are assumed to be infinite in the z direction.

The concept of a shell balance can be used to analyze this problem. The shell balance is an important technique for developing mathematical models to describe the transport of such quantities as momentum, mass, and energy. The shell balance

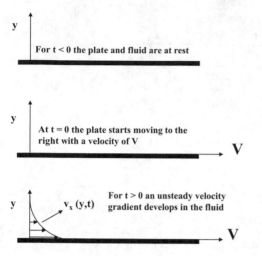

Figure 4.13　Flow of a fluid near a flat plate that is set in motion.

approach is conceptually easy to use and is based on the application of the generalized balance equation (Equation 1.8) to a given finite volume of interest.

Consider a small volume element of the fluid $\Delta x\, \Delta y\, W$, where W is the width of the plate in the z direction and is assumed to be very large. Recall that momentum is (mass) $\times$ (velocity) and we can write the momentum per unit volume of the fluid as $\rho\, v_x$. The rate of accumulation of momentum within this volume element of the fluid is equal to $\rho \dfrac{\partial v_x}{\partial t} \Delta x\, \Delta y\, W$. This term has units of force and is equal to the sum of the forces acting on the the volume element of the fluid. The only forces acting on this volume element of fluid are the shear forces acting on the surfaces at y and $y + \Delta y$, i.e. $\tau_{yx}\,|_y$ and $\tau_{yx}\,|_{y+\Delta y}$. Hence, our momentum shell balance on the volume element can be written as follows:

$$\rho \frac{\partial v_x}{\partial t} \Delta x\, \Delta y\, W = \left(\tau_{yx}\Big|_y - \tau_{yx}\Big|_{y+\Delta y} \right) \Delta x\, W \tag{4.38}$$

Eliminating Δx and W, and then dividing by Δy and taking the limit as $\Delta y \to 0$, we obtain the following partial differential equation.

$$\rho \frac{\partial v_x}{\partial t} = -\frac{\partial \tau_{yx}}{\partial y} \tag{4.39}$$

The above equation is valid for any fluid, Newtonian or non-Newtonian. For the special case of the Newtonian fluid, we can use Equation 4.2 for τ_{yx} and obtain:

$$\frac{\partial v_x}{\partial t} = v \frac{\partial^2 v_x}{\partial y^2} \tag{4.40}$$

where v is the kinematic viscosity and is defined as μ / ρ. The initial condition (IC) and boundary conditions (BC) are as follows:

IC: $t = 0$, $v_x = 0$ for all values of y

BC1: $y = 0$, $v_x = V$ for all values of $t > 0$ (4.41)

BC2: $y = \infty$, $v_x = 0$ for all values of $t > 0$

Equation 4.40 can easily be solved using the Laplace transform technique. Recall that the Laplace transform of a function f(t) is defined by the following equation.

$$L\left[f(t)\right] = \overline{f}(s) = \int_0^\infty e^{-st} f(t) \, dt \tag{4.42}$$

Tables of Laplace transforms may be found in a variety of calculus textbooks and mathematical handbooks. Table 4.5 summarizes some of the more commonly used Laplace transforms.

Taking the Laplace transform of Equation 4.40 results in the equation shown below.

$$s\overline{v}_x - v_x\left(t = 0\right) = v \frac{d^2 \overline{v}_x}{dy^2} \tag{4.43}$$

where $\overline{v}_x$ denotes the Laplace transformed velocity. From the initial condition (Equation 4.41) we have that $v_x(t = 0) = 0$ and the other boundary conditions become:

BC1: $y = 0$, $\overline{v}_x = \dfrac{V}{s}$

BC2: $y = \infty$, $\overline{v}_x = 0$ (4.44)

Equation 4.43 can then be written as:

$$\frac{d^2 \overline{v}_x}{dy^2} - a^2 \, \overline{v}_x = 0 \tag{4.45}$$

with $a^2 = s / v$. The above equation is a homogeneous second order differential equation having the general solution:

$$\overline{v}_x = C_1 e^{ay} + C_2 e^{-ay} \tag{4.46}$$

The constants C_1 and C_2 can then be found from the boundary conditions given by Equation 4.44. Using these boundary conditions we find that $C_1 = 0$ and $C_2 = V / s$. Hence our solution in the Laplace transform space is:

$$\frac{\overline{v}_x}{V} = \frac{1}{s} e^{-ay} = \frac{1}{s} e^{-\frac{y}{\sqrt{v}}\sqrt{s}} \tag{4.47}$$

Now inverting the above equation, we find the function $v_x(t)$ whose Laplace transform

Table 4.5 Some commonly used Laplace transforms (from Arpaci 1966)

ID	Function	Transform
1	$C_1 f(t) + C_2 g(t)$	$C_1 \bar{f}(s) + C_2 \bar{g}(s)$
2	$\dfrac{df(t)}{dt}$	$s\bar{f}(s) - f(0+)$
3	$\dfrac{\partial^n f(x_i, t)}{\partial x_i^n}$	$\dfrac{\partial^n \bar{f}(x_i, s)}{\partial x_i^n}$
4	$\displaystyle\int_0^t f(\tau)\, d\tau$	$\dfrac{1}{s}\bar{f}(s)$
5	$f(\alpha t)$	$\dfrac{1}{\alpha}\bar{f}\left(\dfrac{s}{\alpha}\right)$
6	$e^{-\beta t} f(t)$	$\bar{f}(s + \beta)$
7	$\displaystyle\int_0^t f(t - \tau) g(\tau)\, d\tau$	$\bar{f}(s)\bar{g}(s)$
8	1	$\dfrac{1}{s}$
9	t	$\dfrac{1}{s^2}$
10	$\dfrac{1}{(\pi t)^{1/2}}$	$\dfrac{1}{s^{1/2}}$
11	$-\dfrac{1}{2\pi^{1/2} t^{3/2}}$	$s^{1/2}$
12	$e^{-\alpha t}$	$\dfrac{1}{s + \alpha}$
13	$\dfrac{e^{-\beta t} - e^{-\alpha t}}{\alpha - \beta}$	$\dfrac{1}{(s + \alpha)(s + \beta)}$
14	$t e^{-\alpha t}$	$\dfrac{1}{(s + \alpha)^2}$
15	$\dfrac{(\gamma - \beta)e^{-\alpha t} + (\alpha - \gamma)e^{-\beta t} + (\beta - \alpha)e^{-\gamma t}}{(\alpha - \beta)(\beta - \gamma)(\gamma - \alpha)}$	$\dfrac{1}{(s + \alpha)(s + \beta)(s + \gamma)}$
16	$\dfrac{1}{2} t^2 e^{-\alpha t}$	$\dfrac{1}{(s + \alpha)^3}$
17	$\dfrac{\alpha e^{-\alpha t} - \beta e^{-\beta t}}{\alpha - \beta}$	$\dfrac{s}{(s + \alpha)(s + \beta)}$

18	$\dfrac{(\beta-\gamma)\alpha e^{-\alpha t}+(\gamma-\alpha)\beta e^{-\beta t}+(\alpha-\beta)\gamma e^{-\gamma t}}{(\alpha-\beta)(\beta-\gamma)(\gamma-\alpha)}$	$\dfrac{s}{(s+\alpha)(s+\beta)(s+\gamma)}$
19	$\sin\alpha t$	$\dfrac{\alpha}{s^2+\alpha^2}$
20	$\cos\alpha t$	$\dfrac{s}{s^2+\alpha^2}$
21	$\sinh\alpha t$	$\dfrac{\alpha}{s^2-\alpha^2}$
22	$\cosh\alpha t$	$\dfrac{s}{s^2-\alpha^2}$
23	$\dfrac{x}{2\left(\pi at^3\right)^{1/2}}e^{-x^2/4at}$	$e^{-qx},\ q=(s/a)^{1/2}$
24	$\left(\dfrac{a}{\pi t}\right)^{1/2}e^{-x^2/4at}$	$\dfrac{e^{-qx}}{q},\ q=(s/a)^{1/2}$
25	$erfc\left(\dfrac{x}{2(at)^{1/2}}\right)$	$\dfrac{e^{-qx}}{s},\ q=(s/a)^{1/2}$
26	$2\left(\dfrac{at}{\pi}\right)^{1/2}e^{-x^2/4at}-x\,erfc\left(\dfrac{x}{2(at)^{1/2}}\right)$	$\dfrac{e^{-qx}}{qs},\ q=(s/a)^{1/2}$
27	$\left(t+\dfrac{x^2}{2a}\right)erfc\left(\dfrac{x}{2(at)^{1/2}}\right)-x\left(\dfrac{t}{\pi a}\right)^{1/2}e^{-x^2/4at}$	$\dfrac{e^{-qx}}{s^2}$

---

is the above equation. From Table 4.5 we then find that the inverse of the above equation provides the solution for $v_x(t)$ as shown below.

$$\frac{v_x(y,t)}{V}=erfc\left(\frac{y}{\sqrt{4vt}}\right)=1-erf\left(\frac{y}{\sqrt{4vt}}\right) \qquad (4.48)$$

The above solution is in terms of a new function called the *error function* and the *complementary error function* which are abbreviated as "*erf*" and "*erfc*" respectively. The error function is defined by the following equation:

$$erf(x)=\frac{2}{\sqrt{\pi}}\int_0^x e^{-x^2}dx \quad \text{and} \quad erfc(x)=1-erf(x) \qquad (4.49)$$

Note that the integral of e^{-x^2} cannot be integrated analytically. Since this integral is quite common in the solution of transport problems, this function has been tabulated in mathematical handbooks and mathematical software and can be treated as a known function much like logarithmic and trigonometric functions.

Recall that we defined the boundary layer thickness as that distance y from the surface of the plate where the velocity has decreased to a value of 1% of V. The complementary error function of $\dfrac{y}{\sqrt{4vt}} = 2$ provides a value of v_x / V that is very close to 0.01. Hence we can define the boundary layer thickness, $\delta(y,t)$ as follows:

$$\delta(y,t) = 4\sqrt{vt} \tag{4.50}$$

The value of δ can also be interpreted as the distance to which momentum from the moving plate has penetrated into the fluid at time t.

Example 4.4 Calculate the boundary layer thickness 1 second after the plate has started to move. Assume the fluid is water for which $v = 1 \times 10^{-6}$ m^2 sec^{-1}.

SOLUTION Using Equation 4.50 we can calculate the thickness of the boundary layer as shown below.

$$\delta = 4\sqrt{10^{-6}\,\text{m}^2\,\text{sec}^{-1} \times 1\,\text{sec}} = 0.004\,\text{m} = 4\,\text{mm}$$

4.15.2 Laminar Flow of a Fluid Along a Flat Plate

Now consider the situation shown in Figure 4.14 of the steady laminar flow of a fluid along a flat plate. The plate is assumed to be semi-infinite, and the fluid approaches the plate at a uniform velocity of V. Since the fluid velocity at the surface of the plate is zero, a boundary layer is formed near the surface of the plate, and within this boundary layer, v_x increases from zero to its free stream value of V.

In the following discussion we will obtain an approximate solution for the boundary layer flow over a flat plate. The approximate solution is very close to the exact solution. Details on exact solutions to the boundary layer equations may be found in Schlichting (1979).

Consider the shell volume shown in Figure 4.14 located from x to $x + \Delta x$ and from $y = 0$ to $y = \delta(x)$. We first perform a steady state mass balance on this shell volume which is given by:

$$W \int_0^\delta \rho v_x \, dy \bigg|_x - W \int_0^\delta \rho v_x dy \bigg|_{x+\Delta x} - \rho\, v_y \big|_\delta W \Delta x = 0 \tag{4.51}$$

The first two terms in the above equation provide the net rate at which mass is being added to the shell volume. The third term accounts for the loss of mass from the shell volume at the top of the boundary layer due to flow in the y-direction. Eliminating W, and dividing by Δx, followed by taking the limit as $\Delta x \to 0$ provides the following equation for $v_y|_\delta$.

$$v_y \big|_\delta = -\frac{d}{dx} \int_0^\delta v_x \, dx \tag{4.52}$$

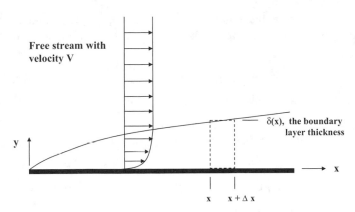

Figure 4.14   Laminar boundary layer flow of a fluid over a flat plate.

In a similar manner we can write an x-momentum balance on the shell volume as shown below.

$$W \int_{0}^{\delta(x)} \rho v_x v_x dy \bigg|_{x} - W \int_{0}^{\delta(x)} \rho v_x v_x dy \bigg|_{x+\Delta x} - \rho v_y \big|_{\delta} VW\Delta x + W\Delta x\, \tau_{yx}\big|_{y=0} = 0 \quad (4.53)$$

The first two terms in the above equation represent the net rate at which x-momentum is being added to the shell volume. The third term represents the rate at which x-momentum at the top of the boundary layer is being lost due to the flow of fluid out of the boundary layer in the y-direction. The last term in the above equation represents the loss of momentum as a result of the shear stress generated by the fluid at the surface of the plate. After eliminating W and dividing by Δx and taking the limit as $\Delta x \to 0$, we can write Equation 4.53 as shown below.

$$-\tau_{yx}\big|_{y=0} = -\frac{d}{dx}\left(\int_{0}^{\delta} \rho v_x v_x dy\right) - \rho v_y\big|_{y=\delta} V \quad (4.54)$$

Now we can use Equation 4.52 to eliminate v_y in the above equation and obtain the result shown below.

$$-\tau_{yx}\big|_{y=0} = \frac{d}{dx}\left(\int_{0}^{\delta} \rho\left(V - v_x\right) v_x\, dy\right) \quad (4.55)$$

For a Newtonian fluid we can use Equation 4.2 once again and obtain that:

$$\mu \frac{\partial v_x}{\partial y}\bigg|_{y=0} = \frac{d}{dx}\left(\int_{0}^{\delta} \rho\left(V - v_x\right) v_x\, dy\right) \quad (4.56)$$

The above equation is also known as the *von Karman* momentum balance equation and forms the basis for obtaining an approximate solution to the boundary layer flow over a flat plate. To obtain an approximate solution, we first need to approximate the shape of the velocity profile within the boundary layer, i.e. $v_x(x,y)$. The simplest function that reasonably approximates the shape of the velocity profile is a simple cubic equation, i.e.

$$v_x(y) = a + by + cy^2 + dy^3 \qquad (4.57)$$

The velocity profile assumed above also has to satisfy the following boundary conditions.

$$BC1: \; y = 0, \, v_x = 0$$
$$BC2: \; y = \delta(x), \, v_x = V$$
$$BC3: \; y = \delta(x), \, \frac{\partial v_x}{\partial y} = 0$$
$$BC4: \; y = 0, \, \frac{\partial^2 v_x}{\partial y^2} = 0 \qquad (4.58)$$

The first boundary condition is referred to as the "no slip" boundary condition and requires that the velocity of the fluid at the surface of the plate be the same as the velocity of the plate which, in this case, is zero. The last boundary condition expresses the fact that the stress at the surface of the plate is a constant. Boundary conditions 2 and 3 state that beyond the boundary layer the velocity is constant and equal to the free stream value V.

When the above boundary conditions are imposed on Equation 4.57 the following expression is obtained for the velocity profile within the boundary layer.

$$\frac{v_x(x,y)}{V} = \frac{3}{2}\left(\frac{y}{\delta(x)}\right) - \frac{1}{2}\left(\frac{y}{\delta(x)}\right)^3 \qquad (4.59)$$

The above equation indicates that v_x depends on $\delta(x)$ which is still not known. However, we can insert the above equation into the the von Karman momentum balance equation (Equation 4.56) and, after some simplification, we obtain:

$$\delta \frac{d\delta}{dx} = \frac{140}{13}\left(\frac{\mu}{\rho V}\right) \qquad (4.60)$$

with the condition that at $x = 0$, $\delta = 0$. Integration of the above equation results in the following expression for the boundary layer thickness, $\delta(x)$.

$$\delta(x) = 4.64\sqrt{\frac{\nu x}{V}} \qquad (4.61)$$

This equation shows that the boundary layer thickness grows in proportion to the square root of the distance from the upstream leading edge of the plate. Also the above equation indicates that the boundary layer thickness is inversely proportional to the square root of the freestream velocity. We can also rearrange Equation 4.61 into the following form:

$$\frac{\delta(x)}{x} = \frac{4.64}{\sqrt{\dfrac{\rho V x}{\mu}}} = 4.64 Re_x^{-\frac{1}{2}} \tag{4.62}$$

In the above equation, $Re_x = \dfrac{\rho V x}{\mu}$ is defined as the local value (i.e. at location x) of the Reynolds number. The *Reynolds number* is a very important dimensionless number in the field of fluid mechanics. Physically the Reynolds number represents the ratio of the inertial forces acting on the fluid to the viscous forces acting on the fluid. A high Reynolds number indicates that the inertial forces dominate the viscous forces. On the other hand, a low Reynolds number means viscous forces are much larger than the inertial forces. The Reynolds number also provides insight into when the fluid transitions from uniform laminar flow to turbulent flow. For example, for boundary layer flow over the flat plate, experiments show that the flow is laminar provided $Re_x < 300,000$.

The approximate velocity profile within the boundary layer is then given by the combination of Equations 4.59 and 4.61 as shown below.

$$\frac{v_x(x,y)}{V} = 0.3233 \left(y\sqrt{\frac{V}{vx}} \right) - 0.005 \left(y\sqrt{\frac{V}{vx}} \right)^3 \tag{4.63}$$

With the velocity profile given by the above equation, we can also calculate the drag force exerted by the fluid on the plate. For a plate of length L and width W, this force that acts on the surface of both sides of the plate in the positive x direction is given by:

$$F_x = 2 \int_0^W \int_0^L \left(\mu \frac{\partial v_x}{\partial y} \Big|_{y=0} \right) dx\, dz = 1.293 \sqrt{\rho \mu L\, W^2\, V^3} \tag{4.64}$$

The exact solution as well as experimental data for the drag force on the flat plate is about 3% greater than that predicted by the above approximate solution to the flat plate boundary layer problem. The constant in Equation 4.64 being 1.328 for the exact solution. Hence we see that this approximate solution to the flat plate boundary layer problem is quite good.

We can also calculate the power associated with this drag force. Recall that power is defined as (force × velocity) so, after multiplying Equation 4.64 by V, the power becomes:

$$P = V\, F_x = 1.293 \sqrt{\rho \mu L\, W^2\, V^5} \tag{4.65}$$

The *friction factor* (*f*) is defined as the ratio of the shear stress at the wall and the kinetic energy per volume of the fluid based on the free stream velocity, i.e.:

$$f = \frac{-\tau_{yx}\big|_{y=0}}{\dfrac{1}{2}\rho V^2} = \frac{\mu \dfrac{\partial v_x}{\partial y}\Big|_{y=0}}{\dfrac{1}{2}\rho V^2} = \frac{3\mu}{\rho V \delta(x)} = \frac{0.646}{\sqrt{Re_x}} \tag{4.66}$$

Example 4.5 Consider a manta ray swimming along in the ocean at 2 meters/second. Assuming the ray approximates a rectangular shape with a length of 0.70 meters and a width of 1.5 meters, how much power is the ray expending to move through the water as a result of the drag force at its surface? Assume the density of the water is 1000 kg/m^3 and the viscosity of the water is 1 cP = 0.001 kg/m/sec.

SOLUTION Assume that the geometry of the manta ray approximates that of a flat plate. Therefore, we can assume that we have laminar flow of the seawater over a flat plate. Using Equation 4.65 we can calculate the power requirement as follows:

$$P = 1.293\sqrt{1000\,\text{kg m}^{-3} \times 0.001\,\text{kg m}^{-1}\,\text{sec}^{-1} \times 0.7\,\text{m} \times 1.5^2\,\text{m}^2 \times 2^5\,\text{m}^5\,\text{sec}^{-5}}$$

$$= 9.18\,\text{kg m sec}^{-2}\,\text{m sec}^{-1} = 9.18\,\text{J sec}^{-1} = 9.18\,\text{W} = 0.012\,\text{HP}$$

4.16 GENERALIZED MECHANICAL ENERGY BALANCE EQUATION

Equation 4.11 provides a useful relationship for describing the laminar flow of a fluid in a cylindrical tube of constant cross section. We will oftentimes need a more generalized relationship that can account for not only the effect of the pressure drop on fluid flow, but also the effect of changes in elevation, tube cross section, changes in fluid velocity, sudden contractions or expansions, pumps, and the effect of fittings such as valves. This general relationship for the flow of a fluid is called the *Bernoulli equation* and is valid for laminar or turbulent flow.

The Bernoulli equation accounts for the effect of changes in fluid pressure, potential energy, and kinetic energy on the flow of the fluid. It also accounts for the energy added to the fluid by pumps and accounts for energy losses due to a variety of frictional effects. For steady state flow of an incompressible fluid (density ρ = constant) such as blood from inlet station "1" to exit station "2", the Bernoulli equation may be written as follows (Bird *et al.* 2002; McCabe *et al.* 1985):

$$\frac{P_1}{\rho} + g Z_1 + \frac{\alpha_1 V_1^2}{2} + \eta W_p = \frac{P_2}{\rho} + g Z_2 + \frac{\alpha_2 V_2^2}{2} + h_{\text{friction}} \tag{4.67}$$

The units of each term in this equation must be consistent with one another and, in general, are expressed in terms of energy per unit mass of fluid. In this equation, P represents the pressure, Z the elevation relative to a reference plane, and V the average fluid velocity. Gravitational acceleration is represented by g and, in SI units, is equal to 9.8 m/sec^2. The α's are kinetic energy correction terms. For laminar flow in a cylindrical tube, i.e. $Re = \dfrac{\rho D_{\text{tube}} V}{\mu} < 2000$, $\alpha = 2.0$, and for turbulent flow $\alpha = 1.0$. W_p represents work done per unit mass of fluid. The pump efficiency is represented by η and accounts for frictional losses and other inefficiencies of the pump. Therefore, $\eta < 1$. Other frictional effects between positions 1 and 2 due to the tube itself, contractions, expansions, and fittings are accounted for by the term h_{friction}.

In addition to the Bernoulli equation, we also need to write a mass balance on the fluid. At steady state, this simply says that the rate of mass leaving the system, or a region of interest (2), must equal the rate at which mass enters the system (1). For steady state flow, this can be expressed by the continuity equation given below.

$$\dot{m} = (\rho V S)_1 = (\rho V S)_2 = \text{constant} \tag{4.68}$$

In this equation, $\dot{m}$ is the mass flow rate and S represents the cross-sectional area of the tube.

The frictional effects are described by the following equation.

$$h_{\text{friction}} = \left(4 \sum_i f_i \frac{L_i V_i^2}{2 D_i} + \sum_j \frac{V_j^2}{2} K_{\text{fitting}_j} \right) \tag{4.69}$$

In this equation, f_i is called the "friction factor" and represents the skin friction contribution in tube segment i of length L_i and diameter D_i. V_i represents the average

velocity $\left(V_i = \dfrac{\text{volumetric flowrate}}{\dfrac{\pi D_i^2}{4}} \right)$ within tube section i. The friction factor (also

known as the Fanning friction factor) is defined as the ratio of the wall shear stress (τ_w) and the average velocity head $\dfrac{\rho V^2}{2}$. For laminar flow, it is easy to show that $f = 16 / Re$. For turbulent flow ($Re > 2000$) in smooth tubes, the friction factor may be evaluated from the following equation (McCabe et $al.$ 1985). If the flow is turbulent and the walls are not smooth, then the friction factor charts in McCabe et $al.$ should be consulted.

$$\frac{1}{\sqrt{f}} = 4.07 \log\left(Re \sqrt{f} \right) - 0.60 \tag{4.70}$$

The term in the second summation of Equation 4.69 represented by K_{fitting} accounts for energy loss in the fluid due to tube contractions, expansions, or fittings such as valves. The average velocity in this summation is for the fluid just downstream of the contraction, expansion, or fitting. K_{fitting} for such items as valves is highly dependent on the valve's degree of openness and on the type of valve used. It is generally best to consult the manufacturer's literature for the particular valve being considered. As examples, a gate valve that is wide open has a K_{fitting} of about 0.2 and a value of about 6 when half open. A globe valve that is wide open may have a value as high as 10. Simple fittings, such as a tee, have a value of 2.0, and a 90° elbow has a value of about 1.0. For sudden contractions and expansions of the fluid, the following equations may be used to estimate K_{fitting}. Once again S refers to the cross-sectional area of the tube.

$$K_{\text{fitting}} = 0.4 \left(1 - \frac{S_{\text{downstream}}}{S_{\text{upstream}}} \right), \text{ sudden contraction, turbulent flow}$$

$$\tag{4.71}$$

$$K_{\text{fitting}} = \left(\frac{S_{\text{downstream}}}{S_{\text{upstream}}} - 1 \right)^2, \text{ sudden expansion, turbulent flow}$$

For the most part these fitting losses tend to be negligible for laminar flow of a fluid.

If the tube through which a fluid flows is not circular, then the equivalent diameter can be used in the above calculations. The equivalent diameter for flow in tubes of non-circular cross section is defined as four times the hydraulic radius (r_H). The hydraulic radius is defined as the ratio of the cross-sectional area of the flow channel to the wetted perimeter of the tube.

$$D_H = 4\,x\,r_H = \frac{4 \times \left(\text{cross sectional area}\right)}{\left(\text{wetted perimeter}\right)} \tag{4.72}$$

The following example illustrates the use of the above relationships.

Example 4.6 The simplest patient infusion system is that of gravity flow from an intravenous (IV) bag. A 500 ml IV bag containing an aqueous solution is connected to a vein in the forearm of a patient. Venous pressure in the forearm is 0 mm Hg (gauge pressure[1]). The IV bag is placed on a stand such that the entrance to the tube leaving the IV bag is exactly one meter above the vein into which the IV fluid enters. The length of the IV bag is 30 cm. The IV is fed through an 18 gauge tube (internal diameter = 0.953 mm) and the total length of the tube is 2 meters. Calculate the flowrate of the IV fluid. Estimate the time needed to empty the bag.

SOLUTION We apply Bernoulli's equation from the surface of the fluid in the IV bag ("1") to the entrance of the vein ("2"). We expect the flow of the fluid through the bag and the tube to be laminar and, therefore, neglect the contraction at the entry of the feed tube and the expansion at the vein. The pressure at the surface of the fluid in the bag will be atmospheric (P_1 = 760 mm Hg absolute) since the bag collapses as the fluid leaves the bag. The venous pressure, or P_2, is 0 mm Hg gauge or 760 mm Hg absolute. Because the fluid takes some length of time to leave the bag, we neglect the velocity of the surface of the fluid in the bag in comparison to the fluid velocity at the exit of the tube. Therefore, we assume V_1 = 0 and $V_2 \gg V_1$. We also set the reference elevation as the entrance to the patient's arm, hence Z_2 = 0. Therefore, Z_1 is equal to the elevation of the bag relative to the position where the fluid enters the patient's arm. This would equal 1 meter plus the 30 cm length of the bag. Since there are no pumps in the system, W_p = 0. We can now write the Bernouilli equation for this particular problem as:

$$g\,Z_1 = V_2^{\,2} + \left(4f\frac{L}{D}\right)\frac{V_2^{\,2}}{2}$$

Recall that the friction factor for laminar flow in a cylindrical tube is equal to 16 / Re. We may substitute this relationship into the above equation to obtain the following quadratic equation that can be solved for the exiting velocity, V_2.

---

[1]Recall that gauge pressure is that pressure relative to the local atmospheric pressure. Absolute pressure is gauge pressure plus local atmospheric pressure.

$$V_2^2 + \left(\frac{32\,\mu\,L}{\rho\,D^2}\right)V_2 - gZ_1 = 0$$

This equation may now be solved for the exit velocity recognizing that this quantity must be positive.

$$V_2 = \frac{-\left(\dfrac{32\,\mu\,L}{\rho\,D^2}\right) + \left[\left(\dfrac{32\,\mu\,L}{\rho\,D^2}\right)^2 + 4\,g\,Z_1\right]^{\frac{1}{2}}}{2}$$

Assuming the IV fluid has the same properties as water, and substituting the appropriate values for the parameters in the above equation, the exit velocity is calculated as follows:

$$V_2 = \frac{-\left(\dfrac{32\times0.001\dfrac{\text{kg m sec}}{\text{sec}^2\,\text{m}^2}\times2.0\,\text{m}}{1000\dfrac{\text{kg}}{\text{m}^3}\times0.000953^2\,\text{m}^2}\right) + \left[\left(\dfrac{32\times0.001\dfrac{\text{kg m sec}}{\text{sec}^2\,\text{m}^2}\times2.0\,\text{m}}{1000\dfrac{\text{kg}}{\text{m}^3}\times0.000953^2\,\text{m}^2}\right)^2 + 4\times9.8\dfrac{\text{m}}{\text{sec}^2}\times1.3\,\text{m}\right]^{\frac{1}{2}}}{2}$$

$$V_2 = 0.18\frac{\text{m}}{\text{sec}} \times \frac{100\,\text{cm}}{\text{m}} \times \frac{60\,\text{sec}}{\text{min}} = 1082\frac{\text{cm}}{\text{min}}$$

$$Q = 1082\frac{\text{cm}}{\text{min}} \times \frac{\pi}{4}(0.0953\,\text{cm})^2 \times \frac{\text{ml}}{\text{cm}^3} = 7.72\frac{\text{ml}}{\text{min}}$$

$$t_{\text{empty}} \approx \frac{V_{\text{bag}}}{Q} = \frac{500\,\text{ml}}{7.72\dfrac{\text{ml}}{\text{min}}} = 65\,\text{minutes}$$

With the exit velocity of the fluid now estimated, we need to check the Reynolds number to see if our assumption of laminar flow is valid.

$$Re = \frac{\rho\,D\,V_2}{\mu} = \frac{1000\dfrac{\text{kg}}{\text{m}^3}\times0.000953\,\text{m}\times0.18\dfrac{\text{m}}{\text{sec}}}{0.001\dfrac{\text{kg m sec}}{\text{sec}^2\,\text{m}^2}} = 172$$

Since the $Re < 2000$, our assumption of laminar flow is correct.

4.17 CAPILLARY RISE AND CAPILLARY ACTION

Numerous processes depend on capillary action, that is the ability of liquids to penetrate freely into small pores, openings, and cracks. Capillary action is responsible for transporting water to the uppermost parts of tall trees and has a variety of applications in the fields of printing, textiles, agriculture, cleaning and sanitation products, and medical devices.

4.17.1 Capillary Rise

Consider the situation shown in Figure 4.15. A small capillary tube is placed within a liquid. The liquid is drawn into the capillary as a result of the surface forces acting on the liquid wetting the inside surfaces of the capillary tube. A meniscus is also formed, and the angle (θ) between the liquid surface and the wall of the capillary is called the *contact angle*. The radius of curvature of the meniscus (r) is then related to the radius of the capillary tube (R) and the contact angle by the following equation.

$$r = \frac{R}{\cos \theta} \tag{4.73}$$

The liquid continues to rise up in the capillary until the forces tending to draw up the liquid are balanced by the downward force of gravity. The pressure at point 3 is lower than the pressure at point 2 by an amount equal to $\rho_L g h$ and the pressure at point 4 is lower than the pressure at point 1 by $\rho_V g h$. From the Laplace-Young equation that we developed in Chapter 2 (i.e. Equation 2.185), we can write at the meniscus interface that:

$$P_4 - P_3 = \frac{2\gamma}{r} \tag{4.74}$$

But $P_1 = P_4 + \rho_V g h$, $P_2 = P_3 + \rho_L g h$, and we also have that $P_1 = P_2$. Therefore, the above equation may be written as shown below recognizing that $\rho_L \gg \rho_V$

$$h = \frac{2\gamma \cos \theta}{R \rho_L g} \tag{4.75}$$

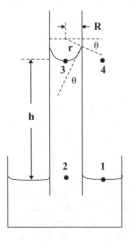

Figure 4.15   Capillary rise of a liquid in a small diameter tube of radius R.

Note that the above equation also provides a simple means to determine the surface tension of a liquid by measuring its capillary rise.

Example 4.7 Calculate the capillary rise for water in a tube with a diameter of 1 mm. Assume that $\cos \theta \approx 1$ and that the surface tension of water is 72 mN/m.

SOLUTION Using Equation 4.75 we can calculate the capillary rise:

$$h = \frac{2 \times 72 \times 10^{-3} \text{ N m}^{-1} \times 1}{0.0005 \text{ m} \times 1000 \text{ kg m}^{-3} \times 9.8 \text{ m sec}^{-2}} = 0.0294 \text{ m} = 29.4 \text{ mm} = 1.16 \text{ inches}$$

4.17.2 Dynamics of Capillary Action

Suppose it is desired to estimate the rate at which the fluid enters the capillary, i.e. find $Q(t)$ and $h(t)$. For the situation shown in Figure 4.15, the liquid is drawn into the capillary by forces arising from surface tension, and this force is retarded by the viscous and gravitational forces acting on the fluid. The goal is then to determine an expression for $Q(t)$ and $h(t)$. In a general sense, this is a very difficult problem, however, an approximate solution can be obtained if we assume that the flow in the capillary tube is laminar ($Re < 2000$) and that the velocity profile maintains the same parabolic shape as the liquid is drawn into the capillary tube. In this case, we can say that the volumetric flowrate of the fluid is then given by Equation 4.11 written as follows:

$$Q(t) = \frac{\pi R^4 (P_2 - P_3)}{8 \mu h(t)} \tag{4.76}$$

where P_2 is the pressure of the liquid as it enters the capillary and P_3 is the pressure of the liquid at the interface. Note that h is changing with time, hence Q also depends on time. This approximate solution should provide reasonable answers provided that $h(t) > R$, the radius of the capillary.

P_3 is also less than P_2 by an amount equal to $\rho_L gh$. We showed in the previous section that at the interface of the liquid within the capillary tube that:

$$P_4 - P_3 = \frac{2\gamma}{r} = \frac{2\gamma \cos \theta}{R} \tag{4.77}$$

Now P_4 represents the atmospheric pressure above the meniscus, and neglecting the small density of the gas phase surrounding the capillary tube, this is the same value as the pressure P_2 at the entrance to the capillary tube. Combining Equation 4.77 with Equation 4.76 we then obtain the following result for the volumetric flowrate of the liquid entering the capillary tube.

$$Q(t) = \frac{\pi R^3 \gamma \cos \theta}{4 \mu h(t)} \tag{4.78}$$

Note that this equation predicts an infinite flowrate at zero time, where $h(t) = 0$. This is an artifact of the assumptions made regarding the velocity profile having the same shape regardless of the time. However, once the fluid has penetrated a short distance into the capillary tube the above equation will become better at approximating the flowrate since the flow will start to approximate the parabolic velocity profile one expects at steady state.

The flowrate $Q(t)$ is also related to the capillary rise $h(t)$, or the amount of liquid in the capillary at any time, by the following equation:

$$Q(t) = \pi R^2 \frac{dh(t)}{dt} = \frac{\pi R^3 \gamma \cos \theta}{4 \mu h(t)} \tag{4.79}$$

Integration of the above equation with the initial condition that $h(t=0) = 0$ provides the following result for the capillary rise as a function of time.

$$h(t) = \sqrt{\frac{R \gamma \cos \theta \, t}{2 \mu}} \tag{4.80}$$

The above equation is known as the *Lucas-Washburn equation*.

Example 4.8 Estimate the time for water at 25°C to reach a height of 15 mm in a capillary tube that has a diameter of 1 mm. Assume that cos $\theta \approx 1$ and that the surface tension of water is 72 mN/m.

SOLUTION Using Equation 4.80 we find:

$$t = \frac{2 \times 0.001 \, \text{N sec m}^{-2} \times 0.015^2 \, \text{m}^2}{0.0005 \, \text{m} \times 72 \times 10^{-3} \, \text{N m}^{-1} \times 1} = 0.0125 \, \text{sec}$$

PROBLEMS

1. Derive Equations 4.10 and 4.21.

2. Derive Equation 4.23

3. Derive Equations 4.27 and 4.28

4. Using Equation 4.21 with a yield stress $\tau_y = 0.289$ dynes/cm^2 and $s = 0.229$ (dynes sec/cm^2)$^{1/2}$, show that this equation provides an excellent fit to the data given in Figure 4.8.

5. Estimate the shear rate at the wall of a capillary.

6. Obtain an expression for the radial dependence of the shear rate for a Newtonian fluid. What is the ratio of the average shear rate to the wall shear rate?

7. In Example 4.6, we obtained a simple estimate of the time needed to drain the IV bag. However, the IV bag has a length of 30 cm, and as the fluid drains from the bag, the potential energy (Z_1) of the remaining fluid that drives the flow will change with time. Therefore, to predict correctly the time to drain the bag, we also need to include how Z_1 changes with time. This can be obtained by combining the expression for the exit velocity (V_2) derived in Example 4.6 with a mass balance on the bag itself. Letting $M(t)$ denote the mass of IV fluid remaining in the bag at any time t, we can write the IV bag mass balance as follows:

$$\frac{dM(t)}{dt} = \rho S_{bag} \frac{dZ_1(t)}{dt} = -\rho \pi \frac{D_{tube}^2}{4} V_2$$

In this equation, S_{bag} is the cross-sectional area of the IV bag and D_{tube} is the diameter of the tube. Use this equation, and the results from Example 4.6, to obtain the time to just drain the bag. How does this time compare with the simple estimate of drain time obtained in Example 4.6?

8. Derive Equation 4.11. Start with Equation 4.10 and the assumption of a Newtonian fluid. Also show that Q may be obtained by integrating $v_z(r)$ in Equation 4.13 using Equation 4.8.

9. The cardiac output in a human is normally about 6 liters/min. Blood enters the right side of the heart at a pressure of about 0 mm Hg gauge and flows via the pulmonary arteries to the lungs at a mean pressure of 11 mm Hg gauge. Blood returns to the left side of the heart through the pulmonary veins at a mean pressure of 8 mm Hg gauge. The blood is then ejected from the heart through the aorta at a mean pressure of 90 mm Hg gauge. Use the Bernoulli equation to obtain an estimate of the total work performed by the heart. Carefully state any assumptions and express your answer in watts (recall that 1 watt = 1 Joule (J) / sec, where a J = Nm = kg m^2/sec^2).

10. Use the Bernoulli equation to describe the expected velocity and pressure changes upstream, within and downstream of an arterial stenosis. A stenosis is a partial blockage or narrowing of an artery by formation of plaque (atherosclerosis).

11. Blood is flowing through a bundle of hollow fiber tubes that are each 50 microns in diameter. There are 10,000 tubes in the bundle. The tube length is 12 cm and the pressure drop across each tube is found to be 250 mm Hg. The hematocrit of the blood is 0.40. The blood flowrate for these conditions is 80 μL/hr in each tube. Estimate the marginal zone thickness.

12. Blood enters a hollow fiber unit that is used as an artificial kidney. The unit consists of 10,000 hollow fibers arranged in a shell and tube configuration. Blood flows from an artery in the patient's arm through a tube and is uniformly distributed to the fibers via an arterial head space region at the entrance of the unit. The blood then leaves each fiber through the venous head space region of the unit and is returned to a vein in the patient's arm. Each hollow fiber has an inside diameter of 250 microns and a length of 15 cm. The fiber void volume within the shell of the unit is 50%. Assuming the maximum available pressure drop across the hollow fiber unit is 90 mm Hg, estimate the total flowrate of blood through the hollow fiber unit.

13. You are designing a hollow fiber unit. The flowrate of blood is assumed to be at a high enough shear rate that the blood behaves as a Newtonian fluid. The fiber diameter is 800 microns and their length is 30 cm. You want a flowrate of 8 ml/min for each fiber. What should the pressure drop (mm Hg, where 760 mm Hg = 101,325 Nm^{-2}) be across the fiber length to achieve this flowrate?

14. Another model proposed to describe the shear stress–shear rate of blood is that of the power law formulation wherein:

$$\tau = K\,\dot{\gamma}^{\,n} + \tau_y$$

The yield stress (τ_y) is considered for practical purposes to be negligible. Show that the power law equation above can be used to represent the data for the apparent viscosity of blood as a function of shear rate that is shown in Figure 4.4. What are the best values of K and n?

15. The following values were obtained for the apparent viscosity of blood ($H = 40\%$) in tubes of various diameters. Estimate the thickness of the marginal layer (δ) in microns from these data. The viscosity of the plasma is 1.09 cP. A little math hint, from the binomial series we know that:

$$(1 - \delta/R)^4 \approx 1 - 4\,(\delta/R) + O\,(\delta/R)^2 \text{ when } (\delta/R)^2 \ll 1$$

Then show that Equation 4.33 simplifies to the following equation:

$$\mu_{apparent} = \frac{\mu_c}{1 + 4\left(\dfrac{\delta}{R}\right)\left(\dfrac{\mu_c}{\mu_p} - 1\right)}$$

Make a plot of $1/\mu_{apparent}$ versus $1/R$ and see what happens in terms of finding μ_c and δ.

Tube Radius (R), microns	Apparent Viscosity, cP
20	1.68
40	2.25
60	2.49
100	2.88
300	3.00

16. A bioartificial liver has a plasma flow of 1000 ml/min through a hollow fiber unit that contains hepatocytes on the shell side. The hollow fiber unit contains 10,000 fibers. The fiber length is 75 cm and the inside diameter of the fibers is 300 microns. What is the pressure drop across each fiber in mm Hg? (1 atm = 101,325 N/m^2 = 760 mm Hg)

17. A design for a novel aortic cannula consists of a smooth thin-walled polyethylene tube 7 mm in diameter and 40 cm in length. For a blood flow rate of 5 liters/min through the cannula, estimate the pressure drop (mm Hg) over the length of the cannula. Carefully state and justify any assumptions that you make.

18. For a cell free layer of 3 microns and a hematocrit of 40%, calculate the apparent viscosity for blood flowing in a 100 micron diameter tube. Assume the plasma viscosity is 1.093 cP and the core viscosity is about 3.7 cP.

19. You are designing a small implantable microfluidic pump for the continuous delivery of a drug. The pump is a two compartment cylindrical chamber; one compartment contains the drug dissolved within a solvent, and the other compartment is the pump engine. These two compartments are separated by a movable piston that pushes on the drug compartment as the pump engine operates. At the other end of the drug compartment there is an exit tube through which the drug solution flows. The exit tube of the pump through which the drug leaves the pump has an internal diameter of 10 microns and a total length of 15 cm. What gauge pressure is needed (mm Hg) within the drug compartment to maintain a flowrate of the drug solution of 350 micrograms/day? You may assume that the drug solution has a density of 1 g/ml and a viscosity of 3 cP. *Also here are some useful unit conversions: 1 cP = 0.01 g/cm/sec; 1 N = 1 k gm/sec^2; 1 Pa = 1 N/m^2; 10^5 Pa = 750.06 mm Hg.*

20. You are part of a team developing an osmotic pump for the delivery of a drug. An osmotic pump has two compartments; one compartment, the osmotic engine, contains an osmotic agent that is retained by a membrane and imbibes water when placed within the body. This compartment also has a piston that expands against another compartment containing the drug solution as water is imbibed from the surroundings. The movement of the piston forces the drug solution out into the body. A question has been raised as to what would be the maximum pressure within the device if, after implantation, the exit tube that delivers the drug becomes blocked. Assume the interstitial fluid pressure is –3 mm Hg and its osmotic pressure is 8 mm Hg. If the concentration of the osmotic agent is 0.05 OsM, what would be the maximum hydrodynamic pressure within the device assuming the delivery tube becomes blocked?

21. A viscometer has been used to measure the viscosity of a fluid at 20°C. The data of the shear stress versus the shear rate when plotted on a log–log graph is linear. Is this fluid Newtonian? Explain your answer.

22. Blood flows through a bundle of hollow fibers at a total flowrate of 250 ml/min. There are a total of 7500 fibers and the diameter of a fiber is 50 microns and the length of a fiber is 45 cm. The viscosity ratio for blood under these conditions is estimated to be 0.70. What is the pressure drop across a fiber in mm Hg?

23. [From "Rheological properties of contraceptive gels," DH Owen, JJ Peters, DF Katz, *Contraception* 62 (2000) 321–326.] Commercially available spermicidal or contraceptive gels have been developed for the purpose of preventing sperm transport and thus blocking fertilization. Current interest has also led to the possible use of contraceptive gels as a means to reduce the spread of sexually transmitted diseases, such as AIDS, and to interest in developing formulations of these gels for both prophylaxis and contraception. The physical properties of these gels must be such that, when applied, they spread to coat the vaginal epithelia and stay in place long enough to provide contraception as well as adequate protection from disease causing agents such as bacteria and viruses. This is accomplished by gel formulations that deliver topically bioactive compounds, such as microbiocides, and also by the physical barrier to infection provided by the coating layer. The spreading and retention of intravaginal contraceptive formulations are fundamental to their efficacy, and these performance characteristics are governed, in part, by their rheological properties.

In vivo, these contraceptive gels will experience a wide range of shear rates as a result of movements of the vaginal epithelial surfaces, gravity, capillary flow, and sex. It is estimated that these shear rates may range from as low as 0.1/sec to as high as 100/sec during sex. The polymeric nature of these gels suggests that they will exhibit non-Newtonian rheological behavior.

A popular model for describing the rheological behavior of non-Newtonian gels is that of the two-parameter power law model. This model is shown by the equation below for flow in a cylindrical tube:

$$\tau_{rz} = -m\,\gamma^{n-1}\,dv_z\,/\,dr$$

In this equation, m and n are constants characterizing the fluid, and γ is the shear rate equal to $-dv_z\,/\,dr$.

From the above equation, what is the relationship for the apparent viscosity's ($\mu_{apparent}$) dependence on the shear rate of the fluid? Using the above equation, derive an expression for the axial velocity profile, $v_z(r)$, and the mass flowrate ($w = \rho\,Q$) of a fluid described by the power law model.

The table below shows data obtained from a rheometer for the commercially available gel called *Conceptrol*.

Apparent viscosity of Conceptrol vs. shear rate

Shear rate (1/sec)	Viscosity (Pa sec)
0.01	6000
0.05	2000
0.1	800
0.5	400
1	100
5	80
10	40
50	15
100	5
500	0.80
1000	0.2

Perform a regression analysis of the data in the above table and find the power law parameters m and n. Compare your model predictions to the data shown in the above table. Carefully state the units these parameters have.

24. Bush, Petros, and Barrett-Lennard (*J. Biomech.* 30, 967–969, 1997) studied the flow of urine in an anatomical model of the human female urethra, as shown below in Figure 4.16.

They obtained the following data shown in the table below from their experimental model.

(A)

(B)

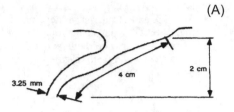

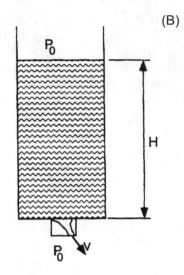

Figure 4.16 (A) A longitudinal profile of the model urethra adapted from a video cystogram. The lateral cross-section is taken to be circular. (B) The experimental apparatus showing the reservoir head, H, and the distal velocity, V. The atmospheric pressure is P_0. The apparatus permitted a maximum head of 65 cm H_2O to be applied to the urethra. (Bush, Petros, and Barrett-Lennard, *J. Biomech.* 30, 967–969, 1997, with permission).

Flow rate (cm³ sec⁻¹)	Pressure difference (ΔP, cm of H_2O)
5	2.0
8	4.5
10	8.0
13	14.0
15	21.0
17	28.0
20	38.0
23	52.0
25	60.0

The pressure difference represents the reservoir head (H) shown in Figure 4.16B, however the total pressure head driving the urine flow also includes the additional 2 cm (see Figure 4.16A) from the bottom of the reservoir to the exit of the urethra. Use the Bernoulli equation (Equation 4.67 in the book) along with the equations for friction to predict the expected pressure difference for each of the flowrates given in the above table. Show your results as a graphical comparison between the data and the model. Assume that the diameter of the urethra is 3.25 mm and that its length is 4 cm. Determine whether or not the flow is laminar or turbulent. If the model does not fit the data, what parameters in the model can you change to improve the fit, for example what is the effect of the urethral diameter?

25. Consider the steady laminar flow of a Newtonian fluid in the thin channel formed between two large parallel and horizontal plates of length L and width W. The plates are separated by a distance of $2H$ and $H \ll L$ and W. Show that the velocity profile $v_z(y)$ and the volumetric flowrate are given by the following expressions:

$$v_z(y) = \frac{(P_0 - P_L)H^2}{2\mu L}\left[1 - \left(\frac{y}{H}\right)^2\right]$$

$$Q = \frac{2}{3}W H^3 \frac{(P_0 - P_L)}{\mu L}$$

with P_0 and P_L the inlet and exit pressures.

26. Using the above result show that the following expression approximates the penetration of liquid, $L(t)$, by capillary action into a slit channel used in a diagnostic device.

$$L(t) = 2\left[\frac{H\gamma \cos \theta}{3\mu}\right]^{1/2} t^{1/2}$$

A diagnostic device makes use of a thin rectangular channel to draw in a sample of blood. Assuming the blood sample has a viscosity of 3 cP and that the plates forming the channel are separated by a distance of 1 mm, estimate the time for the sample of blood to travel a distance of 15 mm in the channel. Assume the blood has a surface tension of 0.06 N/m and that the contact angle is 70°.

27. The following table shows data for the measured velocity profile for the laminar boundary layer flow of a fluid across a flat plate (from H. Schlicting, *Boundary Layer Theory*, McGraw-Hill, New York, 7th ed. 1979). Make a plot of these data and compare with the approximate velocity profile given by Equation 4.63.

$y\sqrt{\dfrac{V}{vx}}$	$\dfrac{v_x(x,y)}{V}$
0.25	0.1
0.50	0.175
1.0	0.34
1.5	0.495
2.0	0.63
2.5	0.745
3.0	0.85
3.5	0.92
4.0	0.955
5.0	0.98

28. The total surface area of the two wings on a new passenger jet (both sides) is about 6000 ft^2 with a width of 150 ft. With a cruising speed of 567 mph, calculate the drag force. How much power is required? Assume the density of air at an altitude of 20,000 feet is 0.036 lb$_m$ ft^{-3} and that the viscosity of air is 0.017 mPa sec.

29. B.V. Zhmud, F. Tiberg, and K. Hallstensson obtained the following data for the capillary rise of dodecane in a 200 micron diameter capillary tube (*J. Colloid Interface Sci.* 228, 263–269 (2000)). Compare these results to those predicted by the Lucas-Washburn equation, i.e. Equation 4.80. The physical properties for dodecane are as follows: viscosity $= 1.7 \times 10^{-3}$ Pa sec, surface tension $= 2.5 \times 10^{-2}$ N m^{-1}, density $= 750$ kg/m^3, and the contact angle $= 17°$. What effect does the contact angle have on the ability of the Lucas-Washburn equation to predict the results that were obtained?

Time, seconds	h(t), mm
0.03	2
0.06	4.3
0.1	6.1
0.13	7.9
0.16	9
0.20	10
0.23	11
0.26	12
0.30	12.8
0.33	13.5
0.36	14.2
0.40	14.8
0.42	15.5
0.46	16.1
0.50	16.3

30. In the same paper by Zhmud *et al.,* the capillary rise for diethyl ether in a 1 mm diameter capillary tube was found to be 8.6 mm at equilibrium. How does this value compare with the value predicted by Equation 4.75? The physical properties for diethyl ether are as follows: surface tension = 1.67×10^{-2} N m^{-1}, density = 710 kg/m^3, contact angle = 26°.

SOLUTE TRANSPORT IN BIOLOGICAL SYSTEMS

5.1 DESCRIPTION OF SOLUTE TRANSPORT IN BIOLOGICAL SYSTEMS

In this chapter we will focus our discussion on the transport rate of solutes in biological systems. Solute transport occurs both through bulk fluid motion (convection) and by solute diffusion due to the presence of solute concentration gradients. In biological as well as synthetic membrane systems, the diffusion of a solute will also be affected by the presence of a variety of heterogeneous structures. For example, solutes will need to diffuse through porous structures such as the capillary wall or a polymeric membrane, around or through cells within the extravascular space, and through the interstitial fluid containing a variety of macromolecules.

5.2 CAPILLARY PROPERTIES

Our primary focus in this chapter concerns solute transport through the capillary wall as a representative porous semipermeable membrane. Therefore, the following development is also generally applicable to synthetic membranes that are used in a variety of biomedical device applications. To begin our discussion of capillary wall solute transport, we first need to define the physical properties of a typical capillary. These properties are summarized in Table 5.1. Note that capillaries are very small, having diameters of about 8–10 microns and a length that is less than 1 millimeter. The residence time of blood in a capillary is also only on the order of 1–2 seconds. Therefore each capillary can only supply nutrients and remove waste products from a very small volume of tissue that surrounds each capillary.

There are three types of capillaries. They are referred to as continuous, fenestrated,

Table 5.1 Capillary characteristics

Property	Value
Inside diameter (D_c)	10 microns
Length (L)	0.1 cm
Wall thickness (t_m)	0.5 microns
Average blood velocity (V)	0.05 cm/sec
Pore fraction	0.001
Wall pore diameter (d_p)	6–7 nm
Inlet pressure	30 mm Hg
Outlet pressure	10 mm Hg
Mean pressure (P_c)	17.3 mm Hg
Colloid osmotic pressure (π_P)	28 mm Hg
Interstitial fluid pressure (P_{if})	–3 mm Hg
Interstitial fluid colloid osmotic pressure (π_{if})	8 mm Hg

and discontinuous. The *continuous capillaries* are found in muscle, skin, lungs, fat, the nervous system, and connective tissue. The capillary lumen lies within a circumferential ring of several endothelial cells, as shown earlier in Figure 3.2. *Fenestrated capillaries* are much more permeable to water and small solutes in comparison to continuous capillaries. These capillaries are found in tissues that are involved in the exchange of fluid or solutes such as hormones. For example, within the kidney they are found in the glomerulus and the tubules. The endothelium is perforated by numerous small holes called *fenestrae*. The fenestrae are sometimes covered by a thin membrane that provides selectivity with regard to the size of solutes that are allowed to pass through. *Discontinuous capillaries* have large endothelial cell gaps that readily allow the passage of proteins and even red blood cells.

5.3 CAPILLARY FLOWRATES

Some of the blood plasma that enters the capillary will be carried or filtered across the capillary wall by the combined effect of the hydrodynamic and oncotic pressure differences that exist between the capillary and the surrounding interstitial fluid. This perfusion of plasma across the capillary wall is also known as *plasmapheresis*. We can determine the total flowrate of this fluid across the capillary wall using the relationships developed in Chapter 3. Recall from Equation 3.7 that the value of the hydraulic conductance, L_P, is given by the following equation:

$$L_P = \left(\frac{A_P}{S} \right) \frac{r^2}{8\mu t_m \tau} \tag{5.1}$$

In this equation, S represents the circumferential surface area of a given capillary, and A_P represents the total cross-sectional area of the pores of radius r in the capillary wall. The ratio, A_P/S, is referred to as the porosity of the capillary wall (or membrane) and is often given the symbol ε.

Example 5.1 Calculate the convective flowrate of plasma across the capillary wall.

SOLUTION Using the capillary properties provided in Table 5.1, and a plasma viscosity of 1.2 cP, the value of the hydraulic conductance, L_p, can be shown as in Example 3.1 to be equal to 0.61 cm^3/hr m^2 mm Hg (or 1.28×10^{-12} m^2 sec/kg). We can then calculate the filtration rate as follows using Equation 3.4.

$$Q_{filtration} = \frac{0.61 \text{cm}^3}{\text{hr m}^2 \text{ mm Hg}} \times \pi \times 10 \times 10^{-6} \text{ m} \times 0.001 \text{ m}$$

$$\times \left[(17.3 - -3) - (28 - 8) \right] \text{ mm Hg} = 5.75 \times 10^{-9} \text{cm}^3 \text{ hr}^{-1} = 5.75 \times 10^{-6} \mu\text{L hr}^{-1}$$

We can compare this value of the plasma filtration flowrate across the capillary wall, to the total flowrate of blood entering the capillary, i.e. $Q_{capillary} = \frac{\pi}{4} D_c^2 V$:

$$Q_{capillary} = \frac{\pi}{4} \left(10 \times 10^{-6} \text{m} \right)^2 \times \frac{0.05 \text{ cm}}{\text{sec}} \times \left(\frac{100 \text{ cm}}{\text{m}} \right)^2 \times \frac{3600 \text{ sec}}{\text{hr}}$$

$$= 1.41 \times 10^{-4} \text{cm}^3 \text{ hr}^{-1} = 0.14 \ \mu\text{L hr}^{-1}$$

We find in the above example that, in comparison to the volumetric flowrate of blood entering the capillary, the total convective flowrate of plasma filtrate across the capillary wall is negligible. We may, therefore, assume that $Q_{capillary}$ is constant along the length of the capillary.

5.4 SOLUTE DIFFUSION

In addition to the convective transport of the solute, for example as carried along by the bulk motion of the fluid, the solute can also diffuse down its own concentration gradient. From Equation 2.70 this is equivalent to stating that a solute diffuses from a region of high chemical potential to a region of lower chemical potential.

5.4.1 Fick's First and Second Laws

Consider the situation shown in Figure 5.1 The surface of a semi-infinite plate of length L contains a solute that maintains a constant concentration along the surface of the plate. At $t = 0$, this surface is contacted with a quiescent fluid or a solid material that initially does not contain the solute. As time progresses the solute diffuses from the surface of the plate into the quiescent fluid or solid material.

Shown in the figure is a thin shell of thickness Δy. The rate at which the solute enters and leaves this thin shell by diffusion at y and $y + \Delta y$ is proportional to the solute concentration gradient at these locations. The diffusion rate of this solute can therefore be described by *Fick's First Law* given by the equation shown below.

$$J_s = -DS \frac{dC}{dy} \tag{5.2}$$

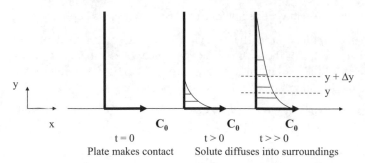

Figure 5.1   Solute concentration in the vicinity of a flat plate of constant surface concentration.

In the above equation, S is the surface area normal to the direction of solute diffusion and D is the *solute diffusivity*. The solute diffusivity generally depends on the size of the solute and the physical properties of the fluid or material in which the solute is diffusing. Since the solute is assumed to be diffusing through a homogeneous medium, this solute diffusivity is sometimes referred to as the bulk diffusivity. C represents the concentration of the solute and typical units are moles/cm^3 or moles/L. The solute diffusion rate is J_S and typically has units of moles/sec, and the diffusivity D has units of cm^2/sec or m^2/sec.

Now if we perform an unsteady solute balance across the shell of thickness Δy we can then write using Equation 1.8 as our guide that:

$$S\Delta y \frac{\partial C}{\partial t} = -DS \frac{\partial C}{\partial t}\bigg|_y - -DS \frac{\partial C}{\partial t}\bigg|_{y+\Delta y} \tag{5.3}$$

We can eliminate S in the above equation and after dividing by Δy, and taking the limit as $\Delta y \to 0$, we obtain the following result that is known as *Fick's Second Law*:

$$\frac{\partial C}{\partial t} = D \frac{\partial^2 C}{\partial y^2} \tag{5.4}$$

Solution of the above equation for the situation shown in Figure 5.1 requires that we state the following initial condition and the boundary conditions for the solute within the fluid or solid material region.

$$\begin{aligned}
\text{IC:} &\quad t = 0, \quad C = 0 \\
\text{BC1:} &\quad y = 0, \quad C = C_0 \\
\text{BC2:} &\quad y = \infty, \quad C = 0
\end{aligned} \tag{5.5}$$

Equations 5.4 and 5.5 are analogous to the problem we examined in Chapter 4 for the flat plate that is set in motion within a semi-infinite fluid that is initially at rest, i.e. Equations 4.40 and 4.41. In that case, we used Laplace transforms to obtain a solution for the unsteady velocity profile within the fluid, i.e. $v_x(y,t)$. We can, therefore, use that result here by simply recognizing that we can replace v_x with C, and V with C_0, in Equation 4.48. Note also in Equation 4.48, that the kinematic viscosity v is replaced by the diffusivity, D. Hence we obtain the following result for the concentration profile within the quiescent fluid or solid material at any location y and time t:

$$\frac{C(y,t)}{C_0} = erfc\left(\frac{y}{\sqrt{4Dt}}\right) = 1 - erf\left(\frac{y}{\sqrt{4Dt}}\right) \qquad (5.6)$$

We can also define a concentration boundary layer thickness, δ_C, as that distance where the concentration has decreased to 1% of the value at the surface of the plate. Recall from Chapter 4 that the complementary error function of $\left(\dfrac{y}{\sqrt{4Dt}}\right) = 2$ provides a value of C/C_0 that is very close to 0.01. Hence we can define the concentration boundary layer thickness, $\delta_C(y,t)$ as follows:

$$\delta_C(y,t) = 4\sqrt{Dt} \qquad (5.7)$$

The value of δ_C can also be interpreted as the distance from the plate to which the solute has penetrated into the fluid at time t.

Example 5.2 Calculate the concentration boundary layer thickness 1 second after the plate has made contact with the fluid. Assume the fluid is water and that the solute diffusivity is $D = 1 \times 10^{-5}$ cm^2 sec^{-1}.

SOLUTION Using Equation 5.7 we can calculate the thickness of the concentration boundary layer as shown below.

$$\delta_C = 4\sqrt{10^{-5}\,cm^2\,sec^{-1} \times 1\,sec} = 0.0126\ cm = 126\ microns$$

The flux of solute ($j_S = J_S/S$) diffusing at any location y is defined as the moles of solute per unit time per unit area normal to the direction of diffusion. The flux at the surface of the plate in Figure 5.1 would therefore be given by the following equation

$$j_S = -D\frac{\partial C}{\partial y}\bigg|_{y=0} \qquad (5.8)$$

Inserting Equation 5.6 into the above equation we find that

$$j_S = \frac{DC_0}{32\sqrt{\pi Dt}} \qquad (5.9)$$

The above equation shows that the solute flux at the surface of the plate is inversely proportional to the square root of the contact time of the plate with the fluid.

Solution of mass transfer problems is often facilitated by defining the mass transfer coefficient, k_m. The mass transfer coefficient can be thought of as the proportionality constant that relates the molar flux of the solute (j_S) to the overall concentration driving force, i.e. ($C_{High} - C_{Low}$). Hence we can write in general that:

$$j_S = -D \frac{\partial C}{\partial y}\bigg|_{y=0} = k_m (C_{High} - C_{Low}) \tag{5.10}$$

For the problem just considered $C_{High} = C_0$ and $C_{Low} = 0$. Comparing Equation 5.10 with Equation 5.9 we see that the mass transfer coefficient is then given by the equation below.

$$k_m = \frac{D}{32\sqrt{\pi D t}} = \frac{D}{8\sqrt{\pi} \, \delta_c(t)} \tag{5.11}$$

From the above equation we also see that the mass transfer coefficient is directly proportional to the solute diffusivity and inversely proportional to the concentration boundary layer thickness.

Example 5.3 Calculate the value of the mass transfer coefficient for the situation described in the previous example.

SOLUTION Using Equation 5.11 and the concentration boundary layer thickness of 0.0126 cm after 1 second of contact obtained in the previous example we can calculate the mass transfer coefficient as shown below.

$$k_m = \frac{10^{-5} \text{cm}^2 \text{sec}^{-1}}{8\sqrt{\pi} \times 0.0126 \, \text{cm}} = 5.6 \times 10^{-5} \frac{\text{cm}}{\text{sec}}$$

5.4.2 Mass Transfer in Laminar Boundary Layer Flow Over a Flat Plate

Figure 5.2 shows the laminar flow of a fluid across a semi-infinite flat plate of length L. The surface of the plate maintains a constant concentration of a solute that diffuses into the fluid. Unlike the situation shown in Figure 5.1, here the solute diffuses from the surface of the plate and is then swept away by the flowing fluid. Hence in this problem solute is transported away from the flat plate by a combination of diffusion and convection.

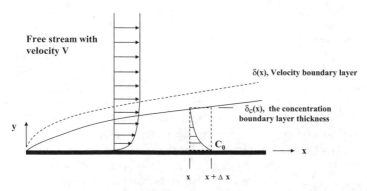

Figure 5.2   Laminar boundary layer flow in the vicinity of a flat plate of constant surface concentration.

In Section 4.15.2 we developed an approximate solution for the steady laminar flow of a fluid along a flat plate and determined the velocity profile within the boundary layer that is formed along the length of the plate. Here we will extend this solution to determine the concentration profile of the solute in the concentration boundary layer that is also formed along the surface of the flat plate.

Consider the shell volume shown in Figure 5.2 and located from x to $x + \Delta x$ and from $y = 0$ to $y = \delta_c(x)$, where δ_c is the concentration boundary layer thickness. We first perform a steady state solute balance on this shell volume which is given by:

$$W \int_0^{\delta_c} C v_x \, dy \bigg|_x - W \int_0^{\delta_c} C v_x \, dy \bigg|_{x+\Delta x} - C v_y \big|_\delta W \Delta x + j_S \big|_{y=0} W \Delta x = 0 \qquad (5.12)$$

The first two terms in the above equation represent the net rate at which solute is being added to the shell volume by flow of the fluid. The third term represents the loss of the solute from the top of the shell volume as a result of flow in the y-direction. The last term represents the rate at which solute is diffusing away from the surface of the flat plate. After eliminating W and dividing by Δx and taking the limit as $\Delta x \to 0$, we can write the above equation as shown below.

$$j_S \big|_{y=0} = \frac{d}{dx} \int_0^{\delta_c} C v_x \, dy + C v_y \big|_{y=\delta_c} \qquad (5.13)$$

Using Fick's First Law, we can insert Equation 5.8 and obtain:

$$-D \frac{dC}{dy} \bigg|_{y=0} = \frac{d}{dx} \int_0^{\delta_c} C v_x \, dy + C v_y \big|_{y=\delta_c} \qquad (5.14)$$

Since we only want to obtain an approximate solution, we can use our previous result for $v_x(y)$, i.e. Equation 4.59:

$$\frac{v_x(x,y)}{V} = \frac{3}{2} \left(\frac{y}{\delta(x)} \right) - \frac{1}{2} \left(\frac{y}{\delta(x)} \right)^3 \qquad (4.59)$$

and in a similar fashion propose that the concentration profile in the concentration boundary layer is approximately described by the following cubic equation:

$$C(y) = a_c + b_c y + c_c y^2 + d_c y^3 \qquad (5.15)$$

The concentration profile described by the above equation also has to satisfy the following boundary conditions.

$$\begin{aligned} &\text{BC1:} \quad y = 0, \ C = C_0 \\ &\text{BC2:} \quad y = \delta_c(x), \ C = 0 \\ &\text{BC3:} \quad y = \delta_c(x), \ \frac{\partial C}{\partial y} = 0 \\ &\text{BC4:} \quad y = 0, \ \frac{\partial^2 C}{\partial y^2} = 0 \end{aligned} \qquad (5.16)$$

The first boundary condition expresses the fact that the concentration of the solute is constant along the surface of the flat plate. The second and third boundary conditions state that beyond the concentration boundary layer the solute is not present. This allows us to eliminate the last term in Equation 5.14 giving:

$$-D \frac{dC}{dy}\bigg|_{y=0} = \frac{d}{dx} \int_0^{\delta_c} C\, v_x\, dy \tag{5.17}$$

The fourth boundary condition is a result of the solute flux being constant along the surface of the flat plate.

After imposing the above boundary conditions on Equation 5.15, the following expression is obtained for the concentration profile within the concentration boundary layer in terms of the concentration boundary layer thickness, which at this time, still needs to be determined.

$$\frac{C(x,y)}{C_0} = 1 - \frac{3}{2}\left(\frac{y}{\delta_c(x)}\right) + \frac{1}{2}\left(\frac{y}{\delta_c(x)}\right)^3 \tag{5.18}$$

We now can substitute Equations 4.59 and 5.18 into Equation 5.17. We also assume that the ratio of the boundary layer thicknesses, i.e. $\Delta = \delta_c(x) / \delta(x)$, is a constant. The algebra is a bit overwhelming but one can obtain the following differential equation for the thickness of the concentration boundary layer:

$$(0.15\Delta - 0.0107\Delta^3)\frac{d\,\delta_c}{dx} = \frac{3}{2}\frac{D}{V\,\delta_c} \tag{5.19}$$

with the boundary condition that at $x = 0$, $\delta_C = 0$. The solution of the above equation is then given by

$$\delta_c(x) = \sqrt{\frac{3}{(0.15\Delta - 0.0107\Delta^3)}\frac{(D\,x)}{V}} \tag{5.20}$$

Recall from Equation 4.61 that the boundary layer thickness δ is given by:

$$\delta(x) = 4.64\sqrt{\frac{v\,x}{V}} = 4.64\sqrt{\frac{\mu\,x}{\rho\,V}} \tag{4.61}$$

Dividing Equation 5.20 by Equation 4.61 and simplifying results in the following equation for $\Delta = \delta_c(x)/\delta(x)$.

$$1.0765\Delta^3 - 0.0768\Delta^5 = \frac{\rho\,D}{\mu} = \frac{1}{Sc} \tag{5.21}$$

where $Sc = \mu/\rho\,D$ is a dimensionless number known as the *Schmidt number*. For solutes diffusing through liquids the Schmidt number is generally much greater than unity hence from the above equation we have that $\Delta < 1$ or $\delta > \delta_c$. For solutes diffusing through gases the Schmidt number is on the order of unity, and Δ is approximately equal to unity or $\delta \approx \delta_C$. For solutes diffusing through liquid metals, the Schmidt number is much less than unity and then $\Delta > 1$ and $\delta_C > \delta$.

In general, if one knows the value of the Schmidt number for the mass transfer problem being considered, then Equation 5.21 can be solved for the value of Δ. The boundary layer thickness would then be given by Equation 4.61 and the concentration boundary layer thickness would be equal to $\Delta \, \delta(x)$.

For most mass transfer problems of interest to biomedical engineers the Schmidt number is often much larger than unity. This makes the right-hand side of Equation 5.21 much smaller than unity. The only way this can happen is if Δ is also much smaller than unity. This means that the concentration boundary layer lies well within the boundary layer for the velocity. Hence for $\Delta << 1$, Equation 5.21 simplifies to the following result.

$$\Delta = \frac{\delta_c(x)}{\delta(x)} = Sc^{-1/3} \tag{5.22}$$

Combining this result with Equation 4.61 results in the following equation for the thickness of the concentration boundary layer.

$$\frac{\delta_c(x)}{x} = 4.64 \left(\frac{\mu}{\rho V x} \right)^{1/2} \left(\frac{\rho D}{\mu} \right)^{1/3} = 4.64 \left(\frac{v}{V x} \right)^{1/2} \left(\frac{D}{v} \right)^{1/3} = 4.64 \left(\frac{1}{Re_x} \right)^{1/2} \left(\frac{1}{Sc} \right)^{1/3} \tag{5.23}$$

The above equation has an extremely important result. Note that the concentration boundary layer thickness depends on two dimensionless parameters that describe the nature of the flow (Re_x) and the physical properties of the solute and the fluid (Sc).

We can also calculate the local value of the mass transfer coefficient defined earlier by Equation 5.10. Here $C_{High} = C_0$ and $C_{Low} = 0$. Using Equation 5.8 and 5.10 we find that the mass transfer coefficient is given by the following expression.

$$k_m = -\frac{D}{\left(C_{High} - C_{Low} \right)} \frac{\partial C}{\partial y} \bigg|_{y=0} = -\frac{D}{C_0} \frac{\partial C}{\partial y} \bigg|_{y=0} \tag{5.24}$$

After substituting Equation 5.18 into the above expression, and replacing $\delta_c(x)$ with Equation 5.23, we obtain the following expression for the local mass transfer coefficient:

$$\frac{k_m x}{D} = Sh_x = 0.323 \left(\frac{\rho V x}{\mu} \right)^{1/2} \left(\frac{\mu}{\rho D} \right)^{1/3} = 0.323 Re_x^{1/2} Sc^{1/3} \tag{5.25}$$

where Sh_x is another dimensionless number known as the local *Sherwood number* at location x. In the above expression, k_m is the local value of the mass transfer coefficient since as one progresses down the length of the plate, the concentration boundary layer thickness increases and, by the above equation, we see that k_m decreases in inverse proportion to $x^{1/2}$. For a plate of length L, the average mass transfer coefficient is given by:

$$\overline{k}_m = \frac{1}{L} \int k_m(x) dx \tag{5.26}$$

Substituting Equation 5.25 into Equation 5.26 gives the following result which is the length averaged mass transfer coefficient.

$$\frac{\bar{k}_m L}{D} = Sh = 0.626 \left(\frac{\rho V L}{\mu} \right)^{1/2} \left(\frac{\mu}{\rho D} \right)^{1/3} = 0.626 Re^{1/2} Sc^{1/3} \qquad (5.27)$$

Equations 5.26 and 5.27 describe the mass transfer of a solute from a flat plate for laminar boundary layer flow provided the local value of the Reynolds is less than 300,000. We also see that at a given location, the average mass transfer coefficient is twice the local value.

Example 5.4 Blood is flowing across the flat surface of a polymeric material coated with an anticoagulant drug. Equilibrium studies of the drug coated polymer show that the drug has a low solubility of 100 mg/L of blood, hence the concentration of the drug at the surface of the plate is, for practical purposes, the same as this equilibrium value. Find the following:

a.   the distance at which the laminar boundary layer for flow over the surface of the polymeric material ends;
b.   the thickness of the velocity and concentration boundary layers at the end of the polymeric material;
c.   the local mass transfer coefficient at the end of the polymeric material and the average mass transfer coefficient for the drug;
d.   the average elution flux of the drug from the polymeric material assuming the concentration of the drug in the blood is much smaller than at the surface.

Assume that the blood traveling across the polymeric material has a free stream velocity of 40 cm/sec. The polymeric material has a length in the direction of the flow of 30 cm. The diffusivity of the drug in blood is 4.3×10^{-6} cm^2/sec.

SOLUTION For part (a) we use the fact that the transition to turbulence begins at a $Re_x = \dfrac{\rho V x}{\mu} = 300,000.$ Hence, we can solve this equation for the value of x at this transition as follows:

$$x = \frac{300,000 \times 3\,\text{cP} \times 0.01\,\text{g cm}^{-1}\,\text{sec}^{-1}\,\text{cP}^{-1}}{1.056\,\text{g cm}^{-3} \times 40\,\text{cm sec}^{-1}} = 213\,\text{cm}$$

Note that this distance for the transition to turbulence is much greater than the actual length of the flat polymeric material, hence the flow is laminar over the region of interest. For part (b) we use Equations 4.61 and 5.22 to calculate the boundary layer thicknesses at the end of the polymeric material as shown below. The thickness of the velocity boundary layer is therefore:

$$\delta = 4.64 \sqrt{\frac{3\,\text{cP} \times 0.01\,\text{g cm}^{-1}\,\text{sec}^{-1}\,\text{cP}^{-1} \times 30\,\text{cm}}{1.056\,\text{g cm}^{-3} \times 40\,\text{cm sec}^{-1}}} = 0.67\,\text{cm}$$

Next we calculate the Schmidt number.

$$Sc = \frac{\mu}{\rho D} \frac{3cP \times 0.01 \, g \, cm^{-1} \, sec^{-1} \times cP^{-1}}{1.056 \, g \, cm^{-3} \times 4.3 \times 10^{-6} \, cm^2 \, sec^{-1}} = 6607$$

We see that the $Sc \gg 1$ and the concentration boundary layer will lie well within the velocity boundary layer calculated above as shown below.

$$\delta_C = \delta \, Sc^{-1/3} = 0.67 \, cm \times 6607^{-1/3} = 0.036 \, cm$$

For part (c) we use Equations 5.25 and 5.27 to calculate the local and average mass transfer coefficients respectively. The Reynolds number at the end of the polymeric surface is:

$$\left(\frac{\rho V x}{\mu}\right) = \frac{1.056 \, g \, cm^{-3} \times 40 \, cm \, sec^{-1} \times 30 \, cm}{3cP \times 0.01 \, g \, cm^{-1} \, sec^{-1}} = 42240$$

and

$$k_m = 0.323 \frac{4.3 \times 10^{-6} \, cm^2 \, sec^{-1}}{30 \, cm} \times 42240^{1/2} \times 6607^{1/3} = 1.78 \times 10^{-4} \, cm \, sec^{-1}$$

Since the average mass transfer coefficient at a given location is twice the local value, we then have that $\bar{k}_m = 3.57 \times 10^{-4}$ cm sec^{-1}. To calculate the average elution flux of the drug from the polymeric surface we use Equation 5.10 which can be rearranged as shown below.

$$\bar{j}_s = \frac{1}{L}\int_0^L j_s \, dx = \bar{k}_m C_0 = 3.57 \times 10^{-4} \, cm \, sec^{-1} \times 100 \, mg \, L^{-1} \times 1 \, L \times 10^{-3} \, cm^3 \, L^{-1}$$

$$\bar{j}_s = 3.57 \times 10^{-5} \, mg \, cm^{-2} \, sec^{-1}$$

5.4.3 Mass Transfer from the Walls of a Tube Containing a Fluid in Laminar Flow

Next we consider the situation shown in Figure 5.3. A fluid in laminar flow is flowing through a tube and comes into contact with a section of the tube that maintains a constant concentration of a solute that diffuses into the fluid. Much like we did with

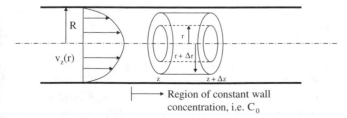

Figure 5.3   Laminar flow of a fluid in a tube with constant concentration of the solute along the surface.

laminar flow over a flat plate, we want to determine the elution rate of the solute from the surface of the tube in terms of dimensionless quantities such as the Reynolds number and the Schmidt number. In order to obtain an analytical solution to this rather complex problem, we will make some simplifying assumptions. First, we will only be interested in obtaining what is known as a short contact time solution. This means that the solute does not penetrate very far from the surface of the tube. Hence the change in concentration of the solute occurs only near the surface of the tube wall.

From the previous chapter, we know that the velocity profile for laminar flow in a tube is given by the following expression, i.e. Equation 4.13:

$$v_z(r) = \frac{(P_0 - P_L)R^2}{4\mu L}\left[1 - \left(\frac{r}{R}\right)^2\right]$$

(4.13)

Next, we perform a steady state solute balance on the cylindrical shell volume shown in Figure 5.3.

$$2\pi r \Delta r \left(v_z C\big|_z - v_z C\big|_{z+\Delta z}\right) + 2\pi r \Delta r \left(D\frac{\partial C}{\partial z}\bigg|_{z+\Delta z} - D\frac{\partial C}{\partial z}\bigg|_z\right) +$$
$$2\pi r \Delta z \left(D\frac{\partial C}{\partial r}\bigg|_{r+\Delta r} - D\frac{\partial C}{\partial r}\bigg|_r\right) = 0$$

(5.28)

The first parenthetical term in the above equation represents the net rate at which solute is added to the shell volume by bulk flow of the fluid. The second term in parentheses represents the net rate at which solute is added to the shell volume by diffusion of solute in the z direction. The final term in the above equation represents the net rate at which solute enters the shell volume by diffusion in the r direction. Now if we divide through by $2\pi r \Delta r\,\Delta z$ and then take the limit as Δr and $\Delta z \to 0$, we then obtain the following partial differential equation that describes the transport of solute at any value of r and z within the tube.

$$v_z\frac{\partial C}{\partial z} = D\left[\frac{1}{r}\frac{\partial}{\partial r}\left(\frac{\partial C}{\partial r}\right) + \frac{\partial^2 C}{\partial z^2}\right]$$

(5.29)

The term on the left represents the solute transport in the flowing fluid by convection, or bulk flow, of the fluid. The right-hand side represents the transport of solute by diffusion. Generally, more solute is transported in the z direction by convection than is transported by axial diffusion, so at any given location in the fluid we expect that $v_z\dfrac{\partial C}{\partial z} \gg D\dfrac{\partial^2 C}{\partial z^2}$.

Next we let $s = R - r$ represent the distance from the wall of the tube. Since we are only interested in a solution near the wall, s will be small relative to R, and we can neglect the effect of the curvature of the tube. With these assumptions the above equation becomes:

$$v_z \frac{\partial C}{\partial z} = D \frac{\partial^2 C}{\partial s^2} \qquad (5.30)$$

Next we recognize that near the tube wall the velocity profile is nearly linear and $v_z \approx \left(\frac{\Delta P\, R}{2\mu\, L} \right) s$. The above equation may then be written as:

$$\left(\frac{\Delta P\, R}{2\mu\, L} \right) s \frac{\partial C}{\partial z} = D \frac{\partial^2 C}{\partial s^2} \qquad (5.31)$$

and we can state the following boundary conditions:

$$\begin{aligned} &\text{BC1:}\quad z = 0,\ \text{and all } s,\ C = 0 \\ &\text{BC2:}\quad s = 0,\ \text{and } z > 0,\ C = C_0 \\ &\text{BC3:}\quad s = \infty,\ \text{and } z > 0,\ C = 0 \end{aligned} \qquad (5.32)$$

The last boundary condition results from the short contact time assumption meaning that the solute does not penetrate very far into the fluid.

Solution of Equations 5.31 and 5.32 is facilitated by rewriting these equations in terms of dimensionless variables defined as follows. We let $\theta \equiv \dfrac{C}{C_0}$, $\varepsilon \equiv \dfrac{z}{R}$, and $\sigma \equiv \dfrac{s}{R}$, where θ is defined as a dimensionless concentration, ε is a dimensionless axial position, and σ is a dimensionless distance from the tube wall. If these dimensionless variables replace C, z, and s in Equations 5.31 and 5.32, the following result is obtained.

$$N\sigma \frac{\partial \theta}{\partial \varepsilon} = \frac{\partial^2 \theta}{\partial \sigma^2} \qquad (5.33)$$

where $N \equiv \dfrac{R^3\, \Delta P}{2\mu\, D\, L}$ and the boundary conditions become

$$\begin{aligned} &\text{BC1:}\quad \varepsilon = 0,\ \text{and all } \sigma,\ \theta = 0 \\ &\text{BC2:}\quad \sigma = 0,\ \text{and } \varepsilon > 0,\ \theta = 1 \\ &\text{BC3:}\quad \sigma = \infty,\ \text{and } \varepsilon > 0,\ \theta = 0 \end{aligned} \qquad (5.34)$$

Now to solve Equations 5.33 and 5.34 the *similarity transform technique* or *combination of variables* approach can be used. In this approach one defines a new independent variable η that is a suitable combination of ε and σ that converts Equation 5.33 from a partial differential equation into an ordinary differential equation that only depends on η. If $\eta \equiv \left(\dfrac{N\sigma^3}{9\varepsilon} \right)^{1/3}$, then it can be shown, with a little bit of effort, that Equations 5.33 and 5.34 become:

$$\frac{\partial^2 \theta}{\partial \eta^2} + 3\eta^2 \frac{\partial^2 \theta}{\partial \eta} = 0 \qquad (5.35)$$

and the boundary conditions become:

$$BC1': \eta = \infty, \; \theta = 0$$
$$BC2': \eta = 0, \; \theta = 1 \tag{5.36}$$

Integration of Equations 5.35 and 5.36 then gives the following result for the solute concentration profile as a function of $\eta \equiv \left(\dfrac{N\sigma^3}{9\varepsilon} \right)^{1/3}$

$$C(\eta) = C_0 \frac{\displaystyle\int_\eta^\infty e^{-\eta^3} \, d\eta}{\Gamma\left(\dfrac{4}{3}\right)} \tag{5.37}$$

Obtaining the above solution involves the use of the *gamma function* which is defined as $\Gamma(j) \equiv \displaystyle\int_0^\infty e^{-x} x^{j-1} \, dx$ along with the gamma function property that $j\,\Gamma(j) = \Gamma(j+1)$.

We can also use the above result to define the concentration boundary layer thickness at any position z as that distance away from the tube wall where $C = 0.01\,C_0$. Hence, letting $C/C_0 = 0.01$, we can solve Equation 5.37 for η and obtain:

$$\delta_C = 2.89 \left(\frac{R^2 z}{N} \right)^{1/3} \tag{5.38}$$

The solute flux from the tube surface at any value of z is given by:

$$j_S = -D \left. \frac{\partial C}{\partial r} \right|_{r=R} = -\frac{D\,C_0}{R} \left. \frac{\partial \theta}{\partial \sigma} \right|_{\sigma=0} = \frac{D\,C_0}{R\,\Gamma\left(\frac{4}{3}\right)} \left(\frac{N}{9\varepsilon} \right)^{1/3} \tag{5.39}$$

and the local value of the mass transfer coefficient may then be written as:

$$k_m = \left. j_S \middle/ C_0 \right. = \frac{D}{\Gamma\left(\frac{4}{3}\right)} \left(\frac{4\,v_{AVG}}{9\,D\,Rz} \right)^{1/3} \tag{5.40}$$

where $v_{AVG} \equiv \dfrac{\Delta P\,R^2}{8\mu\,L}$ is the average velocity of the fluid in the tube which is also equal to the volumetric flowrate of the fluid divided by the tube cross-sectional area, i.e. $v_{AVG} = \dfrac{Q}{\pi R^2}$.

Using Equation 5.40, the total amount of solute eluted from the surface over a length of tube equal to L is given by:

$$J_S = \int_0^L j_S \big|_{r=R} \, 2\pi R \, dz = \frac{2\pi R^{2/3}\,D^{2/3}\,C_0}{\Gamma\left(\frac{4}{3}\right)} \left(\frac{4}{9} v_{AVG} \right)^{1/3} \int_0^L z^{-1/3} \, dz \tag{5.41}$$

Integration of the above equation provides the following result:

$$J_S = \frac{3\pi R \, D \, C_0}{\Gamma\left(\frac{4}{3}\right)}\left(\frac{2}{9}\right)^{1/3} Re^{1/3} Sc^{1/3} \left(\frac{L}{R}\right)^{2/3} \tag{5.42}$$

where the Reynolds number in the above equation is defined as $Re = \frac{\rho \, D_{tube} V_{AVG}}{\mu}$. J_S is also equal to the average mass transfer coefficient multiplied by the concentration driving force, i.e. $(C_0 - 0)$. Hence the average mass transfer coefficient is given by the following equation:

$$Sh = \frac{\bar{k}_m D_{tube}}{D} = \frac{3}{\Gamma\left(\frac{4}{3}\right)}\left(\frac{2}{9}\right)^{1/3} Re^{1/3} Sc^{1/3} \left(\frac{R}{L}\right)^{2/3} \tag{5.43}$$

As in the case for laminar flow over a flat plate we see that the mass transfer coefficient once again is a function of the Reynolds number and the Schmidt number.

5.4.4 Mass Transfer Coefficient Correlations

The above mass transfer equations that we developed for specific situations also provide valuable insight as to how to develop correlations for mass transfer coefficients in more complicated flows and geometries. In many cases we find that mass transfer coefficient data can be correlated with general functions that are very similar in form to these equations. For example, for problems involving forced convection or the flow of a fluid, one can propose a functional dependence on the Re and Sc that is very similar to Equations 5.25 and 5.43, i.e. the $Sh = A \, Re^m \, Sc^n$. The values of A, m, and n can then be found by fitting this equation to experimental data. Table 5.2 provides a summary (Cussler 1984) of some useful mass transfer coefficient correlations for a variety of flow situations.

For the laminar flow of a fluid (i.e. not turbulent) within a cylindrical tube, $\bar{k}_m$ may be estimated from the following equation (Thomas 1992).

$$Sh = 3.66 + \frac{0.104 \dfrac{Re \, Sc}{L/D_{tube}}}{1 + 0.016\left(\dfrac{Re \, Sc}{L/D_{tube}}\right)^{0.8}} \tag{5.44}$$

where the $Sh \; (= \dfrac{\bar{k}_m D_{tube}}{D})$ number in the above equation is the value averaged over the length L.

The above equation is for the case where both the velocity and the concentration profiles are not fully developed. This means that the velocity and concentration profiles are changing with axial position. When the flow is fully developed in terms of the

Table 5.2 A selection of useful mass transfer coefficient correlations (from Cussler 1984)

Physical situation	Correlation
Laminar flow over a flat plate	$\dfrac{k_m L}{D} = Sh = 0.626\left(\dfrac{\rho V L}{\mu}\right)^{1/2}\left(\dfrac{\mu}{\rho D}\right)^{1/3} = 0.626\,Re^{1/2}\,Sc^{1/3}$
Laminar flow in a circular tube, short contact time solution	$Sh = \dfrac{k_m D_{tube}}{D} = \dfrac{3}{\Gamma\left(\dfrac{4}{3}\right)}\left(\dfrac{2}{9}\right)^{1/3} Re^{1/3}Sc^{1/3}\left(\dfrac{R}{L}\right)^{2/3}$
Laminar flow in a circular tube, undeveloped flow and concentration profiles	$Sh = 3.66 + \dfrac{0.104\dfrac{ReSc}{L/D_{tube}}}{1 + 0.016\left(\dfrac{ReSc}{L/D_{tube}}\right)^{0.8}}$
Laminar flow in a circular tube, fully developed flow and concentration profiles	$Sh = 3.66$
Turbulent flow within a horizontal slit $[d = \left(\dfrac{2}{\pi}\right)(\text{slit width})]$	$\dfrac{k_m d}{D} = 0.026\left(\dfrac{d\,v_{AVG}}{v}\right)^{0.8}\left(\dfrac{v}{D}\right)^{1/3}$ with $\bar{k}_m$ the average mass transfer coefficient
Turbulent flow through a circular tube (D_{tube} is the tube diameter)	$\dfrac{k_m d}{D} = 0.026\left(\dfrac{d\,v_{AVG}}{v}\right)^{0.8}\left(\dfrac{v}{D}\right)^{1/3}$ with $\bar{k}_m$ the average mass transfer coefficient
Laminar flow in a circular tube (D_{tube} is the tube diameter	$\dfrac{k_m D_{tube}}{D} = 1.86\left(ReSc\dfrac{D_{tube}}{L}\right)^{1/3}$ with $\bar{k}_m$ the average mass transfer coefficient
Spinning disc (d is the disc diameter and ω is the disc rotation rate in radians/sec)	$\dfrac{k_m d}{D} = 0.62\left(\dfrac{d^2\omega}{v}\right)^{1/2}\left(\dfrac{v}{D}\right)^{1/3}$ with $\bar{k}_m$ the average mass transfer coefficient
Packed beds (d is the particle diameter and v_0 is the superficial velocity defined as the volumetric flowrate divided by the unpacked tube cross section)	$\dfrac{k_m}{v_0} = 1.17\left(\dfrac{d\,v_0}{v}\right)^{-0.42}\left(\dfrac{v}{D}\right)^{-0.67}$
Falling film (z is position along the length of the film and v_{AVG} is the average film velocity)	$\dfrac{k_m z}{D} = 0.69\left(\dfrac{z\,v_{AVG}}{D}\right)^{1/2}$ where $\bar{k}_m$ is the local mass transfer coefficient

velocity and concentration profiles, the Sh number attains its asymptotic value of 3.66. For the hydrodynamic flow to be fully developed at any axial position, $z/D_{tube} > 0.05\,Re$, and for the concentration field to be fully developed, we have that $z/D_{tube} > 0.05\,ReSc$.

The Re for flow in capillaries is on the order of 0.001 to 0.005, and the Sc is on the order of 1000. Since L/D_{tube} for a capillary is on the order of 100, we see that the criteria for fully developed hydrodynamic flow and concentration fields in a capillary are satisfied soon after the blood enters the capillary. Therefore, for blood flow in a capillary we may assume that $Sh = 3.66$.

The length averaged mass transfer coefficient for undeveloped laminar flow in cylindrical tubes and channels can also be estimated from the following equation (Thomas 1992).

$$Sh = 1.86 \left(\frac{Re\ Sc}{L/D_{tube}} \right)^{1/3} \quad \text{for} \quad \frac{L/D_{tube}}{Re\ Sc} \leq 0.01 \tag{5.45}$$

For non-cylindrical flow channels, the diameter (D_{tube}) in these mass transfer coefficient equations is equal to the *equivalent diameter*, D_H, which is defined by the following equation.

$$D_H = 4 \times \frac{\text{channel cross sectional area}}{\text{wetted perimeter}} \tag{5.46}$$

5.4.5 Determining the Diffusivity

Figure 5.4 presents diffusivity data (Renkin and Curry 1979) for a variety of solutes in dilute aqueous solutions at 37°C as a function of solute molecular weight. The solid line through the data is the result of a linear least squares regression. The following

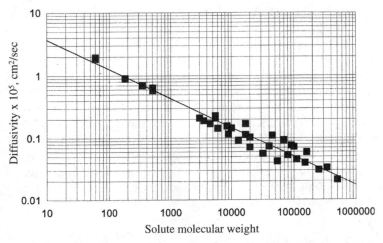

Figure 5.4   Solute diffusivity in water at 37°C (based on data from Renkin and Curry 1979).

empirical equation based on the data in Figure 5.4 provides a relationship between the solute diffusivity and its molecular weight.

$$D = 1.013 \times 10^{-4} (MW)^{-0.46}, \ cm^2 \ sec^{-1} \tag{5.47}$$

Cussler (1984) as well as Tyn and Gusek (1990) provide additional data on the diffusivity for a variety of solutes.

The diffusivity of solutes in dilute solutions can also be estimated from the Stokes–Einstein equation (Bird $et\ al.$ 2002; Cussler 1984):

$$D = \frac{RT}{6\pi\mu\, a\, N_A} \tag{5.48}$$

In this equation, R is the ideal gas constant (8.314 J gmole^{-1} K^{-1}), T is the temperature in K, (a) is the solute radius, N_A is Avogadro's number (6.023 × 10^{23} molecules gmole^{-1}), and μ is the solution viscosity. The viscosity of water at 37°C is 0.76 cP and that of plasma is 1.2 cP (1cP = 1 centipoise = 0.01 g cm^{-1} sec^{-1}).

If the diffusivity for a solute is known, then Equation 5.48 may be used to obtain an estimate of the molecular radius (a) of the solute. If the diffusivity and size of the solute are not known, the solute size can first be estimated from the following equation. This equation assumes that the solute of molecular weight (MW) is a sphere with a density ($\rho \approx 1$ g cm^{-3}) equal to that of the solute in the solid phase.

$$a = \left(\frac{3\, MW}{4\pi\rho\, N_A} \right)^{1/3} \tag{5.49}$$

5.5 SOLUTE TRANSPORT BY CAPILLARY FILTRATION

The perfusion of plasma across the capillary wall will carry with it a variety of solutes that are present in the blood. Because of the size of the pores in the capillary wall, this filtration of the plasma by the capillary wall will also tend to separate the solutes on the basis of their size. For example, smaller solutes like ions, glucose, and amino acids will readily pass through the capillary pores. However, larger proteins will be inhibited to varying degrees in their passage across the capillary wall by the size of the pores. The solute selectivity during this plasma filtration is described by the sieving coefficient (S_a), which is defined as the ratio of the solute concentration in the filtrate (C_f) to the solute concentration at the surface of the capillary on the blood side (C_{bs}). Theoretical expressions based on the motion of a spherical solute moving through a cylindrical pore have been developed to estimate the value of the sieving coefficient (Anderson and Quinn 1974; Deen 1987). The development of these expressions neglects secondary effects between the solute and the membrane pore, for example electrostatic, hydrophobic, and van der Waals interactions. The following expression can be used to estimate the sieving coefficient.

$$S_a = \frac{C_f}{C_{bs}} = (1-\lambda)^2 [2 - (1-\lambda)^2][1 - \frac{2}{3}\lambda^2 - 0.163\lambda^3] \tag{5.50}$$

In this equation, λ is defined as the ratio of the solute radius (a) to the capillary pore radius (r).

Under some conditions, the high filtration rate of the plasma leads to the formation of a layer of retained proteins at a higher concentration near the non-filtrate side of the membrane. This effect, shown in Figure 5.5, is also called *concentration polarization*, and the layer of retained proteins will affect the convective transport of the solutes. Concentration polarization is particularly important at the higher filtration rates used in commercial membrane-based plasmapheresis systems used for separating plasma from the cellular components of blood as well as in membrane systems used to purify protein solutions.

The sieving coefficient for the case of concentration polarization (S_0) can be developed as follows. The flux of solute at any position y in the concentration polarization region shown in Figure 5.5 has to equal the rate at which the solute is diffusing back towards the bulk solution plus the amount of solute that is carried across the membrane by the filtration flow. This can be expressed by the following equation:

$$q\, C = -D\frac{dC}{dy} + q\, C_f \tag{5.51}$$

where q is defined as the filtration flux (Q/S), and Q is given by Equation 3.4. S is the surface area normal to the direction of the filtration flow. This equation can then be rearranged and integrated across the concentration polarization region as shown below.

$$q\int_0^{\delta c} dy = -D\int_{C_{bs}}^{C_b} \frac{dC}{C - C_f} \tag{5.52}$$

After performing the integration we obtain:

$$q = -\frac{D}{\delta_C}\ln\left(\frac{C_b - C_f}{C_{bs} - C_f}\right) = -k_m \ln\left(\frac{C_b - C_f}{C_{bs} - C_f}\right) \tag{5.53}$$

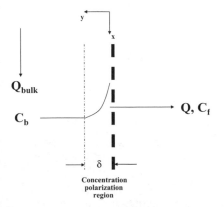

Figure 5.5 Sieving and concentration polarization.

In the above equations δ_C is defined as the concentration boundary layer thickness that also defines the thickness of the concentration polarization region. It should be apparent from our discussion of mass transfer in boundary layers that the mass transfer coefficient k_m is the same as D/δ_C.

The sieving coefficient for the case of concentration polarization (S_0) is defined as the ratio of the solute concentration in the filtrate (C_f) to that of the solute concentration in the bulk fluid (C_b). Equation 5.53 can then be rearranged and after some algebra we obtain the following expression for the the the sieving coefficient.

$$S_0 = \frac{C_f}{C_b} = \frac{S_a}{(1-S_a)\exp(-q/k_m)+S_a} \tag{5.54}$$

S_a in the above equation is given by Equation 5.50. k_m is the mass transfer coefficient and as discussed in the previous sections is dependent on the fluid properties and the nature of the flow field (Table 5.2).

Example 5.5 Consider the filtration of a fluid flowing within a hollow fiber. Assume the length of the hollow fiber is given by L and the radius of the hollow fiber is R. Develop the differential equation that describes the change with axial position (z) of the volumetric flowrate of the fluid (F) as a result of filtration across the walls of the hollow fiber. Assume steady state flow.

SOLUTION Perform a shell balance across a finite axial shell volume of the hollow fiber equal to $\pi R^2 \Delta z$. The amount of the fluid filtered across the wall of the hollow fiber within the shell volume is equal to $2\pi R\Delta z q$, where $2\pi R\Delta z$ is the circumferential surface area of the shell volume. Our shell balance may then be written as:

$$0 = F\mid_z - F\mid_{z+\Delta z} - 2\pi R\Delta z\, q$$

Dividing by Δz, and taking the limit as $\Delta z \to 0$, results in the following differential equation that describes the axial variation of F due to filtration.

$$\frac{dF(z)}{dz} = -2\pi R q(z)$$

If $q(z)$ is constant, then the above equation can be integrated to give:

$$F(z) = F_0 - 2\pi R q z$$

where F_0 is the volumetric feed rate to the fiber.

Example 5.6 Consider a hollow fiber module that is being used to separate a protein solution. The module contains 10,000 fibers and each fiber has a diameter of 400 microns and a length of 20 cm. The wall thickness of the hollow fibers is 75 microns. The nominal molecular weight cutoff (NMWCO) for the hollow fibers is 100,000 and the porosity of the hollow fibers (A_p/S) is equal to 0.40. The protein solution has a viscosity of 0.001 Pa sec. The protein solution enters the module at a flowrate of 250 ml min^{-1}. The composition of the protein solution is protein A (4 g L^{-1}, MW = 20,000), protein B (7 g L^{-1}, MW = 150,000), and protein C (6 g L^{-1}, MW = 300,000). Determine the total filtration flow rate across the hollow fibers assuming the pressure drop across the fibers is 750 mm Hg. What is the composition of the fluid leaving on the filtrate side? What is the fractional removal of protein A from the feed solution to the hollow fiber module? What is the composition of the fluid leaving the hollow fibers?

SOLUTION We can calculate the filtration flux from Equations 3.3 and 3.4. Since proteins B and C are retained by the hollow fiber membranes we find that their osmotic pressure contribution is 1.29 mm Hg. In comparison to the pressure drop across the membranes of 750 mm Hg the osmotic pressure is rather small hence we can assume for the most part that the filtration flux is constant along the length of the hollow fibers. Next we calculate the hydraulic conductance from Equation 3.7 assuming the radius of the pores can be found from the NMWCO using Equation 5.49.

$$r = \left(\frac{3 \times 100,000 \text{ g mole}^{-1}}{4\pi \times 1\text{g cm}^{-3} \times 6.023 \times 10^{23} \text{ mole}^{-1}} \right)^{1/3} = 3.41 \times 10^{-7} \text{ cm}$$

and

$$L_p = \frac{0.40 \times (3.41 \times 10^{-7} \text{ cm})^2}{8 \times 0.001 \text{ Pa sec} \times 0.0075 \text{ cm}} = 7.75 \times 10^{-10} \text{ cm Pa}^{-1} \text{ sec}^{-1}$$

Next we can calculate the total filtration flow for the hollow fiber module as shown below using Equation 3.4.

$$Q = 7.75 \times 10^{-10} \text{ cm Pa}^{-1} \text{ sec}^{-1} \times 10,000 \times 2\pi \times 0.02 \text{ cm} \times 20 \text{ cm}$$

$$\times (750 - 1.29) \text{ mm Hg} \times \frac{1 \text{ atm}}{760 \text{ mm Hg}} \times \frac{101,325 \text{ Pa}}{1 \text{ atm}} = 1.94 \text{ cm}^3 \text{ sec}^{-1}$$

The filtration flowrate is almost 50% of the total flowrate entering the hollow fiber module which is 250 cm^3 min^{-1} or 4.17 cm^3 sec^{-1}. The flowrate of fluid leaving the fibers of the hollow fiber module is therefore $F(L) = (4.17 - 1.94)$ cm^3 sec^{-1} = 2.23 cm^3 sec^{-1}. The filtration flux (q) equals Q divided by the total circumferential surface area of the fibers. Therefore, q = 1.94 cm^3 sec^{-1}/25,133 cm^2 = 7.72 $\times$ 10^{-5} cm sec^{-1}. Next, we perform a solute balance on protein A over the length of a fiber from z to $z + \Delta z$ as shown below.

$$FC_b \big|_z - FC_b \big|_{z+\Delta z} = 2\pi R \, \Delta z \, C_b S_0 q$$

After dividing by Δz and taking the limit as $\Delta z \to 0$ we obtain that:

$$\frac{dF C_b}{dz} = F \frac{dC_b}{dz} + C_b \frac{dF}{dz} = -2\pi R C_b S_0 q$$

From the previous example we also have that $\dfrac{dF}{dz} = -2\pi Rq$, and the above equation may then be written as:

$$\frac{dC_b}{dz} = -\frac{2\pi R(1-S_0)q}{F}C_b = \frac{2\pi R(1-S_0)q}{F_0 - 2\pi Rqz}C_b$$

The above equation can then be integrated as shown below to obtain an expression for how the bulk solute concentration for protein A changes with position in the hollow fiber module.

$$\int_{C_{b0}}^{C_b} \frac{dC_b}{C_b} = 2\pi R(1-S_0)q \int_0^z \frac{dz}{F_0 - 2\pi Rqz}$$

$$C_b(z) = C_{b0}\left[1 - \left(\frac{2\pi Rq}{F_0}\right)z\right]^{-(1-S_0)}$$

To find $C_b(z)$, we now need to calculate S_0 which is also dependent on k_m. Based on the average of the entrance and exit flowrates, the average velocity of the fluid in a fiber is:

$$V = 3.2 \text{ cm}^3 \text{ sec}^{-1} \times \frac{1}{10,000 \text{ fibers}} \times \frac{4}{\pi(0.04 \text{ cm})^2} = 0.25 \text{ cm sec}^{-1}$$

and the average Reynolds number of the fluid in a fiber is:

$$Re = \frac{1 \text{ g cm}^{-3} \times 0.04 \text{ cm} \times 0.25 \text{ cm sec}^{-1}}{0.01 \text{ g cm}^{-1}\text{ sec}^{-1}} = 1.02$$

So the flow of the fluid within the fibers is laminar. From Figure 5.4 we can estimate the diffusivity of protein A as $6.42 \times 10^{-7} \text{ cm}^2 \text{ sec}^{-1}$. The Schmidt number is then calculated as follows:

$$Sc = \frac{0.01 \text{ g cm}^{-1}\text{ sec}^{-1}}{1 \text{ g cm}^{-3} \times 6.42 \times 10^{-7} \text{cm}^2 \text{ sec}^{-1}} = 15,576$$

Previously, we discussed that the flow is fully developed if $z > D_{\text{tube}} \times 0.05\ Re$. In this case the flow is fully developed if $z > 0.002$ cm. For the concentration field to be fully developed $z_C > D_{\text{tube}} \times 0.05\ Re\ Sc$ which, in this case, gives that $z_C > 31$ cm which is longer than the fibers. Hence, the concentration profile is not fully developed and we should use Equation 5.44 to calculate the value of k_m as shown below.

$$Sh = 3.66 + \frac{0.104 \times 1.02 \times 15,576 \times \left(\dfrac{0.04}{20}\right)}{1 + 0.016\left(1.02 \times 15,576 \times \dfrac{0.04}{20}\right)^{0.8}} = 6.29$$

The mass transfer coefficient is then calculated as shown below.

$$k_m = \frac{D}{D_{tube}} Sh = 6.29 \times \frac{6.42 \times 10^{-7}\, cm^2\, sec^{-1}}{0.04\, cm} = 1.01 \times 10^{-4}\, cm\, sec^{-1}$$

Next we can use Equation 5.50 to find the value of S_a. The radius of protein A (a) would be given by Equation 5.49:

$$a = \left(\frac{3 \times 20,000\, g\, mole^{-1}}{4\pi \times 1\, g\, cm^{-3} \times 6.023 \times 10^{23}\, mole^{-1}} \right)^{1/3} = 1.99 \times 10^{-7}\, cm$$

and the value of $\lambda = a/r$ is then equal to 0.585.

$$S_a = (1-0.585)^2 \left[2 - (1-0.585)^2 \right] \left[1 - \frac{2}{3} \times 0.585^2 - 0.163 \times 0.585^3 \right] = 0.233$$

From Equation 5.54 we can then calculate the value of S_0:

$$S_0 = \frac{0.233}{(1-0.233)\exp\left(-\dfrac{7.72 \times 10^{-5}\, cm\, sec^{-1}}{1.01 \times 10^{-4}\, cm\, sec^{-1}} \right) + 0.233} = 0.395$$

Now we can calculate the concentration of protein A exiting ($z = L$) the hollow fiber module.

$$C_{b\,out} = 4g\, L^{-1} \left[1 - \left(\frac{2\pi \times 0.02\ cm \times 7.72 \times 10^{-5}\, cm\, sec^{-1}}{4.17 \times 10^{-4}\, cm^3\, sec^{-1}} \right) \times 20\ cm \right]^{-(1-0.395)} = 5.84\ g\ L^{-1}$$

An overall solute balance can then be used to find the concentration of protein A in the filtrate.

$$Q C_{f\,out} = C_{b0} F_0 - C_{b\,out} F$$

$$C_{f\,out} = \frac{C_{b0} F_0 - C_{b\,out} F}{Q} = \frac{4\ g\ L^{-1} \times 4.17\ cm^3\, sec^{-1} - 5.84\ g\ L^{-1} \times 2.23\ cm^3\, sec^{-1}}{1.94\ cm^3\, sec^{-1}}$$

$$C_{f\,out} = 1.88\ g\ L^{-1}$$

The % removal of protein A from the stream entering the hollow fiber module is then given by the following.

$$\% \text{ removal of A} = \frac{C_{b0} F_0 - C_{b\,out} F}{C_{b0} F_0} =$$

$$\frac{4\ g\ L^{-1} \times 4.17\ cm^3\, sec^{-1} - 5.84\ g\ L^{-1} \times 2.23\ cm^3\, sec^{-1}}{4\ g\ L^{-1} \times 4.17\, cm^3\, sec^{-1}} \times 100 = 22\%$$

The concentrations of protein B and C in the fluid exiting the hollow fibers may be calculated as shown below.

$$C_{\text{protein B}} = \frac{7 \text{ g L}^{-1} \times 4.17 \text{ cm}^3 \text{ sec}^{-1}}{2.23 \text{ cm}^3 \text{ sec}^{-1}} = 13.1 \text{ g L}^{-1}$$

$$C_{\text{protein C}} = \frac{6 \text{ g L}^{-1} \times 4.17 \text{ cm}^3 \text{ sec}^{-1}}{2.23 \text{ cm}^3 \text{ sec}^{-1}} = 11.2 \text{ g L}^{-1}$$

5.6 SOLUTE DIFFUSION WITHIN HETEROGENEOUS MEDIA

A unique aspect of biological systems is their heterogeneous nature. We therefore need to expand our understanding of solute diffusion to see how heterogeneous materials affect solute diffusion. Although the following discussion focuses a lot on the transport of a solute from blood across the capillary wall, the concepts are general and may be applied to systems involving synthetic membranes as well.

Figure 5.6 illustrates the situation for diffusion of a solute from blood, through the cell free plasma layer (see Chapter 4), across the capillary wall, and into the surrounding tissue space. Shown in the figure are the types of solute diffusivity we need to consider. These are the solute diffusivity in blood or tissue (D_T), the diffusivity in plasma (D_{plasma}), the diffusivity within the pores of the capillary wall (D_m), diffusivity in the interstitial fluid (D_0), and the diffusivity through cells (D_{cell}). The following discussion will relate these various solute diffusivities to the diffusivity of the solute in water (D).

We can use the Stokes–Einstein equation to estimate the solute diffusivity in plasma by adjusting the solution viscosity. The diffusivity of a solute in plasma at 37°C would therefore be given by the following relation: $D_{\text{plasma}} = D_{\text{water}} \times \left(\frac{0.76 \text{ cP}}{1.2 \text{ cP}} \right) = 0.63 \, D_{\text{water}}$.

For solute transport across the capillary wall, we must consider that the available surface area is not the total surface area S, but the pore area A_P. This assumes that the

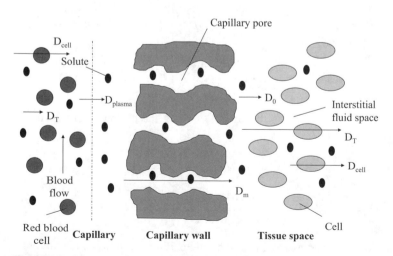

Figure 5.6   Solute transport from blood.

solute itself is not soluble in the continuous non-porous phase. Furthermore, the solute must follow a path through the pores that is more tortuous making the diffusion distance greater than the thickness of the capillary wall or the membrane (t_m).

In many cases in pore diffusion, the solute radius is also comparable to that of the pore radius. This leads to two additional effects both reducing the diffusion rate of the solute. The first effect, called *steric exclusion*, is illustrated in Figure 5.7. Steric exclusion restricts the ability of the solute to enter the pore from the bulk solution. In this figure, we see that a solute can get no closer to the pore wall than its radius (a). Therefore, only a fraction of the pore volume is available to the molecule. The fraction of pore volume available to the solute is given by the ratio $K = \pi(r - a)^2/\pi r^2 = (1 - a/r)^2$, which is also known as the *partition coefficient*. Because of steric exclusion, the equilibrium concentration of solute is less within the pore mouth than in the bulk solution. This simple expression for K ignores any secondary attractive or repulsive interactions between the solute and the pore and basically treats the molecule as a hard sphere.

As the solute molecule diffuses through the pore, it experiences hydrodynamic drag caused by the flow of solvent over the surface of the solute. This is shown in Figure 5.8. As the solute radius increases relative to that of the pore radius, this hydrodynamic drag increases, reducing or restricting the diffusion of the solute through the pore compared with its motion in the bulk fluid. This drag effect on the solute in the pore

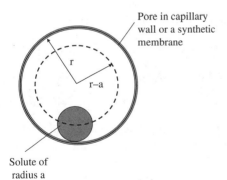

Solute of
radius a

Figure 5.7   Steric exclusion.

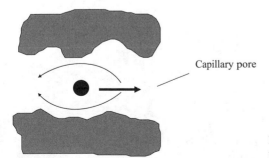

Capillary pore

Figure 5.8   Solvent flow across solute in small pores increases drag.

results in a decrease in the solute diffusivity relative to its value in the bulk solution and is called *restricted* or *hindered diffusion*.

Beck and Schultz (1970) studied the diffusion of a variety of solutes across track-etched mica membranes. The mica membranes were bombarded with U^{235} fission fragments. The etched particle tracks that were formed resulted in relatively straight pores with a narrow pore size distribution. Their data for a number of solutes is shown in Figure 5.9, where the ratio of the pore diffusivity (D_m) and the bulk diffusivity (D) is plotted as a function of the ratio of the solute radius to pore radius ($\lambda = a/r$). Note that when the solute size is only 1/10th of the pore size that the solute diffusivity is reduced by 30%. When a/r is equal to 0.40, the solute diffusivity has been reduced by nearly 90%.

The solid line in Figure 5.9 represents the fit of the Renkin equation (1954) to their data given by the following expression:

$$\frac{D_m}{D} = \left(1 - \frac{a}{r}\right)^2 \left[1 - 2.1\left(\frac{a}{r}\right) + 2.09\left(\frac{a}{r}\right)^3 - 0.95\left(\frac{a}{r}\right)^5\right]$$

$$= K\left(\frac{a}{r}\right) \times \omega_r\left(\frac{a}{r}\right)$$

(5.55)

We see that the Renkin equation describes their data rather well. The first term in this equation is the partition coefficient mentioned earlier, $K\left(\dfrac{a}{r}\right)$, and represents the effect of steric exclusion or the reduction in the solute concentration at the pore mouth compared with its value in the bulk solution. The bracketed term, $\omega_r\left(\dfrac{a}{r}\right)$, accounts for the increase in hydrodynamic drag as the solute diffuses through the pore. If the solute is already within the porous structure, then $K = 1$ and $\dfrac{D_m}{D} = \omega_r\left(\dfrac{a}{r}\right)$. Additional expressions for describing the hindered diffusion of large molecules in liquid filled pores may be found in the review paper by Deen (1987).

With these corrections for the diffusion of a solute in pores of the capillary wall or a synthetic membrane, we can now write Fick's first law as follows:

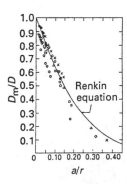

Figure 5.9   Apparent diffusivity in the pores of a membrane (from Beck and Schultz 1970, by permission).

$$J_s = -DA_P \left(\frac{K\omega_r}{\tau} \right) \frac{dC}{dx} \tag{5.56}$$

In this equation, τ is an experimentally defined parameter called the *tortuosity* and accounts for the fact that the solute diffusion distance is greater than that of the capillary wall or membrane thickness. If we divide the above equation by the total surface area (S) available for mass transfer of the solute then we have that the solute flux j_s is given by:

$$j_s = -D \frac{A_P}{S} \left(\frac{K\omega_r}{\tau} \right) \frac{dC}{dx} = -D_e \frac{dC}{dx} \tag{5.57}$$

where D_e is called the *effective* diffusivity of the solute in the heterogeneous material. D_e is then given by the following equation:

$$D_e = D \frac{A_P}{S} \left(\frac{K\omega_r}{\tau} \right) = \frac{\varepsilon D}{\tau} K\omega_r \tag{5.58}$$

where $\varepsilon = A_P/S$ is known as the porosity, or void volume fraction of the capillary wall or the membrane.

5.6.1 Diffusion of a Solute from a Polymeric Material

Biocompatible polymers are widely used for the controlled release of a variety of drugs. Consider the situation shown in Figure 5.10. A drug is uniformly distributed within a microporous polymeric disc of radius R and thickness L. All of the surfaces of the disc are coated with an impermeable material except for the surface located at $x = L$. Since the drug release occurs over many days or weeks, the rate limiting process for release of the drug from the polymeric support at $x = L$ is diffusion of the drug through the pores of the polymeric material. An unsteady solute mass balance over the region from x to $x + \Delta x$ shows that Fick's second law describes the diffusion of the solute within the polymeric material.

$$\frac{\partial C}{\partial t} = D_e \frac{\partial^2 C}{\partial x^2} \tag{5.59}$$

Note that we now use the effective diffusivity (i.e. Equation 5.58) to describe the diffusion of the solute through the microporous structure of the polymeric material. The initial concentration of the drug is C_0 and the drug concentration at $x = L$ is zero since the drug is immediately taken up by the surroundings, a process that is much faster than

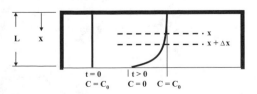

$$t = 0 \qquad |\ t > 0$$
$$C = C_0 \qquad C = 0 \quad C = C_0$$

Figure 5.10   Diffusion of a solute within a microporous polymeric disk.

the diffusion of the drug through the polymeric material. The initial and boundary conditions may then be written as shown below.

$$IC: \quad t = 0, C(x,t) = C_0$$

$$BC1: \quad x = 0, \frac{dC}{dx} = 0 \tag{5.60}$$

$$BC2: \quad x = L, C = 0$$

Boundary condition one expresses the additional fact that the solute cannot diffuse out through the top surface of the polymeric disc since this surface is coated with an impermeable material.

A solution to Equations 5.59 and 5.60 can be obtained using the *separation of variables* technique (check out any advanced calculus book for more information on this technique). We let $C(x, t)$ be given by the product of a function that depends only on x, i.e. $F(x)$, and another function that depends only on t, i.e. $T(t)$. Therefore, $C(x, t) = F(x) T(t)$. We substitute this equation into Equation 5.59 and obtain:

$$\frac{1}{T D_e} \frac{dT}{dt} = \frac{1}{F} \frac{d^2 F}{dx^2} = -\lambda^2 \tag{5.61}$$

Note that the term on the left side of the above equation depends only on t and the term in the middle only depends on x. The only way they can then be equal to each other is if they equal the same constant, i.e. λ^2. The minus sign is in front of λ^2 in order to obtain a solution that remains finite as time increases. Equation 5.61 can then be rearranged to give the following two ordinary differential equations:

$$\frac{dT}{dt} = -\lambda^2 D_e T \quad \text{and} \quad \frac{d^2 F}{dx^2} + F\lambda^2 = 0 \tag{5.62}$$

These equations may then be integrated to provide the results shown below.

$$T = C_1 e^{-\lambda^2 D_e t} \text{ and } F = C_2 \sin \lambda x + C_3 \cos \lambda x \tag{5.63}$$

Applying boundary conditions one and two on the expression for F above we obtain $C_2 = 0$ and $C_3 \cos \lambda L = 0$. C_3 cannot be equal to zero since that provides a trivial solution. Hence we then find that there are an infinite number of λ's that can satisfy the condition that the $\cos \lambda L = 0$. It is easy to then show that these λ's are given by the following equation.

$$\lambda_n = \left(\frac{2n+1}{2} \right) \frac{\pi}{L} \quad \text{for } n = 0, 1, 2, ... \infty \tag{5.64}$$

The λ_n are also known as *eigen values*, and $\cos \lambda_n x$ is known as the *eigen functions*. Our solution now is $C_n(x,t) = C_{3n} e^{-\lambda_n^2 D_e t} \cos \lambda_n x$ for $n = 0, 1, 2, ... \infty$. Each value of n gives us a solution to the original partial differential equation, i.e. Equation 5.59 that satisfies boundary conditions one and two. We can also sum all of these n solutions and obtain the following result:

$$C(x,t) = \sum_{n=0}^{\infty} C_n e^{-\lambda_n^2 D_e t} \cos \lambda_n x \tag{5.65}$$

The C_n's can then be obtained by requiring that the above equation satisfies the initial condition, i.e.

$$C_0 = \sum_{n=0}^{\infty} C_n \cos \lambda_n x \tag{5.66}$$

Next we multiply the above equation by $\cos \lambda_m x$ and integrate from $x = 0$ to $x = L$.

$$C_0 \int_0^L \cos \lambda_m x \, dx = \int_0^L \cos \lambda_m x \sum_{n=0}^{\infty} C_n \cos \lambda_n x \, dx \tag{5.67}$$

The above equation can also be written as:

$$\frac{C_0}{\lambda_n} \sin \lambda_n L = \sum_{n=0}^{\infty} C_n \int_0^L \cos \lambda_m x \cos \lambda_n x \, dx = C_n \int_0^L \cos^2 \lambda_n x \, dx \tag{5.68}$$

where $\int_0^L \cos \lambda_m x \cos \lambda_n x \, dx = 0$ except when $m = n$. Hence we then can solve Equation 5.68 for the C_n's which are given by the next expression:

$$C_n = \frac{4 C_0 (-1)^n}{(2n+1)\pi} \tag{5.69}$$

Combining Equations 5.65 and 5.69 then provides the solution for the concentration distribution of the drug or solute within the polymeric material.

$$C(x,t) = \frac{4 C_0}{\pi} \sum_{n=0}^{\infty} \frac{(-1)^n}{2n+1} e^{-\frac{(2n+1)^2 \pi^2 D_e t}{4 L^2}} \cos \frac{(2n+1)\pi x}{2 L} \tag{5.70}$$

Of particular interest would be the flux of drug leaving the polymeric disk, i.e. Equation 5.57 at $x = L$. After finding dC/dx from the above equation, we can solve for the elution flux of the drug given by the equation below.

$$j_S \big|_{x=L} = \frac{2 D_e C_0}{L} \sum_{n=0}^{\infty} e^{\frac{-(2n+1)^2 \pi^2 D_e t}{4 L^2}} \tag{5.71}$$

The above equation can also be combined with a pharmacokinetic model for drug distribution in the body to predict how the drug concentration in the body changes with time. This is discussed later in Chapter 7.

The amount of drug or solute remaining in the polymeric material at any time t is also of interest. If we let S represent the total surface area of the polymeric material normal to the diffusion direction, then the total amount of drug (D_0) initially present in the polymeric material is equal to $D_0 = C_0 SL$. Note that C_0 is defined in terms of the total volume of the polymeric material which includes the void space and the polymer. The amount of drug remaining in the polymeric material at time t, i.e. $D(t)$, is then given by integrating the concentration distribution at any time t, i.e. Equation 5.70, from $x = 0$ to $x = L$

$$D(t) = S \int_0^L C(x,t)\, dx = \frac{8}{\pi^2} D_0 \sum_{n=0}^{\infty} \frac{1}{(2n+1)^2} e^{-\frac{(2n+1)\pi^2 D_e t}{4L^2}} \tag{5.72}$$

We can also define the cumulative fraction of the drug (f_R) that has been released as the amount released (i.e. $D_0 - D(t)$) divided by the amount of drug originally present, i.e. D_0.

$$f_R = 1 - \frac{8}{\pi^2} \sum_{n=0}^{\infty} \frac{1}{(2n+1)^2} e^{-\frac{(2n+1)\pi^2 D_e t}{4L^2}} \tag{5.73}$$

5.6.1.1 A solution valid for short contact times For small times, the drug concentration change occurs only over a thin region near where $x = L$. In this case, Equation 5.59 still describes the diffusion of the drug within the polymeric material, however, we can replace the boundary conditions with those shown below where now y represents the distance from the exposed surface.

$$\begin{aligned} &\text{IC:} &&t = 0,\ C(y,t) = C_0 \\ &\text{BC1:} &&y = 0,\ C = 0 \\ &\text{BC2:} &&y = \infty,\ C = C_0 \end{aligned} \tag{5.74}$$

Boundary condition two expresses the fact that at distances far from the exposed surface the drug concentration within the polymeric material is still equal to its initial value. Equations 5.59 and 5.74 can be solved using the Laplace transform technique. If we let $C' \equiv C_0 - C$, then Equation 5.59 remains the same with the exception that C is replaced by C' and the initial and boundary conditions above become:

$$\begin{aligned} &\text{IC:} &&t = 0,\ C'(y,t) = 0 \\ &\text{BC1:} &&y = 0,\ C' = C_0 \\ &\text{BC2:} &&y = \infty,\ C' = 0 \end{aligned} \tag{5.75}$$

The solution to this problem for C' using Laplace transforms is identical to the solution we obtained earlier for Equations 5.4 and 5.5 which is given by Equation 5.6, that is:

$$C(y,t) = C_0\, erf\left(\frac{y}{\sqrt{4D_e t}} \right) \tag{5.76}$$

The flux of drug from the exposed surface would also be given by Equation 5.9.

Example 5.7 Bawa *et al.* (*J. Controlled Release*, 1, 259–267, 1985) investigated the release of macromolecules from small slabs of a polymeric material made of ethylene-vinyl acetate copolymer. The polymeric material was formed into slabs that were 1 cm × 1 cm and 1 mm in thickness. All surfaces of the slab were coated with paraffin except one 1 cm × 1cm face which was exposed to the surroundings. During the formation of the polymeric slabs, bovine serum albumin (BSA) particles were incorporated within the polymer forming a system of interconnected pores after the BSA became solubilized. In one case using BSA particles that ranged in size from 106–150 microns, the cumulative fraction of BSA released as a function of the square root of the time is shown in the table below.

Time$^{1/2}$, hrs$^{1/2}$	Cumulative fraction of BSA released (f_R)
2.5	0.05
5.0	0.1
7.0	0.12
11.0	0.20
13.0	0.22
15.5	0.28
17.5	0.30
20.0	0.32
22.0	0.35

If the initial concentration of BSA within the slab is 300 mg cm^{-3} and the porosity is 0.26, estimate the effective diffusivity of BSA within the slab.

SOLUTION Equation 5.73 applies, and the only adjustable parameter in this equation is the effective diffusivity, D_e. Our goal is then to find the value of D_e that minimizes the error between the experimental values of f_R and those values of f_R predicted by Equation 5.73. For this type of problem, one typically uses as the objective function the sum-of-the-square-of-the-error (SSE) defined by the following equation:

$$SSE = \sum_{i=1}^{N} (f_{R\,\text{experimental}_i} - f_{R\,\text{theory}_i})^2$$

where $f_{R\,\text{experimental}}$ are the values in the above table and $f_{R\,\text{theory}}$ are the values predicted by Equation 5.73. N represents the total number of data points. The goal is then to find the value of D_e that minimizes SSE, and this value then represents the best fit of Equation 5.73 to the experimental data. There are a variety of mathematical software programs that one may use to solve these types of problems which are also known as non-linear regression problems. Figure 5.11 shows the results of fitting Equation 5.73 to the data in the above table. As shown in the figure, we obtain an excellent fit of Equation 5.73 to the experimental data. The value of D_e is found to be 2.22×10^{-6} cm^2 hr^{-1}.

5.6.2 Diffusion in Blood and Tissue

In heterogeneous regions like blood or that in the tissue space, the solute can diffuse through both the continuous fluid space as well as through the cells themselves. The transport mechanism in blood or the tissue space is by diffusion and D_T represents the "effective" diffusivity of the solute through the heterogeneous region. The simplest approach for estimating D_T is based on a model developed by Maxwell (1873).

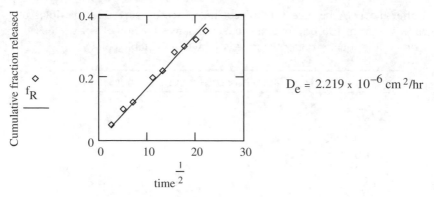

Square root of time in hrs

Figure 5.11   Cumulative fraction of BSA released as a function of time$^{1/2}$.

$$\frac{D_T}{D_0} = \frac{2D_0 + D_{cell} - 2\phi(D_0 - D_{cell})}{2D_0 + D_{cell} + \phi(D_0 - D_{cell})} \tag{5.77}$$

In this equation, D_0 is the diffusivity of the solute through the interstitial fluid space, or in the case of blood, $D_0 = D_{plasma}$. D_{cell} is the diffusivity of the solute in the cells. ϕ represents the volume fraction of the cells. Interestingly, this equation does not depend on the size of the discrete particles, i.e. the cells, but only on their volume fraction.

More recently, Riley *et al.* (1994, 1995a,b,c, 1996) have developed an empirical relation based on Monte Carlo simulations for D_T/D_0 that shows good agreement with available data throughout a wide range of cell volume fractions, $0.04 < \phi < 0.95$.

$$\frac{D_T}{D_0} = 1 - \left(1 - \frac{D_{cell}}{D_0}\right)\left(1.727\phi - 0.8177\phi^2 + 0.09075\phi^3\right) \tag{5.78}$$

The interstitial fluid that lies between the cells has a gel-like consistency due to the presence of a variety of macromolecules. Therefore, the solute must diffuse around this random network of macromolecular obstacles. The reduction in the solute diffusivity relative to its value in pure water (D) due to the network of macromolecules within the interstitial fluid gel may be described by the following equation developed by Brinkman (1947).

$$\frac{D_0}{D} = \omega_r = \frac{1}{1 + \kappa a + \tfrac{1}{3}(\kappa a)^2} \tag{5.79}$$

In this equation, the ratio of D_0 to D depends on only one parameter, κ. κ is a function of the gel's microstructure and may be fitted to experimental data for solute diffusion in the gel.

The steric exclusion, or equilibrium partitioning, of a solute upon entrance from a bulk solution into a gel-like material such as the interstitial fluid may be described by the following equation developed by Ogston (1958).

$$K = \exp\left[-\phi\left(1+\frac{a}{a_f}\right)^2\right]$$ (5.80)

In this equation, a_f represents the radius of the macromolecules which are assumed to form very long cylindrical fibers. If the length of these fibers per volume of gel is given by L, then the volume fraction of these macromolecular fibers is given by $\phi = \pi$ $a_f^2 L$ (Tong and Anderson 1996). Tong and Anderson (1996) showed that Equations 5.79 and 5.80, i.e. $K D_0$, provided excellent representation of the partitioning and diffusion of two representative globular proteins (albumin and ribonuclease-A) in a polyacrylamide gel.

Nugent and Jain (1984a,b) examined the effective diffusion of various sized solutes through normal and tumor tissue. Their results are shown in Figure 5.12 where we see the ratio of the effective diffusivity of the solute in the tissue (D_T) to its bulk diffusivity (D) plotted as a function of the solute radius. This figure may be used to provide reasonable estimates of the solute diffusivity within tissue. We see that normal tissue results in a greater reduction in the solute diffusivity than that observed for tumor tissue. This reflects the smaller interstitial volume of normal tissue in comparison to tumor tissue.

The solid and dashed lines in Figure 5.12 represent the corresponding fits of two diffusion models to their data. The solid lines represent a fiber matrix model (Curry and Michel 1980) wherein the diffusivity ratio is given by this equation.

$$\frac{D_T}{D} = \exp[-(1+a/a_f)v^{1/2}C_F^{1/2}]$$ (5.81)

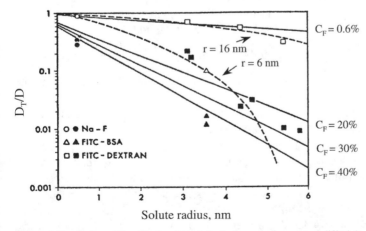

Figure 5.12   Ratios of the effective diffusivity in tissue to the aqueous diffusivity as a function of the solute radius. Open and closed symbols represent data for tumor and normal tissue, respectively. Solid lines represent the fiber matrix model and dashed lines represent the Renkin pore model equation (from Nugent and Jain 1984a, with permission).

The tissue interstitial space is modeled as a random matrix of straight fibers of radius a_f, and the solute is considered to be a rigid sphere of radius a. C_F is the fiber concentration in the interstitial fluid and v the specific volume of the fibers. The fibers, once again, are assumed to be the macromolecules that form the gel-like consistency of the interstitial fluid. The quantity $C_F v$ is related to the fiber volume fraction in the gel by the following expression $\phi/(1 - \phi)$. The dashed lines in Figure 5.12 represent the predictions of the diffusivity ratio (D_T/D) based on the hydrodynamic pore model as given by $\omega_r\left(\dfrac{a}{r}\right)$ in Equation 5.55.

5.7 SOLUTE PERMEABILITY

Sometimes the term *permeability* (P_m) is used to describe solute transport $(N_S,$ for example in gmoles/sec) across a membrane. Permeability is defined by the following expression using the <u>total</u> membrane surface area S.

$$N_s = P_m S (C_{high} - C_{low}) \tag{5.82}$$

$(C_{high} - C_{low})$ is the difference in the surface concentrations of the solute on either side of the membrane. In the absence of external diffusion effects, this concentration difference would equal the difference in the bulk concentration of the solution on either side of the membrane. The permeability is also simply equal to the effective diffusivity of the solute in the membrane divided by the thickness of the membrane (D_e/t_m). The permeability may be shown to be given by the following relationship.

$$P_m = \frac{D}{t_m}\left(\frac{A_P}{S}\right)\left(\frac{K\omega_r}{\tau}\right) \tag{5.83}$$

For a membrane with cylindrically-shaped pores, the quantity $K \omega_r$ would be given by Equation 5.55. The tortuosity (τ) would need to be estimated separately, perhaps by comparison to experimental permeabilty data. On the other hand, for diffusion through a gel-like membrane, $\left(\dfrac{A_P}{S}\right)\left(\dfrac{K\omega_r}{\tau}\right)$ would be given by the product of Equations 5.79 and 5.80. It is important to note that in both cases, if the solute is already within the membrane that $K = 1$.

In Equation 5.82, we see the product of the solute permeability and the membrane surface area, i.e. P_mS. Although we just developed a framework for describing P_m, i.e. Equation 5.83, there is still a lot of uncertainty in the values that go into estimating the permeability for a given solute by this equation. In some cases, for example for solute diffusion through the capillaries in a large region of tissue, it is also difficult to estimate the surface areas of the capillaries involved. In cases like this, it is more convenient to work with the product of P_m and S, rather than specific values of P_m or S. This is because of the uncertainty in accurately defining either the permeability or the capillary surface area in biological systems.

Renkin and Curry (1979) summarized the results of a variety of experiments that defined continuous capillary P_mS values for solutes of various sizes. These data are

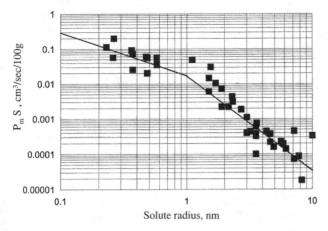

Figure 5.13   Solute permeability of capillaries (based on data from Renkin and Curry 1979).

shown in Figure 5.13. Note that in some cases for a given size solute, the value of $P_m S$ can vary by almost an order of magnitude. The $P_m S$ value is highly dependent on the nature of the capillaries that exist in the tissue of interest. In addition, fenestrated and discontinuous capillaries can have $P_m S$ values that are several order of magnitudes higher than those given in Figure 5.13.

The solid lines in Figure 5.13 represent a linear least square regression of the data. The following empirical equations may be used to estimate capillary $P_m S$ values for a given solute of radius (a). Note that the units on $P_m S$ are cm³/sec/100 g of tissue.

$$P_m S = 0.0184 \, a^{-1.223}, \, a < 1\text{nm} \tag{5.84a}$$

$$P_m S = 0.0287 \, a^{-2.92}, \, a > 1\text{nm} \tag{5.84b}$$

5.8   THE IRREVERSIBLE THERMODYNAMICS OF MEMBRANE TRANSPORT

The transport of solute across membranes can occur, as discussed in the previous sections, by a combination of bulk flow or convection, and by diffusion. In some cases, the relative magnitudes of these two transport rates for a given solute are comparable, and we must consider the fact that these processes are, therefore, interdependent. This interaction makes the description of the solute transport rate much more complicated, and to arrive at the proper result, requires an understanding of the thermodynamics of irreversible processes.

The application of irreversible thermodynamics to membrane processes was developed by Staverman (1948) and Kedem and Katchalsky (1958). For dilute solutions, the theory of irreversible thermodynamics states that the filtration rate of the solvent (Q), and the filtration rate due to the solute relative to that of the solvent (J_V), both depend in a linear manner on the driving forces ΔP (pressure for flow) and $RT\Delta C$

(concentration for diffusion). This linear combination of the driving forces is then given by the following equations:

$$Q = SL_P\Delta P + SL_{PS}RT\Delta C \tag{5.85a}$$

$$J_V = SL_{SP}\Delta P + SL_SRT\Delta C \tag{5.85b}$$

Q and J_V are vectors and must have the correct sign sense for proper interpretaton. This is easily accomplished if we simply assume a vertical orientation for the membrane. Flow of solvent and solute from left to right is considered positive. The "Δ" sign in this case also represents the value of the property (P or C) on the left side of the membrane minus the corresponding value on the right side of the membrane.

The cross coefficients (L_{PS} and L_{SP}) represent secondary effects that are caused by the primary driving forces. For example, the primary ΔP driving force generates the filtration flow (Q) and also induces a relative flow between solute and solvent represented by $L_{SP}\Delta P$. This relative flow between the solute and solvent is capable of producing a separation of solute and solvent by the sieving mechanism discussed earlier and is referred to as *ultrafiltration*. Along the same line of reasoning, the $RT\Delta C$ primary driving force is responsible for solute diffusion. This diffusive transport produces an additional contribution to the filtration flow that is given by $L_{PS}RT\Delta C$. From Equation 3.1, van't Hoff's equation, we recognize this as an osmotic flow.

The coefficient L_P is the hydraulic conductance of the membrane defined earlier by Equation 3.7. We will soon show that L_S is related to the permeability of the membrane. Another basic theorem from the work of Onsager is that the cross coefficients L_{SP} and L_{PS} are equal. We can then simplify Equation 5.85 as follows:

$$Q = SL_p[\Delta P - \sigma RT\Delta C] \tag{5.86a}$$

$$J_V = SL_p\left[-\sigma\Delta P + \frac{L_S}{L_p}RT\Delta C\right] \tag{5.86b}$$

The parameter σ, defined as $-L_{SP}/L_P$ or $-L_{PS}/L_P$, is called the Staverman *reflection coefficient*.

Note the similarity between the filtration flow rate (Q) in Equation 5.86a and that given by Equation 3.4, recognizing that $RT\Delta C$ is equivalent to ($\pi_C - \pi_{IF}$). These equations are identical when $\sigma = 1$. This implies that the pores of the membrane are impermeable to the solute as required by the derivation of Equation 3.4, the solute is therefore completely "reflected" by the membrane and the $RT\Delta C$ term is in fact the osmotic pressure difference. We then recover Starling's equation. When $\sigma = 0$ in Equation 5.86a, then from our previous discussion we have no secondary effect of osmotic flow caused by the primary concentration difference driving force. This means that the solute flows through the pores completely unimpeded, just as easily as the solvent, hence, the solute is not "reflected". The concept of the reflection coefficient is illustrated in Figure 5.14. Most solutes have a reflection coefficient that lies between these two extremes.

The total rate of transfer of solute (N_S) through the pores of the capillary wall is given by the product of the solute concentration (C) and the combined flowrate of the

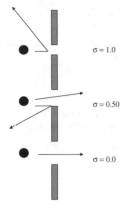

$\sigma = 1.0$

$\sigma = 0.50$

$\sigma = 0.0$

Figure 5.14 The reflection coefficient.

solution due to the applied pressure and concentration differences, i.e. $C(Q + J_V)$. Combining Equations 5.86a and 5.86b results in the following equation for the transport rate of the solute.

$$N_S = CL_P S \left[(1-\sigma)\Delta P + \left(\frac{L_S}{L_P} - \sigma \right) RT \Delta C \right]$$ (5.87)

This equation can be rewritten by using Equation 5.86a to express ΔP as a function of Q and ΔC.

$$N_S = C(1-\sigma)Q + CSL_P \left(\frac{L_S}{L_P} - \sigma^2 \right) RT \Delta C$$ (5.88)

The second term in the above equation has the following factor $(L_S/L_P - \sigma^2)$ in it that must be determined. Recognizing that when $Q = 0$, no filtration flow, diffusion is the only solute transport mechanism, then by Equation 5.86a we have:

$$(\Delta P)_{Q=0} = (\sigma RT \Delta C)_{Q=0}$$ (5.89)

Substituting this expression into Equation 5.87 for $Q = 0$ gives the following equation, where we must recognize that in the absence of Q, the flowrate of the solute across the membrane must be equal to our previous result given by Equation 5.82.

$$(N_S)_{Q=0} = CL_P S (RT \Delta C)_{Q=0} \left(\frac{L_S}{L_P} - \sigma^2 \right) = P_m S(\Delta C)_{Q=0}$$ (5.90)

We can then solve the above equation for $(L_S/L_P - \sigma^2)$ which is given by the following equation:

$$\left(\frac{L_S}{L_P} - \sigma^2 \right) = \frac{P_m}{L_P RTC}$$ (5.91)

where P_m is the membrane permeability defined earlier by Equation 5.82. Equation 5.88 can now be simplified using Equation 5.91 to give our final result for the solute transport rate through the pores of the capillary wall, recognizing that Q is the filtration flowrate defined earlier by Equation 5.86a.

$$N_s = C(1-\sigma)Q + P_m S \Delta C \tag{5.92}$$

The first term in the above equation represents the amount of solute transported by bulk flow (convection) of the solvent, and the second term represents the contribution of diffusion. C represents the solute concentration on the side of the membrane where the filtration flow (Q) originates.

5.8.1 Finding L_P, P_m, and σ

We must know the three parameters (L_P, P_m, and σ) in Equations 5.86a and 5.92 to calculate the transport rate of a solute. L_P is the hydraulic conductance, and it can be estimated using Equation 3.7. It can also be measured using a solute whose value of $\sigma = 0$, and measuring the flowrate Q across the membrane for a given ΔP. Then by Equation 5.86a, $L_P = \dfrac{J}{S \Delta P}$.

Recall that P_m is the permeability of the solute in the membrane and may be estimated by Equation 5.83. Alternatively, in the absence of any bulk flow across the membrane ($Q = 0$), the permeability can be obtained by measuring the solute transport rate (N_S) for a given concentration difference. Then, from Equation 5.92, the permeability would be given by $P_m = \dfrac{N_S}{S \Delta C}$.

The reflection coefficient, σ, can be measured by one of two ways. In the first method, one employs a pressure drop across the membrane and measures the filtration flowrate Q, as well as the transport rate of the solute N_S, across the membrane in the absence of a solute concentration difference across the membrane. Then from Equation 5.92, we obtain the fact that $(1-\sigma) = \dfrac{N_S}{CJ}$. Alternatively, we could measure the filtration flowrate Q, in the absence of a pressure drop across the membrane, but under the control of a concentration difference across the membrane (osmotic flow). Then from Equation 5.86a, the reflection coefficient would be given by $\sigma = \dfrac{-Q}{SL_P RT \Delta C}$.

Durbin (1960) has obtained the data shown in Figure 5.15 for the reflection coefficient of a variety of solutes in dialysis tubing, cellophane, and a wet gel. One could use this figure directly to estimate the reflection coefficient for a solute of known (a/r).

The reflection coefficient can also be estimated using theories developed to describe the motion of particles in small pores. Anderson and Quinn (1974) reexamined the basic hydrodynamic equations that describe hindered particle motion in small pores. Under the assumptions of a rigid and spherical solute, and no electrostatic interactions with the pore wall, which is assumed to be inert, they showed that the sieving

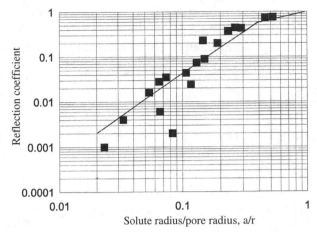

Figure 5.15 Reflection coefficient (based on data from Durbin 1960).

coefficient, $S_a \equiv N_S/C\,Q$, is the same as $(1 - \sigma)$. In this expression, C is the concentration of the solute on the non-filtrate side of the membrane. Recall that their expression for S_a is given by Equation 5.50. This equation can be rewritten to give the following predictive equation for the reflection coefficient.

$$\sigma = 1 - \left(1 - \frac{a}{r}\right)^2 \left[2 - \left(1 - \frac{a}{r}\right)^2\right] \times \left[1 - \frac{2}{3}\left(\frac{a}{r}\right)^2 - 0.163\left(\frac{a}{r}\right)^3\right] \qquad (5.93)$$

The solid line in Figure 5.15 represents a comparison of Equation 5.93 to the experimental data obtained by Durbin (1960) for the reflection coefficient. Equation 5.93 provides a reasonable comparison to Durbin's data at higher values of a/r, where the a/r ratio would be expected to be the dominant factor controlling the value of the reflection coefficient. At lower values of (a/r), other effects not accounted for in the derivation of Equation 5.93 dominate. In the absence of experimental data, Equation 5.93 can be used to provide an estimate of the reflection coefficient.

5.8.2 Multicomponent Membrane Transport

Equations 5.86a and 5.92 describe the filtration flowrate and the solute transport rate in a system composed of only a single solute and a solvent. However, similar equations can be derived for a multicomponent solute and solvent system by recognizing that now Equations 5.85a,b become:

$$Q = SL_P \Delta P + SRT \sum_i L_{PS_i} \Delta C_i \qquad (5.94a)$$

$$J_V = S \Delta P \sum_i L_{SP_i} + SRT \sum_i L_{S_i} \Delta C_i \qquad (5.94b)$$

The summations are over each solute in the system. Using the approach outlined earlier for the derivation of Equations 5.86a and 5.92, the following equations are obtained for the filtration rate and the transport rate of solute i across the membrane in multicomponent systems:

$$Q = SL_P \left[\Delta P - RT \sum_i \sigma_i \Delta C_i \right] \tag{5.95a}$$

$$N_i = C_i \left(1 - \sigma_i \right) Q + P_{m_i} S \Delta C_i \tag{5.95b}$$

5.9 TRANSPORT OF SOLUTES ACROSS THE CAPILLARY WALL

We will continue to use the capillary wall as a representative membrane to illustrate in the following examples the calculation of solute transport rates using the relationships developed in the previous sections. We will find it convenient in our discussion to classify solutes into three categories. The first category concerns the transport of small water soluble but lipid insoluble substances such as Na^+ and K^+ ions, glucose, amino acids, and other metabolites. The second category of solutes are lipid soluble substances such as oxygen and carbon dioxide, and the third category we will discuss is the transport of large lipid insoluble substances such as proteins (macromolecules).

Example 5.8 Calculate the rate of transport by filtration and diffusion of a small water soluble, but lipid insoluble solute, such as glucose. The *diffusivity of glucose (D)* is 0.91×10^{-5} cm^2 sec^{-1} and its molecular radius 0.36 nm. The average concentration of glucose in plasma is about 5 μmole ml^{-1}, and it is assumed that all the glucose transported to the extracapillary space is consumed instantly by the cells. Assume that all of the plasma proteins are retained by the capillary wall. Also, assume that the tortuosity of the capillary wall is equal to 2.

SOLUTION The filtration flowrate of plasma (Q) across the capillary wall is given by Equation 5.86a with $\sigma = 1.0$ and was shown earlier in Example 5.1 to be equal to 5.75×10^{-9} cm^3 hr^{-1}. Equation 5.92 may be used to calculate the solute transport rate of glucose. We first need to estimate the reflection coefficient for glucose. Equation 5.93 may be used to obtain an estimate assuming the diameter of the capillary pores is 7 nm.

$$\sigma = 1 - \left(1 - \frac{0.36}{3.5} \right)^2 \left[2 - \left(1 - \frac{0.36}{3.5} \right)^2 \right] \times \left[1 - \frac{2}{3} \left(\frac{0.36}{3.5} \right)^2 - 0.163 \left(\frac{0.36}{3.5} \right)^3 \right]$$

$$\sigma = 0.05$$

The permeability of glucose in the capillary wall may be estimated using Equation 5.83. The quantity $K \omega_r$ may be estimated from Equation 5.55 assuming the pores in the capillary wall are cylindrical in shape.

$$P_m = \frac{0.91 \times 10^{-5} \, \text{cm}^2 \, \text{sec}^{-1}}{2 \times 0.5 \times 10^{-6} \, \text{m}} \times \frac{1 \, \text{m}}{100 \, \text{cm}} \times 0.001 \times \left(1 - \frac{0.36}{3.5}\right)^2$$

$$\times \left[1 - 2.1\left(\frac{0.36}{3.5}\right) + 2.09\left(\frac{0.36}{3.5}\right)^3 - 0.95\left(\frac{0.36}{3.5}\right)^5\right] = 5.76 \times 10^{-5} \, \text{cm} \, \text{sec}^{-1}$$

Renkin (1977) reports values of the glucose permeability ranging from 1.6×10^{-6} cm/scc for capillaries in the dog brain to as high as 9.6×10^{-5} cm/sec for cat leg capillaries. On average, the capillary glucose permeability is about 2.65×10^{-5} cm/sec, thus the value calculated above is of the correct order of magnitude. We may now calculate by Equation 5.92 the transport rate of glucose across the capillary wall.

$$N_S = 5\frac{\mu\text{mole}}{\text{cm}^3}(1 - 0.05) \times 5.75 \times 10^{-9} \frac{\text{cm}^3}{\text{hr}} + 5.76 \times 10^{-5} \frac{\text{cm}}{\text{sec}} \times \pi$$

$$\times 10 \times 10^{-6} \, \text{m} \times 0.001 \, \text{m} \times (5 - 0)\frac{\mu\text{mol}}{\text{cm}^3} \times \frac{(100 \, \text{cm})^2}{\text{m}^2} \times \frac{3600 \, \text{sec}}{\text{hr}}$$

$$N_S = 2.73 \times 10^{-8} \frac{\mu\text{mole}}{\text{hr}} + 3.26 \times 10^{-4} \frac{\mu\text{mole}}{\text{hr}} = 3.26 \times 10^{-4} \frac{\mu\text{mole}}{\text{hr}}$$

The first term in the calculation of N_S by Equation 5.92 represents the transport rate of glucose by filtration. The second term represents the transport rate by diffusion. Comparing the above result, we find that the transport rate of small water soluble but lipid insoluble solutes such as glucose across the capillary wall is several thousand times higher by diffusion than by convection! We may therefore neglect the convective transport of these substances across the capillary wall.

Example 5.9 Now consider the transport rate of lipid soluble solutes such as oxygen. In this case the entire surface area of the capillary wall is available for transport. Oxygen has considerable solubility in the cell membrane and will readily diffuse through the endothelial cells of the capillary wall. Oxygen can also diffuse through the pores in the capillary wall. The *diffusivity of oxygen (D)* is 2.11×10^{-5} cm^2/sec and its molecular radius is 0.16 nm. The average concentration of oxygen in the capillary is 0.09 μmole/cm^3 Once again, assume that the tortuosity of the capillary wall is 2.

SOLUTION The filtration flowrate of plasma (Q) across the capillary wall was shown earlier to be equal to 5.75×10^{-9} cm^3/hr. Equation 5.92 may be used to calculate the solute transport rate of oxygen through the pores in the capillary wall. We first need to estimate the reflection coefficient for oxygen. Equation 5.93 may be used to obtain an estimate assuming the diameter of the capillary pores is 7 nm.

$$\sigma = 1 - \left(1 - \frac{0.16}{3.5}\right)^2 \left[2 - \left(\frac{0.16}{3.5}\right)^2\right] \times \left[1 - \frac{2}{3}\left(\frac{0.16}{3.5}\right)^2 - 0.163\left(\frac{0.16}{3.5}\right)^3\right]$$

$$\sigma = 0.01$$

The permeability of oxygen in the capillary pores may now be determined using Equations 5.55 and 5.83.

$$P_m = \frac{2.11 \times 10^{-5}\,\text{cm}^2\,\text{sec}^{-1}}{2 \times 0.5 \times 10^{-6}\,\text{m}} \times \frac{1\,\text{m}}{100\,\text{cm}} \times 0.001 \times \left(1 - \frac{0.16}{3.5}\right)^2$$

$$\times \left[1 - 2.1\left(\frac{0.16}{3.5}\right) + 2.09\left(\frac{0.16}{3.5}\right)^3 - 0.95\left(\frac{0.16}{3.5}\right)^5\right] = 1.74 \times 10^{-4}\,\text{cm sec}^{-1}$$

We may now calculate by Equation 5.92 the transport rate of oxygen through the pores of the capillary wall.

$$N_S = 0.09\frac{\mu\text{mole}}{\text{cm}^3}(1 - 0.01) \times 5.75 \times 10^{-9}\frac{\text{cm}^3}{\text{hr}} + 1.74 \times 10^{-4}\frac{\text{cm}}{\text{sec}}$$

$$\times \pi \times 10 \times 10^{-6}\,\text{m} \times 0.001\,\text{m} \times (0.09 - 0)\frac{\mu\text{mol}}{\text{cm}^3} \times \frac{(100\,\text{cm})^2}{\text{m}^2} \times \frac{3600\,\text{sec}}{\text{hr}}$$

$$N_S = 5.12 \times 10^{-10}\frac{\mu\text{mole}}{\text{hr}} + 1.77 \times 10^{-5}\frac{\mu\text{mole}}{\text{hr}} = 1.77 \times 10^{-5}\frac{\mu\text{mole}}{\text{hr}}$$

Comparing the above result, we again find that the transport rate of small molecules like oxygen through the capillary pores is several thousand times higher by diffusion than by convection.

The permeability of oxygen through the capillary wall itself is given by the following equation.

$$P_{oxygen} = \frac{K_{oxygen}D}{t_m} = \frac{0.25 \times 2.11 \times 10^{-5}\frac{\text{cm}^2}{\text{sec}}}{0.5 \times 10^{-6}\,\text{m}} \times \frac{1\,\text{m}}{100\,\text{cm}} = 0.11\frac{\text{cm}}{\text{sec}}$$

K_{oxygen} is the oxygen partition coefficient and represents the ratio of the oxygen concentration in the capillary wall to that in the bulk solution. The value of 0.25 was used in order to match the oxygen permeability of the capillary wall with the accepted value of about 0.1 cm/sec. The oxygen transport rate from the capillary is then equal to:

$$N_{oxygen} = 0.11\frac{\text{cm}}{\text{sec}} \times \pi \times 10 \times 10^{-6}\,\text{m} \times 0.001\,\text{m} \times (0.09 - 0)\frac{\mu\text{mol}}{\text{cm}^3} \times \frac{(100\,\text{cm})^2}{\text{m}^2} \times \frac{3600\,\text{sec}}{\text{hr}}$$

$$= 0.011\frac{\mu\text{mol}}{\text{hr}}$$

This is essentially the total oxygen transport rate from the capillary, since the contribution due to the capillary pores is negligible in comparison. The smaller value of the oxygen transport rate through the pores (N_S) in comparison to the value given above for the entire capillary wall (N_{oxygen}) represents the fact that the pore surface area is 1/1000th that of the capillary wall. Since the permeability of oxygen through the pores of the capillary and the capillary wall itself need to be multiplied by the same values of S and ΔC to calculate the transport rate of oxygen, it is seen that regardless of these values, the diffusive transport of oxygen is 1000-fold higher through the endothelial cells lining the capillary walls than that transported by diffusion through the capillary pores. Thus we can conclude that the dominant mode for transport of oxygen is simple diffusion across the entire surface area of the capillary walls.

Example 5.10 Our final category of solute to consider is lipid insoluble macromolecules such as proteins. These solutes are characterized by a molecular radius that is comparable to the pores in the capillary wall. Calculate the rate of transport by filtration and diffusion through the capillary wall of a protein of molecular weight equal to 50,000 Daltons. Assume the concentration of the protein in the plasma is 0.05 micromoles/ml.

SOLUTION The filtration flowrate of plasma (Q) across the capillary wall is equal to 5.75×10^{-9} cm^3/hr. Since we only know the molecular weight of the protein, we need to first estimate its radius and its diffusivity. From Equation 5.49 we find that the radius of the solute is equal to 2.71 nm. We can also use Equation 5.47 to estimate the diffusivity. Therefore, we find that $D = 7 \times 10^{-7}$ cm^2/sec. Equation 5.92 may be used to calculate the solute transport rate of the protein. We first need to estimate the reflection coefficient for the protein. Equation 5.93 may be used to obtain an estimate assuming the diameter of the capillary pores is 7 nm.

$$\sigma = 1 - \left(1 - \frac{2.71}{3.5}\right)^2 \left[2 - \left(1 - \frac{2.71}{3.5}\right)^2\right] \times \left[1 - \frac{2}{3}\left(\frac{2.71}{3.5}\right)^2 - 0.163\left(\frac{2.71}{3.5}\right)^3\right]$$

$$\sigma = 0.95$$

The permeability of the protein in the capillary wall may be estimated using Equation 5.83. The quantity $K \omega$, may be estimated from Equation 5.55 assuming the pores in the capillary wall are cylindrical in shape.

$$P_m = \frac{7.0 \times 10^{-7} \text{cm}^2/\text{sec}}{2 \times 0.5 \times 10^{-6} \text{m}} \times \frac{1 \text{ m}}{100 \text{ cm}} \times 0.001 \times \left(1 - \frac{2.71}{3.5}\right)^2$$

$$\times \left[1 - 2.1\left(\frac{2.71}{3.5}\right) + 2.09\left(\frac{2.71}{3.5}\right)^3 - 0.95\left(\frac{2.71}{3.5}\right)^5\right] = 2.85 \times 10^{-8} \text{cm/sec}$$

This value of the permeability is representative of the experimental values for macromolecules of this size reported by Renkin (1977). We may now calculate by Equation 5.92 the transport rate of the protein across the capillary wall.

$$N_S = 0.05 \frac{\mu mole}{cm^3}(1 - 0.95) \times 5.75 \times 10^{-9} \frac{cm^3}{hr} + 2.85 \times 10^{-8} \frac{cm}{sec}$$

$$\times \pi \times 10 \times 10^{-6} m \times 0.001 m \times (0.05 - 0) \frac{\mu mol}{cm^3} \times \frac{(100 \, cm)^2}{m^2} \times \frac{3600 \, sec}{hr}$$

$$N_S = 1.44 \times 10^{-11} \frac{\mu mole}{hr} + 1.61 \times 10^{-9} \frac{\mu mole}{hr} = 1.61 \times 10^{-9} \frac{\mu mole}{hr}$$

Once again we observe that even for a macromolecular solute with a molecular weight of 50,000 Daltons, that diffusive transport across the capillary wall is still significantly higher than the amount of solute transported by convection.

Example 5.11 Repeat the previous example assuming that the mean capillary pressure is increased by 20 mm Hg.

SOLUTION The filtration flowrate of plasma (Q) across the capillary wall will be significantly increased as a result of the increase in mean capillary pressure. The net filtration pressure will increase from 0.3 to 20.3 mm Hg. We can calculate the filtration rate of the plasma following the procedure outlined in Example 5.1. We can then show that the filtration rate equals 3.89×10^{-7} cm^3 hr^{-1}. This represents a nearly 68-fold increase in the plasma filtration rate and would lead to edema in the interstitium. The reflection coefficient and solute permeability would still have the same values as calculated in the previous example, i.e. $\sigma = 0.95$ and $P_m = 2.85 \times 10^{-8}$ cm sec^{-1}.

We may now calculate by Equation 5.92 the transport rate of the protein across the capillary wall for the special case of increased mean capillary pressure.

$$N_S = 0.05 \frac{\mu mole}{cm^3}(1 - 0.95) \times 3.89 \times 10^{-7} \frac{cm^3}{hr} + 2.85 \times 10^{-8} \frac{cm}{sec}$$

$$\times \pi \times 10 \times 10^{-6} m \times 0.001 m \times (0.05 - 0) \frac{\mu mol}{cm^3} \times \frac{(100 \, cm)^2}{m^2} \times \frac{3600 \, sec}{hr}$$

$$N_S = 9.73 \times 10^{-10} \frac{\mu mole}{hr} + 1.61 \times 10^{-9} \frac{\mu mole}{hr} = 2.58 \times 10^{-9} \frac{\mu mole}{hr}$$

The first term in the calculation of N_S by Equation 5.92 represents the transport rate of the protein by filtration. The second term represents the transport rate by diffusion. Unlike the calculations for solute transport in the previous examples, we now observe that the increased filtration pressure has significantly increased the transport rate of the solute by convection. Convection of the solute now accounts for 38% of the total solute transport rate and is responsible for the 60% increase in the solute transport rate as a result of the increased mean capillary pressure.

5.10 TRANSPORT OF A SOLUTE BETWEEN A CAPILLARY AND THE SURROUNDING TISSUE SPACE

Now that we have developed an understanding of solute transport across a semipermeable membrane like the capillary wall, it is possible to develop a model for the spatial distribution and consumption or production of a particular species within the tissue space surrounding a given capillary. The following equations also describe the transport of a solute from within a hollow fiber to the cylindrical space surrounding the fiber. In the case of oxygen, the situation is somewhat more complex since the red blood cell provides an additional source or sink of oxygen through interactions with hemoglobin. We will discuss oxygen transport later in Chapter 6.

5.10.1 The Krogh Tissue Cylinder

Microscopic studies support somewhat the notion that a bed of capillaries in tissue can be represented as a repetitive arrangement of capillaries surrounded by a cylindrical layer of tissue. Figure 5.16 shows an idealized sketch of the capillary bed and the corresponding cylindrical layer of tissue of thickness equal to r_T. As mentioned at the beginning of this chapter, the residence time for blood in the capillary is on the order of 1 second. The diffusivity of a solute in tissue is on the order of 10^{-5} cm^2/sec. The characteristic time for diffusion in the Krogh tissue cylinder can be expressed by the following relation: $t_{\text{diffusion}} = \dfrac{r_T^2}{2 D} \ln\left(\dfrac{r_T}{r_C}\right)$ (Lightfoot 1995). Since the blood residence time and the diffusion time must be of the same order of magnitude, we can use this simple relationship to estimate that the Krogh tissue cylinder radius is on the order of 30 microns. Krogh (1919) used this cylindrical capillary tissue model to study the supply of oxygen to muscle. This approach is now known as the *Krogh tissue model*.

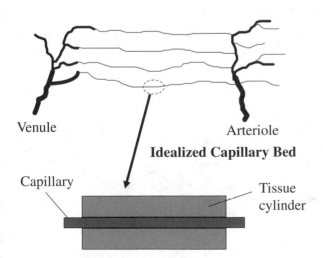

Venule   Arteriole

Idealized Capillary Bed

Capillary   Tissue cylinder

Figure 5.16   The Krogh tissue cylinder.

A steady state mathematical model for the transport of a solute can be developed using the Krogh tissue model. We will treat the tissue space surrounding the capillary as a continuous phase and ignore the fact that it is composed of discrete cells. Therefore, the diffusion of the solute in the tissue space may be described by an effective diffusivity (D_T) as defined earlier by Equations 5.77, 5.78, or 5.81. The driving force for diffusion of the solute is the consumption (or production) of the solute by the cells within the tissue space.

We also need to have an expression that relates the consumption or production of the solute to the concentration of the solute in the tissue space ($\bar{C}$).

The Michaelis–Menten equation can be used to describe the metabolic consumption (or production) of the solute in the tissue space. The Michaelis–Menten equation may be written as follows:

$$R(\bar{C}) = \frac{V_{max}\bar{C}}{K_m + \bar{C}} \tag{5.96}$$

For consumption of the solute, $R(\bar{C})$ will be assumed to have a positive value, whereas if the solute is produced, then $R(\bar{C})$ will have a negative value. In this equation, V_{max} represents the maximum reaction rate. The maximum reaction rate occurs when $\bar{C} \gg K_m$, and the reaction rate is then said to be zero-order in the solute concentration (i.e. $\bar{C}^0$). K_m is called the Michaelis constant and represents the value of $\bar{C}$ for which the reaction rate is one-half the maximal value. When $\bar{C} \ll K_m$, then $R(\bar{C}) = \frac{V_{max}}{K_m}\bar{C}$, and the reaction is first-order in the solute concentration (i.e. $\bar{C}^1$). In many cases for biological reactions, $\bar{C} \gg K_m$, and we may reasonably assume that the reaction is zero order and that $R(\bar{C}) = V_{max} = R_0$. This will considerably simplify the following mathematical analysis.

5.10.2 A Model of the Krogh Tissue Cylinder

Figure 5.17 illustrates a finite volume section of the Krogh tissue cylinder extending from z to $z + \Delta z$. Within the capillary, the solute is transported primarily by convection in the axial direction and by diffusion in the radial direction. Diffusion of the solute

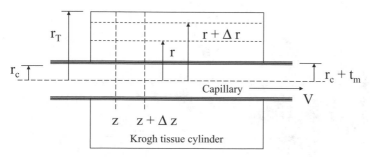

Figure 5.17   Geometry of the Krogh tissue cylinder.

from the blood and through the capillary wall is proportional to the concentration difference of the solute and is represented by an overall mass transfer coefficient (K_0). The overall mass transfer coefficient represents the combined resistance of the fluid flowing through the capillary (i.e. k_m) and the permeability of the solute in the capillary wall (P_m). The overall mass transfer coefficient K_0 can be evaluated by recognizing that the solute flux at any value of z is constant (neglecting the curvature of the capillary wall) from the blood to the capillary wall at r_C, and across the capillary wall to the beginning of the tissue space at $r_c + t_m$. The transport rate across the capillary wall is given by Equation 5.92, recognizing that in most cases, as shown in the previous examples, the convective term $C(1 - \sigma)Q$ is negligible in comparison to the amount of solute transported by diffusion. We may then write at a particular axial location along the capillary that:

$$N_s = 2\pi r_C \Delta z\, k_m \left(C - \hat{C}\big|_{r_C} \right) = 2\pi r_C \Delta z\, P_m \left(\hat{C}\big|_{r_C} - \bar{C}\big|_{r_C + t_m} \right)$$

$$= 2\pi r_C \Delta z\, K_0 \left(C - \bar{C}\big|_{r_C + t_m} \right) \tag{5.97}$$

where $\hat{C}$ now denotes the solute concentration at the surface of the capillary wall on the blood side. Equation 5.97 may then be rearranged to obtain the following expression for the overall mass transfer coefficient.

$$K_0 = \cfrac{1}{\cfrac{1}{k_m} + \cfrac{1}{P_m}} \tag{5.98}$$

P_m is the solute permeability across the capillary and k_m is the blood side film mass transfer coefficient defined earlier in this chapter.

The transport properties for the blood, as represented by the viscosity and the solute diffusivity, are based on their respective plasma values. This occurs because for blood the Schmidt number, $Sc = \mu/\rho D$, is much greater than unity. Laminar boundary layer theory (Bird *et al.* 2002) tells us that the concentration boundary layer, the distance from the capillary wall over which the solute concentration changes from its bulk value to its value at the capillary wall, is much thinner for $Sc \gg 1$ than the momentum or velocity boundary layer, the distance from the capillary wall over which the axial velocity changes from its bulk value to zero at the capillary wall. The red blood cells also have a tendency, as discussed in Chapter 4, to accumulate along their axis of flow. This creates a cell-free plasma layer adjacent to the capillary wall. Because of the thinness of the concentration boundary layer, it lies well within the cell-free layer making it appropriate to base the transport properties on the plasma values.

We also treat the blood as a continuous phase and ignore the finite size of the red blood cells. We will assume that the blood flows through the capillary with an average velocity of V. With these assumptions, a steady state shell balance on the solute in the blood from z to $z + \Delta z$ provides the following equation.

$$V \pi r_c^2\, C\big|_z - V \pi r_c^2\, C\big|_{z + \Delta z} = 2\pi r_c \Delta z\, K_0 \left(C - \bar{C}\big|_{r_c + t_m} \right) \tag{5.99}$$

Dividing by Δz, and taking the limit as $\Delta z \to 0$, results in the following differential equation for the solute concentration in the blood.

$$-V\frac{dC}{dz} = \frac{2}{r_c}K_0\left(C - \bar{C}\big|_{r_c+t_m}\right) \tag{5.100}$$

A steady state shell balance at a given value of z from r to $r + \Delta r$ may also be written for the solute concentration in the tissue space.

$$-D_T\frac{d\bar{C}}{dr}2\pi r\Delta z\Big|_r + D_T\frac{d\bar{C}}{dr}2\pi r\Delta z\Big|_{r+\Delta r} = R(\bar{C})2\pi r\Delta r\Delta z \tag{5.101}$$

This equation neglects axial diffusion in comparison to radial diffusion within the tissue space. After dividing by $2\pi\, r\, \Delta r$, and taking the limit as $\Delta r \to 0$, the following differential equation results for the solute concentration in the tissue space.

$$\frac{D_T}{r}\frac{d}{dr}\left(r\frac{d\bar{C}}{dr}\right) - R(\bar{C}) = 0 \tag{5.102}$$

The boundary conditions for differential Equations 5.100 and 5.102 are:

$$\text{BC1}: \ z = 0, \ \ C = C_0$$
$$\text{BC2}: \ r = r_c + t_m, \ \ \bar{C} = \bar{C}\big|_{r_c+t_m} \tag{5.103}$$
$$\text{BC3}: \ r = r_T, \ \ \frac{d\bar{C}}{dr} = 0$$

It should be noted that even though we neglect axial diffusion in the tissue space in comparison to radial diffusion, the solute concentration in the tissue space will still have an axial dependence. This axial dependence of the tissue space solute concentration results from the boundary condition at $(r_c + t_m)$ and the solute flux term, $K_0(C - \bar{C}\big|_{r_c+t_m})$, which couples the solutions for the solute concentration within the capillary and the tissue space. In this way, the axial variation of the solute concentration in the capillary is impressed on that of the tissue space.

The solute concentration in the tissue space is then easily found by solving Equation 5.102 subject to boundary conditions 2 and 3 in Equation 5.103. Here we assume the consumption of solute is zero-order as discussed earlier, hence $R(\bar{C}) = R_0$ is a constant. The following equation is then obtained.

$$\bar{C}(r,z) = \bar{C}(z)\big|_{r_c+t_m} + \frac{R_0}{4D_T}\left[\left(r^2 - (r_c + t_m)^2\right)\right] - \frac{R_0 r_T^2}{2D_T}\ln\left(\frac{r}{r_c+t_m}\right) \tag{5.104}$$

This equation provides no information on the value of $\bar{C}(z)\big|_{r_c+t_m}$. However, at any axial position z from the capillary entrance, we must have at steady state that the change in solute concentration within the blood must equal the consumption of solute in the tissue space. Therefore, we can express this requirement by writing the following equation:

$$V \pi r_c^2 C_0 - V \pi r_c^2 C \big|_z = \pi \left[r_T^2 - \left(r_c + t_m \right)^2 \right] z \, R_0 \tag{5.105}$$

This equation may be rearranged to give the following equation that provides for the axial variation of the solute concentration in the capillary.

$$C(z) = C_0 - \frac{R_0}{V r_c^2} \left[r_T^2 - \left(r_c + t_m \right)^2 \right] z \tag{5.106}$$

Equation 5.106 can now be used to find $\dfrac{dC}{dz}$ in Equation 5.100 with the result that we can solve for $\overline{C}(z) \big|_{r_c + t_m}$ which is given by the equation below.

$$\overline{C}(z) \big|_{r_c + t_m} = C(z) - \frac{R_0}{2 r_c K_0} \left(r_T^2 - \left(r_c + t_m \right)^2 \right) \tag{5.107}$$

Note that $C(z)$ in Equation 5.107 makes $\overline{C}(z) \big|_{r_c + t_m}$ depend on z, and by Equation 5.104, the tissue space solute concentration then depends on both r and z as discussed earlier. Equations 5.104, 5.106, and 5.107 can be combined to give the following equation for the solute concentration in the tissue space.

$$\overline{C}(r,z) = C_0 - \frac{R_0}{V r_c^2} \left(r_T^2 - \left(r_c + t_m \right)^2 \right) z$$

$$- \frac{R_0}{2 r_c K_0} \left[r_T^2 - \left(r_c + t_m \right)^2 \right] + \frac{R_0}{4 D_T} \left[\left(r^2 - \left(r_c + t_m \right)^2 \right) \right] - \frac{R_0 r_T^2}{2 D_T} \ln \left(\frac{r}{r_c + t_m} \right) \tag{5.108}$$

Under some conditions either the delivery of the solute to the capillary may be limited by the capillary flowrate, or the transport rate of the solute across the capillary wall is limited, or the consumption of the solute by the tissue is very rapid. Any one of these conditions may lead to regions of the tissue that have no solute. We can then define a *critical radius* in the tissue, $r_{critical}(z)$, defined as the distance beyond which no solute is present in the tissue. For this situation we need to modify boundary condition 3 in Equation 5.103 to the following:

$$\text{BC3}' : r = r_{critical}(z), \quad \frac{d\overline{C}}{dr} = 0 \text{ and } \overline{C} = 0 \tag{5.109}$$

Under these conditions, the solute concentrations in the capillary, i.e. $C(z)$, at the interface between the capillary and the tissue space, i.e. $\overline{C}(z) \big|_{r_c + t_m}$, and in the tissue space itself, i.e. $\overline{C}(r,z)$, would still be given respectively by Equations 5.106, 5.107, and 5.108, however, the Krogh tissue cylinder radius, r_T, is replaced with $r_{critical}(z)$ once the solute concentration in the tissue at a particular location has reached zero. The critical radius may be obtained by recognizing that at $r_{critical}(z)$, $\overline{C}(r, z) = 0$. Thus we may use Equation 5.108, with $r_T = r_{critical}(z)$ and $\overline{C}(r, z) = 0$, to obtain the following expression for the critical radius.

$$\left(\frac{r_{critical}(z)}{r_c+t_m}\right)^2 \ln\left(\frac{r_{critical}(z)}{r_c+t_m}\right)^2 - \left(\frac{r_{critical}(z)}{r_c+t_m}\right)^2 + 1 = \left(\frac{4D_TC_0}{R_0(r_c+t_m)^2}\right)$$

$$-\frac{4D_T}{Vr_c^2}\left[\left(\frac{r_{critical}(z)}{r_c+t_m}\right)^2 - 1\right]z - \frac{2D_T}{r_cK_0}\left[\left(\frac{r_{critical}(z)}{r_c+t_m}\right)^2 - 1\right] \quad (5.110)$$

This is a non-linear algebraic equation that may be solved for the critical radius. This is illustrated in Example 5.13 below. Of particular interest are the conditions under which the solute just begins to be depleted within the tissue space. The depletion of the solute will start at the downstream corner of the Krogh tissue cylinder represented by the coordinates $z = L$ and $r = r_T$. Equation 5.108 may be used to explore the conditions when the solute depletion just begins by recognizing that $\overline{C}(r_T,L) = 0$.

The calculation of the solute concentration within the capillary and the tissue space surrounding the capillary is illustrated in the following examples.

Example 5.12 Calculate the glucose concentration profiles for an exercising muscle capillary. Assume the capillary properties summarized in Table 5.1 apply and that the glucose consumption rate, R_0, is equal to 0.01 μmole/sec/cm³. Also assume that the Krogh tissue cylinder radius, r_T, is 40 microns.

SOLUTION Using Equations 5.106 and 5.108, we obtain the solution shown in Figure 5.18. Shown in the figure are the axial solute concentrations within the capillary and in the tissue space at the Krogh tissue cylinder radius (r_T). As shown in the figure, we observe that the axial solute concentration within the capillary and in the tissue space decreases in a linear fashion. We also observe very little difference between the gluose concentration at the outer surface of the capillary wall and at the Krogh

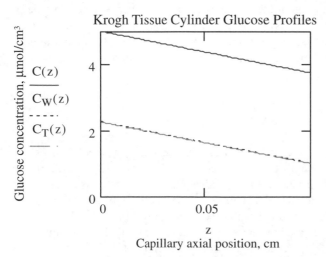

Figure 5.18 Glucose concentration as a function of axial position in a capillary.

tissue cylinder radius. This indicates that there is little mass transfer resistance to glucose within the tissue space. It is also important to note for glucose transport that the bulk of the mass transfer resistance is the capillary wall as reflected by the value of the glucose permeability, P_m. We see that the glucose permeability (5.76×10^{-5} cm/sec) is about two orders of magnitude smaller than the film mass transfer coefficient k_m (0.033 cm/sec) and for practical purposes we can ignore the blood side mass transfer effect since it offers little resistance in comparison to that of the capillary wall. We also see that because of the large mass transfer resistance of the capillary wall, that a significant concentration difference must exist between the capillary and the tissue space to obtain the required mass transfer rate. Transport of glucose for these conditions is, therefore, said to be diffusion limited.

Example 5.13 Calculate the critical radius for the capillary conditions used in the previous example. Assume that the flowrate in the capillary has been reduced by a factor of ten to $V = 0.005$ cm/sec.

SOLUTION Figure 5.19 shows the solution for the critical radius as a function of the axial position in the capillary. The calculation shows that the glucose concentration at the Krogh tissue cylinder radius is non-zero until 0.018 cm into the capillary of total length 0.1 cm. The region above the curve in the critical radius figure represents the region of the tissue where the glucose concentration is zero. We also see that the exiting glucose concentration for the reduced flow conditions in this example, i.e. 0.887 μmole cm^{-3}, is significantly reduced in comparison to the result obtained in the previous example, i.e. 3.8 μmole cm^{-3}. Although the exiting glucose concentration in the capillary is lower, this does not necessarily mean that the total amount of glucose transferred is higher. In fact the amount of solute transferred in this example is about one-third lower than the previous example because of the

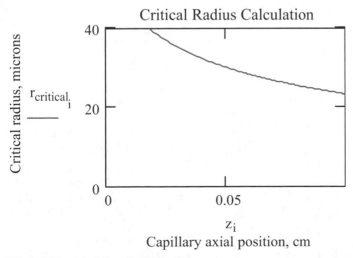

Figure 5.19 Calculation of critical radius.

reduced velocity of the blood. For solutes with reasonably high capillary permeability, i.e. lipophilic solutes like oxygen or small lipophobic solutes like glucose, the solute transport rate as in this example can be flow limited. The transport rate of the solute can, therefore, be increased to the tissue by increasing the blood flow rate as in the previous example. However, for larger lipid insoluble solutes, the capillary solute permeability is significantly reduced, and the solute transport rate is diffusion limited. In this case, the blood flow rate has little effect on the solute transport rate.

5.10.2.1 A comparison of convection and diffusion effects Recall that in developing Equations 5.106 and 5.108 that we ignored axial diffusion of the solute within the capillary in comparison to axial convection of solute, that is the solute carried along by the flowing blood. We can easily check the validity of this assumption through the following arguments. The amount of solute carried by convection is on the order of:

$$\text{solute transport by axial convection} = \frac{\pi}{4} D_C^2 V C_0 \tag{5.111}$$

and the amount of solute carried by axial diffusion is on the order of:

$$\text{solute transport by axial diffusion} = \frac{\pi}{4} D_C^2 \frac{D C_0}{L} \tag{5.112}$$

By taking the ratio of Equations 5.111 and 5.112 we obtain what is called the *Peclet (Pe) number*, whose magnitude represents the importance of axial convection in comparison to axial diffusion. The criterion for ignoring axial diffusion is then given by the fact that $Pe = V\,L/D \gg 1$. For the example problems presented above:

$$Pe = VL/D = (0.05 \text{ cm sec}^{-1} \times 0.1 \text{ cm})/0.91 \times 10^{-5} \text{ cm}^2 \text{ sec}^{-1} = 550 \gg 1$$

and we conclude that we can ignore axial diffusion.

A similar line of reasoning may be applied to the tissue space to support the neglect of axial diffusion in comparison to radial diffusion. The solute transport by radial diffusion would be on the order of:

$$\text{solute transport by radial diffusion} = \left(\frac{D_T C_0}{r_T - r_C} \right) 2\pi r_C L \tag{5.113}$$

and the amount of solute transport by axial diffusion would be on the order of:

$$\text{solute transport by axial diffusion} = D_T \frac{C_0}{L} \pi \left(r_T^2 - r_C^2 \right) \tag{5.114}$$

The ratio of Equation 5.113 and 5.114 is a measure of the relative importance of radial diffusion in comparison to axial diffusion. If this ratio is $\gg 1$, then axial diffusion can be neglected. For the example problem we then have:

$$\frac{2 r_C L^2}{(r_T - r_C)(r_T^2 - r_C^2)}$$

$$= \frac{2 \times 0.5 \times 10^{-3} \text{cm} \times (0.1 \text{ cm})^2}{(4 \times 10^{-3} - 0.5 \times 10^{-3}) \left[\left(4 \times 10^{-3} \right)^2 - \left(0.5 \times 10^{-3} \right)^2 \right] \text{cm}^3} = 181 \gg 1$$

and we conclude that we can ignore axial diffusion in the tissue space.

5.10.2.2 The Renkin–Crone equation The above analysis can also be simplified considerably for the special case where the solute concentration in the tissue space is zero or much smaller than the concentration in the blood. We can also assume, as shown in the above examples that $K_0 \approx P_m$. In other words, the blood offers little resistance to the solute transport rate in comparison to that of the capillary wall. For these conditions, Equation 5.100 simplifies to:

$$\frac{dC}{dz} = -\frac{2}{Vr_c} P_m C = -\frac{2\pi r_c}{Q} P_m C \qquad (5.115)$$

Here we have replaced the capillary blood velocity (V) with the volumetric flowrate of blood in the capillary (Q). This equation is easily integrated to obtain the solute concentration in the capillary at any axial position z.

$$E = \frac{[C_0 - C(z)]}{C_0} = 1 - \exp\left(-\frac{2\pi r_c P_m z}{Q}\right) \qquad (5.116)$$

This equation is known as the Renkin–Crone equation and describes how the solute concentration varies along the length of the capillary for the special case where the solute concentration in the tissue space is very small in comparison to its value in the blood. The solute extraction is represented by E. Since the quantity $Q C_0$ represents the maximum amount of solute that can be transported across the capillary, then $E \times Q C_0$ represents the actual solute transport rate for the given conditions. The solute extraction is a measure of the removal efficiency of the solute from the capillary. This equation readily shows that solute transport from the capillary is strongly dependent on the ratio of the permeability of the capillary wall ($2\pi r_c P_m z$) to the blood flow rate (Q). This ratio can be used to define conditions under which the solute transport is either flow limited or diffusion limited as follows:

$$\text{if } \frac{2\pi r_c P_m z}{Q} \gg 1, \text{ then flow limited}$$

$$\text{if } \frac{2\pi r_c P_m z}{Q} \ll 1, \text{ then diffusion limited} \qquad (5.117)$$

For regions of tissue containing many capillaries, we can replace the quantity $2\pi r_c P_m z$ with the group $P_m S$, where S represents the total surface area of the capillaries within the tissue region of interest.

5.10.2.3 Determining the value of $P_m S$ The multiple tracer indicator diffusion technique may be used to obtain a test solute's permeability ($P_m S$) across the capillary wall in organs and large tissue regions. This technique can also be used to evaluate the solute permeability in medical devices that employ membranes, for example in hemodialysis and bioartificial organs.

In this technique, a solution containing equal concentrations of a permeable test solute and a non-permeable reference solute (typically labeled albumin), is rapidly injected into a main artery leading to the region of interest (Levick 1991). Venous blood samples are then collected over the next few seconds and analyzed for the concentratons of the test and reference solutes. Because the test solute is permeable, its concentration in the venous blood will initially fall below that of the reference solute due to its diffusion into the surrounding interstitium. This is shown in Figure 5.20. Because of the transient nature of this technique and the mass transfer resistance of the capillary wall, the concentration of the test solute in the interstitium may be assumed to be close to zero or much smaller than the value within the blood. The concentration of the reference solute in the venous blood provides an estimate of what the test solute concentration would have been at a particular time if no transport of the test solute out of the capillaries had occurred. After several seconds, the reference solute concentration in the blood will fall below that of the test solute since the reference solute has been nearly washed out of the capillary and the test solute concentration in the interstitium is now larger than that in the blood. Hence, the test solute starts to diffuse back into the blood.

The concentrations of the test (i.e. C) and reference solutes (i.e. C_0) in the venous blood at a particular time may then be used to calculate, using Equation 5.116 (with $2\pi r_c P_m z = P_m S$), the test solute extraction (E) and the value of $P_m S$ for a given blood flowrate (Q). It is important to note that accurate results require that the test be repeated at increasing blood flow rates (Q) to ensure that the solute transport is not flow limited.

The term *clearance* is often used to describe the removal of a solute from a flowing fluid such as in a capillary. Clearance, given the symbol CL, is the volumetric flowrate of the fluid that has been totally cleared of the solute.

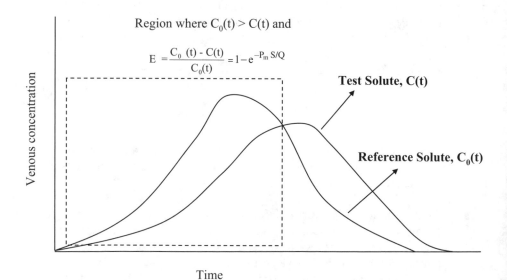

Figure 5.20   Multiple tracer indicator diffusion technique.

Clearance is then defined by the following expression:

$$CL(z) \equiv \frac{Q(C_0 - C(z))}{C_0} = QE = Q\left(1 - e^{-\frac{P_m S}{Q}}\right) \qquad (5.118)$$

This expression shows that capillary solute clearance is linearly dependent on the flowrate (i.e. $CL = Q$) under flow limited conditions where $P_m S \gg Q$ and $E \approx 1$. However, under diffusion limited conditions where $P_m S \ll Q$, the parenthetical term in the above equation becomes equal to $P_m S/Q$ and we obtain the result that the clearance becomes independent of the flowrate (Q) and is equal to the capillary wall permeability (i.e. $CL = P_m S$).

Example 5.14 Blood is flowing through a hollow fiber based bioartificial organ at a flowrate (Q) of 8 ml min^{-1} for each fiber. Surrounding the fibers in the shell space are living cells obtained from a pig's liver. Each fiber has a diameter of 800 microns and a length of 30 cm. At time equal to zero, equal concentrations of a permeable (relative to the fiber wall) test solute and labeled albumin (impermeable) are injected into the blood entering the device. Exiting blood samples are collected over the next few seconds and analyzed for the concentrations of the test solute (C) and the reference solute, albumin (C_0). At a particular time, the value of C/C_0 was found to equal 0.60. Estimate the solute permeability (P_m) of the hollow fiber membrane (cm sec^{-1}).

SOLUTION We can use Equation 5.118 first to calculate the clearance of a given hollow fiber for the test solute as shown below.

$$CL = Q\left(1 - \frac{C}{C_0}\right) = 8 \text{ ml min}^{-1} \times (1 - 0.6) = 3.2 \text{ ml min}^{-1}$$

Next we calculate from Equation 5.118 the value of P_m.

$$P_m = -\frac{Q}{S}\ln\left(1 - \frac{CL}{Q}\right) = -\frac{8 \text{ ml min}^{-1} 1 \min (60 \sec)^{-1}}{2\pi \times 0.04 \text{ cm} \times 30 \text{ cm}}\ln\left(1 - \frac{3.2 \text{ ml min}^{-1}}{8 \text{ ml min}^{-1}}\right)$$

$$P_m = 0.009 \text{ cm sec}^{-1}$$

5.10.3 Solute Transport in Vascular Beds

The Krogh tissue cylinder model provides a description of solute transport within a single capillary. In many cases, a model is needed to describe the solute transport in a much larger region containing numerous capillaries as illustrated for the capillary bed shown in Figure 5.16. Equation 5.116 with $2\pi\, r_c\, P_m\, z$ replaced with $P_m S$ applies to vascular beds but assumes the solute concentration is zero in the tissue space.

One simple approach to describe solute transport in this situation would be to treat the blood in the capillary bed and the tissue space as separate well-mixed regions. To facilitate the analysis, we introduce two parameters that characterize the degree of vascularization in the region of interest. The first parameter would be the *surface area capillary density* (*s*), which is the capillary surface area per unit volume of tissue. Assuming the geometry of the Krogh tissue cylinder shown in Figure 5.17, the following equation provides an estimate of (*s*) based on given values of the capillary radius, r_c, and the Krogh tissue cylinder radius, r_T.

$$s = \frac{2r_c}{r_T^2} \tag{5.119}$$

The second parameter is the *volume of capillaries per unit volume of tissue* (*v*). In terms of the Krogh tissue model, this may also be expressed in terms of r_c and r_T.

$$v = \left(\frac{r_c}{r_T}\right)^2 \tag{5.120}$$

A steady state solute balance, assuming only diffusion is relevant for solute transport across the capillary wall, may be written for the capillary bed illustrated in Figure 5.16. We may therefore write that the solute carried into the capillary bed by blood flow must equal the amount removed by diffusion across the capillary walls and that which leaves with the blood flow.

$$QC_0\big|_{in} = P_m S(C - \bar{C}) + QC_0\big|_{out} \tag{5.121}$$

Q is the volumetric flowrate of blood to the region of interest. If V_T is the volume of the tissue region considered, then $q_b = \dfrac{Q}{V_T}$ is defined as the *tissue blood perfusion rate*. Assuming a capillary of length L and blood velocity within the capillary of V, it is easily shown that:

$$q_b = \left(\frac{r_c}{r_T}\right)^2 \left(\frac{V}{L}\right) = \frac{v}{\tau} \tag{5.122}$$

where τ is the blood residence time $\left(\tau = \dfrac{L}{V} = 0.1\,\text{cm } 0.05^{-1}\text{ cm}^{-1}\text{ sec}^{-1} = 2\,\text{sec}\right)$.

The diffusion term, $P_m S(C - \bar{C})$, for a zero-order reaction occuring within the tissue space must equal $V_T R_0$. Equation 5.121 is then readily solved to provide the solute concentration in the blood exiting the tissue region of interest.

$$C = C_0 - \frac{V_T R_0}{Q} = C_0 - \frac{R_0}{q_b} \tag{5.123}$$

The solute concentration within the tissue space is then given by the following equation.

$$\bar{C} = C_0 - R_0 \left[\frac{1}{q_b} + \frac{V_T}{P_m S}\right] \tag{5.124}$$

The following example illustrates the use of these equations.

Example 5.15 Consider the transport of glucose from capillary blood to exercising muscle tissue. As a basis consider 1 gram of tissue. The glucose consumption of the tissue is 60 μmol min^{-1} 100^{-1} g^{-1} or 0.01 μmol sec^{-1} g^{-1}. Blood flow to the region is 60 ml min^{-1} 100^{-1} g^{-1} or 0.01 ml sec^{-1} g^{-1}. The arterial glucose concentration is 5 mM (5 μmol cm^{-3}). The value of P_mS based on capillary recruitment during exercise is 0.33 cm^3 sec^{-1} 100^{-1} g^{-1}, or for the 1 gram of tissue considered here, 0.0033 cm^3 sec^{-1}. Calculate the glucose concentration in the tissue space and in the exiting blood (based on data provided in Levick 1991).

SOLUTION

$$C = 5\frac{\mu\text{mol}}{\text{cm}^2} - \frac{0.01\dfrac{\mu\text{mol}}{\text{g sec}}}{0.01 \text{ ml g}^{-1} \text{ sec}^{-1}} = 4.0\frac{\mu\text{mol}}{\text{cm}^3} = 4 \text{ mM, by Equation 5.123}$$

$$\bar{C} = 5\frac{\mu\text{mol}}{\text{cm}^3} - 0.01\frac{\mu\text{mol}}{\text{g sec}}\left[\frac{1}{0.01 \text{ ml g}^{-1} \text{ sec}^{-1}} + \frac{1 \text{ g}}{0.0033 \times \dfrac{\text{cm}^3}{\text{g sec}}}\right], \text{ by Equation 5.124}$$

$$= 5\frac{\mu\text{mol}}{\text{cm}^3} - 4.03\frac{\mu\text{mol}}{\text{cm}^3} = 0.97\frac{\mu\text{mol}}{\text{cm}^3} = 0.97 \text{ mM}$$

PROBLEMS

1. Consider a polymeric membrane that is 30 microns thick, with a total surface area of 1 cm^2. The membrane has pores equivalent in size to a spherical molecule with a molecular weight of 100,000, a porosity of 80%, and a tortuosity of 2.5. On one side of the membrane we have a protein solution that is to be concentrated (protein properties, $a = 3$ nm and $D = 6.0 \times 10^{-7}$ cm^2/sec) at a concentration of 80 g/L. Assume the solution viscosity is 1 cP. On the other side of the membrane is pure water. The pressure on the protein side of the membrane is 20 psi higher than on the filtrate side of the membrane. Determine the convective flowrate of the solution and the rate at which the protein crosses the membrane.

2. Hemoglobin has a molecular weight of 64,460. Estimate its molecular radius and its diffusivity at 37°C. Compare the predicted hemoglobin diffusivity to the data shown in Figure 5.4.

3. Show graphically that Equation 5.48 provides a reasonable estimation for the data shown in Figure 5.4.

4. Tong and Anderson (1996) obtained the following data in a polyacrylamide gel for the partition coefficient (K), as a function of gel volume fraction (ϕ), for bovine serum albumin (BSA). The BSA they used had a molecular weight of 67,000 Daltons, a molecular radius of 3.6 nm, and a diffusivity of 6 × 10^{-7} cm^2/sec. Compare the Ogston equation to their data and obtain an estimate for the radius of the cylindrical fibers (a_f) that comprise the gel.

Gel volume fraction (ϕ)	K_{BSA}
0.00	1.0
0.025	0.35
0.05	0.09
0.06	0.05
0.075	0.017
0.085	0.02
0.105	0.03

5. Tong and Anderson (1996) also obtained the following data for the diffusive permeability of BSA (KD_0) in polyacrylamide gels as a function of the gel volume fraction (ϕ). Show that the combined Brinkman and Ogston equations may be used to represent this data.

Gel volume fraction (ϕ)	Diffusive permeability (KD_0), cm^2/sec
0.00	5.5×10^{-7}
0.02	1.0×10^{-7}
0.025	6.5×10^{-8}
0.05	8.0×10^{-9}
0.065	4.5×10^{-9}
0.075	1.0×10^{-9}

6. Show for a protein with a molecular weight of 50,000 Daltons that $S_0 = S_a$ for blood flowing through a capillary.

7. Investigate the fit of Equations 5.77 and 5.78 to the data presented in Figure 5.12.

8. The following data was obtained by Iwata *et al.* (1996) for the effective diffusivity of several solutes through a 5% agarose hydrogel. Show how well the Brinkman (1947) and Ogston (1958) equations and the Renkin (1954) equation represents this data. What is the effective pore size of the gel based on the Renkin equation?

Solute	Molecular weight	Gel diffusivity (KD_0), cm^2/sec
glucose	180	4.5×10^{-6}
vitamin B$_{12}$	1200	1.7×10^{-6}
myoglobin	17,000	4.0×10^{-7}
BSA	69,000	1.0×10^{-7}
IgG	150,000	1.3×10^{-7}

9. Iwata *et al.* (1996) also measured the sieving coefficient for a variety of solutes using an XM-50 ultrafilter reported to have a nominal molecular weight cutoff of 50,000 Daltons. Compare their measured sieving coefficients to predictions for the sieving coefficient calculated by Equation 5.50. What pore diameter is needed to get a reasonable representation of the data?

Solute molecular weight	Sieving coefficient
180	0.98
1200	0.94
14,300	0.55
17,000	0.12
69,000	0.01

10. Derive Equation 5.104.

11. Membrane plasmapheresis (Zydney 1995; Zydney and Colton 1986) is a technique used to separate plasma from the cellular components of blood. The plasma that is collected may be processed further to yield a variety of substances that are useful for the treatment of a variety of blood disorders. Membrane-based systems employ blood flow that is parallel to the membrane. The filtrate of plasma is, therefore, perpendicular to the membrane. A cell-free filtrate with minimal retention of plasma proteins is obtained since the membrane pores are typically 0.2 to 1.0 microns in diameter. The filtrate flux is limited, however, by the accumulation (concentration polarization) of cells along the surface of the membrane that provides an additional hydraulic resistance.

At any axial position (z) from the device inlet, this layer of cells at steady state represents a balance between the flux of cells carried to the membrane by the filtration flux, i.e. $q(z) C(z)$, and the diffusive flux of cells carried away from the membrane, i.e. $D\dfrac{dC(z)}{dy}$. Here we let $q(z)$ represent the filtration flux ($Q(z)/S$), $C(z)$ is the concentration of the cells in the blood, D is the diffusivity of the red blood cells, and y is the distance from the membrane surface. Show that after equating these flux expressions, and integrating over the thickness of the cell boundary layer (δ), that the following expression may be obtained for the filtrate flux.

$$q(z) = \frac{D}{\delta} \ln\left(\frac{C_w}{C_b}\right) = k_m \ln\left(\frac{C_w}{C_b}\right)$$

In this equation, k_m represents the local film mass transfer coefficient (i.e. D/δ) and C_w and C_b represent the cell concentration at the membrane surface and in the bulk blood, respectively. Here the thickness of the cell boundary layer is much thinner than that of the flow channel. For these conditions, the axial velocity profile will vary linearly across the cell boundary layer and the mass transfer coefficient may be evaluated from Equation 5.45 with the lead constant of 1.86 replaced with the value of 1.03. The shear-induced diffusivity of the red blood cells is given by the following equation (Zydney and Colton 1986).

$$D = 0.03a^2 \, \gamma_w$$

Here (a) represents the radius of the red blood cells, about 4.2 microns, and γ_w is the wall shear rate given by the following equation for flow of blood in a single cylindrical hollow fiber of radius equal to R (discussed in Chapter 4).

$$\gamma_w = \frac{4F(z)}{\pi R^3}$$

Show that Equation 5.45 (with 1.86 replaced with 1.03) may be rewritten as follows for the flow of blood in the hollow fiber membrane.

$$k_m = 0.05\left(\frac{a^4}{z}\right)^{1/3} \gamma_w$$

The above expressions may be substituted into the following equation that was obtained in Example 5.5 for the axial change of the volumetric flowrate as a result of plasma filtration.

$$\frac{dF(z)}{dz} = -2\pi R q(z)$$

This equation may be integrated to find the fractional filtrate yield, defined as the ratio of the total plasma filtrate formed (Q_f) and the inlet blood flow rate $F_B(0)$. However, to obtain an analytical expression that is useful for design calculations, several assumptions are needed.

To obtain an analyical expression, Zydney and Colton (1986) assumed that the bulk cell concentration remained constant at its inlet value ($C_b(0)$) and that the wall shear rate (γ_w) varied linearly with $F(z)$. Using these assumptions, show that the following equation is obtained for $Q_f/F_B(0)$.

$$\frac{Q_f}{F_B(0)} = 1 - \exp\left[-0.90\beta\ln\frac{C_w}{C_b(0)}\right] \text{ where } \beta = \frac{2}{3}\left(\frac{a^2 L}{R^3}\right)^{2/3}$$

β is a dimensionless length. Zydney and Colton (1986) provided the following performance data for a variety of membrane plasmapheresis units. Show that the above equation provides a remarkable prediction of the fractional filtration yield for these units. Use 0.35 and 0.95 for the concentration of the red blood cells entering in the blood and at the wall, respectively.

β	Fractional filtrate yield
0.20	0.12
0.27	0.20
0.27	0.23
0.32	0.22
0.32	0.25
0.35	0.25
0.40	0.30
0.42	0.32
0.43	0.30
0.44	0.39
0.45	0.38
0.49	0.36
0.51	0.35
0.58	0.40
0.60	0.35
0.62	0.37
0.70	0.41

12. If the surface area of capillaries is 7000 cm^2/100 g of tissue, estimate the Krogh tissue cylinder radius.

13. In a discussion of glucose transport through exercising muscle capillaries, Renkin (1977) reports the capillary glucose permeability (P_m) to be on the order of 12×10^{-6} cm/sec and a capillary surface area (S) of 7000 cm^2/100 g of tissue. How does the value of $P_m \times S$ compare with the data shown in Figure 5.13?

14. Using the Renkin–Crone equation, make a graph of capillary solute extraction (E) as a function of capillary position (z) for various values of the ratio ($2\pi r_c P_m z)/Q$. Identify conditions on your plot for which the solute transport is flow limited and diffusion limited.

15. Levick (1991) presented the following data obtained by Renkin (1977) for the capillary clearance of antipyrene and urea as a function of blood flow in skeletal muscle. Based on this data, which of these solutes' transport rate is flow limited and which is diffusion limited? Calculate the value of $P_m S$ for urea. How does this value of $P_m S$ for urea compare with the results presented in Figure 5.13?

Solute	Blood flow $(Q, \text{ml min}^{-1} \ 100^{-1} \ g^{-1})$	Clearance $(Cl, \text{ml min}^{-1} \ 100^{-1} \ g^{-1})$
antipyrene	4.5	4
	4.5	4.5
	6.5	6.5
	10	10
urea	5	3.5
	5.5	3
	7	4
	8	2.5
	8.5	2.6
	8.5	4
	10	2.6
	10.5	2.5
	12	4.5
	13.5	2.6

16. Mann *et al.* (1979) obtained the following concentration versus time data for several solutes in cat fenestrated salivary glands using the multiple tracer indicator diffusion technique. The test solutes were cyanocobalamin (vitamin B_{12}, MW = 1353) and insulin (MW = 5807). Albumin (MW = 69,000) served as the non-permeable reference solute. The perfusion rate for the gland was 8 ml/min/g. The following table summarizes the concentration of these solutes (% dose/0.2 ml of perfusate sample) for various periods of time following their injection into the gland. From this data calculate the value of P_mS for B_{12} and insulin. How do the P_mS values for B_{12} and insulin compare with the results shown in Figure 5.13?

Time, seconds	C, B_{12}	C, Insulin	C, Albumin
0.3	1.5	2.2	2.3
0.66	4.4	6.3	7.3
1.0	7.3	10.0	11.9
1.5	8.6	11.0	12.9
1.9	7.6	10.2	10.7
2.1	6.0	7.3	7.8
2.5	5.0	5.9	5.7
2.8	4.1	4.2	3.9
3.1	3.8	3.7	3.0
3.5	3.0	2.6	2.2
3.8	2.7	2.2	2.0

17. Estimate the diffusivity of a spherically shaped molecule of molecular weight equal to 35,000 Daltons through a membrane containing cylindrical pores that are 8 nm in diameter. Assume that the molecule enters the pores of the membrane from a well-stirred bulk solution.

18. Some folks suggest that the cell walls of living cells have pores that are 30 angstroms in diameter (1 angstrom = 10^{-8} cm). Estimate the diffusivity (cm²/sec) at 37°C for a solute 5 angstroms in diameter through such pores. Assume the pores are filled with water having a viscosity of 0.76 cP (1 cP = 0.01 g/cm/sec).

19. A molecule with a radius of 2.5 nm enters a membrane having pores with a radius of 7.5 nm. For a temperature of 37°C, and assuming the fluid phase is water, what is the apparent diffusivity of the molecule in the pores of the membrane? Express your answer in cm²/sec.

20. A single hollow fiber is placed within a very large glass tube. The hollow fiber is 20 cm in length and has a diameter of 400 microns. The flowrate of a liquid containing a permeable solute through the hollow fiber is 1 ml/min. It is found that the concentration of the permeable solute exiting the hollow fiber is 10% of the concentration of this solute when entering the hollow fiber. Estimate the permeability of the hollow fiber membrane for this solute. Express you answer in cm/sec.

21. Your group has developed a new drug for the treatment of hilariosis. This drug is lipid insoluble and has a molecular weight of 1200. As part of the design of a controlled release system for this drug, you have been asked to estimate the permeability ($cm^3/sec/100$ g) of this drug through the capillary wall.

22. Estimate the reflection coefficient for a molecule that has a radius of 0.75 nm assuming the pores in the membrane have a radius of 2.5 nm.

23. IgG (150,000 Daltons) is diffusing from a bulk solution through a 5% agarose hydrogel. The polymer in the hydrogel material has a radius $a_f = 2.79$ nm. The radius of the IgG molecule is 3.9 nm. The volume fraction of the polymer fibers in the hydrogel is $\phi = 0.05$. The value of κ for the hydrogel is equal to 0.57. Estimate the diffusivity (cm^2/sec) of IgG in the hydrogel.

24. A new photosensitizer drug has a molecular weight of 1400 Daltons. Estimate its diffusivity in water at 37°C.

25. A single hollow fiber is placed within a larger diameter glass tube forming a shell space that surrounds the hollow fiber. The hollow fiber is 20 cm in length and has a diameter of 400 microns. The flowrate of a liquid through the hollow fiber is 0.1 ml/min and in the shell space another liquid also flows through at a flowrate of 100 ml/min. The liquid entering the hollow fiber also contains a permeable solute. It is found that the concentration of the permeable solute exiting the hollow fiber is 20% of the concentration of this solute when entering the hollow fiber in the liquid. Estimate the permeability for this solute in cm/sec.

26. Blood is flowing through a hollow fiber that is 800 microns in diameter and 30 cm in length. The average velocity of the blood within the hollow fiber is 25 cm/sec. The concentration of a drug is maintained at 10 mg/L along the inside surface of the hollow fiber. The diffusivity of the drug in blood is 4×10^{-6} cm^2/sec. Estimate the mass transfer coefficient k_m for the drug. Assuming no drug enters the hollow fiber with the blood, estimate the exiting concentration of the drug in the blood.

27. Derive Equation 5.18.

28. Derive Equation 5.19.

29. Derive Equation 5.35.

30. Derive Equation 5.37.

31. Derive Equation 5.54.

32. Bawa *et al.* (*J. Controlled Release*, 1, 259–267, 1985) measured the drug distribution inside a polymer matrix material. These polymeric slabs of thickness 1 mm were loaded with albumin at an initial concentration of 484 mg/cm^3 of slab. The table below shows two distributions of the drug concentration as a function of x/L after 6.5 hrs and 172 hrs. Find the value of D_e that best fits these data using Equation 5.70. How well does Equation 5.76 fit the 6.5 hr concentrations?

x/L	C @ 6.5 hrs	C @ 172 hrs
0.039	484 mg cm^{-3}	175
0.115	484	200
0.20	484	213
0.32	484	188
0.38	484	175
0.46	484	150
0.58	475	113
0.66	450	100
0.73	275	75
0.82	200	38
0.89	112	25
0.93	80	12.5

33. Bawa *et al.* (*J. Controlled Release*, 1, 259–267, 1985) obtained the following data for the cumulative fraction released for lysozyme from 1 mm thick polymeric slabs. Estimate the value of the D_e that best fits these data using Equation 5.73. The initial concentration of lysozyme is about 400 mg cm^{-3} and the diffusivity of lysozyme in water is 3.74×10^{-3} cm^2 hr^{-1}.

Time$^{1/2}$, hrs$^{1/2}$	f_R
3	0.03
5	0.12
7	0.18
8	0.21
9.5	0.24
13	0.32
15	0.40
16.5	0.45
18.5	0.50
21	0.52

34. Investigate strategies for improving the removal of protein A by membrane filtration in Example 5.6.

SIX

OXYGEN TRANSPORT IN BIOLOGICAL SYSTEMS

6.1 THE DIFFUSION OF OXYGEN IN MULTICELLULAR SYSTEMS

The growth of simple multicellular systems beyond a diameter of a hundred microns or so is limited by the availability of oxygen. The metabolic needs of the cells near the center of the cell aggregate exceed the supply of oxygen available by simple diffusion. This is illustrated in Figure 6.1 for isolated islet organs (Brockman bodies) of the fish, *Osphronemus gorami*. The islet organs, typically 800 microns in diameter, were placed in culture and the oxygen partial pressure in the surrounding media and within the cells was measured radially using an oxygen microelectrode.

For the case with no convection in the media (closed symbols), we see that the oxygen partial pressure drops sharply and has been reduced to nearly zero within a few hundred microns of the islet organ's surface. Those cells that experience reduced oxygen levels suffer from *hypoxia*. If the oxygen level is reduced beyond some critical value the cells may die, which is called *necrosis*. Note that with increased convection (fluid mixing) in the surrounding media (open symbols), the partial pressure of oxygen at the surface and core regions of the islets is increased significantly in comparison to the case without convection. This is a result of the decrease in the resistance to oxygen transport of the fluid surrounding the islets. However, we still observe a large decrease in the oxygen partial pressure within the islet body.

Clearly the development of such complex cellular systems as vertebrates required the development of oxygen delivery systems that are more efficient than diffusion alone. This has been accomplished by the development of two specialized systems that enhance the delivery of oxygen to cells. The first is the circulatory system, which carries oxygen in blood by convection to tiny blood vessels called capillaries in the neighborhood of the cells. Here oxygen is released and then diffuses over much shorter distances to the cells. Secondly, a specialized oxygen carrier protein, called

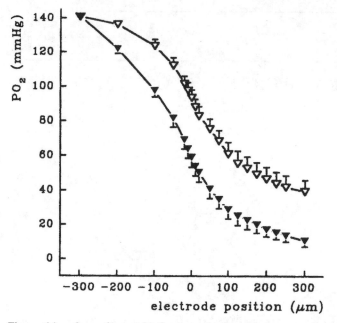

Figure 6.1 pO_2 profiles of 12 Brockman bodies with (open symbols) and without (closed symbols) convection in the medium. Measurements were started 300 microns outside of the tissue (−300). At the surface of the body, the electrode position is 0 microns. (from Schrezenmeir *et al*. 1994, with permission).

hemoglobin, is contained within the red blood cells that are suspended within the blood plasma. The presence of hemoglobin overcomes the very low solubility of oxygen in water which is the major component of plasma. In this chapter, we will investigate the factors that affect oxygen delivery to cells.

6.2 HEMOGLOBIN

Figure 6.2 shows the basic structural features of the hemoglobin molecule. The hemoglobin found in adult humans is called hemoglobin A. Hemoglobin consists of four polypeptide chains (a polymer of amino acids), two of one kind called the α chain and two of another kind called the β chain. These polypeptide chains of hemoglobin A

Heme, oxygen binding site

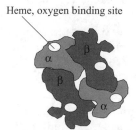

Figure 6.2   Structure of hemoglobin.

are held together by non-covalent attractions. The hemoglobin molecule is nearly spherical with a diameter of about 5.5 nm and a molecular weight of 68,000 Daltons.

The oxygen binding capacity of hemoglobin (Hb) depends on the presence of a non-polypeptide unit called the *heme group*. The iron atom within heme gives blood its distinctive red color. The heme group is also called a *prosthetic group* because it gives the hemoglobin protein its overall functional activity. The heme group consists of an organic part that binds through four nitrogen atoms with the iron atom within the center. The organic part is called the *protoporphyrin* and consists of four pyrole rings linked by methene bridges. The iron atom can form up to six bonds, four of which are commited to the pyrole rings. So we have two additional bonds available, one on each side of the heme plane. A heme group can bind one molecule of oxygen.

The four heme groups in a hemoglobin molecule are located near the surface of the molecule, and the oxygen binding sites are about 2.5 nm apart. Each of the four chains contains a single heme group and, therefore, a single oxygen binding site. A molecule of hemoglobin, therefore, has the potential to bind four molecules of oxygen. Hemoglobin is an *allosteric protein*, and its oxygen binding properties are affected by interactions between these separate and non-adjacent oxygen binding sites. The binding of oxygen to hemoglobin enhances the binding of additional oxygen molecules. The binding of oxygen to hemoglobin is, therefore, said to be *cooperative*.[1]

6.3 THE OXYGEN–HEMOGLOBIN DISSOCIATION CURVE

The binding of oxygen with hemoglobin is described by the oxygen–hemoglobin dissociation curve shown in Figure 6.3. This is an equilibrium curve that expresses, for a given oxygen partial pressure (or pO_2), the fractional occupancy of the hemoglobin oxygen binding sites (Y). The equilibrium partial pressure of oxygen, or pO_2, above a solution of blood is related to the dissolved oxygen concentration by *Henry's law* which is expressed by the following equation.

$$pO_2 \equiv P\, y_{oxygen} = H_{oxygen}\, C_{oxygen} = HC \qquad (6.1)$$

where H_{oxygen} is Henry's constant with a value for normal blood at 37°C of 0.74 mm Hg μM^{-1}. C_{oxygen} is the dissolved oxygen concentration (μM), and y_{oxygen} is the mole fraction of oxygen in the gas phase. It is important to realize that the pO_2 is exerted only by the dissolved oxygen. Oxygen bound to hemoglobin does not affect the pO_2 but serves only as a source or sink for oxygen.

6.4 OXYGEN LEVELS IN BLOOD

Table 6.1 summarizes some typical properties of arterial and venous blood. The concentration of hemoglobin in blood (see Table 4.1) is about 150 g L^{-1} or 2200 μM.

---

[1] An analogy for cooperative binding would be several people entering a life raft from the water. Because of the high sides of the life raft it is very difficult for the first person to get aboard. Once one person is aboard she can help the next person. Two people on board make it even easier to load the next person and so on.

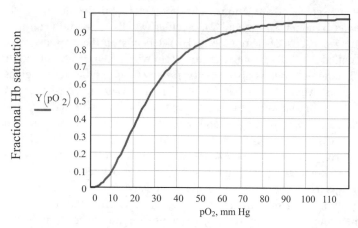

Figure 6.3 Oxygen–hemoglobin dissociation curve.

Since each molecule of hemoglobin can bind at most four molecules of oxygen, the saturated ($Y = 1$) concentration of oxygen bound to hemoglobin is equal to 8800 μM. Comparing this saturated value, or even the arterial or venous values shown in Table 6.1, to the dissolved oxygen value of 130 μM clearly shows that the bulk of the oxygen that is available in the blood is carried by hemoglobin. The P_{50} value represents the value of the oxygen partial pressure (pO_2) at which 50% of the oxygen binding sites are filled, that is $Y = 0.5$. From Figure 6.3, we see this occurs when the $pO_2 \approx 26$ mm Hg.

We can gain a feeling for the importance of the cooperative binding of oxygen through the following simple calculation. Consider an active capillary bed (see Figure 5.16) for which the entering arterial blood has a pO_{2A} of 95 mm Hg and the exiting venous blood has a pO_{2V} of 20 mm Hg. The amount of oxygen delivered to the tissue surrounding the capillary bed would be proportional to the difference in pO_2 between the arterial and venous blood ($pO_{2A}-pO_{2V}$). This difference in pO_2 from Figure 6.3 is equivalent to a 60% change in the hemoglobin saturation. If the oxygen binding was non-cooperative, then $n = 1$, and assuming the same value for P_{50}, the difference in hemoglobin saturation is now only 35%. Therefore, cooperative binding of oxygen to hemoglobin provides for about a 70% increase in oxygen delivery.

Table 6.1 Oxygen levels in blood

Oxygen property	Arterial	Venous
Partial Pressure (tension), pO_2, mm Hg	95	40
Dissolved O_2, μM	130	54
As Oxyhemoglobin, μM	8500	5820
Total Effective (dissolved + oxy – Hb), μM	8630	5874

Other useful data:
$H_{oxygen} = 0.74$ mm Hg/μM
Saturated oxyhemoglobin = 8800 μM
$P_{50} \sim 26$ mm Hg
$n = 2.34$

6.5 THE HILL EQUATION

In 1913 Archibold Hill proposed that the sigmoidal shape of the oxygen dissociation curve could be described by the following equilibrium reaction.

$$Hb(O_2)_n \underset{k_1}{\overset{k_{-1}}{\longleftrightarrow}} Hb + nO_2 \qquad (6.2)$$

In this equation, n represents the number of molecules of oxygen that bind with hemoglobin (Hb) to form oxyhemoglobin, i.e. $[Hb(O_2)_n]$. Writing an elementary rate expression for the appearance of Hb with k_1 and k_{-1} representing the reaction rate constants we obtain:

$$\frac{d[Hb]}{dt} = k_1 \left[Hb(O_2)_n \right] - k_{-1}[Hb][O_2]^n = 0 \qquad (6.3)$$

Assuming the reaction is at equilibrium, we set the above equation equal to zero and solve for the concentration of oxygenated hemoglobin in terms of the hemoglobin and dissolved oxygen concentrations.

$$\left[Hb(O_2)_n \right] = \kappa[Hb][O_2]^n \qquad (6.4)$$

In this equation, κ is defined as k_{-1}/k_1. The fraction of the hemoglobin that is saturated (Y) is be given by the following relationship.

$$Y \equiv \frac{\left[Hb(O_2)_n \right]}{\left[Hb(O_2)_n \right] + [Hb]} = \frac{\dfrac{\left[Hb(O_2)_n \right]}{[Hb]}}{1 + \dfrac{\left[Hb(O_2)_n \right]}{[Hb]}} \qquad (6.5)$$

This equation can be simplified using Equation 6.1 and 6.4 to obtain the following result.

$$Y = \frac{\kappa[O_2]^n}{1 + \kappa[O_2]^n} = \frac{(pO_2)^n}{P_{50}^n + (pO_2)^n} \qquad (6.6)$$

The P_{50}^n value was substituted for H^n_{oxygen}/κ. This equation is called the *Hill equation* and can be used to provide a mathematical relationship between Y and pO_2.

Regression analysis of Y and pO_2 data is facilitated by the following rearrangement of the Hill equation.

$$\ln \frac{Y}{1-Y} = n \ln(pO_2) - n \ln P_{50} \qquad (6.7)$$

A plot of $\ln(Y/(1-Y))$ versus $\ln(pO_2)$ is linear with a slope equal to n and a y-intercept equal to $-n \ln P_{50}$. This is called the *Hill plot*. The following example illustrates the use of the Hill equation to represent the oxygen–hemoglobin dissociation curve.

Example 6.1 The following table presents data that represents the oxygen–hemoglobin dissociation. Show that the Hill equation provides excellent representation of this data.

Oxygen partial pressure, pO_2, mm Hg	Fractional hemoglobin saturation
10	0.12
20	0.28
30	0.56
40	0.72
50	0.82
60	0.88
70	0.91
80	0.93
90	0.95
100	0.96

Source: A.C. Guyton, *Textbook of Medical Physiology*, 8th ed., 1991, W.B. Saunders Co.

SOLUTION Performing a linear regression analysis on the above equation we find that the P_{50} value is 26 mm Hg and the value of n is 2.34. We see in Figure 6.4 that the Hill equation (dashed line) provides excellent representation of the oxygen dissociation curve.

6.6 OTHER FACTORS THAT CAN AFFECT THE OXYGEN DISSOCIATION CURVE

Other molecules such as H^+, CO_2, and organic phosphates (such as 2,3-diphosphoglycerate, DPG) also bind to specific sites on the hemoglobin molecule and greatly affect its oxygen binding ability through *allosteric interactions*. A lowering of the pH shifts the oxygen dissociation curve to the right as shown in Figure 6.5. This decreases for a given pO_2 the oxygen affinity of the hemoglobin molecule. CO_2 at

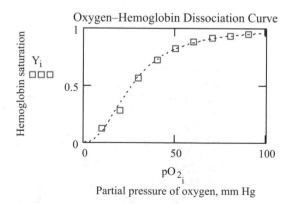

Figure 6.4 The Hill equation.

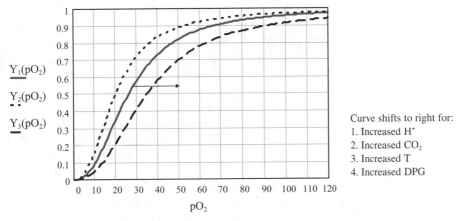

Figure 6.5   Shifts in the oxygen–hemoglobin dissociation curve.

increased levels and constant pH also lowers the oxygen affinity of hemoglobin. In metabolically active tissues such as the muscle, the higher levels of CO_2 and H^+ in the capillaries, therefore, has a beneficial effect by promoting the release of oxygen. This is called the *Bohr effect*.

As shown in Figure 6.5, at high partial pressures of carbon dioxide (pCO_2), CO_2 can reversibly bind to hemoglobin displacing oxygen to form a compound known as *carbaminohemoglobin*. The reverse process also occurs, that is oxygen binding to hemoglobin has the ability to displace CO_2 that is bound to hemoglobin. This is important in both tissues and in the lungs. In the tissue capillaries, oxygen is released from hemoglobin and the higher pCO_2 in the tissue space results in the formation of carbaminohemoglobin. Within the lungs, where the pO_2 is much higher within the gas space of the alveoli, there is displacement of CO_2 from the hemoglobin by oxygen. This is called the *Haldane effect*.

The affinity of hemoglobin for oxygen within the red blood cell is significantly lower than that of hemoglobin in free solution. Within the red blood cell there exists an organic phosphate called 2,3-diphosphoglycerate (DPG) at about the same concentration as hemoglobin. DPG binds to hemoglobin and reduces the oxygen binding capacity as shown in Figure 6.5. Without DPG, the value of P_{50} is reduced to about 1 mm Hg. This makes it very difficult for hemoglobin to release the bound oxygen at low pO_2. Note that without DPG, there would be essentially no release of oxygen in the physiological range of pO_2 values from 40–95 mm Hg. Therefore, the presence of DPG within the red blood cell is vital for hemoglobin to perform its oxygen carrying role.

6.7 TISSUE OXYGENATION

With this understanding of the oxygen binding ability of hemoglobin, we can now focus on tissue oxygenation. Hemoglobin bound oxygen within the red blood cells is transported by the circulatory system to the capillaries where a change in pO_2 levels along the length of the capillary causes a release of oxygen. The released oxygen then diffuses across the capillary wall and into the surrounding tissue space where it is consumed by the cells.

Consider the capillary bed shown earlier in Figure 5.16. What would be the nominal oxygen consumption rate of the tissue lying within this capillary bed? To answer this question, we can develop the following simple mathematical model to describe oxygen usage by the tissue. We assume that the capillary bed and the surrounding tissue space are each well-mixed with respect to oxygen. This assumption allows us to write the following mass balance equations for oxygen. We will let V_T represent the volume of the tissue space (including the capillaries), $\Gamma_{metabolic}$ represents the tissue metabolic volumetric oxygen consumption rate (μM/sec), and q is the tissue blood perfusion rate (ml of blood/cm^3 tissue/min). C represents the dissolved oxygen concentration in the blood (μM) and C' represents the concentration of oxygen in the blood that is bound to hemoglobin (μM). C_T represents the concentration of oxygen dissolved within the interstitial space of the tissue (μM). P_C (cm/sec) is the permeability of oxygen through the capillary wall, and S_C represents the total surface area of the capillaries (cm^2).

blood $\qquad 0 = qV_T(C+C')_A - qV_T(C+C')_V - P_CS_C(C_V - C_T)$ $\qquad$ (6.8)

tissue $\qquad 0 = P_CS_C(C_V - C_T) - V_T\Gamma_{metabolic}$ $\qquad$ (6.9)

Note that the convective terms (containing q) in Equation 6.8 include oxygen transported in the dissolved states as well as that bound to hemoglobin. However, the mass transfer of oxygen across the capillary wall is only based on the difference between the dissolved oxygen concentration in the blood and that within the tissue interstitial space. The above two equations may be added together and solved to give the following equation for the value of the metabolic oxygen consumption rate.

$$\Gamma_{metabolic} = q\left[(C+C')_A - (C+C')_V\right]$$ $\qquad$ (6.10)

To solve this equation for the metabolic oxygen consumption rate of tissue requires a value of the tissue blood perfusion rate, q. Table 6.2 provides representative values for the blood flow to various organs and tissues in the body. The blood perfusion is expressed on the basis of flow in ml per min per 100 grams of tissue or organ. The kidneys are highly perfused because of their role in the filtration of blood. A nominal tissue perfusion rate is on the order of 0.5 ml/cm^3/min (assuming the density of tissue, $\rho_{tissue} \approx 1$ g/cm^3). The nominal arterial and venous pO_2 levels are 95 and 40 mm Hg respectively. We can then use Equation 6.10 to obtain an estimate of the metabolic oxygen consumption rate, $\Gamma_{metabolic}$, as shown in the following example.

Example 6.2 Calculate the metabolic oxygen consumption rate in μM sec^{-1} using the above nominal values for the blood perfusion rate and the arterial and venous pO_2 levels.

Table 6.2 Blood flow to different organs and tissues under basal conditions

Organ	Percent	ml min^{-1}	ml min^{-1} 100^{-1} gm^{-1}
Brain	14	700	50
Heart	4	200	70
Bronchial	2	100	25
Kidneys	22	1100	360
Liver	27	1350	95
Portal	(21)	(1050)	
Arterial	(6)	(300)	
Muscle (inactive state)	15	750	4
Bone	5	250	3
Skin (cool weather)	6	300	3
Thyroid gland	1	50	160
Adrenal glands	0.5	25	300
Other tissues	35.5	175	1.3
Total	100.0	5000	—

Source: A.C. Guyton, *Textbook of Medical Physiology*, 8th edition, 1991, W. B. Saunders Co.

SOLUTION

$$\Gamma_{metabolic} = \frac{0.5\,ml}{cm^3\,min} \times [8630 - 5874]\,\mu M \times \frac{1\,min}{60\,sec}$$

$$\times \frac{1000\,cm^3}{Liter_{tissue}} \times \frac{\mu mol}{Liter_{blood}\,\mu M} \times \frac{1\,Liter_{blood}}{1000\,ml} \times \frac{\mu M\,Liter_{tissue}}{\mu mol} = 22.97\frac{\mu M}{sec}$$

This calculation shows that the tissue metabolic oxygen consumption rate is on the order of 20 μM sec^{-1}. The value for any given tissue will depend on the cells specific requirement for oxygen.

If the value of $\Gamma_{metabolic}$ were known for a given tissue, then Equation 6.10 could be rearranged to solve for the change in blood oxygenation in order to provide the oxygen demands of the tissue.

$$(C+C')_V = (C+C')_A - \frac{\Gamma_{metabolic}}{q} \tag{6.11}$$

To illustrate this type of calculation, where $\Gamma_{metabolic}$ is known, consider an *islet of Langerhans*. The islets of Langerhans are a specialized cluster of cells located within the pancreas. The islets amount to about 1–2% of the pancreatic tissue mass. The cells that lie within the islet are responsible for the control of glucose metabolism through secretion of the hormones insulin and glucagon. Loss of the insulin-secreting β cells as a result of a defect in the immune system results in diabetes.

An islet of Langerhans is only about 150 microns in diameter. The *single islet blood flow* has been determined to be about 7 nl/min (about 4 ml cm^{-3} min) (Lifson et al. 1980). This value is about 4 × the blood perfusion rate of the pancreas itself and is probably related to the hormonal function of the islet.

The oxygen consumption rate for tissue like islets may be described by Michaelis–Menten type kinetics[2]. Here we see that the oxygen consumption rate of the tissue is dependent on the pO_2 in the tissue.

$$\Gamma_{metabolic} = \frac{V_{max}\, pO_2}{K_m + pO_2} \tag{6.12}$$

For islets, the value of K_m is 0.44 mm Hg and the value of V_{max} is 26 μM sec^{-1} when the islets are exposed to basal levels of glucose (100 mg dL^{-1}) and 46 μM sec^{-1} under stimulated glucose levels (300 mg dL^{-1}) (Dionne et al. 1989, 1991).

Because of the small value of K_m, the tissue oxygen consumption rate is generally independent of the tissue pO_2 until the pO_2 in the tissue reaches a value of just a few mm Hg. Cellular metabolic processes are therefore relatively insensitive to the local pO_2 level until it reaches a value of 1–2 mm Hg. Therefore, we can approximate the value of $\Gamma_{metabolic}$ as simply the value of V_{max}. Only at very low pO_2 levels would this approximation no longer be valid.

In Equation 6.11 we know the value of $\Gamma_{metabolic}$, and the value of $(C + C')_A$ would be the nominal arterial value of 8630 μM (from Table 6.1). We now must solve for the value of $(C + C')_V$ recognizing that C' (amount of oxygen bound to hemoglobin) depends on C (the dissolved oxygen concentration) through the Hill equation (Equation 6.6) where we make use of the fact that $pO_2 = H\,C$ (i.e. Henry's law).

$$C' = \frac{C'_{SAT}(HC)^n}{P_{50}^n + (HC)^n} \tag{6.13}$$

Substituting this equation into Equation 6.11 for the value of C', we obtain the equation shown below for the dissolved oxygen concentration in the venous blood.

$$\left[C + \frac{C'_{SAT}(HC)^n}{P_{50}^n + (HC)^n} \right] = \left(C + C'\right)_A - \frac{\Gamma_{metabolic}}{q} \tag{6.14}$$

For given values of q, $\Gamma_{metabolic}$, and $(C + C')_A$, this equation can be solved for the dissolved oxygen concentration (C) in the blood leaving the tissue which, in this case, would be an islet. Once we have obtained the value of C, then the concentration of oxygen bound to hemoglobin, C', can be found from Equation 6.13. The pO_2 of the exiting blood would be given by Henry's law, Equation 6.1. The following examples illustrate these calculations.

Example 6.3 Calculate the change in blood oxygenation within an islet of Langerhans. Perform the calculations under conditions of normal blood perfusion for basal and stimulated levels of glucose. Find the critical blood perfusion rate to maintain the exiting pO_2 at 20 mm Hg under conditions of basal and stimulated levels of glucose.

---

[2] A derivation of the Michaelis–Menten equation can be found in Section 8.8.3

SOLUTION Equation 6.14 provides the solution. Since this equation is non-linear in C, a Newton root finding method is needed to solve for C in the venous blood. For the case of basal glucose stimulation, the venous dissolved oxygen concentration is found to be 102.7 μM. The corresponding value of the exiting pO_2 from Equation 6.1 is 76 mm Hg and the fractional saturation of the hemoglobin from Equation 6.6 or Figure 6.3 is 0.925. The exiting oxygen levels under stimulated glucose conditions are found to be 86.7 μM with a pO_2 of 64 mm Hg and a fractional saturation of 0.89. In both of these cases we observe a modest decrease in oxygen levels in the blood exiting the islet. This is because the islets are highly perfused with blood. The critical perfusion rate is defined as that blood flowrate to the tissue that just maintains a critical pO_2 level, for example 20 mm Hg. Hence we solve Equation 6.14 for the value of the blood perfusion rate, i.e. q. We then find that the critical blood perfusion rate for an islet is 0.28 ml cm^{-3} min^{-1} under basal glucose conditions. A similar calculation shows that the critical blood perfusion rate is 0.50 ml cm^{-3} min^{-1} under stimulated glucose conditions.

6.8 OXYGEN TRANSPORT IN A BIOARTIFICIAL ORGAN

This discussion on oxygen transport is also of critical importance in the design of bioartificial organs (Colton 1995). In one type of device for the treatment of diabetes, therapeutic cells such as the islets of Langerhans or other insulin secreting cells are sandwiched between two permselective membranes that are adjacent to a polymer matrix material that has been vascularized by the host. The permselective membrane has a porous structure at the molecular level such that small molecules like oxygen, key nutrients, and the therapeutic agent are readily permeable, however the components of the host immune system cannot cross the membrane. The cells are therefore said to be *immunoprotected*. We will discuss this in greater detail in Chapter 10, but here our interest is in providing sufficient oxygen to the cells.

Consider the bioartificial organ shown in Figure 6.6. As a first approximation, we will assume that the major resistance to oxygen transport lies in the immunoprotective membrane and the layer of cells. We will, therefore, prescribe the pO_2 level at the outer surface of the membrane to be some average of that of the blood pO_2. A steady state shell balance over a thickness Δx of the cell layer for oxygen may be written as follows:

$$0 = -AD_e \frac{dC}{dx}\Big|_x + AD_e \frac{dC}{dx}\Big|_{x+\Delta x} - \Gamma_{\text{metabolic}} A \, \Delta x (1-\varepsilon) \tag{6.15}$$

Note that the volumetric metabolic oxygen consumption rate of the tissue, $\Gamma_{\text{metabolic}}$ (assumed constant), is multiplied by the islet volume, $A \, \Delta x \, (1-\varepsilon)$, to get the amount of oxygen consumed by the tissue in the shell of thickness Δx. The void volume within the tissue space is given by ε. D_e is the effective diffusivity of oxygen within the combined tissue and void space and was defined earlier in Chapter 5. After dividing by Δx, and taking the limit as $\Delta x \to 0$, the following equation is obtained. Henry's law was used to express the dissolved oxygen concentration in terms of the oxygen partial pressure.

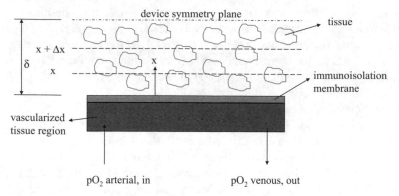

Figure 6.6   Conceptual model for a bioartificial organ.

$$D_e \frac{d^2 pO_2}{dx^2} = \Gamma_{\text{metabolic}} H(1-\varepsilon) \tag{6.16}$$

Since this equation is a second order differential equation, we require the following two boundary conditions in order to obtain the solution for the pO_2 profile within the tissue layer.

$$\text{BC1:} \qquad x = 0, \; pO_2 = pO_2^{x=0}$$
$$\text{BC2:} \qquad x = \delta, \; \frac{dpO_2}{dx} = 0 \tag{6.17}$$

The first boundary condition expresses the pO_2 level at the interface between the membrane and the tissue region. The second boundary condition results from device symmetry and expresses the fact that there is no net flow of oxygen across the symmetry plane. Equation 6.16 can now be integrated twice and Equation 6.17 can be used to find the integration constants. The following equation is then obtained for the oxygen profile within the tissue layer.

$$pO_2(x) = pO_2^{x=0} + \left[\frac{\Gamma_{\text{metabolic}} H(1-\varepsilon)\delta^2}{2D_e} \right] \left[\left(\frac{x^2}{\delta} \right) - 2\left(\frac{x}{\delta} \right) \right] \tag{6.18}$$

The diffusivity of oxygen within the tissue layer (D_e) may be estimated using methods discussed in Chapter 5.

At this point the value of $pO_2^{x=0}$ in Equation 6.18 is unknown. We need to develop an additional equation that relates the transport rate of oxygen across the membrane to the total consumption of oxygen by the islets. This is given by the following equation:

$$\frac{AP_m}{H}\left(pO_2^B - pO_2^{x=0} \right) = A\delta(1-\varepsilon)\Gamma_{\text{metabolic}} \tag{6.19}$$

where P_m is the permeability of the immunoisolation membrane and pO_2^B is the average partial pressure of oxygen in the blood in the vascularized region which is assumed to be known. Rearranging the above equation provides the value of $pO_2^{x=0}$.

$$pO_2^{x=0} = pO_2^{B} - \left[\frac{\Gamma_{met} H(1-\varepsilon)\delta}{P_m} \right] \qquad (6.20)$$

Equations 6.18 and 6.20 can now be used for basic calculations on the effect of oxygen transport on the design of a bioartificial organ. These calculations are illustrated in the following example.

> **Example 6.4** Consider a bioartificial organ such as that shown in Figure 6.6. Assume the value of pO_2^{B} is the mean of the arterial and venous oxygen values, i.e. 68 mm Hg. Let the cells be the islets of Langerhans, assumed spheres with a diameter of 150 microns, with the basal metabolic oxygen consumption rate of 25.9 μM sec^{-1}. The half-thickness (δ) of the islet layer is 200 microns and the void volume fraction (ε) is equal to 0.85. The oxygen permeability of the membrane (P_m) is equal to 4×10^{-3} cm sec^{-1}. Calculate the pO_2 profile in the tissue space of the bioartificial organ.
>
> SOLUTION Using a value for the bulk diffusivity of oxygen as 2.11×10^{-5} cm^2 sec^{-1}, the effective diffusivity by Equation 5.77 for the given void fraction of 0.85 is equal to 1.67×10^{-5} cm^2 sec^{-1}. Equations 6.18 and 6.20 can then be solved for the oxygen profile in the layer of islets. Figure 6.7 shows a plot of the pO_2 profile within the islet layer of the bioartificial organ. The graph shows that for these conditions the tissue layer is fully oxygenated.
>
> **Example 6.5** Malda *et al.* (2004) measured oxygen gradients in tissue engineered cartilaginous constructs. Using a microelectrode, oxygen concentrations were measured as a function of distance into the tissue engineered constructs containing chondrocytes. In addition they also determined the chondrocyte distribution as a

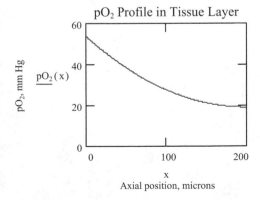

Figure 6.7 Oxygen profile in a bioartificial organ.

function of position with the construct. The average number of chondrocytes per cm^3 of total construct, i.e. the cell density, 28 days after seeding is about 1.5×10^8 cells cm^{-3}. The table below summarizes the pO_2 levels that were measured within the construct as a function of the distance from the exposed surface of the construct.

Depth, microns	pO_2, mm Hg
0	160
250	114
500	95
750	76
1000	61
1250	53
1500	42
1750	38
2000	30
2500	19

They also estimated the diffusivity of oxygen in the construct to be 3.8×10^{-6} cm^2 sec^{-1}. From these data, estimate the metabolic oxygen consumption rate of the chondrocytes in μM sec^{-1} assuming the diameter of a chondrocyte is 20 microns.

SOLUTION Equation 6.18 can also be shown to apply to this situation as well with $pO_2^{x=0} = 160$ mm Hg. The volume of one cell is equal to 4.19×10^{-9} cm^3 $cell^{-1}$. Multiplying the cell volume by the cell density provides the value of $1-\varepsilon = 0.63$. A regression analysis can then be performed to find the best value of $\Gamma_{metabolic}$ in Equation 6.18 that fits the above data. Figure 6.8 shows the results of these calculations. $\Gamma_{metabolic}$ is found to equal 0.037 μM sec^{-1}, and we see that Equation 6.18 provides a good representation of the data. It is also important to note that chondrocytes have very low respiratory activity which helps them survive in the relatively avascular regions of cartilage tissue.

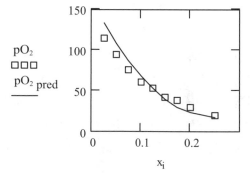

Figure 6.8 Measured and predicted oxygen gradients in engineered cartilaginous constructs. From Example 6.5.

6.9 STEADY STATE OXYGEN TRANSPORT IN A PERFUSION BIOREACTOR

Tilles *et al.* (2001) and Allen and Bhatia (2003) developed steady state microchannel perfusion bioreactor systems for studying the effect of oxygen on a variety of cellular functions. The oxygen concentration within the bioreactor can be controlled by adjusting the flowrate of media through the device.

Figure 6.9 illustrates their parallel-plate microchannel bioreactor. Nutrient media flows through the bioreactor with an average velocity equal to V. On the lower surface there is a monolayer of attached cells that consume the oxygen that is in the flowing stream. The entering oxygen concentration is C_{in}. A steady state shell balance on oxygen from x to $x + \Delta x$ and from y to $y + \Delta y$, assuming the flow is constant across the bioreactor cross section and equal to V, may be written as follows:

$$V\Delta y W \, C\big|_x - V\Delta y W \, C\big|_{x+\Delta x} + D\Delta x W \frac{\partial C}{\partial y}\bigg|_{y+\Delta y} - D\Delta x W \frac{\partial C}{\partial y}\bigg|_y = 0 \qquad (6.21)$$

After dividing by $\Delta x \Delta y$, and taking the limit as Δx and $\Delta y \to 0$, the following differential equation is obtained.

$$V\frac{\partial C}{\partial x} = D\frac{\partial^2 C}{\partial y^2} \qquad (6.22)$$

The term on the left of the above equation represents transport of oxygen by convection, or the bulk flow of the fluid flowing through the bioreactor, and the term on the right-hand side represents oxygen transport by diffusion in the y direction. The boundary conditions are as follows:

$$\text{BC1:} \quad x = 0, \, C(0,y) = C_{in}$$

$$\text{BC2:} \quad y = h, \, D\frac{\partial C}{\partial y}\bigg|_{y=0} = \Gamma_{metabolic}\delta_{cell}\left(1-\varepsilon\right)$$

$$\text{BC3:} \quad y = 0, \, \frac{\partial C}{\partial y}\bigg|_{y=h} = 0$$

$$(6.23)$$

Boundary condition one expresses the fact that fluid flowing through the bioreactor enters at a uniform oxygen concentration, i.e. C_{in}. The second boundary condition

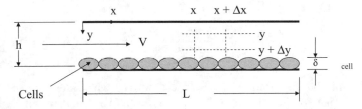

Figure 6.9   Parallel-plate microchannel bioreactor.

expresses the fact that the flux of oxygen at the cellular surface is equal to the rate of oxygen consumption by the cells. δ_{cell} is the thickness of the cellular layer along the lower surface and $(1-\varepsilon)$ represents the volume fraction of the cells along the lower surface. The third boundary condition represents the fact that oxygen is not consumed or lost from the upper surface of the bioreactor.

After introducing the following dimensionless variables into Equations 6.22 and 6.23 (i.e. $\alpha = L/h$, $\hat{x} = x/L$, $\hat{y} = y/h$, $\hat{C} = C/C_{in}$), the following dimensionless equations are obtained:

$$\frac{\partial \hat{C}}{\partial \hat{x}} = \frac{\alpha}{Pe} \frac{\partial^2 \hat{C}}{\partial \hat{y}^2} \tag{6.24}$$

$$\text{BC1:} \quad \hat{x} = 0, \ \hat{C} = 1$$

$$\text{BC2:} \quad \hat{y} = 1, \frac{\partial \hat{C}}{\partial y}\bigg|_{\hat{y}=0} = Da \tag{6.25}$$

$$\text{BC3:} \quad \hat{y} = 0, \frac{\partial \hat{C}}{\partial \hat{y}}\bigg|_{\hat{y}=1} = 0$$

Pe is the Peclet number (Vh/D) and represents the ratio of oxygen transport by convection to that by diffusion. The dimensionless oxygen flux is also known as the Damkohler number $\left(Da = \dfrac{\Gamma_{metabolic}(1-\varepsilon)h\delta_{cell}}{DC_{in}} \right)$. Equations 6.24 and 6.25 can be solved analytically using separation of variables as discussed in Chapter 5. The resulting dimensionless oxygen concentration profile can be shown to be given by the following equation (Allen and Bhatia 2003).

$$\hat{C}(\hat{x}, \hat{y}) = 1 + Da \left[\frac{1-3\hat{y}^2}{6} - \frac{\alpha}{Pe}\hat{x} + \frac{2}{\pi^2} \sum_{n=1}^{\infty} \frac{(-1)^n}{n^2} \exp\left(-\frac{\alpha n^2 \pi^2 \hat{x}}{Pe} \right) \cos(n\pi\hat{y}) \right] \tag{6.26}$$

The above equation provides the oxygen concentration within the bioreactor at any position $(\hat{x}, \hat{y})$. For large Pe numbers (where the convective flow dominates axial diffusion), the summation term in the above equation becomes insignificant a short distance into the bioreactor, and the above solution can then be approximated by the following result.

$$\hat{C}(\hat{x}, \hat{y}) \approx 1 + Da\left(\frac{1-3\hat{y}^2}{6} - \frac{\alpha}{Pe}\hat{x} \right) \tag{6.27}$$

Of particular interest would be the location along the cell surface where the oxygen concentration becomes equal to zero, i.e. $\hat{C}(\hat{x}_{critical}, \hat{y} = 1) = 0$. We can then solve Equation 6.27 for this critical value of x as shown below.

$$\hat{x}_{critical} \approx \frac{Pe}{\alpha}\left(\frac{1}{Da}-\frac{1}{3}\right)$$ (6.28)

We can also calculate the average oxygen concentration at any value of $\hat{x}$ by integrating Equation 6.26 from $\hat{y}=0$ to $\hat{y}=1$. The following result is then obtained for the average oxygen concentration in the bioreactor.

$$\hat{C}_{average}(\hat{x})=1-\left(\frac{\alpha Da}{Pe}\right)\hat{x}$$ (6.29)

Example 6.6 Calculate the distance into the bioreactor where the cell surface pO_2 reaches 0 mm Hg. The flow rate through the bioreactor is 0.5 cm³/min and the inlet pO_2 is 75 mm Hg. Use the bioreactor parameters in the table below from Allen and Bhatia (2003).

Parameter	Value	Units
D, oxygen diffusivity	2×10^{-5}	cm² sec⁻¹
V_{max}, max. O_2 uptake	0.38	nmol sec⁻¹ 10⁻⁶ cells⁻¹
ρ, cell density	1.7×10^5	cells cm⁻²
pO_{2in}, inlet pO_2	75	mm Hg
Q, flowrate	0.5	cm³ min⁻¹
h, height of flow channel	100	microns
W, width of flow channel	2.8	cm
L, length of flow channel	5.5	cm

SOLUTION First we need to calculate α, Da, and Pe. $\alpha = L/h = 550$. For the Da number we also recognize that $\rho V_{max} = \Gamma_{metabolic}(1-\varepsilon)\delta_{cell}$. Therefore we can calculate the Da number as shown below. Note that we have divided the inlet pO_2 by the Henry's constant for oxygen in media (~ 1.04 mm Hg/μM) to estimate the inlet oxygen concentration as 72 nmol/cm³.

$$Da = \frac{\rho V_{max}h}{DC_{in}} = \frac{1.7 \times 10^5 \text{cells cm}^{-2} \times 0.38 \text{ nmol sec}^{-1}\ 10^{-6} \text{ cell}^{-1} \times 0.01 \text{ cm}}{2 \times 10^{-5} \text{ cm}^2 \text{ sec}^{-1} \times 72 \text{ nmol cm}^{-3}} = 0.45$$

The Pe number is then calculated as follows:

$$Pe = \frac{Vh}{D} = \frac{0.5 \text{ cm}^3 \text{ min}^{-1} 1 \text{ min}/(60 \text{ sec})^{-1} \times \dfrac{1}{(0.01 \text{ cm} \times 2.8 \text{ cm})} \times 0.01 \text{ cm}}{2 \times 10^{-5} \text{cm}^2 \text{ sec}^{-1}} = 149$$

Using the above values of α, Da, and Pe, we find that both Equations 6.26 and 6.28 give an $\hat{x}_{critical} = 0.57$ or an $x_{critical}$ of 0.512×5.5 cm $= 2.81$ cm.

Example 6.7 Allen and Bhatia (2003) obtained the following outlet pO_2's (equivalent to the average oxygen concentration of pO_2) for their bioreactor at various flowrates for an inlet $pO_2 = 158$ mm Hg. Compare these results with that predicted by Equation 6.29.

Flow rate, cm³/min	pO_2 out, mm Hg
0.5	42
0.75	73
1.0	92
1.5	112
2.0	122
3.0	133

SOLUTION Once again we first need to calculate α, Da, and Pe. $\alpha = L/h = 550$. We can calculate the Da number as shown below. Note that we have divided the inlet pO_2 by the Henry's constant for oxygen (~1.04 mm Hg μM^{-1}) to estimate the inlet oxygen concentration as 152 nmol/cm³.

$$Da = \frac{\rho V_{max} h}{DC_{in}} = \frac{1.7 \times 10^5 \text{ cells cm}^{-2} \times 0.38 \text{ nmol sec}^{-1} \ 10^{-6} \text{ cell} \times 0.01 \text{cm}}{2 \times 10^{-5} \text{ cm}^2 \text{ sec}^{-1} \times 152 \text{ nmol cm}^{-3}} = 0.21$$

For the Peclet number we have to calculate a value for each of the flowrates:

$$Pe = \frac{Vh}{D} = \frac{Q \text{ cm}^3 \text{ min}^{-1} 1\text{min}/ (60 \text{ sec})^{-1} \times \dfrac{1}{(0.01 \text{cm} \times 2.8 \text{cm})} \times 0.01 \text{cm}}{2 \times 10^{-5} \text{ cm}^2 \text{ sec}^{-1}} = 298 \, Q$$

Figure 6.10 shows the comparison between the measured outlet pO_2 values and those predicted by Equation 6.29. The agreement between the data and the model is quite good.

6.10 OXYGEN TRANSPORT IN THE KROGH TISSUE CYLINDER

Our previous analysis for tissue oxygenation (Equation 6.10) treats the tissue and blood as well-mixed regions, i.e. each region is spatially lumped or averaged. This type of

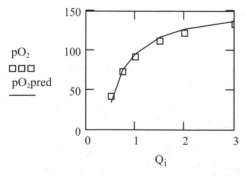

Figure 6.10 Comparison of actual and predicted exiting pO_2 levels from the parallel-plate microchannel bioreactor, Example 6.7.

analysis provides no information about the oxygen diffusion and consumption within the tissue surrounding a given capillary.

Oxygen transport by diffusion to the tissue surrounding a given capillary is usually the rate limiting step as far as maintaining viable tissue is concerned. The effective oxygenation distance within the tissue outside the capillary is quite small, typically only several capillary diameters, thus requiring an extensive network of capillaries to meet the tissue oxygen demand.

We can use the Krogh tissue cylinder approach (Krogh 1919) developed earlier in Chapter 5 to analyze oxygen transport to the tissue surrounding a given capillary. The Krogh tissue model is a good place to start before considering more advanced models that are described in the recent literature (Mirhashemi *et al.* 1987; Tsai *et al.* 1990; Lagerlund and Low 1993; Secomb *et al.* 1993; Intaglietta 1997; Li *et al.* 1997).

Figure 5.17 illustrated a concentric shell of the Krogh tissue cylinder. Considering blood first, the shell balance equation for oxygenated hemoglobin may be written as follows[3]:

$$\left(2\pi r\Delta r\Delta z\right)\frac{\partial C'}{\partial t} = 2\pi r\Delta rVC'\Big|_{z} - 2\pi r\Delta rVC'\Big|_{z+\Delta z} + R_{\mathrm{HbO}}\, 2\pi r\Delta r\Delta z \qquad (6.30)$$

In this equation, we ignore the particulate nature of blood as well as the mass transfer resistance of the red blood cell. The blood is assumed to be in plug flow with an average velocity represented by V. Also note that the hemoglobin is carried along by the red blood cell at the average blood velocity (V) and, accordingly, there is no diffusive transport of hemoglobin. R_{HbO} represents the volumetric production rate of oxygenated hemoglobin. After dividing by $2\pi r\, \Delta r\, \Delta z$, and taking the limit as $\Delta z \to 0$, we obtain the following differential equation that describes the mass balance for oxygenated hemoglobin within the blood flowing through the capillary.

$$\frac{\partial C'}{\partial t} + V\frac{\partial C'}{\partial z} = R_{\mathrm{HbO}} \qquad (6.31)$$

A shell balance for the dissolved oxygen in the blood is given by the next equation.

$$(2\pi r\Delta r\Delta z)\frac{\partial C}{\partial t} = 2\pi r\Delta rVC\,|_{z} - 2\pi r\Delta rVC\,|_{z+\Delta z} + (-2\pi r\Delta z)D\frac{\partial C}{\partial r}\,|_{r}$$

$$-(-2\pi r\Delta z)D\frac{\partial C}{\partial r}\,|_{r+\Delta r} + (-2\pi r\Delta r)D\frac{\partial C}{\partial z}\,|_{z} - (-2\pi r\Delta r)D\frac{\partial C}{\partial z}\,|_{z+\Delta z}$$

$$+R2\pi r\Delta r\Delta z \qquad (6.32)$$

In addition to convective flow of the dissolved oxygen (V C terms), we also have axial and radial diffusion of the dissolved oxygen $(D\dfrac{\partial C}{\partial z}$ and $D\dfrac{\partial C}{\partial r}$ terms). Once again we

---

[3] Unlike the shell balances developed in Chapter 5, we will now leave in the accumulation term so that we can consider, if necessary, unsteady state problems.

divide by $<2\pi r \Delta r \Delta z>$, and take the limit as Δr and $\Delta z \to 0$. We then obtain the following differential equation that describes the mass balance on dissolved oxygen within the capillary.

$$\frac{\partial C}{\partial t} + V \frac{\partial C}{\partial z} = D\left[\frac{1}{r}\frac{\partial}{\partial r}\left(r\frac{\partial C}{\partial r}\right) + \frac{\partial^2 C}{\partial z^2}\right] + R \qquad (6.33)$$

The two reaction rate terms in Equations 6.31 and 6.33 are simply the negative of each other, therefore $R_{HbO} = -R$. This is true since the rate of disappearance of dissolved oxygen must equal the rate of appearance of oxygenated hemoglobin.

We can also make use of the oxygen dissociation curve to relate C' to C as follows:

$$\frac{\partial C'}{\partial t} = \frac{\partial C}{\partial t}\left(\frac{\partial C'}{\partial C}\right) = m\frac{\partial C}{\partial t}$$

$$\frac{\partial C'}{\partial z} = \frac{\partial C}{\partial z}\left(\frac{\partial C'}{\partial C}\right) = m\frac{\partial C}{\partial z} \qquad (6.34)$$

where $m = \dfrac{dC'}{dC}$ is simply related to the slope of the oxygen dissociation curve.

Equations 6.31 and 6.33 can now be added together using Equation 6.34 and Henry's law, Equation 6.1, to express the dissolved oxygen concentration in terms of the pO_2. We then obtain the following equation that expresses the mass balance for dissolved oxygen within the capillary.

$$(1+m)\frac{\partial pO_2}{\partial t} + V(1+m)\frac{\partial pO_2}{\partial z} = D\left[\frac{1}{r}\frac{\partial}{\partial r}\left(r\frac{\partial pO_2}{\partial r}\right) + \frac{\partial^2 pO_2}{\partial z^2}\right] \qquad (6.35)$$

Before we go any further, let's take a closer look at how to evaluate the parameter, m. m can be related to the blood pO_2 using Henry's law and recognizing that $Y = C'/C'_{sat}$. The following equation then provides a way to calculate m, which depends on the local value of the pO_2. Note that m is dimensionless.

$$m = \frac{dC'}{dC} = HC'_{Sat}\frac{dY}{dpO_2} \qquad (6.36)$$

Using Hill's equation (Equation 6.6), we can evaluate $\dfrac{dY}{dpO_2}$ and obtain the following expression for the dependence of m on the local pO_2.

$$m = nP_{50}^n HC'_{sat}\frac{pO_2^{n-1}}{\left[P_{50}^n + pO_2^n\right]^2} \qquad (6.37)$$

A shell balance on the dissolved oxygen within the tissue interstitial space of void volume (ε^T) surrounding the capillary may be written as follows:

$$2\pi r \Delta r \Delta z \varepsilon^T \frac{\partial C^T}{\partial t} = (-2\pi r \Delta z) D^T \left. \frac{\partial C^T}{\partial r} \right|_r$$

$$-(-2\pi r \Delta z) D^T \left. \frac{\partial C^T}{\partial r} \right|_{r+\Delta r} + (-2\pi r \Delta r) D^T \left. \frac{\partial C^T}{\partial z} \right|_z \qquad (6.38)$$

$$-(-2\pi r \Delta r) D^T \left. \frac{\partial C^T}{\partial z} \right|_{z+\Delta z} - 2\pi r \Delta r \Delta z \Gamma_{\text{metabolic}}$$

Within the tissue, D^T for oxygen can once again be estimated using methods discussed in Chapter 5. After dividing by $2\pi r \, \Delta r \, \Delta z$, and taking the limit as Δr and $\Delta z \to 0$, the following differential equation is obtained for the oxygen mass balance within the tissue region. Henry's law was used to express the dissolved oxygen concentration in the tissue, C^T, in terms of the tissue region oxygen partial pressure, $pO_2{}^T$.

$$\varepsilon^T \frac{\partial pO_2^T}{\partial t} = D^T \left[\frac{1}{r} \frac{\partial}{\partial r} \left(r \frac{\partial pO_2^T}{\partial r} \right) + \frac{\partial^2 pO_2^T}{\partial z^2} \right] - \Gamma_{\text{metabolic}} H^T \qquad (6.39)$$

The boundary conditions needed to solve Equations 6.35 and 6.39 are as follows.

<u>blood region</u> $0 \le z \le L$ and $0 \le r \le r_C$ $\qquad\qquad\qquad\qquad\qquad\qquad$ (6.40)

$$\text{BC1:} \quad z = 0, \ pO_2 = pO_2(t)$$

$$\text{BC2:} \quad z = L, \ \frac{\partial pO_2}{\partial z} = 0$$

$$\text{BC3:} \quad r = 0, \ \frac{\partial pO_2}{\partial r} = 0$$

$$\text{BC4:} \quad r = r_c, \ pO_2 = pO_2^T \ \text{and} \ D \frac{\partial pO_2}{\partial r} = D^T \frac{\partial pO_2^T}{\partial r}$$

Boundary condition 1 expresses the fact that the pO_2 of the blood entering the capillary is assumed to be known and may be a function of time. Boundary condition 2 simply states that oxygen cannot leave the capillary in the axial direction by axial diffusion. Boundary condition 3 assumes the oxygen profile is symmetric with respect to the capillary centerline. The last boundary condition assumes the capillary wall has negligible mass transfer resistance[4] and expresses the requirement that the dissolved oxygen concentrations and the oxygen flux are continuous at the interface between the capillary and tissue regions.

The tissue region boundary conditions may be written as follows assuming there are no anoxic regions.

<u>tissue region</u> $0 \le z \le L$ and $r_C \le r \le r_T$ $\qquad\qquad\qquad\qquad\qquad$ (6.41)

$$\text{BC5:} \quad z = 0, \ \frac{\partial pO_2^T}{\partial z} = 0$$

---

[4] Recall that oxygen is lipid soluble and readily permeates the entire surface of the capillary wall.

$$\text{BC6:} \quad z = L, \quad \frac{\partial pO_2^T}{\partial z} = 0$$

$$\text{BC7:} \quad r = r_T, \quad \frac{\partial pO_2^T}{\partial r} = 0$$

Boundary conditions 5 and 6 state that oxygen cannot leave the tissue region at either end by axial diffusion. Boundary condition 7 states that the oxygen profile in the tissue region between capillaries spaced a distance $2 r_T$ (the Krogh tissue cylinder diameter) apart is symmetric.

Under certain conditions, anoxic regions may develop within the tissue region. These regions will be defined by a *critical radius*, r_{anoxic} (z), a distance beyond which there is no oxygen in the tissue. If an anoxic region exists, then boundary condition 7 above becomes:

tissue region – anoxic $\text{BC8:} \quad r = r_{\text{anoxic}}(z), \dfrac{\partial pO_2^T}{\partial r} = 0$ and $pO_2^T = 0$ \hfill (6.42)

6.11 AN APPROXIMATE SOLUTION FOR OXYGEN TRANSPORT IN THE KROGH TISSUE CYLINDER

Solution of the above equations for the oxygen concentrations within the blood and tissue regions is a formidable problem and requires a numerical solution (Lagerlund and Low 1993). However, with some simplifications, a reasonable analytical solution can be obtained that is a good starting point for exploring the key factors that govern oxygenation of the tissue surrounding a given capillary. We can also limit ourselves to a steady state solution, thus eliminating the time derivatives. Another useful approximation is to treat (m) as a constant. Recall that m is related to the slope of the oxygen dissociation curve by Equation 6.37, and an average value can be used in the range of pO_2 levels of interest. Since the capillary is much longer in length than the Krogh tissue cylinder radius, we can also ignore axial diffusion within the tissue region. This greatly simplifies the equation for the tissue region (Equation 6.39) to give:

$$\frac{d}{dr}\left(r\frac{dpO_2^T}{dr} \right) = \frac{r\Gamma_{\text{metabolic}} H^T}{D^T} \tag{6.43}$$

Within the capillary, we can ignore axial diffusion in comparison to axial convection and, from Equation 6.35, we obtain the following equation.

$$(1+m)V\frac{\partial pO_2}{\partial z} = D\left(\frac{1}{r}\frac{\partial}{\partial r}\left(r\frac{\partial pO_2}{\partial r} \right) \right) \tag{6.44}$$

Being a partial differential equation, this equation is still tough to solve. One approximate approach is to eliminate the radial diffusion term by lumping (i.e. integrating the equation) over the r-direction as illustrated below. Radial averaging of the capillary oxygen levels is appropriate here since the bulk of the oxygen mass transfer resistance is not within the capillary. Hence we would not expect steep gradients in the oxygen concentration in the radial direction within the blood.

$$2\pi \int_0^{r_c} (1+m)V \frac{\partial pO_2}{\partial z} r \, dr = 2\pi D \int_0^{r_c} \frac{1}{r} \left(\frac{\partial}{\partial r} \left(r \frac{\partial pO_2}{\partial r} \right) \right) r \, dr \qquad (6.45)$$

This equation may then be integrated and written as follows:

$$2\pi(1+m)V \frac{d}{dz} \int_0^{r_c} pO_2 r \, dr = 2\pi D r_c \frac{\partial pO_2}{\partial r} \bigg|_{r_c} \qquad (6.46)$$

We next recognize that the radially averaged pO_2 level in the blood, $\langle pO_2 \rangle$, at a given axial location, is defined by the following equation:

$$\langle pO_2 \rangle \pi r_c^2 = 2\pi \int_0^{r_c} pO_2 r \, dr \qquad (6.47)$$

This allows Equation 6.46 to be rewritten in terms of the average pO_2 level in the blood.

$$(1+m) \frac{d \langle pO_2 \rangle}{dz} = \frac{2D}{r_c V} \frac{dpO_2}{dr} \bigg|_{r_c} = \frac{2D^T}{r_c V} \frac{dpO_2^T}{dr} \bigg|_{r_c} \qquad (6.48)$$

Note that the solutions for the oxygen level in the blood, i.e. $\langle pO_2 \rangle$ and tissue regions, i.e. pO_2^T, are connected through the oxygen flux terms on the right-hand side of Equation 6.48 that arises from the use of boundary condition 4 in Equation 6.40.

We can now proceed to obtain an analytical solution for the oxygen levels within the capillary and the tissue region. First, we integrate Equation 6.43 two times and use boundary conditions 4 and 7 in Equations 6.40 and 6.41 to obtain the following result.

$$pO_2^T(r,z) = \langle pO_2(z) \rangle - \frac{r_c^2 \Gamma_{\text{metabolic}} H^T}{4D^T} \left[1 - \left(\frac{r}{r_c} \right)^2 \right]$$

$$- \frac{r_T^2 \Gamma_{\text{metabolic}} H^T}{2D^T} \ln \left(\frac{r}{r_c} \right) \qquad (6.49)$$

Although we ignored axial diffusion within the tissue region, note that the axial average capillary pO_2 level, $\langle pO_2(z) \rangle$ impresses an axial dependence on the tissue region pO_2 level. This equation therefore depends on the local capillary oxygen pO_2 and is valid so long as $pO_2^T \geq 0$ throughout the tissue region.

In some cases part of the tissue region may become anoxic and Equation 6.43 must be solved using the anoxic boundary condition given by Equation 6.42. In this case, the tissue pO_2 is given by:

$$pO_2^T(r,z) = \langle pO_2(z) \rangle - \frac{r_c^2 \Gamma_{\text{metabolic}} H^T}{4D^T} \left[1 - \left(\frac{r}{r_c} \right)^2 \right]$$

$$- \frac{r_{\text{anoxic}}(z)^2 \Gamma_{\text{metabolic}} H^T}{2D^T} \ln \left(\frac{r}{r_c} \right) \qquad (6.50)$$

We can get $r_{\text{anoxic}}(z)$ from the additional condition in Equation 6.42 that requires $pO_2^T = 0$ at $r_{\text{anoxic}}(z)$. We then obtain the following non-linear equation that can be solved for $r_{\text{anoxic}}(z)$.

$$\left(\frac{r_{\text{anoxic}}(z)}{r_c}\right)^2 \ln\left(\frac{r_{\text{anoxic}}(z)}{r_c}\right)^2 - \left(\frac{r_{\text{anoxic}}(z)}{r_c}\right)^2 + 1 = \frac{4D^T \langle pO_2(z) \rangle}{r_c^2 \, \Gamma_{\text{metabolic}} H^T} \tag{6.51}$$

Equations 6.49 through 6.51 provide the pO_2 level within the tissue region under both non-anoxic and anoxic conditions for a given capillary pO_2 level.

The capillary pO_2 level changes with axial position and can now be found by solving Equation 6.48. Solution of this equation requires that we know the value of $D^T \dfrac{dpO_2^T}{dr}\Big|_{r_c}$. We can get this by differentiating either Equation 6.49 or 6.50 and evaluating the derivative at r_c. Upon substitution of this result for the non-anoxic case we obtain the following equation.

$$\frac{d\langle pO_2(z) \rangle}{dz} = -\frac{\Gamma_{\text{metabolic}} H^T}{(1+m)V}\left[\left(\frac{r_T}{r_c}\right)^2 - 1\right] \tag{6.52}$$

This equation may be integrated to give the following result for the axial change in the capillary oxygen partial pressure. Note that this equation predicts that the capillary pO_2 level decreases linearly with axial position in the capillary.

$$\langle pO_2(z) \rangle = \langle pO_2 \rangle_{\text{in}} - \frac{\Gamma_{\text{metabolic}} H^T}{(1+m)V}\left[\left(\frac{r_T}{r_c}\right)^2 - 1\right]z \tag{6.53}$$

For anoxic conditions we can simply replace r_T by $r_{\text{anoxic}}(z)$ in Equation 6.53.

At a given value of z, we can substitute the anoxic version of Equation 6.53 into Equation 6.51 to obtain the axial dependence of $r_{\text{anoxic}}(z)$. This would be the radial position for a given z at which the tissue pO_2 would equal zero.

$$\left(\frac{r_{\text{anoxic}}(z)}{r_c}\right)^2 \ln\left(\frac{r_{\text{anoxic}}(z)}{r_c}\right)^2 - \left(\frac{r_{\text{anoxic}}(z)}{r_c}\right)^2 + 1$$
$$= \left[\frac{4D^T \langle pO_2 \rangle_{\text{in}}}{r_c^2 \Gamma_{\text{metabolic}} H^T}\right] - \frac{4D^T}{(1+m)r_c^2 V}\left[\left(\frac{r_{\text{anoxic}}(z)}{r_c}\right)^2 - 1\right]z \tag{6.54}$$

Of particular interest would be those conditions under which anoxia first begins. Anoxia will first start at the corner of the Krogh tissue cylinder represented by the coordinates $z = L$ and $r = r_T$. This critical tissue pO_2 level, equal to zero at the *anoxic corner*, can be readily found by first solving Equation 6.53 for the pO_2 level in the blood exiting the capillary. Equation 6.49 can then be solved with $r = r_T$ to find the value of pO_2^T at the anoxic corner. One can then adjust various parameters such as V and $\Gamma_{\text{metabolic}}$ until the pO_2^T equals zero at the anoxic corner.

The Krogh tissue cylinder approach assumes that we know what the Krogh tissue cylinder radius (r_T) is. In Chapter 5 we showed that the blood perfusion rate could be related to the Krogh tissue cylinder radius, the capillary radius and length, and the average blood velocity. This relationship is given by Equation 5.122, which when solved for r_T, gives the following equation.

$$r_T = r_c \sqrt{\frac{V}{q_b L}}$$

(6.55)

This completes our analysis of tissue oxygenation in the region surrounding a given capillary. Example 6.8 illustrates how these equations may be solved to obtain the pO_2 levels within the capillary and the tissue region that surrounds the capillary.

Example 6.8 Calculate the oxygen profiles within a capillary and its surrounding tissue for an islet of Langerhans. Assume blood enters the capillary at a $pO_2 =$ 95 mm Hg. Also, calculate the critical blood perfusion rate. Determine the anoxic boundary for a blood perfusion rate of 0.25 ml/cm³/min.

SOLUTION We use Equations 6.49, 6.50, 6.53, and 6.54 to obtain the solution for the islet tissue perfusion rate of 4 ml/cm³/min. Note that the value of (m) is must be based on the average pO_2 of the blood which must be found by trial and error. That is, one assumes an exiting pO_2 in the blood to start the calculation. The calculation is then performed, and the calculated exiting pO_2 is compared with the value previously assumed. This process is repeated until convergence is obtained. Figure 6.11 shows the solution. The solid line represents the change in the capillary pO_2 and the dashed line represents the change in the tissue pO_2 at the Krogh tissue cylinder radius. If the tissue perfusion rate is decreased to about 0.62 ml/cm³/min, the pO_2 level becomes zero at the anoxic corner. This is shown in Figure 6.12. If the tissue blood perfusion rate is decreased further, we will develop an anoxic layer of tissue whose axial boundary would be defined by r_{anoxic} (z). This boundary can be found by solving Equation 6.54. Figure 6.13 shows the anoxic boundary for the case where the tissue perfusion rate has been decreased to 0.25 ml/cm³/min.

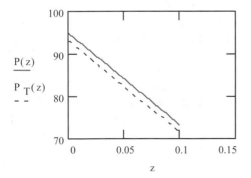

Figure 6.11 Oxygen profile in the Krogh tissue cylinder.

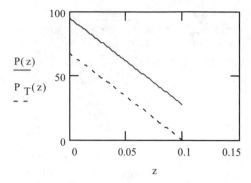

Figure 6.12 Oxygen profile in the Krogh tissue cylinder with an anoxic corner.

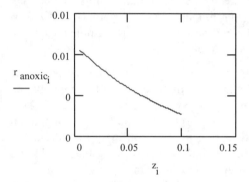

Figure 6.13 Anoxic profile in the Krogh tissue cylinder.

6.12 ARTIFICIAL BLOOD

Blood is crucial for survival and is frequently needed for treatment of life-threatening injuries, to replace blood loss during surgery, and to treat a variety of blood disorders. The world demand for blood amounts to about 100 million units[5] of blood per year. Within the USA, a blood transfusion occurs every 3.75 seconds (Lewis 1997). Blood banking provides, in most cases, an immediate source of blood to meet these needs. However, blood banking requires an extensive infrastructure for collection, testing for disease, cross-matching, and storage. In addition to potential contamination of the blood supply with infectious agents such as HIV and hepatitis, blood itself can cause hemolytic transfusion reactions. The risks associated with blood transfusions range from 0.0004%[6] for HIV infection, 0.001% for a fatal hemolytic reaction, 0.002% to 0.03% for contracting the hepatitis virus, and 1% for a minor reaction (Intaglietta and Winslow 1995).

[5] A unit of blood is 1 pint equal to about 500 ml.
[6] 1% is equal to 1 transfusion incident per 100 units transfused.

Concern in recent years about the safety of the blood supply has resulted in major efforts to develop a substitute for blood. A blood substitute could replace human donor blood and potentially be used for trauma and surgery, as well as for treatment of chronic blood disorders that require frequent transfusions. Blood substitutes are also of interest to the military because of the potential for a significant decrease in the special handling and logistical requirements of human blood on the modern battlefield.

Artificial blood must meet the two most important functions of blood given by transfusion. These are replacement of lost plasma and the ability to transport sufficient amounts of oxygen. There are three approaches being considered that have the potential to be used as artificial blood. These are solutions containing stroma-free[7] hemoglobin, perfluorocarbon (PFC) liquids, and camouflaged RBCs. There are currently about 10 different artificial blood products in development that utilize either some form of hemoglobin, PFCs, or camouflaged RBCs (Winslow 1997; Scott et al. 1997).

Several products are based on the use of stroma-free hemoglobin. The hemoglobin can be derived from a variety of sources. For example, hemoglobin can be of human or bovine origin, it may be derived from bacteria via recombinant DNA techniques, or from transgenic methods in animals and plants. Hemoglobin as discussed earlier consists of four polypeptide chains, two α chains and two β chains. The hemoglobin molecule outside of the RBC is not stable. For example, if hemoglobin is diluted in comparison to its concentration in RBCs, it will spontaneously dissociate into these smaller chains. These lower molecular weight chains are then rapidly removed from the circulation and excreted by the kidneys. Also recall that hemoglobin by itself has a very high affinity for oxygen. Therefore, to be effective as a blood substitute, artificial bloods based on hemoglobin must chemically modify the hemoglobin in order to improve its stability and to decrease its affinity for oxygen. The P_{50} value obtained for artificial blood based on hemoglobin is comparable to that of normal human blood.

PFCs are hydrocarbons in which the bonds between C and H are replaced by the much stronger C and F bonds. These strong C–F bonds give PFCs their chemical inertness (Shah and Mehra 1996). PFCs also exhibit high oxygen solubility, and are, for the most part, biocompatible and low cost. However, they are immiscible with water. In order to be used as a blood substitute, they must first be emulsified into tiny droplets typically in the range of 0.1 to 0.2 microns in diameter. The droplets are also frequently coated with phospholipids derived from egg yolk to stabilize the emulsion. Oxygenated blood carries about 20 ml of oxygen (based on 37°C and 760 mm Hg) per 100 ml of blood. PFCs in equilibrium with pure oxygen at 760 mm Hg have an average oxygen solubility of about 50 ml of oxygen per 100 ml of PFC (Gabriel et al. 1996). PFCs exhibit a linear equilibrium relationship between pO_2 and oxygen solubility. The Henry constant for PFCs is equal to about 0.04 mm Hg/μM. Unlike blood containing hemoglobin, PFCs do not saturate and can, therefore, transport more oxygen by simply increasing the pO_2 level in the gas breathed by the patient. For example, a 40 volume % PFC emulsion when saturated with pure oxygen at 760 mm Hg would have the same oxygen carrying capacity of saturated blood. Because of their relatively high vapor

---

[7] Stroma-free in this case means without the red blood cell and its cellular framework.

pressure, a portion of the PFCs are removed from the circulation by evaporation via respiration.

Artificial bloods based on modified hemoglobin or PFCs still face some significant development problems. Key among these problems are gastrointestinal complaints and vasoconstriction (reduction in blood vessel diameter) when using hemoglobin products and "flu-like" symptoms and thrombocytopenia (very low quantity of platelets) when using PFCs. These artificial bloods also exhibit short lifetimes in the circulation. Modified hemoglobins last on the order of 12 hours to 2 days whereas PFCs are removed in about 12 hours. In comparison, normal RBCs have lifetimes of about 120 days. Until the lifetimes of these products can be significantly improved, they will only be able to be used for acute situations such as trauma and surgery (Winslow 1997).

Recently, a promising technique was described that may overcome some of the problems of artificial blood based on modified hemoglobin or PFCs (Scott *et al.* 1997). This approach involves covalently binding polyethylene glycol (PEG) to the surface of intact red blood cells. The RBCs appear to be unaffected by the presence of the PEG coat. The PEG molecules have the effect of masking the surface antigens on the RBC that lead to the various blood types in the case of human blood or that trigger rejection in the case of animal red blood cells. The result is a universal blood type that could allow the use of unmatched human or animal RBCs. The camouflaged RBCs should have a lifetime that is comparable to that of a normal RBC.

PROBLEMS

1. Consider a very metabolically active tissue region supplied with arterial blood that has a pO_2 = 95 mm Hg. For an exiting venous pO_2 of 40 mm Hg, what values of n and P_{50} would maximize the amount of oxygen delivered to the tissue?

2. Determine the volumetric tissue oxygen consumption rate ($\Gamma_{metabolic}$, $\mu M/sec$) from the data shown in Figure 6.1 for Brockman bodies. Carefully state all assumptions.

3. Calculate the external mass transfer coefficient (k_m, cm/sec) for oxygen from the data shown in Figure 6.1. Recall that the flux of oxygen into the spherical Brockman body by diffusion must equal the oxygen transport rate from the bulk fluid as described by the following equation:

$$-D_{effective} \frac{dC_{tissue}}{dr} \Big|_{surface} = k_m (C_{bulk} - C_{tissue} \Big|_{surface}) \approx \frac{1}{3} R_{Brockman\ body} \Gamma_{metabolic}$$

Show that for the case without convection, the Sherwood number (Sh) is equal to 2.

4. Derive Equation 6.18.

5. For the situation described in Example 6.4, determine the critical loading of islets, i.e. the void volume of the tissue space for which the pO_2 becomes equal to zero at the centerline of the symmetric islet layer.

6. For the results obtained in Example 6.4, determine an estimate of the blood flowrate needed to sustain a total of 750,000 islets. Each islet may be assumed to be 150 microns in diameter. What would be the diameter of such a device?

7. For the situation described in Example 6.4, determine the critical oxygen permeability of the membrane, i.e. the value for which the pO_2 becomes equal to zero at the centerline of the symmetric islet layer.

8. Write a short paper that discusses blood types.

9. Derive Equation 6.49.

10. Derive equation 6.53.

11. Develop a Krogh tissue cylinder model for blood oxygenation using PFCs. Using this model, determine the tissue oxygen profile for an artificial blood perfusion rate of 0.75 ml/cm^3/min.

12. Consider a long and wide planar aggregate or slab of hepatocytes (liver cells) used in a bioartificial liver. The slab is suspended in a growth medium at 37°C with no mass transfer limitations between the slab and the medium. The medium is saturated with air giving a pO_2 of 160 mm Hg or 190 μM. The cell density in the slab is 1.25×10^8 cells cm^{-3}. Each cell is spherical in shape and 20 microns in diameter. The oxygen consumption rate of the cells may be described by the Michaelis–Menten relationship (Equation 6.12) with $V_{max} = 0.4$ nmol 10^{-6} cells^{-1} sec^{-1} and a $K_m = 0.5$ mm Hg. Estimate the maximum half-thickness of the slab of hepatocytes.

13. Photodynamic therapy (PDT) involves the localized photoirradiation of dye sensitized tissue toward the end goal of causing cell death in tumors (Henderson and Doughherty, *Photochem. Photobiol.*, 55, 145–157, 1992). The tumor containing the photosensitizer is irradiated by a laser with light of the proper wavelength (around 630 nm) to generate excited singlet molecules of the sensitizer. In the presence of molecular oxygen, the singlet sensitizer forms a very reactive form of oxygen called singlet oxygen which destroys the tumor. Therefore, sufficient levels of molecular oxygen are needed in the tissue to effect a kill of the tumor by PDT. The action of PDT on tissue oxygen levels is like having an additional sink for oxygen that is dependent on the laser fluence rate, ϕ_0. Foster *et al.* (*Cancer Research*, 53, 1249–1254, 1993) states that this PDT induced oxygen consumption rate may be estimated by the following equation:

$$\Gamma_{PDT} = 0.14 \ (\mu M \ cm^2 \ sec^{-1} \ mW^{-1}) \times \phi_0 \ (mW \ cm^{-2}),$$

where the fluence rate (ϕ_0) ranges between

50 and 500 mW cm^{-2}

Consider a typical capillary within the tumor undergoing PDT treatment. Assume a capillary diameter of 8 microns and a capillary length of 300 microns. The Krogh tissue cylinder radius is assumed to be 40 microns. The metabolic oxygen consumption rate is 11.5 μM sec^{-1}. The average velocity of blood in the capillary is 0.04 cm sec^{-1}. The entering blood pO_2 is 95 mm Hg. Answer the following questions:

a. What is the exiting pO_2 of the blood from the capillary with the laser off?

b. What is the pO_2 in the lethal corner ($z = L$ and $r = r_T$) with the laser off?

c. At what fluence rate would the pO_2 in the lethal corner equal zero? Note this is the point at which PDT would start to be ineffective for those cancer cells at the outer periphery of the Krogh tissue cylinder.

14. Consider the design of a hollow fiber unit for an extracorporeal bioartificial liver. Blood flows through the lumens of the hollow fibers contained within the device at a flow rate of 400 ml/min. The unit consists of 10,000 fibers and the lumen diameter of a fiber is 400 microns. The blood enters the fiber with a pO_2 of 95 mm Hg and must exit the device with a pO_2 no lower than 40 mm Hg. Surrounding each fiber is a multicellular layer of cloned human liver cells. These cells have a void fraction of 80% and they consume oxygen at the rate of 17.5 μM sec^{-1} (based on cellular volume). The hollow fibers are 25 cm in length and the fiber wall provides negligible resistance to the transport of oxygen. Determine the maximum thickness of the layer of cells that can surround each fiber. Carefully state your assumptions.

15. A non-woven mesh of polyglycolic acid (PGA) is proposed to serve as the support for growing chondrocytes in vitro for the regeneration of cartilage. The PGA mesh is a square pad 1 cm × 1 cm and 0.5 mm thick. The PGA mesh has a porosity of 97% and the edges of the mesh are clamped within a support. The chondrocytes were found to consume oxygen at a rate of 2.14

$\times 10^{-13}$ mol cell^{-1} hr^{-1} (based on cellular volume). Assuming the cells are cultured in a well-perfused growth media that is saturated with air, estimate the maximum number of cells that can be grown within each PGA mesh. Assume the diameter of a cell is about 20 microns. Carefully state your assumptions.

16. Blood perfuses a region of tissue at a flow rate of 0.35 ml min^{-1} cm^{-3} of tissue. The pO_2 of the entering blood is 95 mm Hg and the exiting pO_2 of the blood is 20 mm Hg. Calculate the metabolic oxygen consumption rate of the tissue in μM sec^{-1}.

17. Consider a slab layer of cells being grown within an artificial support structure. The layer of cells is immersed in a well-mixed nutrient medium maintained at an oxygen $pO_2 = 150$ mm Hg. The cells are known to consume oxygen at the rate of 40 μM sec^{-1}. Estimate the maximum half-thickness of the cell layer (cm) assuming the void volume fraction in the tissue layer is 0.20.

18. A tumor spheroid 400 microns in diameter is suspended in an infinite and quiescent media at 37°C with a pO_2 of 120 mm Hg. The pO_2 at the surface of the spheroid was measured to be 100 mm Hg. Estimate the oxygen consumption rate of the cells in the spheroid in μM sec^{-1}. Assume H = 0.74 mm Hg μM^{-1}.

19. Blood flows through a membrane oxygenator at a flowrate of 5000 ml min^{-1}. The entering pO_2 of the blood is 30 mm Hg and the exiting blood pO_2 is 98 mm Hg. Calculate the amount of oxygen transported into the blood.

20. Consider a slab layer of cells being grown between two microporous support membranes. The half-thickness of the cell layer is 125 microns and the void volume fraction in the cell layer is 0.90. The permeability of oxygen through the support membrane is estimated to be equal to 1.5 $\times 10^{-3}$ cm sec^{-1}. The layer of cells is immersed in a well-mixed nutrient medium maintained at an oxygen $pO_2 = 150$ mm Hg. An oxygen microelectrode placed at the centerline of the layer of cells gives a pO_2 reading of 15 mm Hg. Estimate the rate at which these cells are consuming oxygen in μM sec^{-1}.

21. Blood is flowing at the rate of 200 ml min^{-1} through the lumens of a hollow fiber unit containing hepatocytes on the shell side. The pO_2 of the entering blood is 95 mm Hg and the exiting pO_2 of the blood is 20 mm Hg. The volume of hepatocytes on the shell side of the device is estimated to be about 600 ml. Estimate the metabolic oxygen consumption rate of the hepatocytes in μM sec^{-1}.

22. Consider a slab layer of cells being grown within an artificial support structure. The layer of cells is immersed in a well-mixed nutrient medium maintained at an oxygen $pO_2 = 130$ mm Hg. The cells are known to consume oxygen at the rate of 10 μM sec^{-1} and the half-thickness of the slab of cells is 35 microns. Estimate the void volume fraction in the tissue.

23. Blood perfuses a region of tissue at a flow rate of 0.50 ml min^{-1} cm^{-3} of tissue. The pO_2 of the entering blood is 95 mm Hg and the exiting pO_2 of the blood is 30 mm Hg. Calculate the metabolic oxygen consumption rate of the tissue in μM sec^{-1}.

24. Consider a slab layer of cells being grown within an artificial support structure. The layer of cells is immersed in a well-mixed nutrient medium maintained at an oxygen $pO_2 = 140$ mm Hg. The cells are known to consume oxygen at the rate of 30 μM sec^{-1}. Estimate the maximum half-thickness of the cell layer (microns) assuming the void volume fraction in the tissue layer is 0.20.

25. In a tissue engineered vascularized tissue structure, pO_2 measurements were taken in vivo in the region equidistant from the capillaries using luminescent oxygen-sensitive dyes. This pO_2 value, which is basically at the Krogh tissue cylinder radius, was found to be 10 mm Hg and the average concentration of oxygen in the blood in the capillaries was 100 μM. Histological analysis of the tissue samples indicated that the capillaries were pretty much parallel to each other. The average distance between the capillaries, measured center to center, was found to be 130 microns. The capillaries themselves are 7 microns in diameter and the rate of oxygen uptake for the cells surrounding the capillaries is estimated to be 30 μM sec^{-1}. From these data, estimate the diffusivity of oxygen through the tissue surrounding the capillaries.

26. Islet of Langerhans are sequestered from the immune system in a device like that shown in Figure 6.6. The pO_2 of the blood in the capillaries adjacent to the immunoisolation membrane is 40 mm Hg. The membrane oxygen permeability is 9.51×10^{-4} cm sec^{-1}. If the islets consume oxygen at the rate of 25.9 μM sec^{-1}, estimate the maximum half-thickness of the islet tissue in cm assuming a void volume in the islet layer of $\varepsilon = 0.95$.

27. A laboratory scale bioartificial liver consists of a single hollow fiber that is 500 microns in diameter and 25 cm in length. The flowrate of blood through the fiber is 0.2 ml min^{-1}. The blood enters the fiber with a pO_2 of 95 mm Hg and exits the fiber at a pO_2 of 35 mm Hg. Surrounding the hollow fiber is a confluent layer of cloned human liver cells that consume oxygen at the rate of 25 μM sec^{-1} (based on cell volume). Assume the hollow fiber wall provides negligible resistance to the transport of oxygen. Determine the maximum thickness of the layer of cells that can surround the hollow fiber.

28. A laboratory scale bioartificial liver consists of 10,000 hollow fibers that are 500 microns in diameter and 25 cm in length. The blood enters the fibers with a pO_2 of 95 mm Hg and exits the fibers at a pO_2 of 35 mm Hg. Surrounding each of the hollow fibers is a single confluent layer of cloned human liver cells that consume oxygen at the rate of 25 μM sec^{-1}. The thickness of the layer of cells is 25 microns. Assume the hollow fiber wall provides negligible resistance to the transport of oxygen. Determine the <u>total</u> flowrate of the blood in ml/min needed to provide these conditions.

29. Design a planar bioartificial pancreas for the treatment of diabetes (what are the dimensions, thickness and radius). Assume that the device will contain a total of 50 million genetically modified human β cells. Each of these cells has a diameter of 15 microns, and when packed into the device, the void volume (ε) must be 0.15. The immunoisolation membrane has a thickness of 50 microns and a porosity of 0.80. Assume the cells consume oxygen at the rate of 25 μM sec^{-1} and that the oxygen pO_2 in the blood adjacent to the immunoisolation membrane is 40 mm Hg.

30. Rework problem 29 assuming there are 500 million cells.

31. Generate the cell surface oxygen concentration profiles in the bioreactor studied by Allen and Bhatia (2003) for an inlet flowrate of 0.5 cm^3 min^{-1} and inlet oxygen concentrations from 90 to 190 nmol cm^{-3}. Other data may be found in Example 6.6.

PHARMACOKINETIC ANALYSIS

7.1 TERMINOLOGY

In this chapter we will focus our attention on some useful techniques in the field of pharmacokinetics. *Pharmacokinetics* is the study of the processes that affect drug distribution and the rate of change of drug concentrations within various regions of the body. These processes are also collectively referred to as ADMET for drug adsorption, distribution, metabolism, excretion, and toxicity. Although pharmacokinetics is of utmost importance in the treatment of diseases using drugs, our major focus in the remainder of this book will be on using this technique as a tool for understanding the transport processes in drug delivery systems and between an artificial device and the body.

We also need to distinguish pharmacokinetics from similar terms. Therefore, *pharmaceutics* concerns the formulation and preparation of the drug to achieve a desired drug availability within the body, *pharmacodynamics* is concerned with the time course of the treatment response that results from a given drug, and the actual physiological response that results from a drug is the subject of *pharmacology*.

7.2 ENTRY ROUTES FOR DRUGS

There are two routes through which a drug can enter the body. The *enteral* route refers to drugs that are given via the gastrointestinal tract (GI tract). All other routes are called *parenteral*. The enteral route includes drugs that are absorbed via one or more of the following components of the GI tract: the buccal cavity (mouth), sublingually (beneath the tongue), gastrically (stomach), intestinally (small and large intestines), and rectally. Once the drug is absorbed from the GI tract, the drug enters the blood and is distributed throughout the body. It is important, however, to point out that only drugs absorbed from the buccal cavity and the lower rectum enter the *systemic circulation* directly. Drugs absorbed from the stomach, intestines, colon, and upper rectum enter the *splanchnic circulation.* The splanchnic circulation then takes the drug to the liver via the portal vein and, after leaving the liver, the drug enters the systemic circulation. Since the liver contains many enzymes capable of degrading the drug (metabolism), a significant portion of the drug may be removed during this first pass through the liver before the drug is available to the general circulation. Table 7.1 summarizes all of the other routes for drug administration that are called parenteral. For the most part, drugs given parenterally enter the systemic circulation directly.

As the drug is being absorbed into the body, its presence will be noticed within the circulation as its concentration in the plasma portion of the blood changes. Recall that plasma refers to the clear supernatant fluid that results from blood after the cellular components have been removed. If the blood sample is simply allowed to clot, then the resulting clear fluid is referred to as serum since the clotting proteins have been removed by the clotting process. For the most part, the concentration of drug in plasma or serum is identical and no distinction is needed. It is usually only the plasma drug concentration in the body that is known.

Figure 7.1 illustrates typical plasma drug concentrations as a function of time after their introduction to the body. Curve I illustrates the case where the drug is first slowly absorbed, resulting in increasing plasma concentrations, followed by a plateau, and then the plasma concentration decreases as the drug is eliminated by a variety of body processes. Curve II represents a very rapid injection (bolus) of the drug, usually intravenously (IV injection), where we see that the drug instantly reaches its peak plasma concentration near the time of injection, and then is slowly eliminated. Curve III results from a continuous infusion, usually intravenous or through a controlled release formulation. We see after a short period of time that the drug reaches a steady state plasma concentration. If the infusion is stopped, then the drug plasma concentration falls due to its elimination as shown by the dashed curve.

Table 7.1 Parenteral routes for drug administration

Route	Result
Intravenous (IV)	Introduced directly into the venous circulation
Intramuscular	Within the muscle
Subcutaneous	Beneath the epidermal and dermal skin layers
Intradermal	Within the dermis, usually a local effect
Percutaneous	Topical treatment applied to the skin
Inhalation	Mouth/nose, pharynx, trachea, bronchi, bronchioles, alveolar sacs, alveoli
Intraarterial	Introduced directly into an artery, regional drug delivery
Intrathecal	Cerebrospinal fluid, subarachnoid space
Intraperitoneal (IP)	Within the peritoneal cavity
Vaginal	Within the vagina

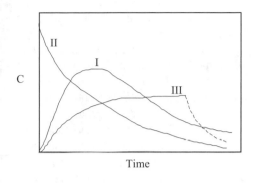

Figure 7.1 Plasma drug concentrations following I: absorption, II: rapid intravenous, and III: continuous infusion.

7.3 MODELING APPROACHES

Our goal is then to develop mathematical models that can be used to describe profiles of plasma drug concentration such as those shown in Figure 7.1. This would allow for the description of the drug concentration in the body as a function of time. Through pharmacology and pharmacodynamics, the effects of the drug on the body can then be related to the plasma drug concentration. These models will incorporate several unknown parameters that we find by fitting our models to experimental drug concentration data. The development of pharmacokinetic models has generally followed three approaches. They are the *compartmental approach*, the *physiological approach*, and the *model independent approach.*

The compartmental approach assumes the drug distributes into one or more "compartments" in the body that usually represent a particular region of the body, an organ, a group of tissues, or body fluids. The compartments are assumed to be well-mixed so that the concentration of the drug within the compartment is spatially uniform. Spatial distribution of the drug within the body is accounted for by the use of multiple compartments. The assumption of well-mixed compartments is usually pretty good since the cardiac output in humans is nominally 5 liters/min, and the blood volume is about 5 liters, giving an effective residence time for one passage through the circulatory system of about one minute. Also, recall that there is a net filtration of several milliliters per minute of the blood plasma forming the interstitial fluid that bathes and circulates around all of the cells in the body. Since drug distribution occurs over periods of hours, and the body's fluids are moving over periods of minutes, and mass transfer occurs on the order of seconds, the well-mixed assumption is appropriate. Movement of the drug between the compartments is usually described by simple irreversible or reversible first order rate processes, i.e. the rate of transport is proportional to difference in the drug concentration between compartments.

In physiological models, the movement of drug is based on the blood flow rate through a particular organ or tissue and includes consideraton of the rates of the mass transport processes within the region of interest. Experimental blood and tissue drug concentrations are needed to define the model parameters. Our Krogh tissue cylinder model developed in Chapter 6 to describe the changes in oxygen levels within the capillary and the surrounding tissue region is an example of a physiological model. In this case however, the "drug" is oxygen. Since the liver plays a major role in the metabolism and excretion of drugs from the body much attention has been focused on the development of physiological models to describe liver function (Bass and Keiding 1988; Niro *et al.* 2003).

The model independent approach does not try to make any physiological connection like the compartmental approach or the physiological approach. Instead, one simply finds the best set of mathematical equations that describes the situation of interest (Notari 1987). These models are considered to be linear if the plasma drug concentration can be represented by a simple weighted summation of exponential decays. For example, the following equation may be used to describe the time course of the plasma drug concentration.

$$C = \sum_{i=1}^{n} C_i e^{-\lambda_i t} \qquad (7.1)$$

Our focus in this chapter will be on the use of compartmental models to describe pharmacokinetic data like that shown in Figure 7.1. Compartmental models are commonly used and provide a convenient framework for building a pharmacokinetic model and are conceptually simple to use (Cooney 1976; Gibaldi and Perrier 1982; Welling 1986; Notari 1987).

7.4 FACTORS THAT AFFECT DRUG DISTRIBUTION

Before we can investigate some simple compartmental models, we first must discuss the following factors that influence how a particular drug is distributed throughout the body. Table 7.2 summarizes these factors, and they are discussed in greater detail in the following discussion.

7.4.1 Drug Distribution Volumes

Recall that Table 3.1 summarized the types of fluids found within the body. Since the drug can only distribute within these fluid volumes, we call these the *true distribution volumes*. Therefore, a drug could distribute within just the plasma volume of the circulatory system, i.e. a true distribution volume of about 3 liters. If the drug readily penetrates the vascular walls, then it could also distribute throughout the extracellular fluid volume for a total true distribution volume of about 15 liters. If the drug can also

Table 7.2 Factors influencing drug distribution

Blood Perfusion Rate
Capillary Permeability
Drug Biological Affinity
Metabolism of Drug
Renal Excretion

permeate the cell wall, then it will also be found within the intracellular fluid spaces as well, giving a total true distribution volume of about 40 liters.

The rate at which the drug is distributed throughout these fluid volumes is controlled by the rate at which drug is delivered to a region of interest by the blood, i.e. the tissue blood perfusion rate, and by the rate at which the drug diffuses from within the vascular system into the extravascular spaces. If the drug is lipid soluble, then its mass transfer across the capillary wall is usually not rate limiting, and equilibrium is quickly reached between the amount of drug in the tissue region and that found in the blood. In this case, the distribution of the drug in a particular tissue region is *perfusion rate limited*. For lipid insoluble drugs, the capillary membrane permeability controls the rate at which the drug distributes between the blood and tissue regions. The distribution of the drug is then said to be *diffusion rate limited*.

Drugs are also capable of binding to proteins found in the plasma and in the extravascular spaces as well. Albumin is perhaps the most common of the blood proteins that will bind with drugs. Recall that the major difference between plasma and interstitial fluid is the amount of protein that is present. For the most part, plasma proteins have a difficult time diffusing through the capillary wall to enter the extravascular space. Therefore, a drug that binds strongly with plasma proteins will, for the most part, be limited to just the volume of the plasma, i.e. about 3 liters. On the other hand, extravascular proteins or cell surface receptors that bind strongly to the drug will tend to accumulate the drug in the extravascular space at the expense of the plasma. In this case, the distribution volume of the drug may appear to be larger than the physical fluid volumes. Protein binding of the drug, therefore, can have a significant effect on the calculated or *apparent distribution volumes* as the following discussion will show.

It is important to point out that most analytical methods measure the total drug concentration (bound and unbound) that is in the plasma. Since this total plasma concentration (C_{total}) is readily measured, the *apparent distribution volume* ($V_{apparent}$) for the drug is simply given by the following equation:

$$V_{apparent} = \frac{A}{C_{total}} \tag{7.2}$$

where A is the total amount of drug within the body. The amount of drug in the body is usually in units of milligrams (mg), and the concentration of drug in micrograms per milliliter of plasma ($\mu g/ml$), which is the same as milligrams per liter of plasma (mg/L), resulting in the apparent distribution volume having units of liters (L).

It is also important to note that the concentration of drug within any particular tissue region may vary significantly from the concentration of the drug in the plasma. Therefore, although the apparent distribution volume seems to have some physical significance, it is really just a factor having units of volume that, when multiplied by the plasma drug concentration provides, the total amount of drug in the body at a particular time.

The apparent distribution volume calculated by Equation 7.2 can also be vastly different from the true distribution volume (V), related to the volume of the various body fluids, because of the effect of protein binding. To examine the effect of protein binding on the apparent distribution volume, consider the conceptual model shown in

Figure 7.2 (Oie and Tozer 1979). In their model, unbound drug is free to distribute between three compartments; the plasma (P), extracellular fluid outside of the plasma (E), and the remainder of the body (R) where the drug may be bound to the surface of the cell by cell surface receptors, or internalized. A mass balance on the total amount of drug in the body (A) at any time may then be written as follows:

$$A = A_P + A_E + A_R \tag{7.3}$$

where A_P, A_E, and A_R represent the total amount of drug in the plasma, the extracellular fluid, and the remainder of the body, respectively.

The amount of drug in any of these three compartments is also related to the corresponding concentration of total drug in a given compartment, i.e. $A_P = C \, V_P$, $A_E = C_E \, V_E$, and $A_R = C_R \, V_R$, where C, C_E, and C_R are the respective total drug concentration in each compartment and V_P, V_E, and V_R are the corresponding compartment volumes. Using these relationships, we can rewrite Equation 7.3 as:

$$V = V_P + V_E \left(\frac{C_E}{C} \right) + V_R \left(\frac{C_R}{C} \right) \tag{7.4}$$

where we also recognize that, by definition, $A = V\,C$.

Ignoring active transport type processes, it is assumed that at distribution equilibrium the unbound concentration of the drug, i.e. C_U, is the same in all of the compartments. In addition, in the extracellular fluid compartment we have $C_E = C_U + C_{EB}$, where C_{EB} is the bound drug concentration in the extracellular fluid. Also, f_{UT} is defined as C_U/C_R and f_U is defined as C_U/C. f_U is also the same as the fraction of unbound drug in the plasma, and f_{UT} is the fraction of unbound drug in the extracellular fluid space. Both f_U and f_{UT} are usually determined experimentally by equilibrating a

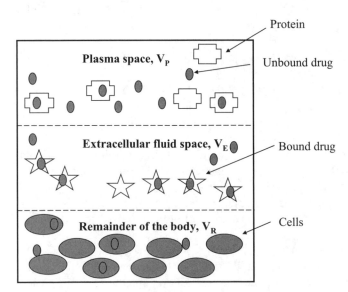

Figure 7.2   Drug distribution in the body with protein binding.

known amount of drug with plasma or extracellular fluid and determining from a drug mass balance the fraction of drug that is unbound. With these relationships Equation 7.4 becomes:

$$V = V_P + V_E\, f_U \left(\frac{C_U + C_{EB}}{C_U} \right) + V_R \frac{f_U}{f_{UT}} \tag{7.5}$$

Unbound drug in the plasma and extracellular spaces can bind with proteins in the plasma (protein concentration in plasma, i.e. P_P) and extracellular spaces (protein concentration in extracellular space, i.e. P_E) as described by the following chemical reactions.

$$C_U + P_P \Leftrightarrow C_{PB} \text{ and } C_U + P_E \Leftrightarrow C_{EB} \tag{7.6}$$

C_{PB} and C_{EB} then represent the concentrations of drug bound protein in the plasma and extracellular spaces. Assuming that the binding sites on these proteins have the same affinity for the unbound drug, then at equilibrium we can write:

$$K_a = \frac{C_{PB}}{C_U\, P_P} = \frac{C_{EB}}{C_U\, P_E} \tag{7.7}$$

K_a is the equilibrium or affinity constant. The total protein concentration in the plasma (P_{PT}) and extracellular spaces (P_{ET}) must be equal to the concentration of the respective proteins that are bound to drug plus the concentration of these proteins that are not bound to drug, i.e.

$$P_P + C_{PB} = P_{PT} \text{ and } P_E + C_{EB} = P_{ET} \tag{7.8}$$

From Equation 7.7 we also have that:

$$C_{EB} = C_{PB} \frac{P_E}{P_P} = C_{PB} \frac{\left(P_{ET} - C_{EB} \right)}{\left(P_{PT} - C_{PB} \right)} \tag{7.9}$$

and this equation can be simplified to give:

$$C_{EB} = C_{PB} \frac{P_{ET}}{P_{PT}} \tag{7.10}$$

Recall that P_{ET} and P_{PT} represent the total protein concentrations in the extracellular and plasma spaces respectively. The total amount of drug binding protein in the plasma and extracellular spaces equals $P_{PT}\, V_P$ and $P_{ET}\, V_E$. $R_{E/l}$ is then defined as the ratio of the total amount of drug binding protein in the extracellular space to that in the plasma space which is equal to $(P_{PT}\, V_P)/ (P_{ET}\, V_E)$. Therefore, Equation 7.10 may be written as shown below.

$$C_{EB} = C_{PB}\, R_{E/l} \frac{V_P}{V_E} \tag{7.11}$$

Now we can substitute Equation 7.11 into Equation 7.5 for C_{EB} and obtain:

$$V = V_P + f_U \left(\frac{V_E \, C_U + C_{PB} \, V_P \, R_{E/I}}{C_U} \right) + V_R \, \frac{f_U}{f_{UT}} \tag{7.12}$$

Recognizing that $\dfrac{C_{PB}}{C_U} = \dfrac{1 - f_U}{f_U}$, we then obtain what is known as the *Oie-Tozer Equation*.

$$V = V_P \left(1 + R_{E/I} \right) + f_U \, V_P \left(\frac{V_E}{V_P} - R_{E/I} \right) + \frac{V_R \, f_U}{f_{UT}} \tag{7.13}$$

From Table 3.1 we have for a 70 kg man that the plasma volume, V_P, is ~3 liters and that the extracellular fluid volume outside of the plasma, i.e. V_E, is ~12 liters. Also, the ratio of drug binding protein in the extracellular fluid space to that in the plasma, i.e. $R_{E/I}$, is ~1.4. Using these values Equation 7.13 simplifies to approximately:

$$V = 7 + 8 f_U + V_R \, \frac{f_U}{f_{UT}} \tag{7.14}$$

The above equation shows us that if a drug is only distributed within the extracellular fluid space and cannot enter the cells, i.e. $V_R = 0$, the apparent distribution volume of the drug then attains its smallest value and is equal to $V_{min} = 7 + 8 f_U$. If, in addition, the drug is not bound to plasma proteins, i.e. $f_U = 1$, then the apparent distribution volume of the drug is limited to 15 liters which is the extracellular fluid volume. Equations 7.13 and 7.14 are also useful for investigating the effect on the apparent distribution volume of a drug for changes in the unbound fraction in the plasma, in the unbound fraction in the extracellular fluids, in the volume of the extracellular fluid, or in the extracellular to plasma ratio of drug binding proteins. Many of these changes occur as a result of age, disease, injury, and interactions with other drugs.

7.4.2 Drug Metabolism

Once the drug is absorbed and distributed throughout the body by the circulatory system, a variety of biochemical reactions will begin to degrade the drug. This breakdown of the drug is part of the body's natural defense against foreign materials. These biochemical reactions, driven by existing enzymes, occur in a variety of organs and tissues. However, the major site of drug metabolism is within the liver. The enzymatic destruction of the drug reduces its pharmacological activity because the active site related to the drug's molecular structure is destroyed. Also, the *metabolites* that result tend to have increased water solubility that decreases their capillary permeability and enhances their removal from the body via the kidneys.

Enzymatic reactions that degrade the drug in the tissues tend to follow the Michaelis–Menten rate model[1] where the reaction rate is given by the following expression:

---

[1]A derivation of the Michaelis-Menten model can be found in Section 8.3.3

$$r_{\text{metabolic}} = -\frac{V_{\text{max}}C}{K_m + C} \tag{7.15}$$

Usually the drug concentration is much smaller than the value of K_m. Therefore, the rate at which the drug is degraded by metabolic reactions can be represented by an irreversible first order rate law, where $k_{\text{metabolic}} = V_{\text{max}}/K_m$ in the above equation.

$$r_{\text{metabolic}} = k_{\text{metabolic}}C \tag{7.16}$$

In the above equation, C represents the total plasma concentration of drug, $k_{\text{metabolic}}$ represents the first order rate constant for the metabolic reactions, and $r_{\text{metabolic}}$ represents the rate at which the drug is degraded on a per unit volume basis. We will use this expression later when we write mass balances for the drug using compartmental models.

7.4.3 Renal Excretion of the Drug

The kidneys also have a major role in the elimination of the drug from the body. Part of this elimination process is enhanced by the enzymatic degradation of the drug by the liver and the formation of more water soluble products. The kidneys receive about 1100 ml/min of blood or about 22% of the cardiac output. On a per mass basis, they are the highest perfused organ in the body. Their primary purpose is to remove unwanted metabolic end products such as urea, creatinine, uric acid, and urates from the blood, and to control the concentrations of such ions as sodium, potassium, chloride, and hydrogen. The concentrated urine that is formed from the kidneys amounts to about 1.5 liters of urine per day.

Figure 7.3 illustrates the basic functional unit of the kidney called the *nephron*.

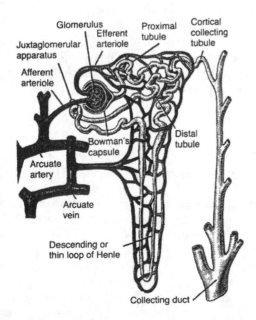

Figure 7.3 The nephron (from A.C. Guyton, 1991, *Textbook of Medical Physiology*. 8th ed. W.B. Saunders, with permission).

Each kidney contains about one million nephrons, and each nephron contributes to the formation of urine. The operation of the kidney can be explained by considering how an individual nephron functions. The nephron consists of the following major components: the *glomerulus* contained within *Bowman's capsule*, the *convoluted proximal tubule*, *Henle's loop*, the *convoluted distal tubule*, and, finally, the *collecting tubule* that branches into the *collecting duct*. The glomerulus, proximal tubule, and distal tubule have a major role in drug elimination from the kidney.

Blood enters the glomerulus via the *afferent arteriole* and leaves by the *efferent arteriole*. The glomerulus consists of a tuft of highly permeable fenestrated capillaries, several hundred times more permeable to solutes than the typical capillary. *Glomerular filtrate* is formed across these capillaries within the Bowman's capsule due to a pressure difference that exists between that within the glomerular capillaries and the Bowman space. The glomerular capillary membrane, for the most part, is impermeable to the plasma proteins retaining those molecules with a molecular weight larger than about 69,000 Daltons. Drugs that are bound to plasma proteins, therefore, cannot be removed from the blood stream by filtration through the glomerulus. However, unbound drugs with a molecular weight less than 69,000 Daltons will readily be filtered out of the bloodstream with the glomerular filtrate. The *glomerular filtration rate* (GFR) is the sum total of all the glomerular filtrate formed per unit time by the nephrons in the two kidneys. This is about 125 ml/min or about 180 liters each day, which is more than twice the weight of the body. Clearly, this volume of fluid must be recovered, and we find that over 99% of the glomerular filtrate is reabsorbed in the tubules.

After the glomerular filtrate leaves Bowman's capsule, it enters the proximal tubule that is primarily concerned with the active reabsorption of sodium ions and water. Other substances, such as glucose and amino acids, are reabsorbed as well. The proximal tubule is important in drug elimination in that it can also actively secrete drugs from the peritubular capillaries surrounding the proximal tubule. This active transport of the drug is sufficiently fast that even drugs that are bound to plasma proteins will dissociate freeing up drug for active secretion across the tubule wall. Within the descending portion of Henle's loop, continued reabsorption of water and other ions occurs primarily by passive diffusion. The ascending loop of Henle is considerably less permeable to water and urea resulting in a very dilute tubular fluid that is rich in urea. Henle's loop does not play a significant role in regard to drug elimination. The first portion of the distal tubule is similar to that of the ascending loop of Henle in that it continues to absorb ions but is impermeable to water and urea. Within the latter portions of the distal tubule and the collecting tubule, ions and water continue to be absorbed along with acidification of the urine thus controlling the acid–base balance of the body fluids.

The reabsorption rate of sodium ions is controlled by the hormone *aldosterone*. The water permeability of these distal segments is controlled by *antidiuretic hormone* providing a means for controlling the final volume of urine that is formed. Drugs can also be reabsorbed within the distal tubule thus affecting the rate of drug elimination from the body. The collecting duct continues to reabsorb water under the control of antidiuretic hormone and does not affect, for the most part, drug elimination. We thus find that the kidney affects drug elimination through the following mechanisms:

filtration through the glomerulus, secretion from the peritubular capillaries into the proximal tubule, and reabsorption within the distal tubule.

7.5 DRUG CLEARANCE

7.5.1 Renal Clearance

The elimination of drug by the kidney is described by the term *renal clearance*. The renal clearance is simply the volume of plasma that is totally "cleared" of the drug per unit time as a result of the drug's elimination by the kidneys. Figure 7.4 illustrates the concept of renal clearance and allows for a mathematical definition for the case where the only elimination pathway for the drug is through the kidneys. As we shall see shortly, however, this figure can be generalized to represent multiple elimination pathways.

   In this figure, a drug is uniformly distributed within an apparent distribution volume, $V_{apparent}$, and has a total plasma concentration now represented by C. CL_{renal} represents the renal plasma flowrate that is totally cleared of the drug. Therefore, renal clearance is given by CL_{renal}, and the rate at which drug is removed from the body would be $CL_{renal} \times C$. We can then write an unsteady mass balance on the drug in the body's apparent distribution volume as given by the next equation.

$$V_{apparent} \frac{dC}{dt} = -CL_{renal} C \tag{7.17}$$

This equation may be integrated to provide an equation for how the drug concentration within the body changes with time starting from an initial plasma drug concentration C_0.

$$C(t) = C_0 e^{-(CL_{renal}/V_{apparent})t} = C_0 e^{-k_{renal}t} \tag{7.18}$$

   We see that drug elimination by the kidneys is a first order process exhibiting an exponential decay in plasma drug concentration with time. The renal elimination rate

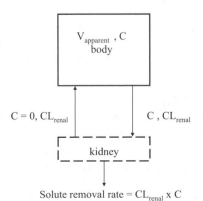

Solute removal rate = $CL_{renal} \times C$ **Figure 7.4**   Concept of renal clearance.

constant k_{renal} is related to the renal clearance and the apparent distribution volume by the following equation.

$$k_{renal} = \frac{CL_{renal}}{V_{apparent}}$$ (7.19)

If Q_{urine} represents the volumetric urine formation rate and C_{urine} the concentration of drug in the urine, then another drug mass balance requires the following relationship.

$$CL_{renal} C = Q_{urine} C_{urine} \quad or \quad CL_{renal} = \frac{Q_{urine} C_{urine}}{C}$$ (7.20)

This equation then provides a connection between the renal clearance and the formation of urine containing the excreted drug.

The polysaccharide *inulin*, with a molecular weight of about 5000 Daltons, is particularly important to mention at this point because its renal clearance can be used to measure kidney function. Inulin is not metabolized by the body, readily passes through the glomerular capillaries, and is neither reabsorbed or secreted within the tubules. Therefore, the rate at which plasma is being cleared of inulin in the kidneys (renal clearance) is the same as the glomerular filtration rate (GFR), i.e. $CL_{renal} \times C_{inulin}$ = GFR $\times C_{inulin}$, where C_{inulin} is the plasma concentration of inulin which is the same as the inulin concentration in the glomerular filtrate.

7.5.2 Plasma Clearance

Plasma clearance (CL_{plasma}) is the term used to represent the sum of all the drug elimination processes of the body. These would include the two primary ones already discussed, that is metabolism and the kidneys, and other secondary processes such as sweating, bile, respiration, and the feces. Each process would have its own characteristic clearance value and a first order elimination rate constant defined as follows:

$$k_i = \frac{CL_i}{V_{apparent}}$$ (7.21)

i = renal, metabolic, sweat, bile, respiration, feces

The plasma clearance would then be related to the sum of all the first order drug elimination rate constants as given by the following relationship. The total first order elimination rate constant is given by k_{te} and is equal to the sum of the elimination rate constants for all elimination processes.

$$k_{te} = \sum_i k_i = \frac{1}{V_{apparent}} \sum_i CL_i = \frac{CL_{plasma}}{V_{apparent}}$$ (7.22)

Similar to Equations 7.17 and 7.18, we can show that the change in drug concentration with time, taking into account all elimination processes, is given by the following equation.

$$C(t) = C_0 e^{-k_{te}t}$$ (7.23)

7.5.3 Biological Half-Life

The *biological half-life* ($t_{1/2}$) of a drug is the time needed for the drug concentration in the plasma to be reduced by one-half. For first order processes, this time is related to the first order elimination rate constant by simply letting C/C_0 equal 0.5 in Equation 7.23. When solved for $t_{1/2}$, the following result is obtained.

$$t_{1/2} = \frac{0.693}{k_{te}} = \frac{0.693 V_{apparent}}{CL_{plasma}} \tag{7.24}$$

7.6 A MODEL FOR INTRAVENOUS INJECTION OF DRUG

Equation 7.23 now represents our first pharmacokinetic model. It is based on a single compartment with all the elimination processes lumped into a single first order elimination rate constant, i.e. k_{te}. This equation predicts that the drug concentration decreases at an exponential rate with a general shape like that of curve II in Figure 7.1. The initial drug concentration is C_0, and this would arise, for example, from a rapid bolus intravenous or intraarterial injection. The initial drug concentration is related to the dose (D), and the apparent distribution volume as given by the following equation.

$$C_0 = \frac{D}{V_{apparent}} \tag{7.25}$$

If the dose and the initial concentration of the drug were known, then Equation 7.25 can be used to solve for the apparent distribution volume of the drug.

The total area under a C versus t curve, such as those shown in Figure 7.1, is given the special symbol $\text{AUC}^{0 \to \infty}$ and is defined mathematically as follows:

$$\text{AUC}^{0 \to \infty} \equiv \int_0^\infty C(t)\,dt \tag{7.26}$$

Equation 7.23 provides a mathematical relationship for the change of the drug concentration with time for the special case of a rapid bolus intravenous injection of a drug that distributes throughout a single compartment. Substituting this equation into Equation 7.26, we obtain the following result for the value of $\text{AUC}^{0 \to \infty}$.

$$\text{AUC}^{0 \to \infty} = \frac{C_0}{k_{te}} = \frac{D}{V_{apparent} k_{te}} = \frac{D}{CL_{plasma}} \tag{7.27}$$

This equation then provides a simple relationship between the total area under the curve, the drug dose, and the plasma clearance. Note that the $\text{AUC}^{0 \to \infty}$ is directly proportional to the dose and inversely proportional to the plasma clearance.

7.7 ACCUMULATION OF DRUG IN THE URINE

As we discussed earlier, some of the drug will be eliminated by the kidneys and show up in the urine. In some cases, it may be convenient to use additional data on the amount of drug (not its metabolites) found in the urine to supplement a pharmacokinetic analysis. The rate at which drug accumulates in the urine would be given by the following equation where M_{urine} is the mass of drug (unchanged) in the urine at any given time.

$$\frac{dM_{\text{urine}}}{dt} = k_{\text{renal}} V_{\text{apparent}} C \tag{7.28}$$

Continuing our discussion with the bolus intravenous injection of drug, we can substitute Equation 7.23 for the value of the drug concentration in the plasma at any time and integrate to obtain the following expression for the mass of drug in the urine at any time.

$$M_{\text{urine}}(t) = \left(\frac{k_{\text{renal}} C_0 V_{\text{apparent}}}{k_{te}} \right) \left(1 - e^{-k_{te} t} \right) \tag{7.29}$$

After a sufficiently long period of time ($t \to \infty$), Equation 7.29 simplifies to provide the total amount of drug collected in the urine, M_{urine}.

$$M_{\text{urine}} = \frac{k_{\text{renal}} C_0 V_{\text{apparent}}}{k_{te}} = D \frac{k_{\text{renal}}}{k_{te}} = \text{AUC}^{0 \to \infty} CL_{\text{renal}} \tag{7.30}$$

Thus, we find a relationship exists between the total amount of drug collected in the urine, the renal clearance, and the total area under the curve. Note that if the only drug elimination pathway is through the kidneys ($k_{\text{renal}} = k_{te}$), then the total amount of drug collected in the urine is equal to the total drug dose, as it should be.

7.8 CONSTANT INFUSION OF DRUG

So far we have discussed the case of first order drug elimination following a rapid bolus intravenous injection of a drug. The drug was assumed to be uniformly distributed throughout a single compartment. In this case, the drug concentration versus time would be expected to follow curve II of Figure 7.1. In many cases, a constant infusion of the drug is given. Here the concentration of the drug increases and reaches a steady state level once a balance is reached between the drug infusion rate and the drug elimination rate. The concentration versus time curve would follow something like curve III of Figure 7.1. This result is similar to the so-called *controlled release* drug delivery systems that are commonly used, for example, to treat motion sickness, for birth control, for extended chemotherapy, and to help people quit smoking by delivering nicotine. The constant infusion of a drug can also be used to provide a steady starting concentration of the drug.

Figure 7.5 illustrates a single compartmental model for continuous infusion of drug. I_0 represents the drug infusion rate (mg/min) assumed here to be constant. However, one could generalize this infusion rate to be any arbitrary time varying function. For example, in Section 5.6.1, we developed an analytical solution describing the diffusion of solute from a polymeric material. In this case, the solute is the drug and

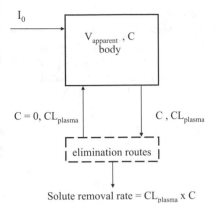

I_0

$V_{\substack{apparent \\ body}}, C$

$C = 0, CL_{plasma}$

C, CL_{plasma}

elimination routes

Solute removal rate $= CL_{plasma} \times C$

Figure 7.5   Single compartment model for continuous infusion with first order elimination.

the polymeric material is a controlled release device. The elution flux of the drug $(j_s|_{x=L})$ is then given by Equation 5.71 (assuming that the concentration of drug in the polymeric material is much larger than that in the body so that BC 2 in Equation 5.60 is satisfied), and I_0 in the equation below is $I_0(t) = j_s|_{x=L} \, S$, where S is the exposed surface area of the device. An unsteady mass balance may then be written for the drug as given by the next equation.

$$V_{apparent} \frac{dC}{dt} = I_0 - CL_{plasma} C = I_0 - V_{apparent} k_{te} C \tag{7.31}$$

This equation may be integrated under the assumption of a constant infusion rate and that no drug is initially present in the body to give:

$$C(t) = \left(\frac{I_o}{k_{te} V_{apparent}} \right) \left(1 - e^{-k_{te} t} \right) \tag{7.32}$$

The general shape of the curve described by this equation is that of curve III in Figure 7.1. For long periods of time, this equation predicts a steady state plasma drug concentration given by the next equation.

$$C_{ss} = \frac{I_0}{k_{te} V_{apparent}} = \frac{I_0}{CL_{plasma}} \tag{7.33}$$

When the infusion is stopped, the drug is eliminated according to Equation 7.23, where the initial drug concentration C_0 is equal to the steady state drug concentration (C_{SS}) arising from the constant infusion. Now the drug concentration decreases as shown by the dotted line of curve III in Figure 7.1.

Example 7.1 The table shown below provides some data for the elimination of radioactive inulin from a 392 gram laboratory rat. The animal was given a bolus injection at $t = 0$ equivalent to 1.01×10^5 cpm (counts per minute, radioactivity directly proportional to the inulin concentration). This was done to hasten the development of a steady state plasma inulin level. Next, a continuous infusion of inulin was started at the rate of 2.76×10^3 cpm min^{-1}. The inulin infusion was then stopped after a total of 80 minutes. Determine from this data the glomerular filtration rate, the renal elimination rate constant, and the apparent distribution volume for inulin.

Inulin elimination from a laboratory rat (Sarver 1994)

Time (minutes)	Inulin concentration (cpm ml^{-1})
0	0
30	849
40	845
60	903
70	888
75	873
80	882 infusion stopped
90	565
100	412
110	271

SOLUTION We see that after about 30 minutes a steady state plasma inulin concentration of about 873 cpm/ml is reached. Once the infusion is stopped, the plasma inulin concentration decreases rapidly. As discussed earlier, inulin is only removed by the kidneys, and its renal clearance is the same as the glomerular filtration rate (GFR). Therefore, for the case of this experiment with inulin, Equation 7.33 can be rearranged to solve for the GFR in terms of the known values of the infusion rate and the steady state plasma concentration.

$$k_{te} V_{apparent} = GFR = \frac{I_0}{C_{ss}}, \text{ for inulin only}$$

For the experimental conditions leading to the data shown in the above table, the GFR is then calculated to be 3.16 ml min^{-1}. Since the GFR is proportional to body weight (BW), this can be expressed as 0.48 ml hr^{-1} gBW^{-1}. The renal elimination rate constant can be found by performing a regression analysis of the data obtained after the infusion is stopped using Equation 7.23. Note that the time in this equation represents the time since the infusion was stopped. The initial concentration is equal to 873 cpm ml^{-1} and k_{te} would be equal to k_{renal} since the only elimination process is that of the kidneys. Note that Equation 7.23 can be linearized by taking the natural logarithm of each side. The regression equation is then given by:

$$\ln C(t) = \ln C_0 - k_{te} t$$

The intercept would be the natural logarithm of the initial concentration, and the slope would be the negative of the elimination rate constant. After performing the linear regression we find that $C_0 = 860.1$ cpm/ml and that $k_{te} = k_{renal} = 0.038$ min^{-1}. Figure 7.6 shows a comparison between the data and the inulin concentrations predicted by Equation 7.23. The apparent distribution volume for inulin is given by:

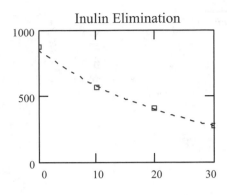

Figure 7.6   Measured and predicted inulin concentrations.

$$V_{apparent} = \frac{GFR}{k_{renal}}$$

For these data, the apparent distribution volume is about 83 ml. Since the distribution volume is also proportional to body weight, the apparent distribution volume for a rat can be expressed as 0.21 ml/gBW.

7.8.1 Application to Controlled Release of Drugs by Osmotic Pumps

There is now considerable interest focused on the development of novel drug delivery systems (Verma *et al.* 2002). Novel drug delivery systems have several advantages to include elimination of first pass metabolism by the liver, minimal discomfort, lower risk of infections, sustained release of the drug, and better patient compliance. For example, osmotic pumps are small implantable devices that can be used to deliver a drug at a constant infusion rate to a specific site in the body for very long periods of time. A recent example is the DUROS® osmotic pump (Wright *et al.* 2001) that is being used to deliver the gonadotropin releasing hormone agonist leuprolide for the palliative treatment of advanced prostate cancer. Serum leuprolide levels of about 1 ng/ml for long periods of time reduces serum testosterone levels in humans to below castration levels which is important in controlling the growth of prostate cancer cells. The device is inserted subcutaneously and is designed to deliver leuprolide at a constant rate for up to one year.

The DUROS® osmotic pump is made from a cylindrically-shaped piece of titanium alloy and measures 4 mm in diameter and 45 mm in length. As shown in Figure 7.7, the device is basically a cylindrical reservoir that is capped at one end by a permselective membrane and at the other end by a flow modulator through which the drug is released to the surroundings.

Within the reservoir are two chambers separated by a piston. In the chamber between the permselective membrane and the piston there is an osmotic agent that increases the osmotic pressure of the fluid within the chamber relative to the osmotic

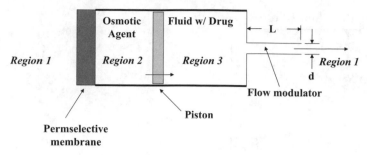

Figure 7.7   Osmotic pump drug delivery system.

pressure of the body fluid that surrounds the device. The osmotic agent in the DUROS® osmotic pump is NaCl at amounts much higher than the NaCl solubility. As water from the surroundings flows across the permselective membrane by osmosis, the NaCl dissolves, and because of being in excess, maintains saturation levels and hence a constant osmotic pressure throughout the lifetime of the implant. As water enters the device by osmosis, the fluid volume containing the osmotic agent increases and exerts pressure on the piston, causing the piston to move. The movement of the piston then acts on the second chamber that contains a fluid with the drug. The piston movement then forces the fluid containing the drug out of the device through the flow modulator at a rate equal to the rate at which water enters the device by osmosis.

The rate at which water enters the device shown in Figure 7.7 from the surroundings is given by Equation 3.4, i.e.

$$Q_{in} = L_p \, S[(\pi_2 - \pi_1) - (P_2 - P_1)] \tag{7.34}$$

where the numerical subscripts refer to the regions shown in Figure 7.7 and S refers to the surface area of the permselective membrane. Since the piston moves very slowly, we can also assume that the pressures in the osmotic agent chamber and the drug chamber are the same, i.e. $P_2 = P_3$.

If the flow modulator is closed or plugged, then water continues to enter the device by osmosis until the hydrodynamic pressure difference, i.e. $(P_2 - P_1)$ balances the osmotic pressure difference $(\pi_2 - \pi_1)$ stopping the influx of water by osmosis, or

$$P_2 = P_1 + (\pi_2 - \pi_1) \text{ equilibrium pressure} \tag{7.35}$$

With the flow modulator open, there must be at steady state some value of the internal pressure, i.e. P_2 or P_3, which gives a flow rate out of the device that is equal to the rate at which water enters the device by osmosis. The value of P_2 can be found by applying Bernoulli's equation, i.e. Equation 4.67, across the flow modulator as shown by the equation below.

$$\frac{P_2}{\rho} = \frac{P_1}{\rho} + \frac{\alpha V_1^2}{2} + h_{\text{friction}} \tag{7.36}$$

In the above equation, h_{friction} represents the frictional losses due to the contraction and expansion of the fluid as it enters and leaves the flow modulator as well as the friction loss as it flows through the flow modulator of length L and diameter d. Recall that $h_{friction}$ is given by Equation 4.69. V_1 is the velocity of the drug containing fluid as it exits the flow modulator. The volumetric flowrate of the drug containing fluid is given by the next equation.

$$Q_{out} = \frac{\pi d^2}{4} V_1 \tag{7.37}$$

At steady state, one needs to find the value of P_2 in Equation 7.36 that makes the Q_{out} calculated by Equation 7.37 equal the osmotic flow into the device as calculated by Equation 7.34. The drug delivery rate, or the drug infusion rate I_0 in Equation 7.33, is then given by the product of Q_{out} and the drug concentration within the osmotic pump, i.e. C_{drug}.

$$I_0 = Q_{out} C_{drug} \tag{7.38}$$

Because the flow rate of the drug from the device is usually so small and occurs over such a long period of time, the pressures within the chambers of the osmotic pump, i.e. P_2 and P_3, are very close to the pressure of the surroundings, i.e. P_1. In addition, the osmotic pressure of the osmotic agent chamber, i.e. π_2, is much larger than the osmotic pressure of the body fluids surrounding the device, i.e. π_1. Hence, for these practical reasons, the value of Q_{in} and Q_{out} are given by the following equation which results from a simplification of Equation 7.34.

$$Q_{in} = Q_{out} = L_P S \pi_2 \tag{7.39}$$

Combining Equations 7.38 and 7.39 results in the following equation for the infusion rate of a drug by an osmotic pump.

$$I_0 = C_{drug} L_P S \pi_2 \tag{7.40}$$

The above equation shows that the infusion rate of the osmotic pump is directly proportional to the drug concentration within the device and the osmotic pressure of the osmotic agent that is used. The proportionality constant is the product of the surface area of the permselective membrane and its hydraulic conductance.

7.8.2 Application to the Transdermal Delivery of Drugs

In transdermal delivery of a drug (i.e. a skin patch) (Panchagnula 1997), the drug molecules diffuse across the skin and enter the systemic circulation. However, the passive transport of the drug across the skin can be a major problem. Skin acts as a major defense barrier for the body and, therefore, most drugs have very low skin permeability, which can limit the ability to achieve therapeutic drug levels in the body. Therefore, transdermal delivery of a drug is best suited for drugs that have high pharmacological activity and good skin permeability.

The bulk of the resistance for the transport of a drug across skin as shown in Figure 7.8 results from the *stratus corneum* (*SC*) (Mitragotri 2003). The stratus corneum is the outermost layer of skin consisting of about 15 layers of keratin-filled cells (keratinocytes). The spaces between these cells are filled with lipid bilayers. The rate of

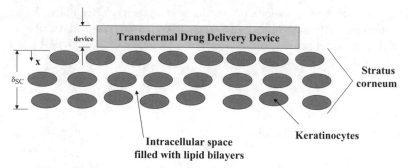

Figure 7.8 Transdermal delivery of drugs.

transdermal transport of a drug across the stratus corneum, i.e. the drug infusion rate, can be described by an equation similar to Equation 5.82, i.e.

$$I_0 = P_{SC} S(C_{drug} - 0) = P_{SC} S C_{drug} \qquad (7.41)$$

where P_{SC} is the permeability of the drug in the stratus corneum, S is the total surface area of the transdermal device, and C_{drug} is the average concentration of the drug within the transdermal device. The permeability of a drug across the stratus corneum needs to be measured experimentally using samples of skin, or it can be estimated as discussed in the next section. It is also assumed that the concentration of the drug at $x = \delta_{SC}$ is zero since the drug is immediately taken up by the blood supply. The stratus corneum permeability, i.e. P_{SC}, is given by an equation similar to Equation 5.83.

$$P_{SC} = \frac{\varepsilon_{SC} D_{SC} K}{\delta_{SC}} \qquad (7.42)$$

where ε_{SC} represents the volume fraction of the lipid bilayers through which the drug diffuses, D_{SC} is the diffusivity of the drug in the lipid bilayer material, and $K = \left. \dfrac{C_{SC}}{C_{drug}} \right|_{x=0}$ represents the equilibrium solubility of the drug in the stratus corneum.

In Equation 7.41 the value of C_{drug} will change with time as the drug is depleted from the device. In the most rigorous case, one could use Equation 5.70 to calculate the average concentration of the drug within the device at a given time. One can also assume that the drug concentration within the device is many times larger than the concentration of drug within the stratus corneum thus satisfying BC 2 in Equation 5.60. However, within the transdermal device we can also make the reasonable assumption that the concentration profile of the drug at any time is flat since the device is very thin, and diffusion of the drug out of the device is a very slow process because the bulk of the mass transfer resistance for the drug lies within the stratus corneum. Therefore, an unsteady mass balance on the amount of drug within the device at any time may be written as:

$$V_{device} \frac{dC_{drug}}{dt} = -SP_{SC} C_{drug} \qquad (7.43)$$

With the initial condition that $C_{drug} = C_{drug0}$, the above equation can be integrated to provide the amount of drug within the transdermal delivery device at any time.

$$C_{drug}(t) = C_{drug0} e^{-\frac{P_{SC}}{\delta_{device}} t} \qquad (7.44)$$

where $\delta_{device} = V_{device}/S$ is the thickness of the transdermal patch. Equations 7.41 and 7.44 can then be combined with Equation 7.31 to obtain the following differential equation that describes the continuous infusion of a drug by a transdermal delivery system.

$$V_{apparent} \frac{dC}{dt} + V_{apparent} k_{te} C = P_{SC} SC_{drug0} e^{-\frac{P_{SC}}{\delta_{device}} t} \qquad (7.45)$$

with the initial condition that at $t = 0$, $C = 0$. The above equation may be easily solved using Laplace transforms (Table 4.5) to give the following equation for the plasma drug concentration as a function of time.

$$C(t) = \frac{P_{SC} SC_{drug0}}{V_{apparent}} \left(\frac{e^{-\frac{P_{SC}}{\delta_{device}} t} - e^{-k_{te} t}}{k_{te} - \frac{P_{SC}}{\delta_{device}}} \right) \qquad (7.46)$$

Note that if the device is very large, i.e. $\delta_{device} \gg 0$, or the stratus corneum permeability is very small, then the above equation simplifies to the previous result given by Equation 7.32 with $I_0 = P_{SC} S C_{drug0}$.

7.8.2.1 Predicting the permeability of skin Often for preliminary design calculations using Equation 7.46, it may be necessary in the absence of experimental values to have an estimate of the permeability of a drug in the stratus corneum. Potts and Guy (1992) developed a relatively simple model for the stratus corneum permeability based on the size of the drug molecule (i.e. its molecular weight, MW) and its octanol/water partition coefficient ($K_{O/W}$). The octanol/water partition coefficient is a commonly used measure of the lipid solubility of a drug. They assumed that the diffusivity of a drug in the stratus corneum depends on the molecular weight of the drug as given by the following equation.

$$\varepsilon_{SC} D_{SC} = (\varepsilon D)^0_{SC} e^{-\beta MW} \qquad (7.47)$$

Substituting the above equation into Equation 7.42 and replacing the equilibrium solubility of the drug in the lipid bilayers (K) with the octanol/water partition coefficient ($K_{O/W}$) we obtain:

$$P_{SC} = K_{O/W} \frac{(\varepsilon D)^0_{SC}}{\delta_{SC}} e^{-\beta MW} \qquad (7.48)$$

After taking the $\log_{10}$ of each side, the above equation becomes:

$$\log P_{SC} = \log \left[\frac{\left(\varepsilon D \right)^0_{SC}}{\delta_{SC}} \right] + \alpha \log K_{O/W} - \beta MW \qquad (7.49)$$

where α is an adjustable constant added to improve the regression analysis and is expected to be on the order of unity. Potts and Guy (1992) then performed a regression analysis using the above equation on a data set of more than 90 drugs for which the stratus corneum permeability is known. The drugs considered in their data set ranged in molecular weight from 18 to 750 Daltons and had octanol/water partition coefficients, i.e. $\log K_{O/W}$, from −3 to +6. The regression analysis found the values of $\left[\dfrac{\left(\varepsilon D \right)^0_{SC}}{\delta_{SC}} \right]$, α

and β that best represented the data set. Their regression analysis resulted in the following equation.

$$\log P_{SC} \ (cm \ sec^{-1}) = -6.3 + 0.71 \log K_{O/W} - 0.0061 \ MW \qquad (7.50)$$

The above equation can be expected to yield predicted values of P_{SC} that are within several fold of the actual values of the stratus corneum permeability for a given chemical.

Example 7.2 Estimate the stratus corneum permeability for caffeine. The molecular weight of caffeine is 194 Daltons, and its octanol/water partition coefficient, $K_{O/W}$, is equal to 1 (Joshi and Raje 2002).

SOLUTION We use the Potts and Guy equation to estimate the stratus corneum permeability for caffeine as shown below.

$$\log P_{SC} = -6.3 + 0.71 \log 1 - 0.0061 \times 194 = -7.483$$

$$P_{SC} = 3.29 \times 10^{-8} \ cm \ sec^{-1} = 1.18 \times 10^{-4} \ cm \ hr^{-1}$$

The value reported by Joshi and Raje for the stratus corneum permeabilty of caffeine is 1×10^{-4} cm hr^{-1} which compares quite well to the value predicted by the Potts and Guy equation.

7.9 FIRST ORDER DRUG ABSORPTION AND ELIMINATION

Many drugs are given orally or by injection. Other routes may include the nasal cavity or by inhalation. In all of these non-intravenous cases, the drug is first absorbed and makes its way into the plasma by diffusion where it becomes distributed throughout the body. The plasma concentration of the drug gradually increases and then finally decreases as the elimination processes begin to overcome the rate of drug absorption. Curve I of Figure 7.1 illustrates the general trend of the plasma drug concentration with time during the absorption and elimination phases. Although the drug absorption rate is strongly affected by the dosage form, i.e. solution, capsule, tablet, suspension, etc., it has been found that, for the most part, the drug absorption process can be adequately described by simple first order kinetics.

Figure 7.9 illustrates the simplest compartmental model for the case of first order drug absorption and elimination. Note a separate mass balance equation is needed for the drug whose amount decreases with time due to its absorption into the body. The factor (f) accounts for the fraction of the dose (D) that is actually absorbable. An amount of the dose equal to $(1-f)$ D is assumed to be removed from the body without ever entering the apparent distribution volume.

A mass balance on the amount of drug within the apparent distribution volume can be written as follows:

$$V_{apparent} \frac{dC}{dt} = k_a A - CL_{plasma} C = k_a A - V_{apparent} k_{te} C \tag{7.51}$$

where A represents the mass of drug remaining to be absorbed, and k_a represents the drug's first order absorption rate constant.

A mass balance on the drug yet to be absorbed is given by the following equation.

$$\frac{dA}{dt} = -k_a A \tag{7.52}$$

When this equation is integrated, recognizing that initially $A_0 = fD$, the amount of drug at anytime waiting to be absorbed is given by:

$$A(t) = f D e^{-k_a t} \tag{7.53}$$

This equation may now be substituted into Equation 7.51 to obtain an expression for the total amount of drug in the plasma as a function of time, assuming the initial concentration of drug is zero.

$$C(t) = \frac{f D}{V_{apparent}} \left(\frac{k_a}{k_a - k_{te}} \right) \left(e^{-k_{te} t} - e^{-k_a t} \right) \tag{7.54}$$

We see that the concentration of drug in the plasma is dependent on the first order rate constants for absorption and elimination, i.e. k_a and k_{te}. The concentration is directly

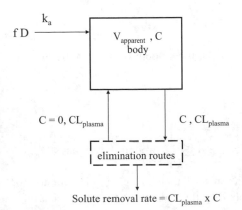

Figure 7.9 Single compartment model with first order absorption and elimination.

proportional to the fraction of the drug dose that is absorbable and inversely proportional to the apparent distribution volume.

An expression for the time at which the concentration peaks can be obtained by finding the derivative of Equation 7.54 with respect to time, setting it equal to zero, and solving for the time. This time, called τ_{max}, is given by the following expression.

$$\tau_{max} = \frac{1}{k_a - k_{te}} \ln\left(\frac{k_a}{k_{te}}\right) \tag{7.55}$$

Note that the time the maximum concentration is reached is independent of the dose. The corresponding concentration at the time of the peak may be found by substituting τ_{max} into Equation 7.54 for the time and simplifying to obtain:

$$C_{max} = \left(\frac{f\,D}{V_{apparent}}\right)\left(\frac{k_a}{k_{te}}\right)^{\frac{k_{te}}{k_{te}-k_a}} \tag{7.56}$$

We can also find the value of $\mathrm{AUC}^{0\to\infty}$ by substituting Equation 7.54 into Equation 7.26. Upon integration, the following results.

$$\mathrm{AUC}^{0\to\infty} = \frac{f\,D}{V_{apparent}}\frac{1}{k_{te}} = \frac{f\,D}{CL_{plasma}} \tag{7.57}$$

It is interesting to note that the value of $\mathrm{AUC}^{0\to\infty}$ does not depend on the absorption rate constant. $\mathrm{AUC}^{0\to\infty}$ is, though, directly proportional to the absorbed dose, $f\,D$, and inversely proportional to the plasma clearance.

To complete our mathematical description of the process shown in Figure 7.9, the rate at which the drug enters the urine is still given by Equation 7.28. Substituting Equation 7.54 for the total plasma drug concentration into this equation gives the following result for the amount of drug in the urine at any given time.

$$M_{urine}(t) = \left(\frac{f\,D\,k_{renal}}{k_{te}}\right)\left[1 - \frac{1}{k_a - k_{te}}\left(k_a\,e^{-k_{te}\,t} - k_{te}\,e^{-k_a\,t}\right)\right] \tag{7.58}$$

After all of the drug has cleared the body ($t \to \infty$), the amount of drug in the urine is given by:

$$M^{\infty}_{urine} = f\,D\,\frac{k_{renal}}{k_{te}} \tag{7.59}$$

Note that if the only elimination pathway for the drug is through the kidneys ($k_{renal} = k_{te}$), then $M_{urine}^{\infty} = f\,D$.

If Equation 7.58 is subtracted from Equation 7.59, then an equation is obtained for the quantity of drug remaining to be excreted by the kidneys. This equation is of a convenient form for determining the absorption and elimination rate constants from urinary excretion data.

$$M^\infty_{urine} - M_{urine}(t) = \frac{M^\infty_{urine}}{(k_a - k_{te})}\left(k_a\, e^{-k_{te}t} - k_{te}\, e^{-k_a t}\right) \tag{7.60}$$

These parameters (k_a and k_{te}) may be determined by performing a non-linear regression on urinary excretion data in a manner similar to that shown in Example 7.3 below. Since M_{urine}^∞ is known from the urinary excretion data, one can then determine the renal elimination rate constant (k_{renal}) from Equation 7.59 given the dose actually absorbed (f D) and the value of k_{te}.

Example 7.3 illustrates the use of the single compartmental model with first order absorption and elimination to describe plasma nicotine levels produced by chewing nicotine gum (McNabb *et al.* 1982). This example requires that a non-linear regression be performed to find the three unknown parameters in Equation 7.54, i.e. $f D/V_{apparent}$, k_a, and k_{te}. Once again we choose as an objective function the sum-of-the-square-of-the-error defined as $SSE = \sum_{i=1}^{N}\left(C_{experimental_i} - C_{theory_i}\right)^2$ where $C_{experimental}$ are the measured plasma drug concentrations and C_{theory} are the values predicted by the pharmacokinetic model, for example Equation 7.54. N represents the total number of data points that are considered. The goal is then to find the set of parameters in the pharmacokinetic model, i.e. $f D/V_{apparent}$, k_a, and k_{te}, that minimizes SSE. These values of the parameters that minimize SSE represent the best fit of the pharmacokinetic model to the data. There are a variety of mathematical software programs that have routines for solving multiple parameter non-linear regression problems. It is important to note that many times when performing a non-linear regression analysis a unique solution may not be found. In other words there can be different values of the regression parameters that can fit the data to the same level of accuracy.

To obtain the values of f and $V_{apparent}$ from the value of $f D/V_{apparent}$ requires additional data such as that obtained from urinary excretion or intravenous injection studies. Other techniques based on graphical methods may also be used to determine pharmacokinetic parameters as discussed in the following references (Cooney 1976; Gibaldi and Perrier 1982; Welling 1986; Notari 1987) and in the problems at the end of this chapter.

Example 7.3 The table below summarizes plasma nicotine levels in a 27-year-old man chewing one piece of gum containing 4 mg of nicotine (McNabb *et al.* 1982). Nicotine gum was developed to help people quit smoking by providing the same amount of plasma nicotine as obtained from cigarettes. Nicotine is absorbed from the gum over a 30 minute period of chewing and is readily absorbed within the buccal cavity. From the data provided in this table, determine the pharmacokinetic parameters that describe the absorption and elimination of nicotine.

Time, minutes	Plasma nicotine concentration, ng/ml
0	0
2	2.5
3	4
6	4
8	5
10	6
12	7
13	8.5
16	10
18	11
20	12
24	13
28	12.5
32	12
35	11
41	10
48	9
52	9
65	7.5

SOLUTION Figure 7.10 shows the results of the non-linear regression. The open symbols are the data points from the above table and the solid line is the resulting best fit of Equation 7.54 to this data. The value of $\dfrac{f\,D}{V_{apparent}}$ = 19.7 ng/ml, k_a = 0.056 min^{-1}, and k_{te} = 0.019 min^{-1}. From Equation 7.55, the predicted maximum occurs after 29.4 minutes, and the peak plasma concentration from Equation 7.56, is 11.3 ng/ml. The fit during the initial absorption phase and the elimination phase is excellent. However Equation 7.54 fails to fit several of the data points in the region around the peak quite as well. This could be because a single compartment model is not sufficient to represent the data or it could represent an error in concentration measurements, or both.

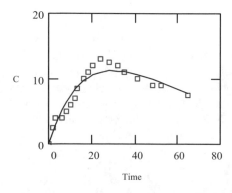

Figure 7.10   A comparison of actual and predicted plasma nicotine levels, Example 7.3.

7.10 TWO COMPARTMENT MODELS

In many cases the single compartment model is not sufficient to represent the time course of the plasma drug concentration. The single compartment model assumes that the drug rapidly distributes throughout the homogeneous apparent distribution volume. In fact, the drug may distribute at different rates in different tissues of the body, and it may require an extended period of time before the drug concentration equilibrates between the different tissue and fluid regions of the body. Clearly, this a complex process, and one could envision building a model in which a central compartment representing the plasma is in direct communication with a multitude of other compartments representing other tissue and fluid regions of the body.

A reasonable simplification of this type of model is the two compartment model shown in Figure 7.11. Drug is introduced into the central compartment that generally represents the plasma and other fluids in which the drug rapidly equilibrates. The drug then slowly distributes into the remaining tissue and fluid spaces represented by a single tissue compartment. The constants k_{pt} and k_{tp} represent the rate constants describing the first order transport between the plasma and tissue compartments. Since most or all of the drug is eliminated by the highly perfused kidneys and liver, the elimination processes (metabolic and renal) are assumed to be associated with the plasma compartment.

The two compartment model has a characteristic biphasic temporal plasma concentration profile that distinguishes it from the single compartment model. During the early period after drug absorption, the drug is eliminated at a more rapid pace from the plasma volume due to the combined effects of the usual elimination processes as well as drug absorption by the tissues and fluids of the secondary compartment. After a period of time, the two compartments equilibrate and the drug elimination rate falls off.

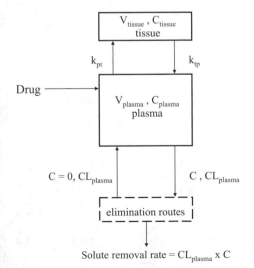

Solute removal rate = CL_{plasma} x C   **Figure 7.11**   A two compartment model.

7.10.1 A Two Compartment Model for an Intravenous Injection

First, let us develop a pharmacokinetic model for the case of a bolus intravenous injection. A separate unsteady mass balance equation for the drug is then written for the plasma and the tissue compartments.

plasma compartment $\quad V_{plasma} \dfrac{dC_{plasma}}{dt} = k_{tp} V_{tissue} C_{tissue} - \left(k_{te} + k_{pt}\right) V_{plasma} C_{plasma}$ (7.61)

tissue compartment $\quad V_{tissue} \dfrac{dC_{tissue}}{dt} = k_{pt} V_{plasma} C_{plasma} - k_{tp} V_{tissue} C_{tissue}$ (7.62)

V_{plasma} and V_{tissue} represent the apparent distribution volumes of the plasma and tissue compartments. C_{plasma} and C_{tissue} represent the concentrations of the drug within these respective compartments. These equations may be solved analytically using Laplace transforms with the following initial condition: $t = 0$, $C_{plasma} = D/V_{plasma}$, and $C_{tissue} = 0$. The equation for the concentration of drug within the plasma compartment is given by:

$$C_{plasma}(t) = \frac{D}{V_{plasma}(A-B)}\left[(k_{tp} - B)e^{-Bt} - (k_{tp} - A)e^{-At}\right]$$ (7.63)

and that for the concentration of drug in the tissue compartment is given by:

$$C_{tissue}(t) = \frac{D k_{pt}}{V_{tissue}(A-B)}\left[e^{-Bt} - e^{-At}\right]$$ (7.64)

The constants A and B are given by the following equations in terms of the first order rate constants.

$$A = \frac{1}{2}\left\{\left(k_{pt} + k_{tp} + k_{te}\right) + \left[\left(k_{pt} + k_{tp} + k_{te}\right)^2 - 4k_{tp}k_{te}\right]^{1/2}\right\}$$
$$B = \frac{1}{2}\left\{\left(k_{pt} + k_{tp} + k_{te}\right) - \left[\left(k_{pt} + k_{tp} + k_{te}\right)^2 - 4k_{tp}k_{te}\right]^{1/2}\right\}$$ (7.65)

A non-linear regression of plasma concentration data using Equation 7.63 can be performed to obtain the constants in this equation, i.e. V_{plasma}, A, B, and k_{tp}. The other first order rate constants, k_{pt} and k_{te}, can then be found through the following process. We first multiply A and B in Equation 7.65 to obtain the following result.

$$k_{te} = \frac{A B}{k_{tp}}$$ (7.66)

Then A and B in Equation 7.65 are added together to give:

$$k_{pt} = (A + B) - (k_{tp} + k_{te})$$ (7.67)

Equations 7.66 and 7.67 may then be solved to obtain the values of k_{te} and k_{pt} from the known values of A, B, and k_{tp}.

The distribution volume of the tissue compartment can be found by assuming that once the two compartments are at equilibrium, then the derivative in Equation 7.62 is equal to zero and the concentrations in the two compartments are the same. Equation 7.62 then simplifies to give the following result for the volume of the tissue space.

$$V_{tissue} = V_{plasma} \left[\frac{k_{pt}}{k_{tp}} \right] \tag{7.68}$$

The overall apparent distribution volume would be equal to $V_{plasma} + V_{tissue}$ and is given by:

$$V_{apparent} = V_{plasma} \left[1 + \frac{k_{pt}}{k_{tp}} \right] \tag{7.69}$$

The data in Example 7.4 illustrates how the pharmacokinetic parameters of the two compartment model for an intravenous dose can be determined. Other techniques based on graphical methods may also be used to determine pharmacokinetic parameters (Welling 1986; Notari 1987). Often, the graphical approaches can provide reasonable starting values for the parameters in a non-linear regression.

Example 7.4 The data shown in the table below was given in Koushanpour (1976) and represents plasma inulin concentrations following a rapid intravenous injection of 4.5 grams of inulin in an 80 kg human. Determine the pharmacokinetic parameters of the two compartment model.

Time, minutes	Plasma inulin concentration, μg/ml
10	440
20	320
40	200
60	150
90	110
120	80
150	60
175	48
210	35
240	25

SOLUTION The logarithmic graph of these data shown in Figure 7.12 shows that the elimination phase in this case is not linear but biphasic, indicating that a two compartment model is needed to provide an adequate representation of the data. Figure 7.13 shows the results of the non-linear regression. The open symbols are the data points from the above table, and the solid line is the resulting best fit of Equation 7.63 to this data. We see that the two compartment model fits this data extremely well. The table below summarizes the results of the regression analysis, and these parameter values provide the best fit of Equation 7.63 to the data in the above table.

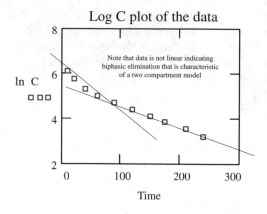

Figure 7.12 Biphasic elimination of inulin, Example 7.4.

Parameter	Value
V_{plasma}	7.8 L
A	0.036 min^{-1}
B	0.0049 min^{-1}
k_{tp}	0.012 min^{-1}
k_{pt}	0.014 min^{-1}
k_{te}	0.016 min^{-1}
V_{tissue}	9.5 L
V_{total}	17.3 L

Since, for inulin, the only elimination pathway is through the kidneys, our earlier discussion showed that the renal clearance is the same as the GFR. Therefore, for this example, the renal clearance (Cl_{renal}), or GFR, is equal to $V_{\text{plasma}} \times k_{te} = 125$ ml min^{-1}. Inulin is known to only distribute throughout the extracellular fluid volume. Hence, inulin can be used to measure the total extracellular fluid. The value of 17 L compares well with the reported value of 15 L for the extracellular fluid volume (Table 3.1).

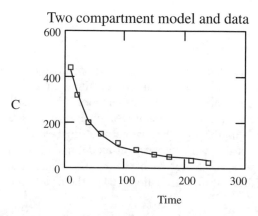

Figure 7.13 A comparison of actual and predicted plasma inulin levels, Example 7.4.

7.10.2 A Two Compartment Model for First Order Absorption

Continuing our discussion of the two compartment model, consider the case now of first order absorption of the drug. Unsteady mass balance equations may once again be written for each of the compartments. The equation for the tissue compartment would still be given by Equation 7.62 and the equation that describes the amount of drug yet to be absorbed (A) would still be given by Equations 7.52 and 7.53. The plasma compartment equation is then given by:

$$V_{plasma} \frac{dC_{plasma}}{dt} = k_a A + k_{tp} V_{tissue} C_{tissue} - (k_{te} + k_{pt}) V_{plasma} C_{plasma} \tag{7.70}$$

The initial conditions are that at time equal to zero there is no drug present in either compartment. These equations may then be solved using Laplace transforms to obtain the following expressions for the concentration of drug within the plasma and tissue compartments. For the plasma concentration we have:

$$C_{plasma}(t) = \frac{f\,Dk_a}{V_{plasma}} \left[\frac{(k_{tp} - A)}{(k_a - A)(B - A)} e^{-A\,t} + \frac{(k_{tp} - B)}{(k_a - B)(A - B)} e^{-Bt} + \frac{(k_{tp} - k_a)}{(A - k_a)(B - k_a)} e^{-k_a t} \right] \tag{7.71}$$

and for the tissue concentration:

$$C_{tissue}(t) = \frac{f\,D\,k_a k_{pt}}{V_{tissue}} \left[\frac{e^{-At}}{(k_a - A)(B - A)} + \frac{e^{-Bt}}{(k_a - B)(A - B)} + \frac{e^{-k_a t}}{(A - k_a)(B - k_a)} \right] \tag{7.72}$$

The constants A and B have the same dependence on k_{pt}, k_{tp}, and k_{te} as the two compartment IV case and are also given by Equations 7.65. Depending on the type of data one has, Equation 7.71 can be fitted to plasma concentration data to obtain the unknown parameters; fD, V_{plasma}, k_a, k_{tp}, A, and B. In some cases, it may not be possible to resolve the group $f\,D/V_{plasma}$ without additional data, perhaps from an IV bolus injection and/or urinary excretion data. Once these parameters have been determined, Equations 7.66 through 7.69 may be used to find the remaining parameters. Other techniques based on graphical methods may also be used to determine pharmacokinetic parameters as discussed in the references (Cooney 1976; Gibaldi and Perrier 1982; Welling 1986; Notari 1987).

Example 7.5 reexamines the nicotine chewing gum data of Example 7.3 using the two compartment model with first order absorption. Example 7.6 uses the two compartment model to examine plasma nicotine levels in a 49-year-old man after smoking one cigarette yielding 0.8 mg of nicotine.

Example 7.5 Reexamine the nicotine chewing gum data used in Example 7.3. Determine the pharmacokinetic parameters of the two compartment model with first order absorption.

SOLUTION Figure 7.14 illustrates the solution. The table below summarizes the results of the regression analysis.

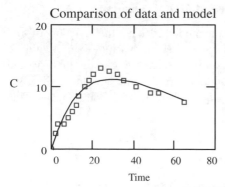

Figure 7.14 A comparison of actual and predicted plasma nicotine levels, Example 7.5.

Parameter	Value
$f D/V_{plasma}$	19.7 ng ml^{-1}
A	0.022 min^{-1}
B	0.0097 min^{-1}
k_{tp}	0.012 min^{-1}
k_{pt}	0.0018 min^{-1}
k_{te}	0.018 min^{-1}

We see that, in this case, the two compartment model offers little improvement to the fit of the data in comparison to the single compartment model shown earlier. This illustrates that the single compartment is sufficient to represent the plasma nicotine concentrations.

Example 7.6 The table below (McNabb *et al.* 1982) presents plasma nicotine levels in a 49-year-old man during and after smoking one cigarette yielding 0.8 mg of nicotine.

Time, minutes	Plasma nicotine concentration, ng/ml
0	0
2	5
3	7
4	11
5	10
6	14
8	12
9	13
11	13
12	12
15	12
21	9
25	9
35	8
41	7.5
49	7
59	7
71	6

Compared with the chewing gum example, there is a much faster absorption of the nicotine by smoking due to the intimate contact with the blood in the lungs. Using the two compartment model, determine the pharmacokinetic parameters and compare them with the chewing gum case of Example 7.5

SOLUTION Figure 7.15 illustrates the solution. We see that the two compartment model fits this data rather well. The table below summarizes the results of the regression analysis.

Parameter	Value
$f\,D/V_{\text{plasma}}$	15.4 ng ml^{-1}
A	0.035 min^{-1}
B	0.0013 min^{-1}
k_{tp}	0.013 min^{-1}
k_{pt}	0.02 min^{-1}
k_{te}	0.0037 min^{-1}

The pharmacokinetic parameters for the case of chewing gum and one cigarette are compared below. One important caveat to consider in comparing the parameters for nicotine delivery by chewing gum or cigarettes in these examples is that the data was obtained from different subjects.

	Chewing gum	Cigarette
Total nicotine available	4 mg	0.8 mg
$f\,D/V_{\text{apparent}}$	19.73 ng ml^{-1}	15.35 ng ml^{-1}
k_a	0.056 1 min^{-1}	0.285 1 min^{-1}
k_{tp}	0.012 1 min^{-1}	0.013 1 min^{-1}
k_{pt}	0.0012 1 min^{-1}	0.02 1 min^{-1}
k_{te}	0.018 1 min^{-1}	0.0037 1 min^{-1}

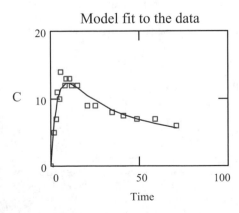

Figure 7.15 A comparison of actual and predicted plasma nicotine levels, Example 7.6.

First, we note that the chewing gum offers nearly five times the potential dose of nicotine than from a single cigarette. Clearly, since the $f D/V_{apparent}$ (like the initial concentration in the plasma) are somewhat comparable, we conclude that only about one-fifth of the nicotine in the chewing gum is actually absorbed. However, the chewing gum at this nicotine loading provides a comparable peak plasma nicotine level to a cigarette. We next notice that the absorption rate constant for the cigarette is about five times that of chewing gum showing that the lungs are very efficient at absorbing nicotine into the blood. Also, note that delivery of nicotine via a cigarette has a marked effect on the rate at which the absorbed nicotine is transported to the tissue compartment. We see a pronounced increase with smoking in the transport rate constant (k_{pt}) of nicotine from the plasma compartment to the tissue compartment. On the other hand, the transport rate constant (k_{tp}) from the tissue compartment to the plasma compartment is about the same for the two cases. Finally, we see that the value of the total elimination rate constant (k_{te}) for the cigarette case is much smaller than that for the chewing gum case. It is difficult to discern without additional data whether this difference is due to the nicotine dosage form or due to individual differences.

PROBLEMS

1. Derive Equation 7.32.

2. Derive Equation 7.54.

3. Derive Equations 7.55 and 7.56.

4. Derive Equation 7.58.

5. The rate constants in Equation 7.54 (k_a and k_{te}) can also be found using a graphical technique called "feathering" (Notari 1987). Note that the plasma concentration in Equation 7.54 is related to the sum of two exponential terms. In most cases, k_a is larger than k_{te}, hence, the second exponential in Equation 7.54 vanishes for long periods of time. Therefore, for long time periods, Equation 7.54 may be approximated as:

$$C(t)_{TL} \approx \frac{f D}{V_{apparent}} \left(\frac{k_a}{k_a - k_{te}} \right) e^{-k_{te} t}$$

On a semi-logarithmic plot ($\log_{10}(C)$ versus t), this equation predicts for long periods of time, that the concentration data should form a straight line. This region is referred to as the terminal elimination phase. The slope of the terminal line (TL) that fits the data during this period of time would then be equal to $2.303 \times k_{te}$, and the intercept would be equal to $\log_{10} \left[\frac{f D}{V_{apparent}} \left(\frac{k_a}{k_a - k_{te}} \right) \right]$.

We can now use the equation for $C(t)_{TL}$ to form the difference or residual, $C(t)_{TL} - C(t)$, with the following result.

$$C(t)_{TL} - C(t) = \frac{f D}{V_{apparent}} \left(\frac{k_a}{k_a - k_{te}} \right) e^{-k_a t}$$

This equation is valid for all values of time starting from time equal to zero. Note for short periods of time, $C(t)_{TL}$ is obtained either by extrapolation of the terminal line or by estimating $C(t)_{TL}$ by the first equation above. For long periods of time this difference is equal to zero. Now, if a plot is made of the $\log_{10}[C(t)_{TL} - C(t)]$ versus time, the slope would equal $2.303 \times k_a$, and the intercept

would once again be equal to $\log_{10}\left[\dfrac{fD}{V_{apparent}}\left(\dfrac{k_a}{k_a-k_{te}}\right)\right]$. With k_a and k_{te} now known, $\dfrac{fD}{V_{apparent}}$
may be obtained from the value of the intercept.

Use this feathering technique to obtain the parameters for the data given in Example 7.3. How do these parameters obtained using the feathering technique compare with those obtained from a non-linear regression? Under what conditions would you expect the feathering technique to be difficult to use?

6. Rework Example 7.4 assuming a single compartment model.

7. You are taking a final exam, it is unlike any exam you have taken before, plus you had a short night of sleep, you had car problems yesterday, you had too much to drink last night, and the phone kept ringing early this morning. The result – you have a terrible headache, unlike any headache you have ever had before. You reach for your bottle of a new headache medicine that claims maximum relief within 30 minutes and contains a well-known pain reliever. An amazing coincidence is that the exam you are taking includes a problem related to the pharmacokinetics of this pain reliever. You are given the following information:

elimination rate constant	=	$0.277\ hr^{-1}$
apparent distribution volume	=	35 liters
fraction absorbed	=	80%
therapeutic range	=	$10\text{–}20\ \mu g\ ml^{-1}$

You plan to take this pain reliever. What dose in milligrams should you take? At what time should you think about taking another dose? How much should you take for the second dose?

8. The pharmacokinetics of nicotinamide has recently been reported in humans (Petley *et al.* 1995. *Diabetes.* 44: 152–155). Nicotinamide, a derivative of the B vitamin niacin, is under investigation as a prevention for insulin dependent diabetes mellitus. The drug was given orally to eight male subjects with a mean body weight of 75 kg. Two formulations were used, a standard powdered form and a sustained release form. The data below summarize the data obtained for the low dose studies (standard = 2.5 mg kg^{-1} BW^{-1} and sustained = 6.7 mg kg^{-1} BW^{-1}). Analyze this data using appropriate pharmacokinetic models. Explain any differences in the pharmacokinetics of the two formulations used.

Time, hours	Nicotinamide plasma concentration, $\mu g\ ml^{-1}$	
	Standard dose	Sustained release dose
0	0	0
0.3	3.3	
0.5	2.45	1.75
0.75	1.90	
1.0	1.45	1.5
1.25	1.00	
1.5	0.80	1.1
1.75	0.55	
2.0	0.50	1.05
2.5	0.30	1.1
3.0	0.20	0.67
3.5		0.50
4.0		0.45
4.5		0.50
5.0		0.55
5.5		0.15

9. The following serum insulin levels were obtained following the intranasal administration of 150 U of insulin to 12 healthy men (Jacobs *et al.* 1993, *Diabetes.* 42: 1649–1655). Analyze this data using an appropriate pharmacokinetic model.

Time, minutes	Serum insulin concentration, $\mu U\ ml^{-1}$
0	10
15	150
20	220
25	230
30	260
35	250
40	210
45	170
60	140
75	105
90	80
105	50
120	45
150	35
180	35
240	35
300	35

10. The following plasma concentrations for the antibiotic imipenem were obtained from six patients suffering multiorgan failure (Hashimoto *et al.* 1997, *ASAIO Journal* 43:84–88). All patients were anuric because of complete renal failure and were placed on continuous venovenous hemodialysis. The antibiotic was delivered by an intravenous infusion pump for 30 minutes resulting in an initial plasma concentration of 32.37 $\mu g\ ml^{-1}$. Following the infusion, blood samples were taken over the next 12 hours. The initial dose was 500 mg and 107.7 mg of imipenem was recovered in the dialysate fluid ("urine"). The table below summarizes the plasma concentrations.

Time, minutes	Plasma concentration, $\mu g\ ml^{-1}$
0	32.47
30	24
60	18.5
120	13
180	10
360	4.6
540	2
720	1.12

Use an appropriate compartment model to analyze the above data.

11. Starting with Equation 7.45, derive Equation 7.46 using Laplace transforms.

12. Estimate the stratus corneum permeability for testosterone and estradiol. The molecular weight of testosterone and estradiol are 288 and 272 respectively. The respective values of their octanol/water partition coefficients are 2070 and 7000. The reported values of the stratus corneum permeability for testosterone is 2.2×10^{-3} cm hr^{-1} and for estradiol the value is 3.2×10^{-3} cm hr^{-1} (Joshi and Raje 2002).

13. Consider the delivery of a drug from a patch applied to the skin. Assume the patch has a surface area for drug transport of 30 cm^2 and a thickness of 0.6 mm, that the concentration of the drug in the patch at any time after it has been applied is given by $C_{drug}(t)$, and that the stratus corneum permeability of the drug is given by P_{SC} which, for this particular drug, has a value of 3.66×10^{-7} cm sec^{-1}. If the initial concentration of the drug in the patch is $C_\infty(0) = 25$ mg ml^{-1}, how long will it take to reduce the concentration of the drug in the patch to $\frac{1}{2} C_\infty(0)$ i.e. 12.5 mg ml^{-1}?

14. Alora is a transdermal system designed to deliver estradiol continuously and consistently over a 3–4 day period of time when placed on the skin. Alora provides systemic estrogen replacement therapy in postmenopausal women. For a 36 cm^2 patch, the release rate of the drug is 0.1 mg of estradiol per day. Clinical trials have provided the following pharmacokinetic data. In one study, following application of the patch, the peak concentration of estradiol was found to be 144 pg ml^{-1} (10^{12} pg = 1 g), and the half-life of the estradiol was found to be 1 hour. Over an 84 hour dosing interval, the plasma clearance for the estradiol from the Alora patch was found to be 61 L hr^{-1}. What is the steady state plasma concentration for estradiol in pg/ml? What is the apparent volume of distribution in L?

15. Deponit® is a nitroglycerin transdermal delivery system for the prevention of angina pectoris. The in vivo release rate of the 16 cm^2 patch is 0.2 mg hr^{-1}. The reported volume of distribution is 3 L kg^{-1} and the observed clearance rate is 1 L kg^{-1} min^{-1}. What is the steady state nicotine concentration (ng ml^{-1}) in a 70 kg human? What is the elimination rate constant in 1 min^{-1}? What is the half-life of nitroglycerine in minutes?

16. A pharmacokinetic study of a particular drug taken orally as a single dose provided an AUC$^{0\rightarrow\infty}$ of 1085.5 ng hr ml^{-1}, a C_{max} equal to 98.5 ng ml^{-1}, and a τ_{max} of 2.1 hrs. From these data, determine the values of $V_{apparent}$ (liters), k_{te} (1 min^{-1}), and k_a (1 min^{-1}). The dose was 80 mg and the entire drug is available, i.e. $f = 1$.

17. Design a skin patch to deliver nicotine. The goal is to have a patch that delivers nicotine for 24 hours, then a new patch is added and so on. Nicotine is somewhat water soluble and has a molecular weight of 162.23 and an octanol–water partition coefficient ($\log K_{o/w}$) of 1.2. The dose of nicotine to be absorbed into the body is 21 mg/24 hrs, and the concentration of nicotine within the patch is 25 mg ml^{-1}. Estimate the surface area of the patch. Assuming that the total nicotine content of the patch is 52.5 mg, how much of this will be absorbed? Also estimate the steady state plasma concentration of nicotine in ng ml^{-1} and then estimate how thick the patch would be. The volume of distribution for nicotine is 2–3 L kg^{-1} of body weight, and the average plasma clearance is 1.2 L min^{-1}.

18. Krewson *et al. Brain Research* 680:196–206 (1995) presented the steady state data in the table below for the distribution of nerve growth factor (NGF) in the vicinity of a thin cylindrical controlled drug release device implanted into the brain of a rat. The polymeric disks containing radiolabeled I^{125}-NGF were 2 mm in diameter and 0.8 mm in thickness. NGF has a molecular weight of 28,000 Daltons. Assuming that the NGF is eliminated from the brain tissue by a first order process (i.e. the rate of elimination is proportional to the concentration of nerve growth factor, i.e. $k_{apparent} C_{NGF}$), develop a steady state reaction–diffusion model to analyze this data. $k_{apparent}$ is the apparent first order elimination rate constant for NGF and accounts for such processes as metabolism, cellular internalization, or uptake by the brain's systemic circulation. The boundary conditions, assuming the origin of the Cartesian coordinate system to be the midline of the polymeric disk, are as follows:

$$C_{NGF} = C_0 \text{ at } x = a \text{ and } C_{NGF} = 0 \; x = \infty$$

where a is the half-thickness of the polymeric disk. Estimate the value of the Thiele modulus $(a\sqrt{k_{apparent} / D}$ that provides the best estimate of the data. D represents the diffusivity of NGF in the brain tissue. After estimating a value of D for NGF, what is the value of $k_{apparent}$? Using Equation 7.24, calculate the half-life of NGF in the brain tissue. How does this value of the NGF half-life compare with the reported half-life of NGF in brain tissue of about 1 hour?

Distance from polymer/tissue interface, mm	NGF concentration, $\mu g\ ml^{-1}$
0	37.0
0.1	31.5
0.2	20.0
0.3	15.0
0.4	10.5
0.5	8.5
0.6	6.5
0.8	3.5
1.0	2.2

EIGHT

EXTRACORPOREAL DEVICES

In this chapter, we will discuss and analyze several examples of extracorporeal devices. These devices lie outside (*extra*) the body (*corporeal*) and are usually connected to the patient by an *arteriovenous shunt*[1]. In some respects, they may be thought of as artificial organs. Their function is based on the use of physical and chemical processes to replace the function of a failed organ or to remove an unwanted constituent from the blood. The patient's blood, before entering these devices, is infused with anticlotting drugs such as *heparin* to prevent clotting. A variety of ancillary equipment may also be present to complete the system. This may include items such as pumps, flow monitors, bubble and blood detectors, as well as pressure, temperature, and concentration control systems. It is important to note that these devices do not generally contain any living cells. Devices containing living cells are called bioartificial organs, and these will be discussed in Chapter 10.

8.1 APPLICATIONS

A variety of extracorporeal devices have been developed. Perhaps the best known are *blood oxygenators* that are used in such procedures as open heart surgery and *hemodialysis* to replace the function of failed kidneys. Other examples include *hemoperfusion* wherein a bed of activated carbon particles are used for cleansing blood of toxic materials, *plasmapheresis*[2] is used to separate erythrocytes from plasma as a first step in the subsequent processing of the plasma, *immobilized enzyme reactors* to

---

[1] A shunt is a means of diverting flow, in this case blood from an artery through a device and then back into the body via a vein.

[2] See problem 11 in Chapter 5 for a discussion and analysis of a membrane plasmapheresis system.

rid the body of a particular substance or to replace lost liver function, and *affinity columns* are used to remove materials such as antibodies that have been implicated as the cause of many *autoimmune diseases*.

The primary functional unit of extracorporeal devices are typically provided as a sterile disposable cartridge. However, in some cases, considering mounting healthcare costs, the functional unit can be reused many times provided it is cleaned and sterilized between applications. Certainly there is no end to the possibilities for extracorporeal devices. However, whatever the function is of the device, there will be some commonality amongst them in terms of their construction, use of membranes, and fluid contacting schemes. In the following discussion, we will examine in greater detail four representative devices; hemodialysis, blood oxygenators, immobilized enzyme reactors, and affinity columns.

8.2 CONTACTING SCHEMES

Most devices, such as blood oxygenators, and dialyzers are typically based on the use of polymeric membranes to create the surface area needed to provide mass transfer between the blood stream and another exchange fluid stream. The membrane is physically retained within the device by a support structure that also creates the flow channels. As illustrated in Figure 8.1 for hemodialyzers, the membranes are typically arranged as stacks of flat sheets, coils of membrane sheets, or bundles of hollow fiber membranes. Hollow fiber membranes are finding increased use because of their low cost, ease of manufacture, and consistent quality.

A variety of contacting patterns between the blood and the exchange fluid are also possible as shown in Figure 8.2. Immobilized enzyme reactors and affinity columns have the active material, i.e. an *enzyme* or other *ligand*, firmly attached to a support particle. Support particles are typically made from a wide variety of polymeric materials such as alginates, agar, carrageenin, and polyacrylamide, or inorganic materials such as glass, silicas, aluminas, and activated carbon. The particles containing the immobilized material may be arranged as a packed bed, or the particles may be suspended in a well-mixed device as shown in Figure 8.3. Complete mixing of the fluid phase and suspended particles can be achieved either by a mechanical impeller or by fluid recirculation.

8.3 MEMBRANE SOLUTE TRANSPORT

In many extracorporeal devices, a solute must be transported through the blood, perhaps across a membrane, and then through an exchange fluid. These transport processes are illustrated in Figure 8.4 for the case of simple diffusion. Concentrations of the solute at the fluid-membrane interface are assumed to be at equilibrium. The partition coefficient, K, is used to describe the solute equilibrium ($C_{mb} = KC_b^s$ and $C_{me} = KC_e^s$) and at the surface of the membrane.

An overall local mass transfer coefficient (K_0) for the situation shown in Figure 8.4 may be written in terms of the individual film mass transfer coefficients on the blood (k_b) and exchange fluid (k_e) sides along with the membrane permeability (P_m) and the

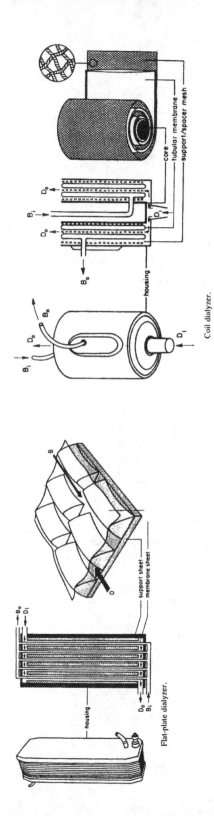

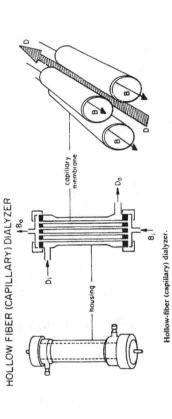

Coil dialyzer.

core
tubular membrane
support/spacer mesh
housing

Flat-plate dialyzer.

housing
support sheet
membrane sheet

HOLLOW FIBER (CAPILLARY) DIALYZER

capillary membrane

housing

Hollow-fiber (capillary) dialyzer.

Figure 8.1 Membrane configurations (from R. Skalak and S. Chien, 1987, *Handbook of Bioengineering*, McGraw-Hill Book Co., with permission).

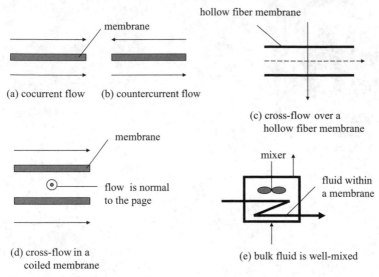

(a) cocurrent flow   (b) countercurrent flow

(c) cross-flow over a
hollow fiber membrane

(d) cross-flow in a
coiled membrane

(e) bulk fluid is well-mixed

Figure 8.2   Fluid contacting patterns.

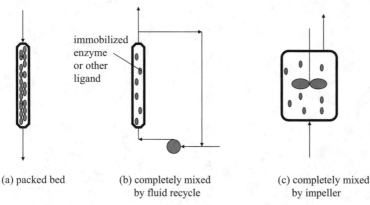

(a) packed bed   (b) completely mixed
by fluid recycle

(c) completely mixed
by impeller

Figure 8.3   Immobilized enzyme and affinity reactor systems.

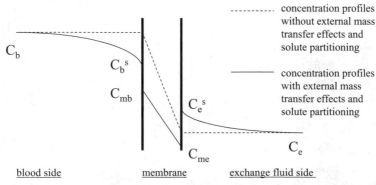

Figure 8.4   Concentration profiles for membrane diffusion.

partition coefficient. Recall that the transport flux is $N_S = K_0(C_b - C_e) = k_b(C_b - C_b^S) = P_m(C_{mb} - C_{me}) = k_e(C_e^S - C_e)$ which can be rearranged and solved for the overall mass transfer coefficient (K_0) as

$$\frac{1}{K_0} = \frac{1}{k_b} + \frac{1}{KP_m} + \frac{1}{k_e} \tag{8.1}$$

This equation states that the total mass transfer resistance $\left(\dfrac{1}{K_0}\right)$ is simply the sum of the individual mass transfer resistances $\left(\dfrac{1}{k_b} + \dfrac{1}{KP_m} + \dfrac{1}{k_e}\right)$. The smallest value of k_b, KP_m, or k_e for a given solute is said to be the *controlling resistance*.

Recall that we discussed in much detail the membrane permeability in Chapter 5. The membrane permeability, although best measured experimentally (Dionne *et al.* 1996; Baker *et al.* 1997), is defined by Equation 5.83. Also as discussed in Chapter 5, the film mass transfer coefficients depend on the physical properties of the fluid and the nature of the flow. They also depend on position due to boundary layer growth, so we usually use length-averaged values of these coefficients in Equation 8.1.

8.4 ESTIMATING THE MASS TRANSFER COEFFICIENTS

Techniques for estimating film mass transfer coefficients for a variety of flow conditions were discussed in Section 5.4.4 in Chapter 5. For laminar flow of fluids in a tube, one may use Equations 5.44 or 5.45 to estimate the film mass transfer coefficient. For other flow situations Table 5.2 provides a summary of useful mass transfer coefficient correlations.

Blood flow through extracorporeal devices is typically laminar and, in some cases, may be fully developed. The Sherwood number (Sh) for blood for these conditions $(k_b h / D_{\text{effective}})$ is equal to 3.66. In the Sherwood number, h = the flow channel thickness in a flat plate membrane arrangement or the tube diameter in a hollow fiber. The effective diffusivity of the solute in blood is given by $D_{\text{effective}}$ and may be quite different from the solute's bulk diffusivity because of the presence of the red blood cells. This is discussed in the next section. Recall from Chapter 5 that the asymptotic limit on the Sh number for fully developed laminar flow is valid when $\dfrac{z}{h} > 0.05 \, Re \, Sc$ (concentration field) or $\dfrac{z}{h} > 0.05 \, Re$ (velocity field). Here, z is defined as the length of the flow path, h is the characteristic dimension of the flow channel, Re is the Reynolds number, and Sc the Schmidt number. For solutes in liquids, the Sc is on the order of 1000 so the velocity field will develop much faster than the concentration field.

The exchange fluid can be either in laminar or turbulent flow. Generally, one operates on the exchange fluid side in such a manner that its contribution to the overall mass transfer resistance would be negligible. Therefore, $k_e \gg k_b$ and KP_m. If estimates of k_e are needed, then one could use Equation 5.44 or 5.45, if the exchange fluid flow is laminar. For turbulent flow of the exchange fluid, one can consult Table 5.2 or these additional references (Cussler 1984; Bird *et al.* 2002; Thomas 1992).

8.5 ESTIMATING THE SOLUTE DIFFUSIVITY IN BLOOD

Transport of a solute through blood can be complicated by the presence of the red blood cells. Estimation of the effective solute diffusivity ($D_{effective}$) needs to account for the effects of the red blood cells on the solute's bulk diffusivity. We need an estimate of solute diffusivity in order to calculate the blood side value of the Sherwood number as discussed in the previous section. Those factors influencing mass transfer within blood are summarized in Table 8.1 (Colton and Lowrie 1981).

Table 8.1 Factors that affect mass transport in blood

Suspension properties
Volume fraction of red blood cells
Volume fraction of proteins
Red blood cell solute permeability

Solute behavior
Protein binding kinetics and equilibrium
Reactions with other solutes

Flow-dependent behavior
Rheological properties
Red blood cell dynamics
   rouleaux formation
   migration from the wall
   rotation and translation

The rheological behavior of blood is not a very important factor in describing solute transport since at the relatively high shear rates encountered in extracorporeal devices, blood behaves as a Newtonian fluid. Mass transport theories for Newtonian fluids are well developed, and we can make use of these for the special case of blood. The rotation and translation of red blood cells can influence mass transport to some degree, however, these effects can usually be neglected unless sufficient data is available to warrant their inclusion.

A complete description of solute transport in blood requires that one knows the solute diffusivity and its red blood cell permeability, its equilibrium distribution between the plasma and the fluid within the red blood cell, and the kinetics of any chemical reaction it may undergo with other solutes, typically proteins, that may be present. Clearly, this is a complex problem. However, for most solutes we can treat the blood as a non-reactive fluid and assume either of the following two cases applies.

First, if the red blood cell permeability for the solute of interest is extremely large, then the solute will diffuse through the blood as if it were a pseudo-homogeneous fluid and the solute diffusivity will not be affected by the presence of the red blood cells. For this case, one only needs to know the solute diffusivity in plasma which can be obtained by estimating the aqueous diffusivity by use of Figure 5.4, followed by a correction for the difference in viscosity between an aqueous solution and plasma using Equation 5.48. Solute diffusion in plasma amounts to about a 40% reduction in comparison to the aqueous solute diffusivity.

For the second case, the red blood cell may be somewhat permeable to the solute. The solute must, therefore, diffuse around or through the red blood cells increasing the solute diffusion path and decreasing the solute effective diffusivity. One could then use relationships such as those presented in Chapter 5, i.e. Equations 5.77 and 5.78. As an example, Colton and Lowrie (1981) investigated the diffusion of urea in both stagnant and flowing blood. They presented data for the ratio of the solute permeability in blood to that in plasma alone as a function of the hematocrit. These results were then used to develop an expression for the effective solute diffusivity in non-reactive blood. The effective diffusion coefficient through blood was defined by the following equation which is similar to Fick's law (Equation 5.2).

$$N_{effective} = D_{effective} \frac{dC_{effective}}{dx} \tag{8.2}$$

The effective concentration ($C_{effective}$) represents the volume fraction weighted sum of the solute concentration in the red blood cell phase (C_{RBC}) and the continuous plasma phase (C_{plasma}) and is given by the next equation.

$$C_{effective} = HC_{RBC} + (1 - H)C_{plasma} \tag{8.3}$$

where H represents the blood hematocrit. Diffusion of solute through just the continuous or plasma phase ($H = 0$) is given by:

$$N = D_{plasma} \frac{dC_{plasma}}{dx} \tag{8.4}$$

Substituting Equation 8.3 into Equation 8.2 for $C_{effective}$, and dividing the result by Equation 8.4, provides a relationship for the effective solute diffusivity of blood in comparison to its value in plasma alone.

$$\frac{N_{effective}}{N} = \frac{D_{effective}}{D_{plasma}} \left[HK_{RBC} + (1 - H) \right] \tag{8.5}$$

In this equation, $K_{RBC} = \dfrac{C_{RBC}}{C_{plasma}}$ represents the equilibrium partition coefficient for the solute between the red blood cells and the plasma. Average values of K_{RBC} for some typical solutes are summarized in Table 8.2.

Table 8.2 Values of K_{RBC} for some typical solutes

Solute	K_{RBC}
Urea	0.86
Creatinine	0.73
Uric acid	0.54
Glucose	0.95

Source: Data from Colton and Lowrie 1981.

Colton and Lowrie (1981) have shown that Equation 8.5 is also equal to the following expression (actually Equation 5.77 with $\phi = H$) assuming the red blood cell is impermeable ($D_{cell} = 0$).

$$\frac{D_{effective}}{D_{plasma}}\left[HK_{RBC} + (1 - H) \right] = \frac{2(1 - H)}{2 + H} \tag{8.6}$$

Equation 8.6 can now be used to estimate the effective diffusivity of a solute in blood. This equation suggests that for a solute such as urea, its effective diffusivity through blood is about 53% of its value in plasma or about 33% of its aqueous diffusivity.

With this background on extracorporeal devices, the following discussion will focus on a description and analysis of four representative devices; hemodialysis, blood oxygenators, enzyme reactors, and affinity columns.

8.6 HEMODIALYSIS

8.6.1 Background

The basic functional unit of the kidney is the nephron (shown earlier in Figure 7.3). Each kidney contains about 1 million nephrons. Only about one third of these nephrons are needed to maintain normal levels of waste products in the blood. If about 90% of the nephrons lose their function, then the symptoms of *uremia* will develop in the patient. Uremia results when waste products normally removed from the blood by the kidneys start to accumulate in the blood and other fluid spaces. For example, water generated by normal metabolic processes is no longer removed and accumulates in the body, leading to *edema*, and in the absence of additional electrolytes, almost half of this water enters the cells rather than the extracellular fluid spaces due to osmotic effects. The normal metabolic processes also produce more acid than base, and this acid is normally removed by the kidneys. Therefore, in kidney failure, there is a decrease in the pH of the body's fluids, called *acidosis*, which can result in *uremic coma*. The end products of protein metabolism include such nitrogenous substances as urea, uric acid, and creatinine. These materials must be removed to ensure continued protein metabolism in the body. The accumulation of urea and creatinine in the blood, although not life threatening by themselves, is an important marker of the degree of renal failure. They are also used to measure the effectiveness of dialysis for the treatment of kidney failure. The kidneys also produce the hormone *erythropoietin* that is responsible for regulating the production of red blood cells in the bone marrow. In kidney failure, this hormone is diminished leading to a condition called *anemia* and a lowered hematocrit. If kidney failure is left untreated, death can occur within a few days to several weeks.

Hemodialysis (HD) has been used for over 50 years (Kolff 1947) to treat patients with degenerative kidney failure or end-stage renal disease (ESRD). Hemodialysis has the ability to replace many functions of failed kidneys. For example, hemodialysis removes the toxic waste products from the body, maintains the correct balance of electrolytes, and removes excess fluid from the body. Hemodialysis can keep patients alive for several years and, for many patients, allows them to survive long enough to receive a kidney transplant. Nearly 500,000 patients are currently being kept alive by hemodialysis.

Figure 8.5 illustrates the basic operation of hemodialysis. Heparinized blood flows through a device containing a membrane. The dialysate or exchange fluid flows on the membrane side opposite the blood. Solutes are exchanged by diffusion between the blood and the dialysate fluid. A variety of membrane configurations are possible and some of the typical ones were shown earlier in Figure 8.1. A variety of flow patterns for the blood and dialysate have been used and these are also summarized in Figure 8.2. The membranes are usually made from such materials as cellulose, cellulose acetate, polyacrylonitrile, and polycarbonate (Zelman 1987; Galletti *et al.* 1995). The membrane surface area is on the order of 1 m^2. Blood flowrates are in the range of several hundred ml min^{-1}, and the dialysate flowrate is about twice that of the blood.

8.6.2 Dialysate Composition

Table 8.3 compares the species present in normal and uremic plasma with those of a typical dialysate fluid. The composition of the dialysate is based on the need to restore the uremic plasma to the normal state. Note in particular the high levels of urea and creatinine in the uremic plasma. For the most part, movement of the species between the uremic plasma and the dialysate is by diffusion. The higher level of HCO$_3^-$ in the dialysate is used to decrease the acidity of the plasma through its buffering action. Because of the reduced concentrations, the osmolarity of the dialysate is about 265 milliosmolar in comparison to about 300 milliosmolar or so for the plasma. Hence, water will tend to leave the dialysate and enter the plasma by osmosis. Since the patient needs to have excess water removed during dialysis, it is then necessary to maintain the dialysate pressure below atmospheric pressure in order to develop a transmembrane pressure gradient that is sufficient to overcome osmosis and remove water from the patient. Removal of waste products from the plasma must also be carefully controlled

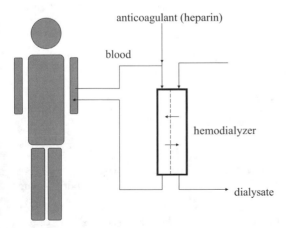

Figure 8.5   Hemodialysis.

Table 8.3 Composition of normal and uremic plasma and dialysate

Species	Normal plasma	Uremic plasma	Dialysate
Electrolytes (mEq^3/liter)			
Na$^+$	142	142	133
K$^+$	5	7	1
Ca^{++}	3	2	3
Mg^{++}	1.5	1.5	1.5
Cl$^-$	107	107	105
HCO$_3^-$	27	14	35.7
Lactate$^-$	1.2	1.2	1.2
HPO$_4^-$	3	9	0
Urate$^-$	0.3	2	0
SO$_4^-$	0.5	3	0
Non-electrolytes (mg/dL)			
Glucose	100	100	125
Urea	26	200	0
Creatinine	1	6	0

Source: Data from A.C. Guyton, *Textbook of Medical Physiology*, 8th ed. (Philadelphia: W.B. Saunders Co., 1991).

in order to avoid the *disequilibrium syndrome*. If waste products are removed from the blood too fast by dialysis, then the osmolarity of the plasma becomes less than that of the cerebrospinal fluid resulting in a flow of water from the plasma and into the spaces occupied by the brain and spinal cord. This increases the local pressure in these areas and can lead to serious side effects.

8.6.3 Role of Ultrafiltration

The rate at which water is removed (ultrafiltration) from the plasma by the pressure gradient across the dialysis membrane can be estimated using Equation 5.95. These equations can be simplified by assuming that only the plasma proteins are impermeable. Hence, the following equation is obtained.

$$Q = SL_P \left(\Delta P_{mean} - \pi_{oncotic\ plasma} \right) \tag{8.7}$$

$\pi_{oncotic\ plasma}$ represents the oncotic pressure of the plasma proteins, about 28 mm Hg, and ΔP_{mean} is the average transmembrane pressure difference which is given by the next equation.

$$\Delta P_{mean} = \left[\frac{P_{b,in} + P_{b,out}}{2} - \frac{P_{d,in} + P_{d,out}}{2} \right] \tag{8.8}$$

Note that to avoid problems with the calculations it is best to use absolute pressures in Equation 8.8, recognizing that since the dialysate is subatmospheric, its reported gauge

---

[3] Equivalents are the amounts of substances that have the same combining capacity in chemical reactions.

pressure will be negative. The pressure drop on the blood side is typically on the order of 20 mm Hg and that on the dialysate side is about 50 mm Hg. The mean transmembrane pressure drop is on the order of a few hundred mm Hg. The hydraulic conductance, L_p, is typically about 3 ml/hr/m²/mm Hg. High flux membranes can have values of the hydraulic conductance as high as 20 ml/hr/m²/mm Hg, however protein deposition on the plasma side of the membrane can reduce this in a linear manner by about 6%/hr (Zelman 1987). Example 8.1 illustrates the calculation of the ultrafiltration rate in a typical membrane dialyzer.

Example 8.1 Using the membrane properties listed above, calculate the ultrafiltration rate of water assuming a membrane surface area of 1 m². How much water would be removed after 6 hours of dialysis? Assume blood enters the device at 120 mm Hg (gauge) and leaves at 100 mm Hg (gauge). The dialysate fluid enters at –150 mm Hg (gauge) and leaves at –200 mm Hg (gauge).

SOLUTION

$$Q = 1\,\text{m}^2 \times \frac{3\,\text{ml}}{\text{hr m}^2\,\text{mm Hg}} \times \left[\frac{(120+760)+(100+760)}{2} \right.$$
$$\left. - \frac{(760-150)+(760-200)}{2} - 28 \right]\text{mm Hg}$$
$$= 771\,\text{ml hr}^{-1}\ (13\,\text{ml min}^{-1})$$

After 6 hours of dialysis, about 10 pounds of water will have been removed.

8.6.4 Clearance and Dialysance

Solute transfer in the dialyzer can be analyzed with the help of the model shown in Figure 8.6 showing cocurrent flow of the blood and dialysate. The Q's represent the volumetric flowrates (ml/min) of the blood (b) and the dialysate (d), and the C's

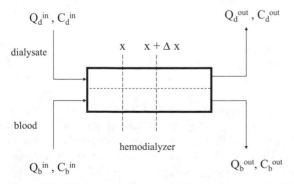

Figure 8.6 Solute mass balance model for a cocurrent hemodialyzer.

represent the species concentration (usually in mg/dL). The mass transfer rate of species i across the dialysis membrane is given by the following equation.

$$M_i = Q_b^{in} C_{b_i}^{in} - Q_b^{out} C_{b_i}^{out} = Q_d^{out} C_{d_i}^{out} - Q_d^{in} C_{d_i}^{in} \tag{8.9}$$

The *clearance of the dialyzer* for a particular solute is defined as the volumetric flowrate of blood (ml/min) entering the dialyzer that is completely cleared of the solute. The dialyzer clearance (CL_D) for solute i is then given by the next equation.

$$CL_D = \frac{M_i}{C_{b_i}^{in}} \tag{8.10}$$

The clearance for such solutes as urea, creatinine, and uric acid is on the order of 100 ml/min. Higher molecular weight solutes show a reduced clearance.

The term *dialysance* is also used to describe the solute removal characteristics of a dialyzer. Dialysance (D_B) is defined by the next equation. We see that the mass transfer rate of species i (M_i) is now divided by the maximum solute concentration difference, i.e. $\left(C_{bi}^{in} - C_{di}^{in} \right)$, between the blood and the dialysate.

$$D_B = \frac{M_i}{C_{b_i}^{in} - C_{d_i}^{in}} \tag{8.11}$$

For a single pass dialyzer, $C_{d_i}^{in} = 0$, and the dialysance is then the same as the clearance.

Because of water removal, the inlet and outlet flowrates of the blood and dialysate streams are not equal. Rather, the following equation holds between the ultrafiltration rate and the flowrates of these streams.

$$Q = Q_b^{in} - Q_b^{out} = Q_d^{out} - Q_d^{in} \tag{8.12}$$

If ultrafiltration is important, then we can use this relationship in combination with Equations 8.9 and 8.10 to derive an expression for the clearance that allows for examination of the importance of ultrafiltration on the clearance of a given solute. Explicitly including ultrafiltration, the solute clearance is given by the following equation.

$$CL_D = Q_b^{in} \left(\frac{C_{b_i}^{in} - C_{b_i}^{out}}{C_{b_i}^{in}} \right) + Q \left(\frac{C_{b_i}^{out}}{C_{b_i}^{in}} \right) \tag{8.13}$$

Since in general the value of $C_{b_i}^{out} \ll C_{b_i}^{in}$, the contribution of the ultrafiltration flow to the solute clearance is less than the value of Q. As shown in Example 8.1, Q is on the order of 10 ml min^{-1}, and earlier it was stated that the clearance of such solutes as urea is on the order of a 100 ml min^{-1}. Therefore, one concludes that the effect of ultrafiltration on the clearance of small molecular weight solutes is negligible. However, for higher molecular weight solutes the effect of ultrafiltration may be significant.

8.6.5 Solute Transfer

We can now develop relationships between solute clearance and the mass transfer characteristics of the membrane dialyzer. In order to perform this analysis, we first must choose one of the contacting patterns shown in Figure 8.2 to describe the flow of blood and dialysate in the dialyzer. Here we will use the cocurrent pattern already shown in Figure 8.6. We will also assume that the overall mass transfer coefficient given by Equation 8.1 is constant. The flowrates of the blood and dialysate are also assumed to be constant, that is, ultrafiltration is ignored.

A shell mass balance for the blood and dialysate can then be written for a given solute as shown next.

$$Q_b C_b \big|_x - Q_b C_b \big|_{x+\Delta x} = K_o W \Delta x (C_b - C_d)$$

(8.14)

$$Q_d C_d \big|_x - Q_d C_d \big|_{x+\Delta x} = -K_o W \Delta x (C_b - C_d)$$

The quantity $W\Delta x$ represents the available mass transfer area within the shell volume ΔV. If L is the length of the dialyzer, then the quantity $A = L \times W$ would be the total membrane area within the dialyzer. After dividing by Δx, and taking the limit as Δx approaches zero, the following two differential equations are obtained that provide the position dependence of the solute concentration in the dialyzer blood and dialysate.

$$Q_b \frac{dC_b}{dx} = -K_o W (C_b - C_d)$$

(8.15)

$$Q_d \frac{dC_d}{dx} = K_o W (C_b - C_d)$$

These equations can be integrated analytically provided we first can relate C_d and C_b. This can be obtained by adding the above two equations, multiplying through by dx, and then integrating between the entrance and any arbitrary value of x. The following result is then obtained for the value of C_d in terms of C_b. Note that $z \equiv \dfrac{Q_b}{Q_d}$.

$$C_d(x) = C_d^{in} - z\left(C_b(x) - C_b^{in}\right)$$

(8.16)

Substituting this equation for C_d in Equation 8.15, the differential equation describing the solute concentration on the blood side can then be written as follows. We see that the only dependent variable is the blood concentration.

$$Q_b \frac{dC_b}{dx} = -K_o W \left[(1+z)C_b - \left(C_d^{in} + zC_b^{in}\right) \right]$$

(8.17)

This equation can now be integrated analytically and rearranged to give the following result. This equation is known as the performance equation for a cocurrent dialyzer.

$$E = \frac{D_B}{Q_b} = \frac{C_b^{in} - C_b^{out}}{C_b^{in} - C_d^{in}} = \frac{1 - \exp\left[-N_T(1+z)\right]}{(1+z)}$$

(8.18)

Two additional dimensionless parameters have been defined in the derivation of this equation, the *extraction ratio*, E, where $0 \leq E \leq 1$; and the *number of transfer units*, $N_T = \dfrac{K_0 A}{Q_b}$. The number of transfer units provides a measure of the mass transfer effectiveness, i.e. the amount of solute tranported across the membrane by diffusion versus the amount of solute that enters the device in the blood. The concentration of solute i in the blood exiting the dialyzer is given by: $C_{b_i}^{\text{out}} = C_{b_i}^{\text{in}}(1-E) + EC_{d_i}^{\text{in}}$.

If the membrane area is infinite for a given value of z (i.e. for a given value of the dialysate and blood flowrates), then we achieve the maximum possible extraction ratio, i.e. $E_{\text{maximum}} = \dfrac{1}{1+z}$. This corresponds to the best possible performance in a dialyzer for a given set of blood and dialysate flowrates, and is independent of the solute concentrations. This also implies that the solute concentration in the exiting blood and dialysate are equal or in equilibrium, i.e. $C_{b_i}^{\text{out}} = C_{d_i}^{\text{out}} = C_{\text{equilibrium}}$. For these conditions, the maximum achievable dialysance for any combination of the blood and dialysate flows is also given by, $D_{B_{\text{maximum}}} = \dfrac{z}{z+1} = \dfrac{Q_b Q_d}{Q_b + Q_d}$.

If $z = 0$, i.e. the dialysate flowrate is significantly higher than that of the blood, and if the membrane area were also infinite, then we would obtain $E = 1$. For these conditions, the solute concentration in the blood exiting the dialyzer would achieve its lowest possible value, i.e. the value of the solute concentration in the entering dialysate fluid, $C_{d_i}^{\text{in}}$.

Performance equations for other fluid contacting patterns such as countercurrent, a well-mixed dialysate, and cross flow are summarized in Table 8.4. The following example illustrates the use of these performance equations.

Table 8.4 Performance equations for hemodialyzers

Contacting pattern	Performance equation
Cocurrent	$E = \dfrac{1 - \exp[-N_T(1+z)]}{1+z}$
Countercurrent	$E = \dfrac{\exp[N_T(1-z)] - 1}{\exp[N_T(1-z)] - z}$
Well-mixed dialysate	$E = \dfrac{1 - \exp(-N_T)}{1 + z[1 - \exp(-N_T)]}$
Crossflow	$E = \dfrac{1}{N_T} \sum\limits_{n=0}^{\infty} \left[S_n(N_T) S_n(N_T z) \right]$
	where
	$S_n(x) = 1 - \exp(-x) \sum\limits_{m=0}^{n} \left[\dfrac{x^m}{m!} \right]$

Source: Colton and Lowrie, 1981.

Example 8.2 Estimate the clearance of urea in a 1 m^2 cocurrent and countercurrent hollow fiber dialyzer. The dialyzer hollow fibers are 25 cm in length and they have an inside diameter of 250 microns. The wall thickness of the fibers is about 40 microns. The void volume (ε) in the dialyzer is 50%. Blood flows within the hollow fibers, and the dialysate flows in a single pass on the shell side of the device parallel to the fibers. The blood flowrate is 200 ml min^{-1}, and the dialysate flowrate is 500 ml min^{-1}. The membrane permeability, KP_m, is estimated to be about 10^{-3} cm sec^{-1} for urea.

SOLUTION For the blood flowing through the fibers the characteristic dimension in the *Sh* number is the internal diameter of the fiber. For flow outside and parallel to the fibers, the characteristic dimension in the dialysate *Sh* number is the equivalent diameter, D_H, defined earlier by Equation 5.46 shown below (Yang and Cussler 1986).

$$D_H = \frac{4 \, (\text{cross sectional area})}{\text{wetted perimeter}}$$

Assuming the fibers are arranged on a square pitch, the following relationship may be developed to express the relationship between the diameter of the hollow fiber (d), the dialyzer void fraction (ε), and the equivalent diameter (D_H).

$$D_H = d\left(\frac{\varepsilon}{1-\varepsilon}\right) \text{ square pitch}$$

The apparent velocity of the dialysate through the void space between the fibers would be equal to the dialysate volumetric flowrate (Q_d) divided by the cross-sectional area of the void space. The cross-sectional area of the void space is equal to the void fraction (ε) times the cross-sectional area of the dialyzer itself. A little algebra provides the following result for the apparent dialysate velocity. N_{fiber} is the number of hollow fibers in the dialyzer.

$$V_d = \frac{4(1-\varepsilon)Q_d}{N_{fiber}\,\varepsilon\pi d^2}$$

After performing the calculations, the predicted urea dialyzer clearance is about 120 ml/min for the cocurrent dialyzer and about 133 ml/min for the countercurrent case. These urea clearance values are typical of what is observed in actual practice. In comparison, the urea clearance of the two normal kidneys is about 70 ml/min. In general, the countercurrent dialyzer will provide the higher clearance everything else being equal. This is because the countercurrent flow pattern maintains a relatively constant mass transfer driving force along the length of the dialyzer. The cocurrent pattern provides a larger driving force at the entrance of the dialyzer, but this difference rapidly decreases along the length of the dialyzer resulting in less overall mass transfer of the solute.

8.6.6 A Single Compartment Model of Urea Dialysis

Urea distributes throughout the total body water for a total distribution volume of about 40 liters. The distribution volume (in liters) can also be estimated to be 58% of body weight (in kg) (Galletti et al. 1995). We can treat the body as a single well-mixed compartment and use the dialyzer performance equations to predict how the amount of urea within the body decreases with time during dialysis. Neglecting the effects of ultrafiltration, the clearance of urea from the body during dialysis is similar to the IV bolus injection examined in Chapter 7, that is Equation 7.23, where the only elimination pathway for urea is the dialyzer. Therefore, k_{te} equals the dialyzer clearance divided by the distribution volume for urea.

$$C_{urea} = C_{urea}^o \, e^{-\left(\frac{CL_D}{V_{apparent}}\right)t} \tag{8.19}$$

Using the urea dialyzer clearance value for a countercurrent unit obtained from Example 8.2, we can show that after 4 hours of dialysis, the urea concentration in the body has been reduced by about 55%. It should be pointed out that during this dialysis period, the amount of urea generated is assumed negligible in comparison to the amount removed by dialysis. However, after dialysis, urea continues to accumulate at the rate of about 10 grams each day (G_{urea}). If we assume the kidney has no residual clearance for urea, then the urea concentration will increase linearly with time after dialysis according to the following equation.

$$C_{urea}(t) = C_{urea}\left(\begin{smallmatrix} end\ of \\ dialysis \end{smallmatrix}\right) + \left(\frac{G_{urea}}{V_{apparent}}\right)t \tag{8.20}$$

This equation can be used to show that dialysis will be required about 3–4 times per week. For some solutes, the removal from the dialyzer is so fast that the rate of solute transfer from the extravascular spaces into the plasma will be the rate limiting step. For these cases a two compartment model may be needed to describe the kinetics of solute removal.

The following references provide additional information on the artificial kidney: (Ramachandran and Mashelkar 1980; Colton and Lowrie 1981; Armer and Hanley 1986, Abbas and Tyagi 1987; Zelman 1987; Lee and Chang 1988; Lee et al. 1989; Shaoting et al. 1990; Cappello et al. 1994; Galletti et al. 1995).

8.6.7 Peritoneal Dialysis

An alternative approach for treating kidney failure is called continuous ambulatory peritoneal dialysis (CAPD) (Lysaght and Farrell 1989; Lysaght and Moran 1995). Unlike the hemodialysis technique that was just described, CAPD has the advantage of the patient not being severely restricted by the regimen of three times per week treatment at a dialysis center. Rather, the patient is responsible for performing a relatively simple maintenance process that can take place at home or even at work. Use of CAPD is growing rapidly. CAPD was used in about 4% of the total dialysis population in 1979 and in 1992 commanded 15% of the dialysis population. Overall, about 90,000 patients used CAPD in 1995 (Lysaght and Moran 1995).

CAPD is based on the addition to the peritoneal cavity of a sterile hypertonic solution of glucose and electrolytes. The peritoneal cavity is a closed space formed by the peritoneum, a membrane-like tissue that lines the abdominal cavity and covers the internal organs such as the liver and the intestines. The peritoneum has excellent mass transfer characteristics for a process like CAPD. The peritoneum has the appearance of a fairly transparent sheath that is smooth and quite strong. The surface area (S) of the peritoneum is about 1.75 m^2. Its thickness ranges from 200 to over 1000 microns. The surface of the peritoneum presented to the CAPD dialysate solution consists of a single layer of mesothelial cells that is densely covered with microvilli or tiny hair-like projections. Beneath this layer of cells is the interstitium that has the characteristics of a gel-like material. Within the interstitium, there is a rich capillary network providing a total blood flow on the order of 50 ml/min.

The CAPD solution is added and later removed from the peritoneal cavity through an in-dwelling catheter. This process can also be automated such that all fluid exchanges are carried out by a pumping unit, even while the patient is asleep. This automated process is called automated peritoneal dialysis (APD).

The intraperitoneal CAPD fluid partially equilibrates across the peritoneal membrane with waste products in the plasma. Excess water from the patient's body is also removed by ultrafiltration across the peritoneal membrane as a result of osmotic gradients. Typically, four 2 liter exchanges are made each day with an additional 2 liters of water removed as a result of ultrafiltration. The average mass removal rate of a particular solute is given by the following equation.

$$\dot{m}_{CAPD} = \frac{V_{CAPD} \, C_{CAPD}}{t_{CAPD}} \tag{8.21}$$

In this equation, V_{CAPD} represents the volume of the CAPD solution, and C_{CAPD} is the final concentration of the solute in this solution. The time the solution was in the peritoneal cavity is given by t_{CAPD}. Recall that the clearance of a solute is equal to its mass removal rate divided by the solute concentration in the blood. Therefore, solute clearance during CAPD is given by the following expression.

$$CL_{CAPD} = \frac{\dot{m}_{CAPD}}{C_b} = \frac{V_{CAPD} \, C_{CAPD}}{t_{CAPD} \, C_b} \tag{8.22}$$

Since the CAPD solution is usually completely equilibrated with urea in the plasma, the urea clearance is then simply $\dfrac{V_{CAPD}}{t_{CAPD}}$. About 10 liters of solution are used each day which equates to a continuous urea clearance of about 7 ml min^{-1}.

The mass transfer charactersitics of CAPD can be described using relationships we have already developed. For example, the ultrafiltration rate is given by Equation 5.95a:

$$Q = S \, L_P \left[\Delta P - RT \sum_i \sigma_i \left(C_{blood} - C_{CAPD} \right)_i \right] \tag{8.23}$$

CAPD usually begins with a solution containing about 4% glucose. The initial ultrafiltration rate for this solution is on the order of 20 ml/min and decreases rapidly with time as the glucose initially in the peritoneal cavity is removed by the blood. Since the ΔP is small during CAPD, and the osmotic pressure of the initial glucose solution is about 4300 mm Hg, then the hydraulic conductance is on the order of 0.15 ml/hr m² mm Hg.

The solute transport rate for species i is given by Equation 5.95b, recognizing that, initially, significant quantities of solute may be transported by ultrafiltration.

$$\dot{m}_{CAPD} = C_i \left(1 - \sigma_i\right) Q + P_{m_i} S \left(C_{blood} - C_{CAPD}\right)_i \qquad (8.24)$$

Typical values of the reflection coefficient and the permeability-surface area product for several solutes are summarized in Table 8.5.

Table 8.5 Transport properties of the peritoneal membrane

Solute	Molecular weight	Reflection coefficient	$P_m S$, cm³/min
Urea	60	0.26	21
Creatinine	113	0.35	10
Uric acid	158	0.37	10
Vitamin B$_{12}$	1355		5
Inulin	5200	0.5	4
β_2 microglobulin	12,000		0.8
Albumin	69,000	0.99	

Source: Lysaght and Moran (1995).

Mathematical models describing CAPD can also be developed to describe the time-course of the treatment process. The patient is considered to be represented by a single well-mixed compartment of apparent distribution volume ($V_{apparent}$) and solute concentration (C_{body}) which is assumed to be the same as the solute concentration in the blood (C_{blood}). The CAPD solution in the peritoneal cavity is in a much smaller variable volume (about 2 liters at the start), represented by V_{CAPD}, and solute concentration, C_{CAPD}. The peritoneal membrane separates these two compartments. The peritoneal lymphatic system also removes fluid from the peritoneal cavity (Q_{lymph}) at a rate of about 0.1 to as high as 10 ml/min. During the initial phases of CAPD, the volume of fluid in the peritoneal cavity will increase due to osmotic flow of water into the cavity, however, as time progresses and glucose is diluted and removed from the peritoneal cavity, the volume of fluid in the peritoneal cavity will decrease as a result of the lymphatic flow.

Models for describing CAPD are reviewed by Lysaght and Farrell (1989). These models range from relatively simple analytical solutions to much more complex models that include selective transport of solutes, ultrafiltration, and the effects of lymphatic flow. These models all require numerical solutions. We will discuss two relatively simple analytical models for CAPD in the following discussion.

The simplest analytical model neglects ultrafiltration and lymphatic flow and assumes that only the peritoneal dialysate concentration changes with time. This diffusion only model, with its assumption that V_{CAPD} is constant, limits the use of this approach to the constant volume phase (isovolemic) that occurs about an hour or so after the exchange begins. For this diffusion only CAPD model, the solute balance equation only needs to be written for the fluid in the peritoneal cavity.

$$\frac{d\left(V_{CAPD}\, C_{CAPD}\right)}{dt} = V_{CAPD}\frac{dC_{CAPD}}{dt} = P_m S\left(C_{body} - C_{CAPD}\right) \tag{8.25}$$

This equation may be easily integrated to obtain the following equations. In these equations, the initial concentration of the solute in the CAPD solution is given by C_{CAPD}^0.

$$P_m S = \frac{V_{CAPD}}{t}\ln\left(\frac{C_{body} - C_{CAPD}^0}{C_{body} - C_{CAPD}(t)}\right) \tag{8.26}$$

$$C_{CAPD}(t) = C_{body} - \left(C_{body} - C_{CAPD}^0\right)\exp\left(-\frac{P_m S t}{V_{CAPD}}\right) \tag{8.27}$$

The first equation above allows for determination of the value of $P_m S$ given information on the solute concentrations at a particular time. The second equation allows for prediction of the solute concentration in the CAPD dialysate solution as a function of time assuming the transport properties of the peritoneal membrane are known.

A slightly more complex model than the one shown above can be developed that includes the effect of ultrafiltration (Garred *et al.* 1983). However, the solute concentration in the body is still assumed to be constant, the peritoneal membrane is not selective, i.e. the reflection coefficients are zero, and lymphatic flow is ignored. The solute balance for the CAPD solution for these conditions may be written as follows.

$$\frac{d\left(V_{CAPD}\, C_{CAPD}\right)}{dt} = P_m S\left(C_{body} - C_{CAPD}\right) + C_{body}\frac{dV_{CAPD}}{dt} \tag{8.28}$$

The second term on the right-hand side $\left(C_{body}\dfrac{dV_{CAPD}}{dt}\right)$ represents the ultrafiltration flow that carries with it the solute in the body. Integration of this equation still requires knowledge of how V_{CAPD} changes with time. This could be obtained from Equation 8.23, however an analytical solution would then be difficult to obtain. If one assumes that the initial and final CAPD volumes are known, then this equation can be integrated using an average value of V_{CAPD} to give the next two equations. $\bar{V}_{CAPD}$ is the average volume defined as the mean of the initial and final CAPD volumes.

$$P_m S = \frac{\bar{V}_{CAPD}}{t}\ln\left[\frac{V_{CAPD}^0\left(C_{body} - C_{CAPD}^0\right)}{V_{CAPD}(t)\left(C_{body} - C_{CAPD}(t)\right)}\right] \tag{8.29}$$

$$C_{CAPD}(t) = C_{body} - \frac{V_{CAPD}^0}{V_{CAPD}(t)}\left(C_{body} - C_{CAPD}^0\right)\exp\left(-\frac{P_m S t}{\bar{V}_{CAPD}}\right) \tag{8.30}$$

Once again, the first equation may be used to estimate the mass transport properties of the peritoneal membrane from given values of the solute concentration and volumes at a particular time. The second equation allows determination of the time course of the solute concentration in the dialysate solution.

8.7 BLOOD OXYGENATORS

8.7.1 Background

Surgery on the heart to repair internal defects or to improve blood flow to the heart muscle by coronary artery bypass grafting may require that the heart be stopped or arrested. Accordingly, blood is no longer pumped throughout the body and is no longer oxygenated by the lungs. Special devices called heart–lung machines or blood pump-oxygenators (Richardson 1987; Makarewicz *et al.* 1993; Galletti and Colton 1995) have been developed and used for over 50 years (Gibbon 1954) to replace the gas exchange function of the lungs and the pumping action of the heart during these cardiac surgical procedures.

Blood flow to the blood-pump oxygenator is usually collected by a pump from the large systemic veins such as the vena cava. This blood is then passed through the oxygenator and returned by a second pump to the aorta. Blood flow to the chambers of the heart itself is stopped providing a dry and bloodless field for surgery. Nearly a million of these cardiopulmonary bypass procedures are performed worldwide each year.

In addition to blood-pump oxygenators for heart surgery, much interest has recently focused on the development of intravascular lung assist devices or ILADs (Vaslef *et al.* 1989; Jurmann *et al.* 1992; Makarewicz *et al.* 1993; Fukui *et al.* 1994). ILADs are being used for the treatment of patients that suffer from adult respiratory distress syndrome (ARDS). ILADs consist of a bundle of hollow fiber membranes that are mounted on a catheter and then inserted into the vena cava. Pure oxygen is supplied within the hollow fibers by an external flow control system. Blood flows externally along the outer surfaces of the hollow fibers, and oxygen and carbon dioxide are exchanged.

8.7.2 Operating Characteristics

There are several important differences between the operational characteristics of the lungs and blood oxygenators. These are summarized in Table 8.6. The blood flow through the blood oxygenator is usually the same as that through lungs since the blood oxygenator must provide the same level of blood oxygenation; that is increasing the pO_2 of the venous blood from about 40 mm Hg to the arterial pO_2 of 95 mm Hg. The capillaries in the lungs present the blood to the alveolar gas as a very thin film of blood about 5–10 microns in thickness. This provides rapid gas transport in a very short time, typically a few tenths of a second. In a blood oxygenator, the blood mass transfer film thickness is governed by the number and thickness of the blood flow channels, typically

Table 8.6 Operational characteristics of the lungs and blood pump-oxygenators

Characteristic	Lungs	Blood Oxygenator
Blood flow rate	5 liters min^{-1}	5 liters min^{-1}
Pressure head	12 mm Hg	0–200 mm Hg
Blood volume	1 liter	1–4 liters
Blood film thickness	5–10 microns	100–300 microns
Length of blood flow channel	100 microns	2–30 cm
Blood contact time	0.7 sec	3–30 sec
Surface area for mass transfer	70 m^2	2–10 m^2
Gas flow rates	7 liters min^{-1}	2–10 liters min^{-1}
pO_2 & pCO_2 blood in	40 & 45 mm Hg	40 & 45 mm Hg
pO_2 & pCO_2 blood out	95 & 40 mm Hg	100–300 & 30–40 mm Hg
Gas pO_2 & pCO_2 in	149 & 0.3 mm Hg	250–713 & 0–20 mm Hg
Gas pO_2 & pCO_2 out	120 & 27 mm Hg	150–675 & 10–30 mm Hg
pO_2 gradient	40–50 mm Hg	650 mm Hg
pCO_2 gradient	3–5 mm Hg	30–50 mm Hg

Source: Cooney (1976); Galletti and Colton (1995).

providing film thicknesses on the order of a few hundred microns and contact times on the order of tens of seconds. Effective gas transport then occurs over greater oxygenator lengths and requires greater pressures to force the blood through the oxygenator.

In the lungs, the pO_2 of the alveolar air is about 104 mm Hg. The average oxygen partial pressure driving force in the lungs is, therefore, on the order of 40–50 mm Hg. Carbon dioxide has a pCO_2 of about 40 mm Hg in the alveolar space, and its level in the blood varies from 40 mm Hg in arterial blood to 45 mm Hg in venous blood. Therefore, the driving force for carbon dioxide transport is only a few mm Hg and considerably smaller than that for oxygen. However, this smaller driving force is offset by the much higher permeability of carbon dioxide through the respiratory membranes which is about 20 times that for oxygen. One significant advantage of the blood oxygenator as compared with the lungs is the driving force for oxygen and carbon dioxide transport. In a blood oxygenator, humidifed oxygen at atmospheric pressure can be used resulting in a pO_2 of about 713 mm Hg [760 mm Hg – 47 mm Hg (water saturation pressure at 37°C) = 713 mm Hg]. Also, little or no carbon dioxide is present in this gas so the driving force for carbon dioxide transport is considerably higher. This results in a considerable reduction in the mass transfer surface area of the blood oxygenator.

8.7.3 Types of Oxygenators

The design goal of a blood oxygenator is to present as large a surface area as possible between the blood and the oxygen carrying gas stream. Several approaches have been developed to accomplish this, and they are summarized in Figure 8.7. *Bubble oxygenators* were the earliest systems developed for cardiopulmonary bypass surgery. Bubble oxygenators develop the required surface area for gas exchange by the production of numerous small bubbles of gas. These gas bubbles are in direct contact with the blood and are typically several millimeters in diameter. Of particular concern

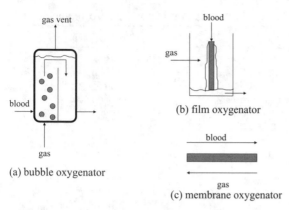

gas vent

blood

gas

(b) film oxygenator

blood

blood

gas

(a) bubble oxygenator

blood

gas

(c) membrane oxygenator

Figure 8.7   Types of blood oxygenators.

in the operation of bubble oxygenators is the removal of foam and gas bubbles prior to the return of the blood to the patient. This is achieved through the use of filters and defoaming sponges as the blood leaves the oxygenator. Design of bubble oxygenators is complicated by the complex nature of bubble motion. For example, bubbles exist in many different sizes and they tend to breakup and coalesce during their passage in the oxygenator.

To avoid the problem with antifoaming and bubble removal, *film oxygenators* have also been developed. In these systems, a film of blood is created on a surface that may be stationary or rotating. The blood film is then exposed directly to the gas. To increase the mass transfer rates, stationary films often employ irregular surfaces to enhance internal mixing of the blood film. Rotating discs whose bottom surface is in contact with a pool of blood create very thin films that are continuously renewed resulting in efficient exposure of the blood to the gas stream.

Membrane oxygenators have, for the most part, now replaced the earlier bubble and film oxygenators. In these systems, the membrane physically separates the blood and gas streams. Membrane systems offer the advantages of minimizing the trauma to the blood that is created in direct contact systems such as bubble and film oxygenators. The membrane and the associated fluid boundary layers, however, do present an additional mass transfer resistance. Recent advances in membrane technology has resulted in membranes with significant oxygen and carbon dioxide permeabilities. The gas permeabilities for various membrane materials are shown in Table 8.7.

Typical units of the thickness normalized membrane gas permeability $\left(\hat{P}_m \right)$ are ml (STP) micron min^{-1} m^{-2} atm^{-1}, where the STP (standard temperature and pressure) conditions are for the pure gas at 0°C and 1 atmosphere. The volumetric transport rate of gas is then given by the following equation:

$$N_V = \frac{\hat{P}_m \times (\text{membrane surface area}) \times \Delta(\text{gas partial pressure})}{\text{membrane thickness}} \quad (8.31)$$

The most common membrane material used in early versions of membrane devices was silicone with wall thicknesses on the order of 50–200 microns. The oxygen

Table 8.7 Oxygen and carbon dioxide permeability of selected membrane materials

Material	O_2 permeability, ml (STP) micron min^{-1} m^{-2} atm^{-1}	CO_2 permeability, ml (STP) micron min^{-1} m^{-2} atm^{-1}
Air	1.27×10^9	1.02×10^9
Polydimethysiloxane (silicone)	27,900	140,000
Water	3810	68,600
Polystyrene	1397	6985
Polyisoprene (natural rubber)	1270	7620
Polybutadiene	1016	7112
Cellulose (cellophane)	635	11,430
Polyethylene	305	1524
Polytetrafluoroethylene (teflon)	203	610
Polyamide (nylon)	2.54	10.2
Polyvinylidene chloride (saran)	0.25	1.52

Source: Galletti and Colton (1995).

permeability for a 130 micron thick silicone membrane is about 215 ml (STP)/min m^2 atm and that for carbon dioxide is about 1100 ml (STP)/min m^2 atm (Cooney 1976; Gray 1981, 1984).

Recent hollow fiber systems employ hydrophobic microporous polypropylene membranes. These membranes allow for free passage of gas molecules by diffusion through the membrane pores, but because of the small size of the pores, there is sufficient surface tension to prevent plasma filtration. The pores in these membranes are typically on the order 0.1 microns in diameter. Note from Table 8.7 that a column of air within the pores of these membranes has a very high gas permeability in comparison to diffusion through the polymeric material itself. Therefore, these membranes have very high gas permeabilities.

These membrane-based systems come in a variety of configurations, for example flat plate, coil, and hollow fiber arrangements as shown earlier in Figure 8.1. Blood and gas stream contacting can include cocurrent, countercurrent, and cross flow. Cross flow systems typically have the gas flowing through the lumen of the hollow fiber with blood flowing at right angles across the outer surface of the fibers. Cross flow of the blood results in significant improvement in performance as measured by gas exchange rates (Catapano et al. 1992; Vaslef et al. 1994; Wickramasinghe et al. 2002a,b, 2005).

8.7.4 Analysis of a Membrane Oxygenator, Oxygen Transfer

A mathematical model for a membrane blood gas oxygenator can be developed as follows. The following model is useful for preliminary design calculations and for exploring the effects of operating conditions on device performance. The model is developed in a manner similar to that used earlier for hemodialysis, however, now we must make use of techniques developed in Chapter 6 to account for the binding of oxygen with hemoglobin.

Figure 8.8 shows the mass balance model where now we assume countercurrent flow of the blood and gas streams. We also assume the total flowrates of these streams

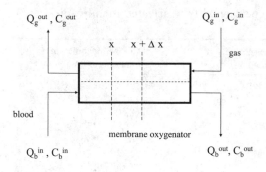

Figure 8.8 Solute mass balance model for a countercurrent membrane blood oxygenator.

are unchanged, hence Q_g and Q_b are constant. C_b and C_g are the bulk concentrations on the blood (b) and gas (g) sides of the membrane. C_b' represents the amount of oxygen in the blood that is bound to hemoglobin. C_{bm} and C_{gm} are the corresponding concentration values at the membrane surface on the blood and gas sides.

A shell balance on oxygen may be written as follows for the blood and gas sides of the membrane of length Δx.

$$Q_b\left(C_b + C_b'\right)\big|_x - Q_b(C_b + C_b')\big|_{x+\Delta x} = k_b W_b\ \Delta x (C_b - C_{bm})$$

$$Q_g C_g\big|_{x+\Delta x} - Q_g C_g\big|_x = k_g W_g\ \Delta x (C_g - C_{gm}) \tag{8.32}$$

In these equations, k_b and k_g are the blood and gas side mass transfer coefficients. W is the membrane area per unit length of membrane with the subscript denoting either the blood (b) or gas (g) sides. This allows application of this discussion to cylindrical fibers where the area normal to the direction of transport changes with radial position. For planar membranes we simply have that $W_b = W_g = W$.

After dividing by Δx, and taking the limit as $\Delta x \rightarrow 0$, we obtain the following differential equations.

$$Q_b \frac{d(C_b + C_b')}{dx} = -k_b W_b (C_b - C_{bm}) \tag{8.33}$$

$$Q_g \frac{dC_g}{dx} = k_g W_g\ (C_g - C_{gm})$$

Once again, we can make use of the fact that $m = \dfrac{dC_b'}{dC_b}$. Recall that m is related to the slope of the oxygen dissociation curve. A value for m may be estimated by Equation 6.36 or 6.37. Assuming m is constant and is evaluated at some suitable combination of the venous and arterial pO_2 values, the blood side equation simplifies to the following result.

$$Q_b(1+m)\frac{dC_b}{dx} = -k_b W_b\ (C_b - C_{bm}) \tag{8.34}$$

It is now convenient to convert from the oxygen concentration to the oxygen partial pressure in our description of the mass transfer process. Recall for blood that the partial pressure of oxygen is related to the dissolved oxygen concentration by Henry's law, $pO_{2b} = HC_b$. The value of H for blood is 0.74 mm Hg/μM. On the gas side, $pO_{2g} = RTC_g$ using the ideal gas law. Our equations for the blood and gas side now may be written as:

$$Q_b(1+m)\frac{dpO_{2b}}{dx} = -k_b W_b (pO_{2b} - pO_{2bm})$$

$$Q_g \frac{dpO_{2g}}{dx} = k_g W_g (pO_{2g} - pO_{2gm})$$

(8.35)

These equations may be rewritten in terms of the overall partial pressure driving force, i.e. $pO_{2g} - pO_{2b}$, through definition of the overall mass transfer coefficient, K_0, defined by the next equation:

$$\frac{1}{K_0} = \frac{H}{k_b} + \frac{W_b}{\rho^{STP} P_m \bar{W}_L} + \frac{RT W_b}{k_g W_g}$$

(8.36)

ρ^{STP} represents the density of the gas at STP conditions. The units on K_0 are typically mol cm^{-2} mm Hg^{-1}sec^{-1}. K_0 is also based on the membrane area on the blood side. The log mean area of the membrane is given by $\bar{W}_L \equiv \dfrac{W_g - W_b}{\ln\left(\dfrac{W_g}{W_b}\right)}$.

In terms of K_0, the blood and gas side equations become:

$$\frac{Q_b(1+m)}{H} \frac{dpO_{2b}}{dx} = K_0 W_b \left(pO_{2g} - pO_{2b}\right)$$

$$\frac{Q_g}{RT} \frac{dpO_{2g}}{dx} = K_0 W_b \left(pO_{2g} - pO_{2b}\right)$$

(8.37)

We can now solve these equations to provide an analytical solution to describe the performance of the oxygenator. First, we subtract the gas side equation from the blood side equation. This allows us to obtain a relationship between pO_{2g} and pO_{2b}.

$$\frac{Q_b(1+m)}{H} \frac{dpO_{2b}}{dx} - \frac{Q_g}{RT} \frac{dpO_{2g}}{dx} = 0$$

(8.38)

This equation may now be integrated from the entrance of the blood stream ($x = 0$) to any arbitrary value of x to give:

$$pO_{2g}(x) = \left(\frac{Q_b}{Q_g}\right)\left(\frac{RT}{H}\right)(1+m)\left[pO_{2b}(x) - pO_{2b}^{in}\right] + pO_{2g}^{out}$$

(8.39)

Now we can substitute this equation for pO_{2g} in the blood side Equation (8.37) to give:

$$-\frac{dpO_{2b}}{dx} = \left(\frac{K_0 W_b H}{Q_b(1+m)}\right) \left\{ \left[\left(\frac{Q_b}{Q_g}\right)\left(\frac{RT}{H}\right)(1+m)-1\right]pO_{2b} \right.$$
$$\left. + \left[pO_{2g}^{out} - \left(\frac{Q_b}{Q_g}\right)\left(\frac{RT}{H}\right)(1+m)pO_{2b}^{in}\right] \right\}$$

(8.40)

This equation may then be integrated to give the following result for the blood side membrane area (A_{oxygen}) required for a given change in blood oxygenation.

$$A_{oxygen} = \frac{\alpha}{\beta} \ln\left(\frac{\beta pO_{2b}^{out} + \gamma}{\beta pO_{2b}^{in} + \gamma}\right)$$

(8.41)

where α, β, and γ are given by:

$$\alpha = \frac{Q_b(1+m)}{K_0 H}$$

$$\beta = \left(\frac{Q_b}{Q_g}\right)\left(\frac{RT}{H}\right)(1+m)-1$$

$$\gamma = pO_{2g}^{out} - \left(\frac{Q_b}{Q_g}\right)\left(\frac{RT}{H}\right)(1+m)pO_{2b}^{in}$$

(8.42)

The value of α will typically have units of cm^2, β is dimensionless, and γ will be in mm Hg.

8.7.5 Analysis of a Membrane Oxygenator, Carbon Dioxide Transfer

We also need to develop a similar set of equations for the transport of carbon dioxide in the oxygenator. Carbon dioxide can exist in blood in a variety of forms, such as a dissolved gas and in combinations with water, hemoglobin, and other proteins. Like oxygen, the total amount of carbon dioxide in the blood depends on its partial pressure, i.e. pCO_2. The carbon dioxide dissociation curve shown in Figure 8.9 provides a relationship between the total amount of carbon dioxide in the blood and its pCO_2. The ordinate expresses the volume of carbon dioxide gas as a percentage of the blood volume. The volume percents are based on body conditions (BTP) of 37°C and 1 atm of pure gas. For example, there is about 50 ml of carbon dioxide at BTP in each 100 ml of blood, i.e. 50 volume per cent, at a pCO_2 of about 42 mm Hg.

The change in the carbon dioxide concentration for normal blood is very narrow. For example, venous blood has a pCO_2 of about 45 mm Hg and a corresponding concentration of 52 volume percent. Arterial blood has a pCO_2 of about 40 mm Hg and a corresponding concentration of 48 volume percent. These values can be used to show that the Henry's constant for carbon dioxide in blood for this range of pCO_2 is about 0.0022 mm Hg/mM.

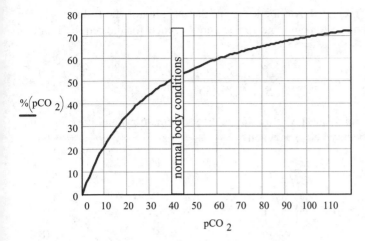

Figure 8.9   Carbon dioxide blood solubility curve.

Figure 8.8 can also be used to write shell balances on the blood and gas sides for carbon dioxide. A set of equations similar to those given by Equations 8.37 are obtained where m is now equal to zero. Henry's law is also assumed to describe the relationship between the carbon dioxide concentration in the blood and its pCO_2. The overall mass transfer coefficient, K_0, is still given by Equation 8.36 recognizing that the physical properties of carbon dioxide are to be used.

$$\frac{Q_b}{H}\frac{dpCO_{2b}}{dx} = K_0 W_b\left(pCO_{2g} - pCO_{2b}\right)$$

$$\frac{Q_g}{RT}\frac{dpCO_{2g}}{dx} = K_0 W_g\left(pCO_{2g} - pCO_{2b}\right)$$

(8.43)

As done before for oxygen, the gas side equation can be subtracted from the blood side equation and the result integrated to provide the following expression relating pCO_{2g} to pCO_{2b}.

$$pCO_{2g}(x) = \left(\frac{Q_b}{Q_g}\right)\left(\frac{RT}{H}\right)\left[pCO_{2b}(x) - pCO_{2b}^{in}\right] + pCO_{2g}^{out}$$

(8.44)

This equation may be used in the blood equation (8.43) to obtain the following differential equation for pCO_{2b}.

$$\frac{dpCO_{2b}}{dx} = \left(\frac{K_0 W_b H}{Q_b}\right)\left\{\left[\left(\frac{Q_b}{Q_g}\left(\frac{RT}{H}\right) - 1\right)\right]pCO_{2b}\right.$$

$$\left. + \left[pCO_{2g}^{out} - \left(\frac{Q_b}{Q_g}\right)\left(\frac{RT}{H}\right)pCO_{2b}^{in}\right]\right\}$$

(8.45)

When this equation is integrated over the length of the oxygenator, the following result is obtained for the blood side membrane surface area needed for the required carbon dioxide removal.

$$A_{\text{carbon dioxide}} = \frac{\alpha'}{\beta'} \ln\left[\frac{\beta' pCO_{2b}^{\text{out}} + \gamma'}{\beta' pCO_{2b}^{\text{in}} + \gamma'}\right] \qquad (8.46)$$

The constants α', β', and γ' are given by the next set of equations and have the same typical units as described earlier for oxygen.

$$\alpha' = \frac{Q_b}{K_0 H}$$

$$\beta' = \frac{Q_b}{Q_g}\left(\frac{RT}{H}\right) - 1$$

$$\gamma' = pCO_{2g}^{\text{out}} - \left(\frac{Q_b}{Q_g}\right)\left(\frac{RT}{H}\right) pCO_{2b}^{\text{in}}$$

$$\qquad (8.47)$$

8.7.6 Example Calculations for Membrane Oxygenators

The following examples illustrate the calculation of the area of a membrane blood oxygenator using the above equations.

Example 8.3 Determine the hollow fiber membrane surface area for a blood oxygenator operating under the following conditions. Assume the fibers are made from microporous polypropylene having a membrane porosity of 40%. The gas permeabilities are then based on diffusion through a stagnant layer of gas trapped within the pores of the membrane. The length of each fiber is 50 cm with a wall thickness of 50 microns. The inside diameter of a fiber is 400 microns. The blood flowrate through the lumen of the fibers in the device is 5000 ml min^{-1}, and the gas flowrate is 5000 ml min^{-1}, both at 37°C and 1 atm. The pO_2 of the entering blood is 40 mm Hg, and the exiting blood must be at a pO_2 of 95 mm Hg. The pCO_2 of the entering blood is 45 mm Hg, and the exiting blood has a pCO_2 of 40 mm Hg. The entering gas contains no carbon dioxide and is saturated with water. The pO_2 of the entering gas is, therefore, 713 mm Hg. Estimate the surface area required to meet these specifications for both oxygen and carbon dioxide. The gas side mass transfer resistance may be considered negligible because of the negligible gas solubility. Hence, the bulk of the mass transfer resistance is a result of the boundary layer formed within the blood.

SOLUTION Using the equations developed above, the membrane area calculated for oxygen uptake is about 12.3 m^2 whereas that for carbon dioxide removal is calculated to be 2.5 m^2. In this example, we must change the blood pO_2 from 40 to 95 mm Hg in order to sustain life for the adult cardiac output of 5000 ml min^{-1}. We find that the total amount of oxygen transported, Q_{oxygen}, is about 250 ml/min based on the required change in blood pO_2 levels. This is the value typically reported for an adult. A proper calculation of the membrane area for oxygen transport requires a value of m based on a representative oxygen partial pressure in the blood, see Equation 6.36 or 6.37. Therefore, we must adjust the value of m, or the reference pO_2, until the amount of oxygen transferred is the same based on the changes in the blood pO_2 and the gas pO_2, thus satisfying the mass balance on oxygen on the blood side and gas sides. For the problem stated in this example, the value of m is calculated to be 25.15 which results in a reference pO_2 of 62 mm Hg which is close to the average of the arterial and venous pO_2 levels of 67.5 mm Hg. The amount of carbon dioxide exchanged is calculated to be 289 ml/min and is the same whether based on the blood or gas sides. This value for an adult should be about 200 ml/min. Therefore, we are removing too much carbon dioxide which will tend to make the blood basic and lead to alkalosis. The *respiratory exchange ratio* (R) is defined as the ratio of carbon dioxide output to oxygen uptake and should be about 0.8. Here we see that R is about 1.17. For a given membrane area, we can control the amount of carbon dioxide removed by adjusting the pCO_2 of the incoming gas. In this example, its value was zero. The next example illustrates the effect of the entering gas pCO_2 on the carbon dioxide removal rate for a given membrane area.

Example 8.4 Use the membrane area calculated for oxygen from Example 8.3. Find the entering gas pCO_2 needed to provide an oxygen uptake of 250 ml min^{-1} and a carbon dioxide removal of 200 ml min^{-1}.

SOLUTION Equations 8.41 and 8.46 are rearranged to solve for the exiting blood partial pressures. By adjusting the entering gas pCO_2 from 0 up to 15 mm Hg, we can decrease the carbon dioxide output to the required value of 200 ml/min, while keeping the oxygen uptake at 250 ml/min, thus achieving the normal respiratory exchange ratio of 0.8. We also find that the membrane area needs to be increased by 0.3 m^2 to maintain the delivery of the same amount of oxygen. This is due to the decreased driving force for oxygen transport as a result of the carbon dioxide that was added to the feed gas.

Hollow fiber membrane blood oxygenators in use today typically have surface areas in the range from about 1.5–5 m^2. In the above examples, it was assumed that blood flowed through the lumen of the hollow fibers in fully developed laminar flow. This assumption provides a conservative, or worst case, estimate of oxygenator performance. Hence, the surface areas calculated in the above examples are several times larger than can be achieved in actual practice. Since the major resistance for oxygen transport resides in the blood, increasing the blood side oxygen mass transfer coefficient can have a significant effect on the size of the oxygenator. This can be accomplished by such methods as pulsating flow, small obstructions in the fiber to cause radial mixing, or a helical arrangement of the hollow fibers to induce secondary flows (swirls). Having the blood flow across the outside of the hollow fibers can also significantly increase the mass transfer rate of oxygen into the blood (Catapano *et al.* 1992; Vaslef *et al.* 1994; Wickramasinghe *et al.* 2005). This is a result of repetitive mixing of the blood as it crosses successive hollow fibers, effectively decreasing the boundary layer resistance to oxygen transport.

Generally, it is recommended that experiments be performed to measure the oxygen transport rates for a given oxygenator design. Mathematical models, such as the one developed above or in the papers by Vaslef *et al.* (1994) and Wickramasinghe *et al.* (2005), can then be used to correlate the results and provide accurate measurements of the blood side mass transfer coefficient. A calibrated model can then be used to explore operations under a wide range of operating conditions and to assist in the optimal development of the device. For example, Vaslef *et al.* (1994) and Wickramasinghe *et al.* (2005) were able to assess the performance of cross flow hollow fiber oxygenators using water and other blood analogues. Subsequent tests using bovine blood provided excellent comparisons between their measurements and model predictions of the oxygen transport rates.

8.8 IMMOBILIZED ENZYME REACTORS

8.8.1 Background

Another example of extracorporeal devices is the use of immobilized enzyme reactors to a species found in the blood. An enzyme is a protein that acts as a biochemical catalyst and offers great specificity in terms of the types of chemical species or substrates it acts on. Enzymes are usually named after the substrate whose reaction is catalyzed. For example, the enzyme that hydrolyzes the substrate urea is called *urease*. Note that the suffix *-ase* is usually added to a portion of the name for the substrate the enzyme acts on.

Figure 8.3 shows several reactor arrangements that are possible when an enzyme is attached to or immobilized within a solid support. Enzyme immobiliziation offers several advantages. First, immobilization keeps the enzyme out of the bulk solution, which in the case of blood returning to the body, could result in an allergic reaction to the foreign protein of the enzyme. Secondly, immobilization offers the potential to reuse the enzyme, which may, in fact, be quite expensive. Finally, in many cases, the enzyme is stabilized (less labile) when immobilized, retaining its activity for longer periods of time.

8.8.2 Examples of the Medical Application of Immobilized Enzymes

One example where immobilized enzyme reactors have been proposed is in the treatment of neonatal jaundice (Lavin *et al.* 1985; Sung *et al.* 1986). Newborns tend to have higher levels of the greenish-yellow pigment *bilirubin* than those found in adults. Bilirubin is a natural product derived from red blood cells after they have lived out their lifespan of about 120 days. It is formed from the heme portion (the four pyrole rings) of the hemoglobin molecule after removal of the iron. The bilirubin binds to plasma albumin for transport to the liver where it is finally excreted from the body in the bile fluids.

The fetus's bilirubin readily crosses the placenta and is removed by the mother's liver. However, in the period after birth, the infant's liver is not fully functional for the first week, resulting in increased levels of plasma bilirubin. In some cases, the infant's bilirubin levels are sufficiently high resulting in a *jaundiced* (yellow) appearance to the skin. High levels of plasma bilirubin can be toxic to a variety of tissues, and in these cases, jaundiced infants are commonly treated by *phototherapy* or blood transfusions. In phototherapy, the infant is placed under a blue light that converts the bilirubin to a less toxic byproduct. Phototherapy through the skin is not capable of controlling cases of severe jaundice. However, blood transfusions can replace the infant's blood with adult blood effectively diluting the infant's plasma bilirubin levels. However, blood transfusions pose their own risk, particularly infectious diseases such as hepatitis and HIV.

An alternative approach to the treatment of neonatal jaundice is the use of a bilirubin specific enzyme for removal of bilirubin from the infant's blood (Lavin *et al.* 1985; Sung *et al.* 1986). The enzyme bilirubin oxidase catalyzes the oxidation of bilirubin according to the following reaction stoichiometry.

$$\text{bilirubin} + 1/2 \ O_2 \rightarrow \text{biliverdin} + H_2O \tag{8.48}$$

Calculations indicate that the amount of oxygen needed to convert all of the bilirubin found in the blood is about 100 times less than the actual oxygen content of blood. Therefore, no external supply of oxygen is needed within the enzyme reactor to carry out this reaction. Biliverdin itself is much less toxic than bilirubin and in fact is further oxidized by bilirubin oxidase to other less toxic substances. Experiments using a water jacketed reactor (much like that in Figure 8.3a) containing bilirubin oxidase covalently attached to agarose beads showed that plasma bilirubin levels in rats decreased by 50% after 30 minutes of treatment. The rat's blood was recirculated through the 6 ml reactor volume at a flow rate of 1 ml/min. Clearly, these results indicate that an immobilized bilirubin oxidase reactor could be a new approach for the treatment of neonatal jaundice. It also shows the feasibility of using immobilized enzyme reactors for the specific removal of a harmful substance present in the blood.

Cells found in the liver and other organs carry out a wide variety of life sustaining enzymatic reactions. There is considerable interest in using these cells or their enzymes to treat liver failure and other enzyme deficiency diseases. The discussion here focuses on the use of just the key enzymes. The use of immobilized cells as bioartificial organs, perhaps using liver cells (hepatocytes), is discussed in Chapter 10. However, it is important to recognize that some of the techniques used in this chapter to design immobilized enzyme reactors are directly applicable to systems that employ immobilized cells.

Another example of an immobilized enzyme reactor that we will look at in considerably more detail is that for the removal of *heparin* (Bernstein *et al.* 1987a,b; Ameer *et al.* 1999a,b). Recall that heparin is used as an anticoagulant in extracorporeal treatments such as hemodialysis and blood oxygenators. Heparin is a large, negatively charged conjugated polysaccharide molecule that is produced by many types of cells in the body. By itself, heparin has little anticoagulant activity at the typical concentrations found in blood. However, in some regions of the body such as the liver and lungs, it is produced in greater amounts. Heparin, therefore, has an important role in preventing blood clots in the slow moving venous blood flow entering the capillaries of the lungs and liver. By combining with *antithrombin III*, it increases, by several orders of magnitude, the ability of antithrombin to remove *thrombin*. Thrombin is an enzyme that converts the plasma protein fibrinogen into fibrin leading to the fibrous mesh-like structure characteristic of a blood clot. Therefore, this synergistic combination of heparin with antithrombin III results in a powerful anticoagulant.

Over 20 million extracorporeal procedures using heparin are performed each year, and in about 15% of these, complications due to heparin arise. Certainly, removal of heparin from the blood before it is returned to the body could significantly improve the safety of these procedures. The enzyme heparinase has the ability to degrade heparin into less harmful byproducts, and one could envision an immobilized heparinase reactor for the removal of heparin from the blood returning to the patient's body.

8.8.3 Enzyme Reaction Kinetics

Our goal now is to develop relationships that can be used to describe the rates or kinetics of chemical reactions that are catalyzed by enzymes. Description of the enzyme reaction kinetics will allow us to develop mathematical models for immobilized enzyme reactors that can be used to analyze experimental data, provide information for scaleup of our devices, and explore the effects of operating conditions on reactor performance. The heparinase reactor described above will be used as a model reaction system for these discussions.

The successful development of an immobilized enzyme reactor requires knowledge of the enzyme kinetics, an understanding of the effects on the observed reaction rate of reactant or substrate diffusion, and a design equation for the specific reactor that is used (Fogler 1992). Each of these is discussed in greater detail in the following discussion.

First, we need to define the kinetics or rate of the enzyme reaction and how it depends on the concentration of the substrate (reactant). This usually entails defining the free enzyme kinetics (enzyme in solution and not on a solid support) and the kinetics after the enzyme has been immobilized on the support material. In some cases, immobilization has no effect on the intrinsic activity of the enzyme, whereas sometimes immobilization can significantly alter the kinetics of the conversion process.

Generally, the kinetics of enzyme reactions are described by the Michaelis–Menten equation. The Michaelis–Menten equation can be derived by assuming that the conversion of substrate to product occurs in two steps. In the first step, the substrate (S) combines with the enzyme (E) to form an enzyme–substrate complex ($E*S$) as shown below.

$$E + S \leftrightarrow E*S \tag{8.49}$$

In the second step, the enzyme–substrate complex ($E*S$) is converted into product (P) and free enzyme (E) which is then available to recombine with substrate.

$$E*S \rightarrow E + P \tag{8.50}$$

The rate controlling step in the above process is assumed to be the conversion of the enzyme–substrate complex to product. Accordingly, it is therefore assumed that the reaction forming the enzyme–substrate complex is at equilibrium. The rate of the enzyme reaction is then given by:

$$r_S = -\frac{dS}{dt} = \frac{dP}{dt} = k_{cat} \, E*S \tag{8.51}$$

where k_{cat} is a first order rate constant that relates the reaction rate (r_S) to the concentration of the enzyme–substrate complex, i.e. $E*S$. S and P are the respective concentrations of the substrate and the product.

From Equation 8.49 we can define the enzyme–substrate dissociation constant (reciprocal of the equilibrium constant) as $K_m = \dfrac{SE}{E*S}$ and then $E*S = SE/K_m$ with the result that Equation 8.51 becomes:

$$r_S = -\frac{dS}{dt} = \frac{dP}{dt} = k_{cat} \frac{SE}{K_m} \tag{8.52}$$

Letting E_0 represent the total concentration of the enzyme, then E_0 must equal the sum of the free enzyme concentration, i.e. E, and the amount of enzyme bound to substrate, i.e. $E*S$. Hence, we can write that $E_0 = E + E*S$. Since $E*S = SE/K_m$ we then have that $E = K_m E_0/(K_m + S)$. Substituting this result into Equation 8.52 then gives us an expression for the rate of the enzyme reaction as shown by the next equation.

$$r_S = \frac{k_{cat} E_0 S}{K_m + S} = \frac{V_{max} S}{K_m + S} \tag{8.53}$$

The reaction rate, r_S, has units of moles/(reaction volume)/time. S represents the substrate or reactant concentration in units of moles/volume. V_{max} represents the maximum reaction rate for a given total enzyme concentration E_0 [Units/(reaction volume)], where $V_{max} = k_{cat} E_0$, and k_{cat} is the reaction rate constant (moles/Units/time). Enzyme activity is commonly expressed in terms of "Units" (U), and for heparinase, a unit of activity is defined as the amount of enzyme required to degrade 1 mg of heparin/hr. The amount or mass of enzyme needed is dependent on the enzyme purity and is usually reported as so many units of activity per mg of enzyme. K_m is the Michaelis constant and may be thought of as that substrate concentration at which the reaction rate is equal to one-half the maximum rate (V_{max}). It is important to note that at high substrate concentrations the reaction rate saturates at V_{max} because in total, there are not enough available free enzyme molecules for the substrate reaction. The reaction rate is then independent of the substrate concentration and is, therefore, said to be zero order. At low substrate concentrations $(S \ll K_m)$, the reaction rate is linearly proportional to the substrate concentration and is, therefore, said to be first order.

Example 8.5 The following table of data was obtained (Bernstein *et al.* 1987b) for the kinetics of heparin degradation by heparinase in solution at 37°C and a pH of 7.4. From these data determine the value of K_m and k_{cat}.

Heparin degradation kinetics

Enzyme loading (Units ml^{-1})	Heparin concentration (mg ml^{-1})	Reaction rate (mg ml^{-1} hr^{-1})	Normalized reaction rate (mg U^{-1} hr^{-1})
24	1	20.83	0.868
	0.5	18.18	0.758
	0.25	13.70	0.571
	0.1	11.11	0.463
	0.05	7.58	0.316
43	1	38.46	0.894
	0.5	32.26	0.750
	0.25	30.30	0.705
	0.1	23.81	0.554
	0.05	16.67	0.388
57	1	50	0.877
	0.5	40	0.702
	0.25	38.46	0.675
	0.1	26.32	0.462
67	1	55.56	0.829
	0.5	52.63	0.786
	0.25	50	0.746
	0.1	35.71	0.533

Source: Bernstein et al. (1987b).

SOLUTION Note that it is convenient for this type of analysis to divide the enzyme reaction rate by the corresponding enzyme concentration, thus expressing the rate on a per unit amount of enzyme that is present in the reactor. One unit of heparinase (1 U) is defined as the amount of enzyme required to degrade 1 mg heparin/hr. Hence, we can rewrite Equation 8.53 as follows:

$$R = \frac{r_S}{E_0} = \frac{k_{cat} S}{K_m + S}$$

and this equation can be rearranged to give:

$$\frac{1}{R} = \left(\frac{K_m}{k_{cat}} \right) \frac{1}{S} + \frac{1}{k_{cat}}$$

The above equation shows that if we plot the above data as $1/R$ vs. $1/S$, the plot should be linear with an intercept equal to $1/k_{cat}$ and a slope equal to K_m/k_{cat}. This is called the *Lineweaver–Burke method* for analyzing enzyme reaction data. Using this method on the data in the above table for heparin degradation by heparinase, we find that $K_m = 0.078$ mg/ml and $k_{cat} = 0.891$ mg/U/hr. As shown in Figure 8.10, we find that the Michaelis–Menten model provides excellent representation of the heparin degradation kinetics.

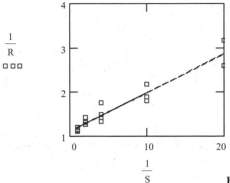

Figure 8.10 Lineweaver–Burke plot for Example 8.5.

8.8.4 Reaction and Diffusion in Immobilized Enzyme Systems

When an enzyme is immobilized within a supporting structure, the substrate has to diffuse into the material in order to come into contact with the immobilized enzyme. Because the substrate diffuses and then reacts within the support structure, a concentration gradient is established within the support structure and the reaction rate depends on location. A reaction–diffusion model is, therefore, needed to describe how the substrate diffuses through the porous support particle and reacts through the action of the immobilized enzyme. We also need to approximate the support geometry, and here, we will assume it is spherical. Extensions to other support geometries is straightforward once a given geometry has been examined in detail. In some cases, we may even need to know how the enzyme is distributed throughout the support particle (Bernstein *et al.* 1987a). In the absence of such information, we simply assume the enzyme is uniformly distributed.

Diffusion through the support material also requires that we know the substrate diffusivity (D) in the bulk or free solution (see Figure 5.4). We also need to know how the solute diffusivity is affected by the properties of the support itself, specifically the porosity, pore size, and tortuosity. For substrates that have molecular dimensions that are comparable to the pore size, we will also need to include the effects of steric exclusion and hindered diffusion as discussed earlier in Section 5.6. These factors combine to define the effective diffusivity (D_e) of the substrate in the support particle as given by the next equation.

$$D_e = \frac{\varepsilon D}{\tau} \omega_r \qquad (8.54)$$

In this equation, ε represents the porosity of the support, τ is the tortuosity of the pores, and ω_r represents the reduction of the substrate diffusivity due to the proximity of the pore wall (i.e. the bracketed term in Equation 5.55). In this discussion, the substrate is already in the pore so there is no steric exclusion. We will see that steric exclusion shows up in the surface boundary condition of the differential equation that describes the reaction–diffusion phenomena occurring in the porous support.

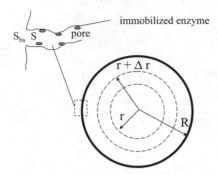

Figure 8.11   Shell balance on a spherical immobilized enzyme particle.

Figure 8.11 illustrates a spherical shell of thickness Δr lying within the support particle containing a uniformily distributed and immobilized enzyme. The particle itself is porous and we can write the following steady state shell balance on the substrate.

$$4\pi r^2 D_e \frac{dS}{dr}\bigg|_{r+\Delta r} - 4\pi r^2 D_e \frac{dS}{dr}\bigg|_r = 4\pi r^2 \Delta r \left(\frac{V_{max} S}{K_m + S} \right) \tag{8.55}$$

This equation simply states that the amount of substrate reacted within the shell volume is equal to the difference in the amount of substrate entering and leaving the shell volume by diffusion. After dividing by $4\pi r^2 \Delta r$, and taking the limit as Δr approaches zero, we obtain the reaction–diffusion equation for the substrate in the particle.

$$D_e \frac{d^2 S}{dr^2} + \frac{2 D_e}{r} \frac{dS}{dr} = \frac{V_{max} S}{K_m + S} \tag{8.56}$$

The boundary conditions are:

$$\text{BC1:} \quad r = 0, \ \frac{dS}{dr} = 0$$
$$\text{BC2:} \quad r = R, \ S = KS_{bs} \tag{8.57}$$

Once again, K represents the partition coefficient which may include effects other than simple steric exclusion, for example electrostatic attraction or repulsion due to the presence of surface charges on the substrate and the particle. It is defined as the ratio of the substrate concentration on the pore side of the support surface (S) to that on the surface on the bulk side (S_{bs}) as shown in Figure 8.11. The substrate concentration at the surface of the support (S_{bs}) may also differ from the bulk substrate concentration (S_b) because of external mass transfer effects. This is discussed later.

Solution of Equations 8.56 and 8.57 is facilitated through the definition of the following dimensionless variables: $s' = S/KS_{bs}$ and $r' = r/R$. Upon substitution of this dimensionless substrate concentration and radial dimension there is obtained:

$$\frac{d^2 s'}{dr'^2} + \frac{2}{r'} \frac{ds'}{dr'} = 9\phi^2 \frac{s'}{1 + \beta s'} \tag{8.58}$$

The dimensionless boundary conditions are given by:

$$BC1: \quad r' = 0, \frac{ds'}{dr} = 0$$
$$BC2: \quad r' = 1, s' = 1 \tag{8.59}$$

Two additional parameters ϕ and β also result.

$$\phi = \frac{R}{3}\left[\frac{V_{max}}{K_m D_e}\right]^{1/2}$$

$$\beta = \frac{S_{bs} K}{K_m} \tag{8.60}$$

The quantitity ϕ is especially important in reaction–diffusion problems and is known as the *Thiele modulus*. The square of this quantity represents the ratio of the substrate reaction rate to its diffusion rate. Its magnitude allows one to determine whether the overall reaction rate is *reaction limited* (small ϕ, minimal intraparticle substrate concentration gradient) or *diffusion limited* (large ϕ, significant intraparticle substrate concentration gradient). The factor $R/3$ is the ratio of the support volume to its external surface area. For non-spherical geometries, one can replace $R/3$ with this ratio, i.e. V/A. For diffusion in a planar material, $R/3$ is equal to L, the thickness. β is a dimensionless Michaelis constant. Large values of β indicate a zero order reaction, whereas $\beta \to 0$ indicates a first order reaction.

8.8.5 Solving the Immobilized Enzyme Reaction–Diffusion Model

The solution to Equation 8.58 for the dimensionless substrate concentration profile in the immobilized enzyme support is usually defined in terms of the effectiveness factor represented by the symbol η. Because of substrate reaction and diffusional effects, the substrate concentration will generally decrease in the radial direction as one enters the support. This radial decrease in substrate concentration in the support will result in a decrease in the volume averaged reaction rate for the support in comparison to the reaction rate that would be possible if the substrate concentration was uniform throughout the support and equal to its surface concentration. The effectiveness factor is used to describe this decrease in observed reaction rate and is defined by the following expression.

$$\eta = \frac{\text{observed particle rate}}{\substack{\text{rate obtained with no} \\ \text{concentration gradient} \\ \text{within the particle}}} = \frac{4\pi R^2 D_e \left.\frac{ds}{dr}\right|_{r=R}}{\frac{4}{3}\pi R^3 \left(\frac{V_{max} S|_R}{K_m + S|_R}\right)} = \frac{\left.\frac{ds'}{dr'}\right|_{r'=1}}{\left(\frac{3\phi^2}{1+\beta}\right)} \tag{8.61}$$

The effectiveness factor (η) varies from 0 (diffusion limited) to 1 (reaction limited). Determination of the effectiveness factor using the non-linear Michaelis–Menten rate equation requires a numerical solution of Equations 8.58 and 8.59 for various values of β. It is also important to point out that η, which depends on S_{bs}, will depend on position within the reactor. So the solution for η will be dependent on the local value of S_{bs} as found by the solution to the reactor design equation to be discussed below.

Example 8.6 Determine the effectiveness factor for the case where the Thiele modulus and the dimensionless Michaelis constant are both unity. Assume the particle is spherical and the kinetics are described by the Michaelis–Menten model.

SOLUTION For this case Equation 8.58 must be solved numerically. This can be accomplished by first letting $y = \dfrac{ds'}{dr'}$ and $\dfrac{dy}{dr'} = \dfrac{d^2 s'}{dr'^2}$. Substituting these equations into Equations 8.58 and 8.59 reduces the original second order differential equation into two first order differential equations as shown below.

$$\frac{dy}{dr'} = -\frac{2}{r'}y + \frac{9\phi^2 s'}{1+\beta s'} \text{ and } \frac{ds'}{dr'} = y$$

$$\text{BC1'}: \ r' = 0, \ y = 0$$
$$\text{BC2'}: \ r' = 1, \ s' = 1$$

Although there are many mathematical software packages that can easily solve this problem, a relatively simple algorithm to implement is based on *Euler's method*. First, we divide the integration interval from $r' = 0$ to $r' = 1$ into N equally sized segments, i.e. $\Delta r' = \dfrac{1}{N}$. Then, at the i-th point we have that $r_i' = i\Delta r'$, and we can approximate the derivatives as follows; $\dfrac{dy}{dr'} \approx \dfrac{y_{i+1} - y_i}{\Delta r'}$ and $\dfrac{ds'}{dr'} \approx \dfrac{s_{i+1}' - s_i'}{\Delta r'}$. After substituting these relationships into the above differential equations and solving for the values of y_{i+1} and s_{i+1}', we obtain:

$$y_{i+1} = y_i + \left(\frac{9\phi^2 s_i'}{1+\beta s_i'} - \frac{2}{i\Delta r'} \right)\Delta r' \text{ and } s_{i+1}' = s_i' + y_i \Delta r_i'$$

When i is equal to zero, we are at the center of the spherical pellet and we use the known value of s' and y' from the boundary conditions. The above two equations can then be solved for the values of y_1 and s_1'. These values are then inserted into the above two equations, and the values of y and s' are then calculated at the next position, and this process is repeated until we reach the surface of the spherical particle. However, one complication should be pointed out about this particular problem which is also known as a two-point boundary value problem. Note that we do not know explicitly what the value of s' is at the center of the particle, we only know that $\dfrac{ds'}{dr'}\bigg|_{r'=0} = 0$ and that $s' = 1$ at $r' = 1$. Hence, to start the solution using the algorithm described above we need to make a guess of what s' is at $r' = 0$. Then we implement the algorithm described above and march through the solution to $r' = 1$. Then, the resulting value of s' calculated at $r' = 1$ is compared to the required value of $s' = 1$ at $r' = 1$. If the calculated value of s' is greater than one, then the value of s' assumed at $r' = 0$ is too high and we must decrease the value assumed at $r' = 0$. If s' is less than one, then the value of s' assumed at $r' = 0$ is too low and we must increase the value assumed at $r' = 0$. We then repeat this process until convergence is obtained. After implementing the algorithm described above to solve this problem, we find that the value of η for these conditions is 0.86. Figure 8.12 shows the calculated substrate concentration profile within the spherical immobilized enzyme particle.

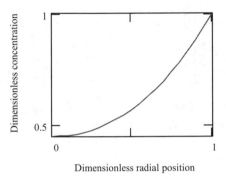

Figure 8.12   Dimensionless concentration profile, Example 8.6.

8.8.6 Special Case of a First Order Reaction

If $K_m > K\,S_{bs}$, then the Michaelis–Menten kinetic model becomes that of a first order reaction. For a first order reaction, Equations 8.58 and 8.59 may be solved analytically resulting in the following relationship between η and ϕ. This relationship between the effectiveness factor and the Thiele modulus for a first order reaction is also shown in Figure 8.13. Note that this solution is independent of the particle surface concentration and, hence, position in the reactor.

$$\eta = \frac{1}{\phi}\left(\frac{1}{tanh\ 3\phi} - \frac{1}{3\phi}\right) \tag{8.62}$$

For large values of ϕ, the value of η approaches $1/\phi$.

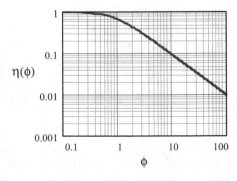

Figure 8.13   Effectiveness factor for a first order reaction.

Example 8.7 For the conditions of the previous example, calculate the effectiveness factor assuming the enzyme reaction described by Michaelis–Menten kinetics can be approximated as a first order reaction.

SOLUTION   The Thiele modulus is equal to unity. From Figure 8.13, we see that the effectiveness factor is about 0.70 or so. From Equation 8.62, we can calculate the value of the effectiveness factor to be 0.67.

8.8.7   Observed Reaction Rate

With the effectiveness factor known, we can write the observed reaction rate for the immobilized enzyme as follows:

$$r_s = \eta \frac{V_{max} \, KS_{bs}}{K_m + KS_{bs}} \tag{8.63}$$

where S_{bs} is the substrate concentration at the support surface and η accounts for the reduction in the reaction rate due to the internal substrate concentration gradient. It is important to note that the above expression for the reaction rate is not explicitly dependent on the substrate concentration within the immobilized enzyme particle.

8.8.8   External Mass Transfer Resistance

Flow of blood or other fluids around the immobilized enzyme support results in the formation of a thin boundary layer of fluid along the surface of the enzyme particle. This boundary layer provides an additional resistance for mass transfer that must be accounted for. An external mass transfer coefficient (k_m) is defined to account for this flow-induced external mass transfer resistance.

Conservation of mass for a given particle requires that the following steady state relationship hold between the external mass transfer rate and the reaction rate within the support.

$$4\pi R^2 k_m \left(S_b - S_{bs} \right) = \frac{4}{3} \pi R^3 \eta \left(\frac{V_{max} \, KS_{bs}}{K_m + KS_{bs}} \right) \tag{8.64}$$

This equation may be solved for the substrate concentration on the surface of the particle (S_{bs}) in terms of the bulk substrate concentration (S_b). Recall it is this surface concentration that is used in the calculation of the effectiveness factor as discussed above. From Table 5.2 the following equation describes the external mass transfer coefficient in packed beds.

$$\frac{k_m}{v_0} = 1.17 \left(\frac{dv_0}{v} \right)^{-0.42} \left(\frac{v}{D} \right)^{-0.67} \tag{8.65}$$

This equation is also equivalent to $Sh = 1.17 \, Re^{0.58} \, Sc^{0.33}$, where $Sh = k_m d/D$, $Re = dv_0/v$, $Sc = v/D$, and d is the particle diameter. Mass transfer correlations for other situations may be found in Cussler (1984) or Bird et al. (2002). In Equation 8.65, v_0 is the superficial velocity of the fluid, i.e. the volumetric flowrate of the fluid divided by the cross-sectional area of the vessel that contains the support particles of diameter d. The kinematic viscosity (v) is defined as the ratio of the fluid viscosity and the fluid density, i.e. μ/ρ. Generally, one would want to operate the reactor in such a manner that the external mass transfer resistance is negligible, hence $S_{bs} = S_b$. This can be accomplished by high bulk flowrates in the case of a plug flow reactor or intense mixing in a well-mixed reactor.

8.8.9 Reactor Design Equations

Finally, we need a reactor design equation that, in concert with the above information on the enzyme kinetics and diffusional effects, provides a relationship as to how the substrate concentration varies within the reaction volume. One usually assumes that the reaction mixture either flows through the reactor in plug flow, leading to a change in substrate concentration in the direction of flow, or that the reaction volume is well-mixed, resulting in a uniform substrate concentration throughout the reaction volume.

8.8.9.1 Packed bed reactor Figure 8.14 shows a packed bed immobilized enzyme reactor. Q_b is the entering flowrate of the fluid, which could, for example, be blood or plasma. The substrate concentration entering the reactor is represented by S_b^{in}. A steady state shell balance may also be performed on a section of the reactor volume ($A_{xs}\Delta x$) where A_{xs} is the cross-sectional area of the reactor. It is assumed that the fluid flows through the reactor in plug flow, hence the velocity does not change with radial position.

$$Q_b S_b \big|_x - Q_b S_b \big|_{x+\Delta x} = (1-\varepsilon_R)\Delta x\, A_{xs}\, \eta \left(\frac{V_{max} K S_{bs}}{K_m + K S_{bs}} \right) \tag{8.66}$$

In this equation, ε_R represents the void volume in the reactor, i.e. the volume of the reactor not occupied by the immobilized enzyme particles. Dividing by Δx and taking the limit as $\Delta x \to 0$, results in the following differential equation:

$$Q_b \frac{dS_b}{dx} = -(1-\varepsilon_R) A_{xs}\, \eta \left(\frac{V_{max} K S_{bs}}{K_m + K S_{bs}} \right) \tag{8.67}$$

$$BC1: \ x = 0, \ S_b = S_b^{in}$$

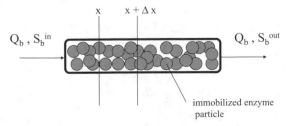

Figure 8.14 Shell balance for a packed bed immbobilized enzyme reactor.

This equation may be solved numerically along with Equation 8.64 which provides the relationship between S_b and S_{bs} due to the effects of external mass transfer resistance. Since β also depends on S_{bs} (see Equation 8.60) which depends on x, we also need to evaluate the value of η as a function of x, as well. For the special case where one can assume that η is approximately constant, and external mass transfer effects are negligible, i.e. $S_{bs} = S_b$, the above equation can be integrated analytically for a reactor of length L to give the following result.

$$K_m \ln\left(\frac{S_b^{out}}{S_b^{in}}\right) + K\left(S_b^{out} - S_b^{in}\right) = -\frac{(1-\varepsilon_R)A_{xs}LKV_{max}\eta}{Q_b} \tag{8.68}$$

Often the operation of a chemical reactor is defined in terms of the residence time (τ), i.e. the time the reacting fluid spends in the reactor. For a packed bed plug flow reactor, the residence time defined in terms of the reactor void volume is given by:

$$\tau = \frac{A_{xs}L\varepsilon_R}{Q_b} \tag{8.69}$$

In addition, we can define the conversion of the substrate, i.e. X_S, as $1 - \dfrac{S_b^{out}}{S_b^{in}}$. With these additional definitions, Equation 8.68 may be written as:

$$K_m \ln\left(1 - X_S\right) - K S_b^{in} X_S = -\frac{\left(1-\varepsilon_R\right)KV_{max}\eta}{\varepsilon_R}\,\tau \tag{8.70}$$

For given values of the enzyme kinetics and size of the reactor or residence time, Equations 8.68 and 8.70 can be solved for the exiting substrate concentration and the substrate conversion.

8.8.9.2 Well-mixed reactor The other type of commonly used immobilized enzyme reactor is the well-mixed reactor shown in Figure 8.15. Perfect mixing results in a uniform substrate concentration throughout the reactor volume. The exiting concentration of the substrate is also the same as that within the reactor, hence the concentration everywhere in the bulk fluid of the reactor is S_b^{out}. A steady state substrate mass balance provides the following equation that can be solved for the exiting substrate concentration.

$$Q_b S_b^{in} = (1-\varepsilon_R)V_{reactor}\eta\left(\frac{V_{max}KS_{bs}^{out}}{K_m + KS_{bs}^{out}}\right) + Q_b S_b^{out} \tag{8.71}$$

This equation may be solved, along with Equation 8.64, and a value of the effectiveness factor evaluated at S_{bs}^{out}, to determine the exiting substrate concentration S_b^{out}.

The substrate conversion is also defined by $X_S \equiv 1 - \dfrac{S_b^{out}}{S_b^{in}}$. In the absence of external mass transfer effects, i.e. $S_{bs}^{out} = S_b^{out}$, the above equation simplifies to:

$$S_b^{in} = \frac{\left(1-\varepsilon_R\right)}{\varepsilon_R}\,\eta\left(\frac{V_{max}KS_b^{out}}{K_m + KS_b^{out}}\right)\tau + S_b^{out} \tag{8.72}$$

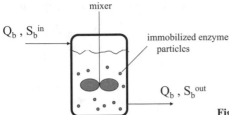

mixer

Q_b , S_b^{in}

immobilized enzyme
particles

Q_b , S_b^{out}

Figure 8.15   A well-mixed immobilized enzyme reactor.

Example 8.8 The following experimental data were obtained for an immobilized heparinase reactor (Bernstein *et al.* 1987b) for two different inlet heparin concentrations and different enzyme loadings. Heparinase was immobilized on a cross-linked agarose support. The flow pattern in the reactor was well-mixed.

Experimental data for a heparin reactor

Total support volume (ml)	Enzyme loading (U ml^{-1}) in the support	Substrate conversion (X_s) S_b^{in} = 0.2 or 0.5 mg ml^{-1}
80	100	0.58 or 0.54
85	150	0.66 or 0.62
96	120	0.69 or 0.55
90	130	0.63 or 0.58

Source: Bernstein *et al.* (1987b).

The diffusivity of heparin is 1.2×10^{-6} cm^2 sec^{-1}. The porosity of the agarose support was 0.92, and the tortuosity was unity. The radius of a heparin molecule is 1.5 nm, and the pores in the agarose support have a radius of 35 nm. The partition coefficient, K, was found from experiments to be equal to 0.36. The agarose support had a radius of 0.0112 cm. The flowrate through the reactor was 120 ml min^{-1}. The external mass transfer coefficient was estimated to be 0.154 cm min^{-1}. Using the kinetic parameters obtained in Example 8.5, use the mathematical model for a well-mixed immobilized enzyme reactor to predict the substrate conversions. Compare the results with the experimental conversions in the above table.

SOLUTION The effective diffusivity of heparin within the enzyme particle is calculated to be $D_e = 1.01 \times 10^{-6}$ cm^2 sec^{-1}. The Thiele modulus, given by Equation 8.60, depends on the enzyme loading since $V_{max} = k_{cat} E_0$. The table below summarizes the calculated Thiele modulus for each of the four enzyme loadings.

Enzyme loading (U ml^{-1} of particle)	Thiele modulus (ϕ)
100	2.098
150	2.569
120	2.298
130	2.392

At this point, we need to calculate the effectiveness factor which also depends on $\beta = \dfrac{S_{bs}\,K}{K_m}$ and the unknown particle surface concentration, i.e. S_{bs}, that is also related to the unknown bulk or exiting substrate concentration, i.e. S_b^{out}. Since we need S_{bs} to start this process and it is unknown, we first assume that the reaction is first order in order to start the calculation. We can then calculate the effectiveness factor, i.e. η, using Equation 8.62, or we can obtain these values from Figure 8.13. From Figure 8.13, we see that these effectiveness factors assuming a first order reaction are on the order of $\eta \sim 0.45$. Next, we solve Equation 8.71 for the exiting bulk substrate concentration, i.e. S_b^{out}, and substitute that result into Equation 8.64 thus providing the following relationship between the inlet substrate concentration, i.e. S_b^{in}, and the substrate surface concentration, i.e. S_{bs}^{out}.

$$\left(S_b^{in} - S_{bs}^{out}\right) - \left[\frac{\eta\, k_{cat}\, E_0\, K\, S_{bs}^{out}}{K_m + K\, S_{bs}^{out}}\right]\left[\frac{V_{reactor}\left(1 - \varepsilon_R\right)}{Q_b} + \frac{R}{3\,k_m}\right] = 0$$

The above is a non-linear algebraic equation that can then be solved using Newton's method (see Example 2.10) for the given conditions to find the value of the substrate concentration at the surface of the immobilized enzyme particle. Once the value of S_{bs}^{out} has been found, then the bulk exiting substrate concentration, i.e. S_b^{out}, can be found from Equation 8.71. Recall that the first time through this process, the effectiveness factors were found assuming the reaction is first order. Once new values of S_{bs}^{out} have been found, the effectiveness values can be updated using the method discussed in Example 8.6 for Michaelis–Menten kinetics. This process is then repeated until convergence is obtained. The table below summarizes the converged values of the effectivenss factors for the two inlet substrate concentrations and the four enzyme loadings.

Calculated effectiveness factors

Enzyme loading (U ml^{-1} particle^{-1})	$S_b^{in} = 0.2$ mg ml^{-1}	$S_b^{in} = 0.5$ mg ml^{-1}
100	0.52	0.62
150	0.43	0.51
120	0.47	0.56
130	0.46	0.54

The table below summarizes the resulting values obtained for the substrate surface and exiting bulk concentrations for the two inlet substrate concentrations and the four enzyme loadings.

Calculated surface and bulk substrate concentrations

Enzyme loading (U ml^{-1} particle^{-1})	$S_b^{in} = 0.2$ mg ml^{-1}		$S_b^{in} = 0.5$ mg ml^{-1}	
	Surface	Bulk	Surface	Bulk
100	0.07	0.074	0.197	0.208
150	0.056	0.061	0.154	0.166
120	0.057	0.061	0.156	0.166
130	0.057	0.061	0.159	0.169

The table below compares the predicted conversions to the experimental substrate conversions for the two inlet substrate concentrations and the four enzyme loadings. Figure 8.16 also shows a parity plot of the predicted conversions versus the experimental conversions. We see that the model's predicted conversions compare rather well with the experimental conversions.

Comparison of predicted and experimental substrate conversions

Enzyme loading (U ml^{-1} particle^{-1})	$S_b^{in} = 0.2$ mg ml^{-1}		$S_b^{in} = 0.5$ mg ml^{-1}	
	Predicted	Experimental	Predicted	Experimental
100	0.63	0.58	0.59	0.54
150	0.70	0.64	0.67	0.62
120	0.70	0.69	0.67	0.55
130	0.69	0.63	0.66	0.58

8.9 AFFINITY ADSORPTION

Adsorption is the process where a substance in a solution literally sticks to the surface of a solid. Adsorption is not the same as absorption. In absorption, the substance goes into the interior of another substance which can be a solid or another liquid. There are two types of adsorption known as either physical or chemical adsorption. In physical adsorption, the substance binds to the solid surface through relatively weak physical interactions between the substance and the solid. The adsorbed substance retains its chemical structure. In chemical adsorption, chemical bonds are actually formed between the adsorbed substance and the solid material.

Adsorption processes can be very selective in terms of the interaction between the solid surface and the substances found in the solution. For example, in affinity adsorption, another substance known as a *ligand* is attached to the surface of a solid material by chemical means. The ligand has a very specific binding site for a substance that is found in the solution as shown in Figure 8.17. Examples of various ligands and the substances they act on that are used in biomedical applications are summarized in Table 8.8.

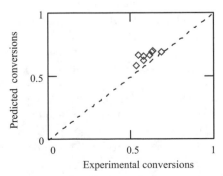

Figure 8.16   A comparison of experimental conversions ($X_{s\,exp}$) to the predicted heparin conversions (X_s).

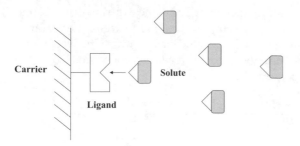

Figure 8.17   Ligand and solute binding.

Table 8.8 Examples of ligands and the substances they bind with

Ligand	Substance that binds with the ligand
Enzyme inhibitor	Enzyme
Antigen	Antibodies
Complementary nucleotide sequence	Nucleic acids
Cellular receptors	Hormones
Carrier protein	Vitamin

An interesting example of the use of affinity adsorption is described in Karoor *et al.* (2003). As mentioned in their paper, there is great interest in using animal organs as a means to address the shortage of human donor organs. This use of animal organs is also known as *xenotransplantation*, and pigs are receiving the most attention because of their adequate supply, and also the size of their organs are comparable to those in humans.

However, transplantation of animal organs into humans leads to a severe antibody-mediated reaction to the foreign tissue, the first phase of which is also known as hyperacute rejection.[3] For this discussion, antibodies are proteins produced by certain cells of the immune system that bind with high specificity to molecules known as antigens. An antigen is just a chemical entity found in the foreign tissue that the immune system's antibodies recognize and then bind to.

These antibodies found in the recipient of a xenotransplant, also known as preformed natural antibodies, bind to the endothelial cells of the xenograft. These antibodies that bind to the endothelial cells then activate the complement system which destroys the transplanted organ within several hours of the transplant. The hyperacute rejection process, for the most part, involves a particular type of antibody known as IgM. A few percent of the IgM antibody population consists of xenoreactive IgM antibodies that bind with high specificity to α-galactosyl structures (sugar-like molecules) that are expressed by non-primate mammalian and New World monkey cells. The selective removal of these α-galactosyl specific IgM antibodies prior to xenotransplantation could eliminate the hyperacute rejection of the transplanted organs.

Figure 8.18 illustrates an affinity adsorption system for removing α-galactosyl reactive IgM antibodies (Karoor *et al.* 2003). Blood, either from the body, or from a

---

[3] A brief discussion on the immune system can be found in Section 10.2.

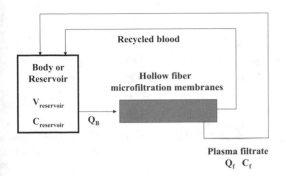

Plasma filtrate
Q_f   C_f   **Figure 8.18**   Affinity adsorption system.

reservoir in the case of in vitro experiments, flows through a microfiltration hollow fiber membrane cartridge. As the blood flows through the inside of the hollow fibers, plasma is filtered across the hollow fiber membrane as a result of the transmembrane pressure difference. Recall that this filtration of blood is also known as plasmapheresis. The exiting blood and the plasma filtrate are then returned to the body or the reservoir. The pores of the hollow fiber membranes have diameters in the range from 0.2–0.5 microns. The diameters of these pores are many times larger than the IgM molecule. The ligand, which in this case is α-galactosyl, was immobilized on all accessible surfaces of the hollow fiber microfiltration membrane, i.e. the inner, outer, and internal surfaces. As the blood and filtered plasma flows through the hollow fiber membranes, the reactive IgM antibodies bind with the α-galactosyl ligand that are bound to the membrane surfaces as shown in Figure 8.17.

We can use a simple well-mixed single compartment model to describe how the concentration of reactive IgM changes with time in the body or the reservoir (Karoor *et al.* 2003). Letting $V_{reservoir}$ denote the reservoir volume or the IgM distribution volume in the body, and $C_{reservoir}$ the IgM concentration, we can write the following mass balance on IgM:

$$V_{reservoir}\frac{dC_{reservoir}}{dt}=-Q_f\left(C_{reservoir}-C_f\right) \tag{8.73}$$

For the case where all of the IgM is bound to the α-galactosyl ligand as it passes through the hollow fiber membranes, then $C_f = 0$ and the above equation can be integrated to give the relative concentration of IgM (χ) in the reservoir as a function of time.

$$\chi=\frac{C_{reservoir}}{C_{reservoir}^0}=e^{-\frac{Q_f t}{V_{reservoir}}} \tag{8.74}$$

where $C_{reservoir}^0$ is the initial concentration of IgM. The filtration flow or plasmapheresis, i.e. Q_f, can be estimated from the following equation (Zydney and Colton 1986, also see problem 5.11 for its derivation):

$$\frac{Q_f}{Q_{b\,in}}=1-\exp\left[-0.90\,\beta\ln\frac{C_w}{C_b\left(0\right)}\right] \tag{8.75}$$

where $\beta = \dfrac{2}{3}\left(\dfrac{a^2 L}{R^3}\right)^{\frac{2}{3}}$, $Q_{b\text{ in}}$ is the inlet blood flow rate, $C_b(0)$ is the bulk red blood cell volume fraction at the inlet (0.40), C_W is the red blood cell volume fraction at the membrane surface ($= 0.95$), a is the red blood cell radius (four microns), L is the length of the hollow fibers, and R is the inside radius of the hollow fibers.

The characteristic time for the IgM adsorption process is defined as $\tau_s \equiv \dfrac{V_{\text{reservoir}}}{Q_f}$, and the number of reservoir volumes that have been filtered (RF) in a period of time (t) through the affinity adsorption membrane system is, therefore, given by the next equation.

$$RF = \text{reservoir volumes filtered} = \frac{Q_f t}{V_{\text{reservoir}}} \qquad (8.76)$$

Results of the studies by Karoor et al. (2003) show that greater than 90% removal of the IgM antibodies (i.e. $\dfrac{C_{\text{reservoir}}}{C^0_{\text{reservoir}}} < 0.10$) can be obtained after three volumes of the plasma have been filtered through the affinity adsorption system.

In designing an affinity adsorption system, one must be certain that the amount of ligand or solute to be removed, i.e. $V_{\text{reservoir}} C^0_{\text{reservoir}}$, is less than the maximum adsorbed solute concentration that is possible for the membrane system, i.e. $V_m Q_m$, where V_m is the membrane volume and Q_m is the solute binding capacity of the membrane. For a human scale device to remove xenoreactive IgM antibodies, the reservoir volume would be the volume of plasma in the body, which is 3000 ml. The IgM concentration is about 0.033 mg/ml so the total amount of IgM to be removed is on the order of 100 mg. IgM binding capacities for the systems studied by Karoor et al. (2003) had total binding capacities on the order of several hundred mg of IgM.

Example 8.9 Karoor et al. (2003) obtained the following data shown in the table below for the relative concentration of IgM as a function of the reservoir volumes filtered. Six hundred nylon hollow fiber membranes were used having an inner and outer diameter of 330 and 550 microns respectively. The active length of the hollow fibers was 10.2 cm, and the total luminal surface area was 718 cm². The filtration flowrate (Q_f) was 50 ml min⁻¹, and the reservoir volume was 1200 ml. The initial IgM concentration was 0.019 mg ml⁻¹. Compare the above model developed for an affinity adsorption system with the results obtained by Karoor et al. (2003). Also, calculate the predicted fractional filtrate yield, i.e. $Q_f/Q_{b\text{ in}}$.

Reservoir volumes filtered, $Q_f t/V_{reservoir}$	Relative concentration, $\dfrac{C_{reservoir}}{C^0_{reservoir}}$
0.025	0.89
0.13	0.68
0.25	0.58
0.55	0.41
0.83	0.30
1.1	0.25
1.38	0.22
1.63	0.21
1.9	0.20
2.20	0.19
2.5	0.16

SOLUTION From Equation 8.75, we calculate that the fractional filtrate yield $(Q_f/Q_{b\ in})$ is equal to 0.23. From Equation 8.74, we can calculate the relative concentration of the IgM antibody in the reservoir. Figure 8.19 shows a comparison between the affinity adsorption model developed above and the results from the above table. The comparison between the model and the data is quite good.

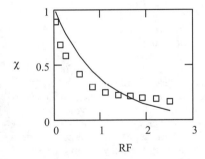

Figure 8.19   Relative reservoir concentration as a function of reservoir volumes filtered.

PROBLEMS

1. Estimate the clearance and fractional % removal of urea from a flat-plate dialyzer for both cocurrent and countercurrent flow under the following conditions:

urea diffusivity	=	2×10^{-5} cm^2 sec^{-1}

4 flat plate membranes each with the following dimensions:

dialyzer length	=	85 cm
dialyzer width	=	15 cm
blood flow rate	=	200 ml min^{-1} (total)
dialyzer flow rate	=	500 ml min^{-1} (total)
mass transfer resistances		
membrane	=	25 min cm^{-1}
blood-side	=	8 min cm^{-1}
dialysate side	=	35 min cm^{-1}

2. Estimate the urea flux [$(NH_2)_2CO$] (g m^{-2} hr^{-1}) through a cellulosic membrane that is 0.001 inches in thickness. The blood at the membrane surface has a urea concentration of 1 mg ml^{-1}. Assume the dialysis fluid on the other side of the membrane surface is free of urea. The void volume of the cellulosic membrane is 0.40 and the membrane tortuosity is 2.0. The pores in the membrane have a diameter such that they exclude all molecules larger than a molecular weight of 20,000 Daltons.

3. The following data for the mass transfer coefficient of water flowing within hollow fibers was reported by Yang and Cussler (1986). Compare their results with the correlation given by Equation 5.45. Show first that the $Pe = Re\ Sc$.

$Pe = \dfrac{V D^2_{\text{hollow fiber}}}{DL}$	$Sh = \dfrac{k_m D_{\text{hollow fiber}}}{D}$
30	4.8
50	6.0
100	7.6
300	9.8
500	13.0
1000	18.0
3000	30.0

4. The following data for the antibiotic imipenem were obtained from six patients suffering multiorgan failure (Hashimoto *et al.* 1997. *ASAIO Journal.* 43: 84–88, see also problem 10 from Chapter 7). All patients were anuric because of complete renal failure and were on continuous venovenous hemodialysis. The antibiotic was first delivered by an intravenous infusion pump for 30 minutes resulting in an initial plasma concentration of 32.37 μg ml^{-1}. Following the infusion, blood samples were taken over the next 12 hours. In addition to the blood samples, drug concentrations in the blood exiting the hemodialysis unit were also taken. The dialysate entering the hemodialyzer was free of drug, and the concentration of the drug in the dialysate leaving the dialyzer was also measured. The blood flowrate through the hemodialyzer was set at 60 ml min^{-1}, and the dialysate flow was set at 20 ml/min. The hemodialyzer had a total surface area of 0.5 m^2. Using this data, develop a model that describes the drug removal process by the patient's body and the hemodialyzer. Carefully state all of your assumptions. Determine key parameters such as the total body clearance, the dialyzer clearance, and the overall mass transfer coefficient for the dialyzer. Compare your model with the actual data. The table below summarizes the drug concentrations in the blood entering the dialyzer, the blood leaving the dialyzer, and the exiting dialysate.

Time, Minutes	Blood drug concentration entering dialyzer, μg ml^{-1}	Blood drug concentration leaving dialyzer, μg ml^{-1}	Drug concentration in dialysate leaving dialyzer, μg ml^{-1}
0	32.47	23	21
30	24	20.5	14
60	18.5	16	12
120	13	11	8.8
180	10	8	6.4
360	4.6	3.8	2
540	2	1.6	1.4
720	1.12	0.9	0.6

5. At the start of CAPD using a 2.5 wt % glucose solution, the filtration rate of water from the plasma was determined to be about 6 ml min^{-1}. Estimate the hydraulic conductance (L_p in ml hr^{-1} m^{-2} mm Hg^{-1}) of the peritoneal membrane.

6. The following data were presented by Lysaght and Farrell (1989) for urea (60 Daltons) and inulin (5200 Daltons) transport during CAPD. The initial glucose concentration in the dialysate solution was equal to 1.5 wt %. Assuming a constant peritoneal fluid volume of 2.4 liters, estimate the permeability of the peritoneal membrane for each solute. What thickness of a stagnant film of water would be required to give an equivalent permeability for urea and inulin? Explain the difference in the observed peritoneal membrane permeabilties based on your understanding of solute diffusion through hydrogels.

Time, minutes	Ratio of dialysate to plasma concentration for urea	Ratio of dialysate to plasma concentration for inulin
0	0.03	0
50	0.33	0.085
100	0.57	0.16
150	0.69	0.21
200	0.77	0.27
300	0.91	0.36
400	0.96	0.42

7. The following data were presented by Lysaght and Farrell (1989) for urea (60 Daltons) transport during CAPD. The urea concentration in blood was 100 mg/dL. From this data, estimate the urea permeability of the peritoneal membrane.

Time, minutes	Peritoneal volume, ml	Dialysate urea concentration, mg dL^{-1}
0	1983	10
15	2304	30
30	2543	42
45	2938	55
60	3146	62

8. Analyze the performance of an oxygenator that uses an ultrathin (25 microns) membrane made of polyalkylsulfone cast on microporous polypropylene. The oxygen permeability of this membrane is reported (Gray 1981) to be 1100 ml(STP)/min^{-1} m^{-2} atm^{-1} and that for carbon dioxide is reported to be 4600 ml(STP)/min^{-1} m^{-2} atm^{-1}. What membrane area is need to deliver 250 ml/min of oxygen and remove 200 ml min^{-1} of carbon dioxide? Carefully state your assumptions.

9. Analyze the performance of an oxygenator in which the hollow fibers have a length of 20 cm. Consider whether or not the assumption of fully developed flow is valid for these conditions. What membrane area is need to deliver 250 ml min^{-1} of oxygen and remove 200 ml min^{-1} of carbon dioxide? Carefully state your assumptions.

10. Consider an oxygenator where the gas flows through the lumen of the hollow fibers and the blood flows on the outside across and perpendicular to the hollow fibers. Assume that the gas flow is sufficiently high that a reasonable assumption is that the gas concentrations are constant along the length of the hollow fiber. Develop a mathematical model that describes the performance of this oxygenator. Carefully state your assumptions. What membrane area is needed to deliver 250 ml min^{-1} of oxygen and remove 200 ml min^{-1} of carbon dioxide? The external mass transfer coefficient for blood flowing across a bank of hollow fibers is given by the following expression (Yang and Cussler 1986).

$$Sh = 1.38 \, Re^{0.34} \, Sc^{0.33}$$

The characteristic diameter for this equation is the external diameter of the hollow fibers. The velocity of the blood (V_b) between the fibers may be calculated by the following equation.

$$V_b = \frac{Q_b}{\varepsilon_{unit} A_{unit}}$$

Here, Q_b is the volumetric flowrate of blood, ε_{unit} is the void volume in the hollow fiber unit, and A_{unit} is the total cross-sectional area of the unit that is perpendicular to the blood flow.

11. Yang and Cussler (1986) considered the design of a human gill using hollow fibers. To support a man, they assumed the human gill would have to provide up to 2500 ml min^{-1} of oxygen at 1 atm and 37°C for moderate levels of activity. Develop your design for the human gill and determine the hollow fiber membrane area that is required. How big would your human gill be? What other issues need to be considered for the design of the human gill?

12. Calculate the effectiveness factor for an enzyme reaction that follows Michaelis–Menten kinetics. Assume the Thiele modulus is 4 and $\beta = 2$. Compare your result with the value obtained assuming the reaction is first order.

13. The following kinetic data (Sung et al. 1986) were obtained for the enzyme bilirubin oxidase at a pH of 7.4 and a temperature of 37°C. The substrate was bilirubin. The enzyme loading in the reactor was equivalent to 0.15 units L^{-1}, where a "unit" of enzyme activity is defined as the amount of enzyme that catalyzes the conversion of 1 μmol min^{-1} of bilirubin. Assuming the data follows the Michaelis–Menten rate law, determine the values of V_{max}, K_m, and k_{cat}.

Bilirubin concentration, μM	Reaction rate, μM bilirubin min^{-1}
1	0.015
5	0.055
10	0.089
15	0.130
25	0.185
40	0.230
50	0.244

14. Suppose the bilirubin oxidase is immobilized within an agarose gel (Sung et al. 1986). The enzyme loading in the gel equals 10 Units ml^{-1} of wet gel. If external and internal mass transfer effects are negligible, calculate the conversion of bilirubin in a well-mixed reactor containing a total gel volume of 15 ml. The void fraction in the reactor is 70%. Assume the flowrate through the reactor is equal to 1.2 ml min^{-1} and that the entering serum bilirubin concentration is 200 μM. The value of k_{cat} for these conditions is equal to 4.67×10^{-4} mmol min^{-1} unit^{-1}. Assume the value of K_m is unaffected by immobilization or other substances found in serum.

15. Fukui et al. (1994) measured the in vitro oxygen transfer rates using an intravascular lung assist device, or ILAD. The membrane oxygenator consisted of microporous polypropylene hollow fibers that were placed within a flexible polyvinyl chloride (PVC) tube. The PVC tube was 30 cm in length with a 20 mm inside diameter. The total surface area of the hollow fibers were 0.3 m^2. The blood flows along a straight path outside of the fibers, and the gas flows within the hollow fibers. For blood flow rates of 1, 2, and 3 L min^{-1}, the oxygen transfer rates were 60, 120, and 140 ml min^{-1} respectively. Estimate the number and the diameter of hollow fibers needed to reproduce their results.

16. A series of experiments was performed using various sizes of particles containing an immobilized enzyme in order to determine the importance of diffusion within the particles. The substrate concentration is much less than the value of K_m so the reaction may be assumed to be first order. In addition, the bulk solution was well-mixed, and the concentration of the substrate at the surface of the particles is 2×10^{-4} mol/cm^3. The partition coefficient was also found to be $K = 1$. The table below summarizes the data that was obtained:

Diameter of the particle (cm)	Observed reaction rate (mol/hr/cm^3)
0.050	0.22
0.010	0.98
0.005	1.60
0.001	2.40
0.0005	2.40

From the data in the above table, determine values of the first order rate constant (k) and the effective diffusivity of the substrate within the particle (D_e). To assess the validity of the assumptions you make to find k and D_e, also calculate the effectiveness factor (η), the Thiele modulus (ϕ), and the predicted reaction rate for each particle size. Make a graph that compares the predicted reaction rate to the observed reaction rate for each particle size. How does it look?

17. A biotech company is designing an immobilized enzyme reactor for the removal of a toxin from blood. The device will be connected to the patient by an arteriovenous shunt. Blood will flow over a packed bed of particles (100 microns in diameter) containing the immobilized enzyme. The following data has been obtained for the toxin enzyme kinetics:

$$K = 1$$

$$V_{max} = 1 \times 10^{-9} \text{ mol cm}^{-3} \text{ sec}^{-1}$$

$$K_m = 1 \times 10^{-9} \text{ mol cm}^{-3}$$

$$D_e = 5 \times 10^{-7} \text{ cm}^2 \text{ sec}^{-1}$$

The maximum concentration of the toxin entering the reactor is estimated to be 1×10^{-11} mol cm^{-3}. Estimate the blood residence time within the reactor to obtain a 60% per pass conversion of the toxin in the blood.

18. Design a packed bed heparinase reactor to achieve a 69% conversion of heparin. Assume an enzyme loading of 120 U ml^{-1} of immobilized enzyme and a feed substrate concentration of 0.2 mg ml^{-1}. The flowrate of plasma to your reactor is 120 ml min^{-1}. Use the heparinase kinetics from Example 8.5, neglect external mass transfer effects, and base the effectiveness factor on that of a 1st order reaction. Other information you may need can be found in Example 8.8. Compare your enzyme support volume for the same case solved using a well-mixed reactor in Example 8.8 (i.e. 96 ml).

TISSUE ENGINEERING

9.1 INTRODUCTION

Over eight million surgical procedures are performed each year in the USA for the treatment of lost tissue or organ function as a result of disease or injury (Holder *et al.* 1997). Organ transplantation has also become a routine and successful treatment method for the replacement of a failed organ. Organs routinely transplanted include the kidneys, heart, lungs, liver, and pancreas. Part of this success is due to better therapies to prevent rejection episodes as well as improved techniques for the procurement and storage of whole organs prior to transplant. Organ transplantation has saved and improved the quality of many lives, however, its widespread application is severely limited by a shortage of whole organ donors. In the USA, only a few thousand organ donors are available each year. Yet tens of thousands are on waiting lists for an organ transplant, and most of these will die waiting for an organ. Certainly, extracorporeal devices such as the hemodialyzer, bioartificial liver (to be discussed in Chapter 10), and artificial heart (Jarvik 1981; Rosenberg 1995; Spotnitz 1987) can be used until an organ is available. These devices, therefore, serve as a "bridge" to a transplant. The continued development and improvement of these devices will help alleviate the pain and suffering while awaiting an organ transplant. However, these devices cannot be used indefinitely and, by themselves, are not a perfect solution.

Many diseases do not involve an entire organ but only certain tissues or cell types that have lost their function. In these cases, it may be simpler to restore the lost function of the tissue or cells than to replace an entire organ. A good example is the treatment of insulin dependent diabetes mellitus. The islets of Langerhans are a specialized cluster of cells scattered throughout the pancreas. Their mass represents only about 1–2% of the pancreas. They secrete a variety of hormones and one of these, insulin, is responsible for the regulation of blood glucose levels. Loss of insulin secretion through an autoimmune disease process (Atkinson and MacLaren 1990; Notkins 1979) results in diabetes. It is estimated that perhaps only about 500 thousand–1 million islets (Warnock *et al.* 1988; Friedman 1989; Smith *et al.* 1991) would be needed to treat a patient with diabetes. With an islet averaging about 150 microns in diameter, this translates to an islet volume of 1 to 2 cm^3, a volume considerably less than that of a whole pancreas. Transplanting just the islets would, therefore, result in a far simpler surgical procedure.

9.2 CELL TRANSPLANTATION

The selective transplantation of just those cells necessary to replace the lost function of a tissue or an organ is receiving considerable attention (Cima *et al.* 1991a,b; Langer and Vacanti 1993; Brown *et al.* 2000; Ma and Choi 2001; Palsson and Bhatia 2004). Selective cell transplantation is being developed for the treatment of a variety of diseases and medical problems, some of which are outlined in Table 9.1. The cited references are also a good resource for additional information concerning specific applications of cell transplantation.

Selective cell transplantation forms the basis of the relatively new fields of *tissue engineering* and *bioartificial organs*. Tissue engineering is defined as a developing multidisciplinary field that utilizes life science and engineering principles to construct biological substitutes containing viable and functioning cells for the restoration, maintenance, or improvement of tissue function.

The transplanted cells can be harvested from a variety of sources. For example, the cells can be obtained from a human donor (*allograft*), animals (*xenograft*), the patient themselves (*autograft*), or from a genetically engineered cell line (Morgan *et al.* 1994; Chang 1997; Zalzman *et al.* 2003). The donor tissue is normally dissociated using

Table 9.1 Opportunities for cell transplantation and tissue engineering

Liver	Plastic/reconstructive surgery
Diabetes	Dental
Parkinson's disease	Nerve regeneration
Kidney	Site-specific delivery of growth factors
Chronic pain	Cornea
Esophagus	Skin
Ureter	Intestine
Bladder	Trachea
Cartilage	Bone
Muscle regeneration	Blood
Artificial blood vessels	Destruction of tumors
Gene therapy	Parathyroid
Cell production	In vitro hematopoiesis
Alzheimer's disease	Neurological diseases
Heart valves	Lymphocytes
Spine	Intevertebral discs

proteolytic enzymes (Ricordi *et al.* 1988) into single cells or small cellular aggregates that are then infused as is, incorporated into a support matrix or scaffold, or encapsulated within a device. The supporting structure or device containing the cells is then implanted within the patient's body. Figure 9.1 illustrates the basic steps of the tissue engineering process.

There is considerable overlap between the two emerging fields of bioartificial organs and tissue engineering. In some cases they are synonymous. For our purposes, we will categorize the field of bioartificial organs as a subset of the larger field of tissue engineering. The major distinction between the two being that in a bioartificial organ, the goal is to use a cellular transplant to replace a lost organ function, for example the use of the islets of Langerhans or genetically engineered beta cells for the treatment of diabetes or *hepatocytes* for treatment of liver failure. Bioartificial organs also isolate the cells from the host's immune system through the use of an *immunoisolation membrane* (Colton 1995). This membrane is permeable to small molecules, such as required nutrients and the therapeutic agent released by the cells, but impermeable to the larger molecules (antibodies and complement) and cells of the host's immune system. The immunoisolation concept is shown in Figure 9.2. Bioartificial organs, therefore, do not

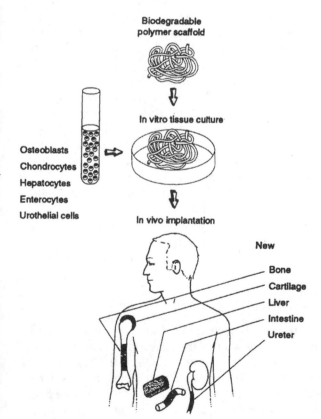

Figure 9.1   The tissue engineering process (from Langer *et al.* 1995, with permission).

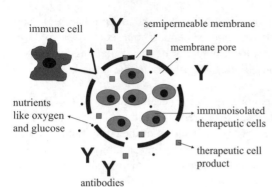

Figure 9.2 Concept of immunoisolation.

require the use of immunosuppressive drugs or *immunomodification* (Lanza and Chick 1994b) of the cells prior to transplant. Because of this immunoprotection, there is the potential to use cells and tissues from animals (xenografts), vastly increasing the availability of donor tissue. Several examples of bioartificial organs will be examined in greater detail in Chapter 10.

Isolated cells or cellular aggregates can be simply infused into several locations within the body. Typical sites include the liver via the portal vein, within regions of fat, the spleen, under the kidney capsule, and the omentum. Isolated cells have the potential to grow and form larger tissue structures when implanted in the vicinity of existing mature tissue. However, a potential problem is their loose association which provides no guide for attachment, restructuring, and formation of larger cellular aggregates. Additionally, cells can not survive if they do not have an adequate supply of blood to provide oxygen, nutrients, and a means for waste product removal. Furthermore, any therapeutic product released by the transplanted cells needs to have access to the host's vasculature in order for it to be effective. This requires that each transplanted cell needs to be within a hundred microns or so of a capillary. Initially, the existing capillary supply in the tissue surrounding the cellular transplant may not be optimal for the sustained growth of the transplanted cells. This severely limits the size of the cellular aggregates that can form, unless a means is provided for the formation of a vascular capillary bed (angiogenesis) within the transplanted cells. The goal of tissue engineering is, therefore, to develop methods and techniques that can enhance the success of cellular transplants.

9.3 THE EXTRACELLULAR MATRIX (ECM)

Most of the effort in tissue engineering is now focusing on the use of polymeric support structures for guiding the growth and organization of the transplanted cells (Cima *et al.* 1991a,b; Goldstein *et al.* 1999; Oerther *et al.* 1999; Petersen *et al.* 2002; Leach *et al.* 2003; Radisic *et al.* 2003; Pratt *et al.* 2004). To understand better how polymeric support structures can enhance the growth of transplanted cells, it is necessary first to examine how cells organize themselves within natural tissue structures.

Tissues are not made up entirely of cells, but also consist of an extracellular gel-like fluid containing a variety of macromolecules collectively referred to as the *extracellular matrix* (ECM) (Alberts *et al.* 1989; Rubin and Farber 1994; Long 1995; Naughton *et al.* 1995; Anderson 1994; Hubbell 1997; Parsons-Wingerter and Sage 1997, Dee *et al.* 2002). The ECM consists of glycosaminoglycans (GAGs) and fibrous proteins that are in contact with the cells and hold them together forming an organized cross-linked mesh-like structure. This is shown in Figure 9.3 for the case of *epithelial cells* that are supported by a layer of *connective tissue*. Connective tissues (also called the *stroma*) provide the framework for the formation and organization of most of the larger structures found within the body. The specific functional cells that define a tissue, for example the hepatocytes found in the liver or the islets of Langerhans in the pancreas, are called *parenchymal cells*.

The ECM not only serves to provide three-dimensional organization of cells and adjacent layers of tissue, but also provides a mechanism for intercellular communication and controls such cellular processes as proliferation, cell migration, attachment, differentiation, and repair. The major components of the ECM are the glycosaminoglycans (GAGs), proteoglycans, collagens, elastic fibers, structural glycoproteins, and the basement membrane.

Most mammalian cells are anchorage dependent and possess cell surface receptors for a variety of the macromolecules, or so-called *attachment factors,* found within the ECM that are responsible for the development, growth, and metabolic functions of cells. The ECM macromolecules are secreted locally, for the most part, by specialized cells called *fibroblasts*. Several of the most important types of macromolecules found within the ECM are summarized in Table 9.2.

9.3.1 Glycosaminoglycans

GAGs (glycosaminoglycans) form from long linear polymers consisting of a repeating disaccharide unit. Molecular weights range from several thousand to over a million Daltons. GAGs tend to have a large number of negative charges because of the presence of carboxyl and sulfate groups. They are highly extended forming random coil-like

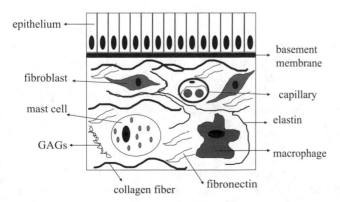

Figure 9.3   Connective tissue.

Table 9.2 Major macromolecules of the extracellular matrix

Glycosaminoglycans (GAGs) – forms the ECM gel
Collagen – provides strength and organization to the ECM
Elastin – provides resilience or elasticity
Fibronectin – adhesion of fibroblasts and other cells
Laminin – adhesion of cells to the basal lamina

structures. Because they are also hydrophilic, they form in combination with the interstitial fluid, the hydrated gel-like material throughout which are found the other macromolecules of the ECM. There are four types of GAGs depending on what types of sugars they are made from. They are *hyaluronic acid, chondroitin sulfate* and *dermatan sulfate, heparin sulfate* and *heparin*, and *keratin sulfate*. With the exception of hyaluronic acid, they also form what are called *proteoglycans* through covalent bonds with proteins.

9.3.2 Collagens

Collagens are fibrous proteins that are made from three polypeptide chains that form the tough, triple-stranded helical structure of the collagen molecule. Tensile strength in tissues is provided by collagen. Although about 10 types of collagen molecules have been identified, the most important types of collagen molecules are the *fibrillar collagens* (type I, type II, and type III) and type IV collagen. The first three types of collagen molecules assemble themselves into much larger structures called collagen fibrils that aggregate further to form collagen fibers. Type I collagen makes up about 90% of the collagen found in the body. It is found mostly in the skin, tendons, ligaments, various internal organs, and bone. Type II collagen is primarily found in cartilage, and type III collagen is found in blood vessels as well as the skin. Type IV collagen organizes itself to form sheets within the basal lamina of the basement membrane. Fibroblasts have the ability to organize the collagen fibrils they secrete forming sheets or rope-like structures, therefore, they can affect the spatial organization of the matrix they produce.

9.3.3 Elastin

Elastin is a hydrophobic glycoprotein molecule that, through cross-links with other elastin molecules, can form a network of sheets and filaments with the unique property of being elastic, allowing for recoil after periods of stretch. The elasticity is a result of the random coil-like structure of the elastin molecule. The elastic nature of elastin is important in blood vessels, skin, the lungs, and the uterus. Inelastic collagen fibrils are interwoven with elastin to limit and control the degree of stretching.

9.3.4 Fibronectin

Fibronectin is the principal adhesive glycoprotein found within the ECM. Fibronectin binds to other ECM molecules, such as collagen, and to cell surface receptors. It, therefore, has a principal role in the attachment of cells to the ECM. Fibronectin is a dimer made up of two subunit chains that are bound together at one end by a pair of disulfide bonds. Along the length of the chains are a series of functional domains that can bind to specific types of molecules (for example collagen or heparin) or cell surface receptors. The cell binding domain has a specific tripeptide sequence consisting of the amino acids: arginine (R), glycine (G), and aspartic acid (D), often referred to as the *RGD sequence*. Peptides containing this sequence will inhibit the attachment of cells to fibronectin through their competition for the RGD binding site on the cell surface.

In addition to fibronectin, there are several other ECM adhesion proteins that express the RGD sequence. These include *vitronectin* found primarily in blood cells. *Thrombospondin* is secreted by a variety of cells involved in the development of the ECM. It serves to bind together other components of the ECM like fibronectin. The *von Willebrand factor* is made by megakaryocytes (platelet generating cells found in the bone marrow) and it is stored in circulating platelets. The von Willebrand factor is released by platelets as a result of injury to blood vessels and this factor then binds to collagen. The platelets then bind to the von Willebrand factor that is also bound to collagen. Fibrinogen is another of the clotting proteins found in blood. During blood clot formation, fibrinogen is converted to fibrin that forms a mesh-like structure that traps red blood cells and platelets.

9.3.5 Basement Membrane

The *basal lamina* is a continuous mat-like structure of ECM materials that separates specific cells, such as epithelial, endothelial, or muscle, from the underlying layer of connective tissue. The basal lamina consists of two distinct layers, the *lamina rara*[1] directly beneath the basal membrane of the specific cells above, and the *lamina densa*[2] just below the lamina rara. Below the two layers of the basal lamina is found the collagen containing *lamina reticularis* that connects the basal lamina to connective tissue that lies below it. All three of these layers together constitute what is known as the *basement membrane* (see Figure 9.3). The basal lamina consists primarily of type IV collagen, proteoglycans such as those formed from heparin sulfate, and the glycoprotein *laminin*. Laminin is an extremely large protein (850,000 Daltons) made from three long polypeptide chains that form the shape of a cross. It also contains a number of functional domains throughout its structure with sites for binding to type IV collagen, heparin sulfate, and laminin cell surface receptors.

9.4 CELLULAR INTERACTIONS

There are three types of interactions involving cells. These interactions may be classified as cell–ECM, cell–cell, and cell growth factor (Long 1995). Figure 9.4 illustrates the various interactions of cells with their surrounding environment. Cells

---

[1]under the electron microscope, this layer is translucent to electrons.
[2]under the electron microscope, this layer is somewhat opaque to electrons.

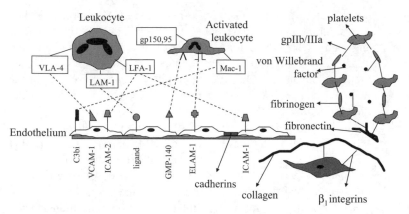

Figure 9.4 Cellular interactions.

interact with other cells, the ECM, and growth factors through cell surface receptor proteins that are an integral part of the cell membrane. Binding of cell surface receptors to surrounding ECM components such as collagen, fibronectin, and laminin enables the cell to link up to the surrounding matrix. Binding of cell surface receptors between cells allow them to organize into larger structures such as tissues and organs. Cell surface receptor interactions with growth factors provide control over a variety of cellular functions through control of gene expression.

There are four types of cell adhesion receptors (Hubbell 1997). Three of these receptors are mainly involved in cell–cell interactions and the fourth is involved in both cell–cell and cell–ECM interactions. The adhesion receptors used in cell–cell interactions include the *cadherins*, the *selectins*, and the *CAMs* (cell adhesion molecules). The receptors involved in cell–ECM interactions belong to a general family of adhesion receptors known as *integrins* (Horwitz 1997).

9.4.1 Cadherins

Cadherins take part in what is known as homophilic binding, that is the binding of a cadherin molecule on one cell with an identical cadherin on another cell of the same type. They do not bind with ECM components. Cadherin molecule binding is strongly dependent on the presence of extracellular Ca^{++}.

9.4.2 Selectins

The selectins are a family of cell adhesion molecules that are found on the surfaces of endothelial cells, leukocytes, and platelets. They exhibit heterophilic binding between the blood cells and the endothelial cells that line the blood vessels. They are important in the localization of leuokocytes to sites of inflammation and tissue injury. The three most important selectins are called GMP-140, endothelial leukocyte adherence molecule-1 (ELAM-1), and leuokocyte adhesion molecule-1 (LAM-1). Like the cadherins, their activity is dependent on the presence of extracellular calcium.

9.4.3 Cell Adhesion Molecules (CAMs)

The CAMs are members of the immunoglobulin gene superfamily. These receptors bind independently of extracellular calcium levels. CAMs are involved in recognition of antigens (foreign agents) and cell–cell interactions. Antigen recognition involves both the T-lymphocyte receptor and two special antigen presenting molecules called major histocompatibility class I and II (MHC I and MHC II). These specialized CAMs will be discussed further in Chapter 10 in the section on immunology. Two other CAMs are important factors in localizing leukocytes to regions of tissue injury. For example, intercellular adhesion molecule-1 (ICAM-1) is found on the surfaces of both endothelial cells and leukocytes and binds to the integrin called leukocyte function antigen-1 (LFA-1) that is found on macrophages and neutrophils. ICAM-2, found on endothelial cells, also binds to LFA-1 and helps to localize neutrophils to sites of injured tissue. VCAM-1 (vascular cell adhesion molecule) is expressed by endothelial cells as a response to injury or inflammation and also promotes adherence of leukocytes to endothelial cells.

9.4.4 Integrins

The integrins are a family of adhesion molecules that are involved in both cell–cell interactions as well as binding with molecules found in the extracellular matrix. Integrin molecules are heterodimers consisting of two protein subunits, the α chain and the β chain. There are at least 15 variants of the α subunit and eight variants of the β subunit. Overall, there are at least 20 different heterodimer $\alpha\beta$ integrins. The β subunits provide the functional aspects of the dimeric integrin molecule, hence, integrins are classified according to the type of their β chain. The subclasses having the β_1, β_2, and β_3 chains are the most common. The β_1 and β_3 subclasses are primarily involved in interactions between the cell and the ECM. The β_2 subclass involves mostly cell–cell interactions involving leukocytes.

β_1 integrins are also referred to as VLA (very late after) antigens (Margiotta et al. 1994). VLA antigens are numbered according to the number of the alpha subunit that forms the heterodimer with the β_1 subunit. These subclasses bind with a variety of ECM proteins such as collagen, fibronectin, and laminin. The primary recognition site for β_1 integrins is the RGD sequence described earlier. Therefore, this group of integrins are important for the overall organization and function of a variety of cell types. VLA-4 is also a receptor found on leukocytes for VCAM-1 expressed by endothelial cells.

Cell–cell interactions, mainly leukocyte to leukocyte, involve mostly the β_2 group of integrins. Hence, this subclass is known as the "leukocyte" integrins. The common β_2 subunit is also known as CD18 (cluster designation), a 95 kDalton protein. The most important alpha subunits, in combination with β_2, form the leukocyte integrins known as LFA-1 (leukocyte function antigen-1), Mac-1 (macrophage antigen-1), and gp150,95 (a glycoprotein with an alpha subunit of molecular weight 150 kD along with the β_2 subunit protein of molecular weight 95 kD). These alpha subunits are also respectively known by the cluster designations CD11a, CD11b, and CD11c. The expression of these integrins on leukocytes is particularly important for defense against bacterial infections.

The β_3 integrins comprise the platelet glycoprotein receptor known as gpIIb/IIIa and the vitronectin receptor. Both of these receptors also recognize the RGD sequence of ECM matrix proteins such as vitronectin, fibronectin, thrombospondin, fibrinogen, and von Willebrand factor. Aggregation of platelets is enhanced by the release of von Willebrand factor. gpIIb/IIIa is calcium dependent and found on the surface of activated platelets. This integrin is important for platelet aggregation and adherence of platelets to the subendothelium. This occurs through binding of this receptor to fibrinogen, fibronectin, vitronectin, and von Willebrand factor.

9.4.5 Cytokines and Growth Factors

A variety of cells also secrete soluble proteins known as *cytokines*. Cytokines serve as intercellular chemical messengers. Many of these proteins are required for the normal development, growth, and proliferation of cells. They are also involved in such processes as inflammation, wound healing, and the varied responses of the immune system. Many are mitogens, that is they stimulate the proliferation of specific types of cells and are often then referred to as growth factors.

Growth factors bind to specific receptors on the surface of their target cell and have the ability to induce or direct the action of specific genes in the targeted cell. A variety of growth factors have been identified and some of the more important of these are summarized in Table 9.3. All of these growth factors play a key role during the wound healing process and many are involved in the growth of new blood vessels. Growth factors can also be used to stimulate and accelerate the development of blood vessels in polymeric scaffolds used in tissue engineering applications.

Platelet derived growth factor (PDGF) is a mitogen for such cells as smooth muscle, fibroblasts, and neuroglial cells. PDGF is stored in platelets and is released after platelet aggregation at the site of injury to blood vessels. Damaged endothelial and smooth muscle cells also secrete PDGF. PDGF is also a chemotactic signal for such

Table 9.3 Growth factors

Growth factor	Property
Platelet derived growth factor (PDGF)	Mitogen for smooth muscle cells, fibroblasts, and neuroglial cells; chemotactic signal for immune system cells
Epidermal growth factor	Mitogen for a variety of cells types, accelerates wound healing
Fibroblast growth factor (aFGF, bFGF) and vascular endothelial cell growth factor (VEGF)	Stimulates growth of blood vessels, mitogen for endothelial cells, fibroblasts, smooth muscle cells
Tumor necrosis factor (TNF)	Stimulates growth of blood vessels
Insulin-like growth factor (INF-I)	Stimulates growth of blood vessels
Transforming growth factor (TGF-β)	Controls cellular response to other growth factors
Nerve growth factor (NGF)	Stimulates axon growth
Interleukin-2 (IL-2)	Stimulates proliferation of T lymphocytes
Interleukin-6 (IL-6)	Stimulates proliferation and differentiation of B lymphocytes

inflammatory cells as macrophages and neutrophils. Epidermal growth factor (EGF) stimulates the growth of a variety of cell types. EGF accelerates the healing of many wounds and stimulates fibroblasts to deposit collagen.

Fibroblast growth factor (FGF) comes in two forms, acidic (aFGF) and basic (bFGF). bFGF is an order of magnitude more potent than aFGF. FGF and vascular endothelial cell growth factor (VEGF) stimulates the growth of capillaries, a process called angiogenesis (Folkman 1985; Folkman and Klagsbrun 1987). FGF also promotes the growth of fibroblasts, endothelial cells, and smooth muscle cells. These processes affected by FGF also accelerate wound healing.

Another angiogenic growth factor is tumor necrosis factor-α (TNF-α) which is secreted by activated macrophages during the initial stages of inflammation. Insulin-like growth factor-I (IGF-I) not only plays a major role in growth, but also stimulates angiogenesis. Transforming growth factor-β (TGF-β) can either promote or inhibit the growth of cells by controlling their response to other growth factors. Several other important growth factors include nerve growth factor (NGF) which enhances function of neurons and stimulates axon growth, interleukin-2 (IL-2) which stimulates the proliferation of T lymphocytes, and interleukin-6 (IL-6) stimulates the proliferation of B lymphocytes and their differentiation to antibody secreting plasma cells. There are also several hematopoietic[3] growth factors that are specific to the development and differentiation of blood cells.

9.5 POLYMERIC SUPPORT STRUCTURES

With this background on how cells organize themselves within the body, we can now address the properties required for an artificial polymeric support structure for transplanted cells. Table 9.4 summarizes some of the characteristics that are important to consider in selecting materials to serve as support structures for transplanted cells.

A variety of biological and synthetic polymers are used as biomaterials and many of these can be used to provide the support structure for transplanted cells. Table 9.5 summarizes some of the materials that are being used or considered for use as support structures in tissue engineering.

The biomaterials listed in Table 9.5 are grouped as to whether or not they are biodegradable. Long-term implantation of non-biodegradable biomaterials poses the risk of localized infection, chronic inflammation, and development of fibrous encapsulation of the implant. This can compromise the proper functioning of transplanted cells if they were to reside within the porous structure of these materials. Non-biodegradable materials such as PTFE, PU, and Dacron, are preferred for those applications where their structural integrity is more important than the requirement for them to serve as a scaffold for transplanted cells; for example, as vascular grafts or structural components of artificial organs. These materials can also be made into a variety of three-dimensional configurations.

Biodegradable materials are now attracting significant interest for tissue

---

[3] hematopoiesis is the process of blood cell development and differentiation.

Table 9.4   Desirable properties of support structures for transplanted cells

Biocompatible
Non-immunogenic
Negligible toxicity
  locally and systemically
  polymer and any degradation products
Chemically and mechanically stable
Processable into a variety of shapes
  hollow tubes, sheets, arbitrary 3D shapes
  open foam or sponges
  woven or non-woven mesh-like structures
  high porosity
  high internal surface area to volume ratio
  controllable pore size
Promotes cell attachment
Promotes angiogenesis
Favorable interaction/mobilization of host cells
Ability to release active compounds such as growth factors
Ability of the cells and ECM materials to interact with the support structure
Ability to obtain the desired cellular response

engineering applications. Biodegradable materials allow the transplanted cells sufficient time to organize the desired three-dimensional structure and develop their own blood supply. The biodegradable polymers are easily hydrolyzed by the body's fluids and slowly disappear wthout leaving behind any foreign residues. Examples of biologically derived biodegradable polymers include collagen, GAGs, and chitosan. Collagen is one of the ECM materials and can be prepared as fibers (Cavallaro et al. 1994) or as a gel. Collagen is being used in such applications as tissue repair and artificial skin (Morgan and Yarmush 1997). PLA, PGA, and PLGA are naturally occurring hydroxy acids. They are biodegradable and have been approved for use by the Food and Drug Administration (FDA) for use in sutures, controlled drug release, and surgical support fabrics (Cima et al. 1991a,b). A great deal of effort is focusing on these materials for tissue engineering applications (Holder et al. 1997; Mooney et al. 1994, 1995a,b, 1996, 1997; Kaufmann et al. 1997). They degrade by hydrolysis and form natural byproducts. Their absorption rate can also be controlled from months to years (Agrawal et al. 1995; Mooney et al. 1994, 1995a,b). The degradation rate of these materials is dependent on their initial molecular weight, the surface area that is exposed, degree of crystallinity, and for the case of co-polymer blends, the ratio of the hydroxy acid monomers used (Pachence and Kohn 1997).

The resulting structure formed from a biomaterial must be mechanically strong enough to support the growth of the transplanted cells and chemically compatible with the intended duration of use. For example, cells secrete a variety of enzymes that may degrade the polymer used. For a biodegradable material, this cell-induced degradation rate must be considered during formulation of the biomaterial to achieve the desired degradation rate. In the case of non-biodegradable materials, they must withstand the cellular attack and not form byproducts that are inflammatory or immunogenic and,

Table 9.5 Materials for tissue engineering and other biomedical applications

Non-biodegradable polymer	Applications
Polydimethylsiloxane or silicone (PDMS)	Breast, penile, testicular prostheses, catheters, drug delivery, heart valves, membrane oxygenators, shunts, tubing, orthopedics
Ceramics	Bone repair
Polyurethanes (PEU)	Artificial hearts and ventricular assist devices, catheters, intraaortic balloons, wound dressings
Polytetrafluoroethylene (PTFE)	Heart valves, vascular grafts, reconstruction, shunts, membrane oxygenators, catheters, sutures, coatings
Polyethylene (PE)	Artificial hips, catheters, shunts, syringes, tubing
Polysulfone (PS)	Heart valves, penile prostheses, artificial heart
Polycarbonate	Hard contact lenses
Poly(methyl methacrylate) (PMMA)	Bone cement, fracture fixation, intraocular lenses, dentures, plasmapheresis membranes
Poly(2-hydroxyethylmethacrylate) (PHEMA)	Controlled drug release, contact lenses, catheters, coatings, artificial organs
Polyacrylonitrile (PAN)	Hemodialysis membranes
Polyamides (nylon)	Hemodialysis membranes, sutures
Polyethylene terephthalate (Dacron)	Vascular grafts, tissue patches, shunts
Polypropylene (PP)	Valve structures, plasmapheresis membranes, sutures
Polyvinyl chloride (PVC)	Tubing, plasmapheresis membranes, blood bags
Poly(ethylene-co-vinyl acetate)	Drug delivery devices
Polystyrene (PS)	Tissue culture flasks
Poly(vinyl pyrolidone) (PVP)	Blood plasma extender
Polyvinyl alcohol (PVA)	Dental, tissue repair, scaffolds for bioartificial organs

Biodegradable polymers	Applications
Poly(L-lactic acid), poly(glycolic acid), and poly(lactide-co-glycolide) (PLA, PGA, and PLGA)	Controlled release drug delivery devices, sutures, scaffolds for cell transplantation and tissue engineering
Collagen	Artificial skin, hemostasis, tissue regeneration scaffold
GAGs (hyaluronan)	Tissue repair, viscoelastic, wound care
Chitosan	Scaffolds for tissue engineering, inhibitors of blood coagulation, cell encapsulation, membrane barriers, contact lens materials
Polyhydroxyalkanoates (PHA)	Controlled drug release, sutures, artificial skin
Poly(ε-caprolactone) (PCL)	Implantable contraceptive devices, controlled drug release, surgical staples

Source: Friedman *et al.* (1994), Marchant and Wang (1994), Saltzman (1997), Pachence and Kohn (1997).

therefore, capable of compromising the function of the implant. The material used to form the support structure will also generally be in direct contact with the host's immune system and connective tissue such as fibroblasts. For any biomaterial, activation of the host's immune system, as well as a fibrotic connective tissue response, will tend to wall off and isolate the transplanted support structure blocking its intended function. The host reaction and formation of a fibrotic capsule around the implant is a function of not only the materials used, but also on the shape and microstructure of the implant.

Unlike in vitro tissue culture of cells wherein the cells are usually grown in only two dimensions, the transplanted support structure containing the cells will be three dimensional. Furthermore, the resulting three-dimensional polymer scaffold must, in general, have high porosity, significant internal surface area, and a controlled pore size distribution. All of these properties must not compromise the structural strength of the material.

Polymer scaffolds used in tissue engineering applications must be processable into whatever shape is required for the implant. A variety of techniques may be used to process polymer scaffolds. These include such methods as fiber bonding, solvent casting, particulate leaching, membrane lamination, melt molding, polymer/ceramic fiber composite foams, phase separation, and in situ polymerization (Lu and Mikos 1996; Thomson *et al.* 1997; Ma and Choi 2001).

For example, the biodegradable open sponge or mesh-like structures shown in Figure 9.5 exhibit high porosity (> 80%) and large pores. The poly(DTE carbonate) shown on the left in the figure was prepared by a salt leaching technique. The resulting structure has pores in the range of 200 to 500 microns and is being used for bone regeneration. The PGA non-woven mesh has fibers 10 to 50 microns in diameter and is being considered for use in organ regeneration. These materials interact favorably with the surrounding tissue allowing for ingrowth of tissue and the formation of a vascular network throughout the support structure. The high porosity of these structures is also important for nutrient and product transport as well. To avoid the presence of non-vascular regions and to optimize implant size, it may be important to be able to control the pore size distribution. A high internal surface area to volume ratio will also provide numerous sites for cell adhesion to the polymer support structure. A high internal surface area also allows for surface treatment of the polymer structure with ECM materials and/or growth factors in order to enhance interaction of the support with the transplanted cells and promote their proliferation and differentiation.

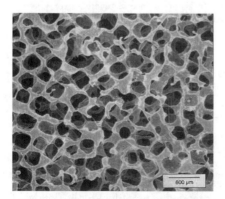 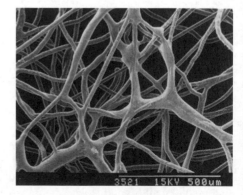

Figure 9.5   Scanning electron micrographs of polymeric scaffolds for tissue engineering. On the left, a poly(desaminotryosyl-tyrosine ethyl ester carbonate) polymer scaffold; on the right, a bonded poly(glycolic acid) non-woven fiber mesh (from James and Kohn, 1996, and Lu and Mikos, 1996, respectively, with permission).

In the best of situations, the implanted support structure becomes vascularized by the host during the period of time the transplanted cells are increasing in number. However, the metabolic demands of the transplanted cells, in many cases, cannot be met by the vasculature that is also developing at the same time within the implant. It may be better first to implant the support structure, allowing it to become vascularized, and then to transplant or seed the cells at a later point in time (Takeda *et al.* 1995). This gives the transplanted cells the opportunity to start in a well-vascularized region thus eliminating any transport limitations due to the metabolic requirements of the transplanted cells. An example of this is shown later in Section 9.11.

Prevascularization of the implant can also be enhanced by first seeding the implant with the host's endothelial cells (Holder *et al.* 1997). The endothelial cell loaded support matrices are believed to enhance vascularization by any or all of the following mechanisms. The endothelial cells can form new capillaries, they may provide chemical signals (ECM and growth factors) for growth of blood vessels, and they may merge with the host's own vascular ingrowth. Immobilization of growth factors on the support material, or within controlled release microspheres, can also be used to facilitate vascularization and improve survival of the transplanted cells (Thompson *et al.* 1988, 1989; Mooney *et al.* 1996). Figure 9.6 shows the tissue region in a polyester support matrix 28 days after implantation that was pretreated with collagen and aFGF. We see a well-vascularized tissue region within the fiber space as evidenced by the presence of numerous capillaries some of which contain red blood cells.

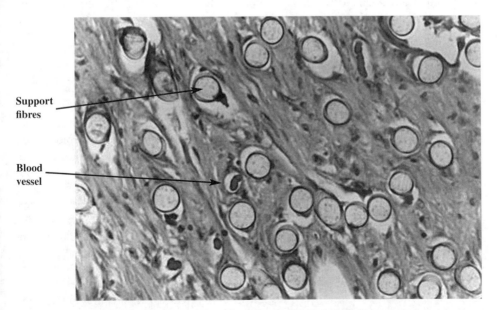

Support
fibres

Blood
vessel

Figure 9.6   Vascularized tissue within a polyester support matrix 28 days after implantation.

9.6 INITIAL RESPONSE TO AN IMPLANT

The first response after implant of the support structure will be the infiltration of the support by a variety of cell types. This process is very similar to what occurs during wound healing and involves three somewhat overlapping phases referred to as *inflammation*, *proliferation*, and *maturation* (Hammar 1993; Arnold and West 1991).

The *inflammation period* lasts for several days and involves the arrival of *platelets* and *neutrophils*. Activation of the clotting process and release of growth factors (Alberts *et al.* 1989) (PDGF, EGF, TGF) leads to the formation of a collagen-free fibrin network that serves as a scaffold for the inflammatory cells. The neutrophils release factors that attract *monocytes* and also ingest foreign material by phagocytosis. The monocytes also enter the site and are transformed into *macrophages*, and when fused together, they become *foreign body giant cells*. They continue to clean up the site removing dead tissue, removing bacteria, and walling off large debris.

During the *proliferative phase*, the macrophages release a variety of factors (PDGF, TGF, EGF, FGF) that activate the migration and proliferation of fibroblasts and endothelial cells. The fibroblasts release a variety of ECM materials and begin to form a collagen network. The generally low oxygen levels within the site also stimulates the movement and growth of capillary sprouts from the surrounding vasculature. These capillary sprouts are formed from endothelial cells and invade the site forming a vascular bed. Over a period of weeks, the entire site becomes vascularized and this marks the end of the proliferative phase.

The *maturation phase* involves final remodeling of the site resulting in contraction of the wound and organization of the collagen matrix. This is also a critical time for tissue engineering applications since the reduced oxygen demand of the cells involved in wound healing can result in regression of the vascular supply. Seeding of the transplanted cells at this time and their need for oxygen may promote the continued growth of the vascular supply.

9.7 TISSUE INGROWTH IN POROUS POLYMERIC STRUCTURES

The rate of tissue ingrowth is an important factor in the development of tissue engineered structures. Presently, the slow rate of ingrowth of vascularized tissue from the host severely limits the thickness of most tissue engineered structures (Holder *et al.* 1997). Much effort, is therefore, being focused on optimizing the rate of tissue ingrowth into support materials so that the thickness of these devices can be increased from millimeters to centimeters (Galban and Locke 1997). These studies on the rate of ingrowth of new tissue in the support structure are being performed in laboratory animals. For example, Mikos *et al.* (1993) prepared highly porous polymer disks 13.5 mm in diameter and about 5 mm thick. The polymers used were made either from PLA, PGA, or PLGA. In forming the polymer disks, NaCl particles on the order of several hundred microns were incorporated in the polymer solution at 80–90 wt % to form a composite membrane. By leaching out the salt particles, a highly porous foam-like support structure was formed. These structures typically had porosities in the range of 64–90% with mean pore sizes ranging from as low as 36 to 179 microns. Porosity and pore size was controllable by adjusting, respectively, the relative amount and particle size of the salt used. The devices were then implanted in the *mesentery* of male rats.

Figure 9.7 illustrates some typical results for a prewetted PLA implant (83% porosity and 166 micron mean pore size) at two sites in the mesentery. Tissue ingrowth was determined from histological sections of implants removed at 5, 15, and 25 days. The optimal time for subsequent cell transplantation corresponds to the time required for 100% tissue ingrowth and assumes that vascularization has also occurred. For these results, it appears that 25 days is sufficient to fill the void space of the support structure with vascularized tissue. However, the regions occupied by tissue were not completely filled. The percent tissue area provides a measure of the device total volume fraction occupied by tissue. At day 25, this volume fraction of the tissue is 79%. Since the device total porosity (ε_p, recall it is 83%) is the sum of the tissue volume fraction ε_T and the remaining void volume (ε_v), we conclude that the remaining void fraction is 4% at day 25. This would be the volume remaining for the transplanted cells.

Also determined from the histological sections of the implant, were the number of capillaries and their average cross-sectional surface area within a 1×1 mm^2 field of

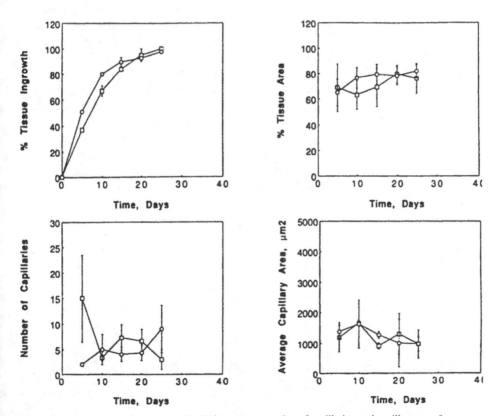

Figure 9.7   Normalized tissue ingrowth, % tissue area, number of capillaries, and capillary area for prewetted PLLA devices (from Mikos *et al.* 1993, with permission).

view. The number of capillaries (N_c) after 25 days is about five per field, whereas the average capillary cross-sectional area ($A_{capillaries}$) is about 1000 μm^2. This gives a mean capillary radius of 8 microns. The volume fraction of capillaries within the tissue (v) would then be given by the next equation:

$$v = \frac{A_{capillaries}}{\varepsilon_T \times 1 \, mm^2} \tag{9.1}$$

This gives a capillary tissue volume fraction of 0.13%. Assuming the idealized Krogh tissue cylinder model shown in Figure 5.16, we can calculate the Krogh tissue cylinder radius using Equation 5.120.

$$v = \left(\frac{r_C}{r_T}\right)^2$$

With a capillary radius (r_C) determined above of 8 microns, we obtain a Krogh tissue cylinder radius of about 220 microns.

In a subsequent study, Wake et al. (1994) focused on disks with the same geometry as the earlier Mikos et al. (1993) study, forming an amorphous PLA device with pore sizes ranging from 91–500 microns. Figure 9.8 illustrates the tissue ingrowth for various implantation times out to 35 days. Once again, the device is ready for cell transplantation once 100% tissue ingrowth has been achieved. Here we see the strong effect that mean pore size has on the rate of tissue ingrowth. A 500 micron pore size provides 100% ingrowth after only 5 days, compared with a requirement of 25 days for the 179 micron mean pore size. For the 91 micron pore size device and after 25 days, tissue ingrowth is only about 90%. Once again however, the void space available for cell transplant after tissue ingrowth was quite small.

Sarver et al. (1995) have examined the tissue ingrowth and vascularization of a polyester surgical support felt placed within the test chamber shown in Figure 9.9. The support material itself was 1 cm in diameter and consisted of two layers for a total thickness of 3 mm (δ_T). The void fraction of the support material (ε_m) was 0.84. The

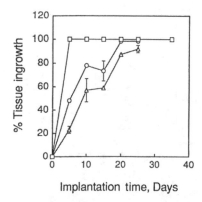

Figure 9.8 Normalized tissue ingrowth in PLLA devices of different pore sizes: 500 microns, square; 179 microns, circle; 91 microns, triangle (from Wake et al. 1994, with permission).

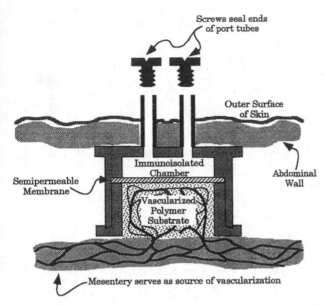

Figure 9.9 A device for cell transplantation (from Sarver *et al.* 1995, with permission).

polymer support material was pretreated with type I collagen and acidic fibroblast growth factor (aFGF) according to the method outlined by Thompson *et al.* (1988, 1989). The combination of the type I collagen and aFGF is a potent promoter of angiogenesis (Folkman 1985; Folkman and Klagsbrun 1987). The test chambers were implanted within the abdominal cavities of rats for periods of time ranging from 1 to 28 days.

Figure 9.10 (Sarver *et al.* 1995) shows the increase in capillary volume fraction of the support matrix as a function of time. Also shown is the fraction of the support matrix (f_T) that has vascularized tissue which, in this case, is equal to the fractional depth of tissue penetration. We see that after 28 days, the capillary volume fraction (based on the total matrix volume) is nearly 0.23% and that about 72% (0.72×3 mm = 2.2 mm of penetration) of the support structure contained vascularized tissue. The tissue ingrowth in this case occurs, for the most part, in one direction only and is comparable to that observed by Mikos *et al.* (1993). The average capillary diameter was also found to be about 7.2 microns, and the Krogh tissue cylinder radius, after 28 days, was about 68 microns.

Similar studies by Holder *et al.* (1997) showed significantly higher numbers of new capillaries during the first 3 weeks for implants that were initially seeded with endothelial cells. They also reported cellular ingrowth rates into their matrices of about 0.5–1 mm/week. These rates of ingrowth are comparable to the values obtained by Mikos *et al.* (1993) and Sarver *et al.* (1995).

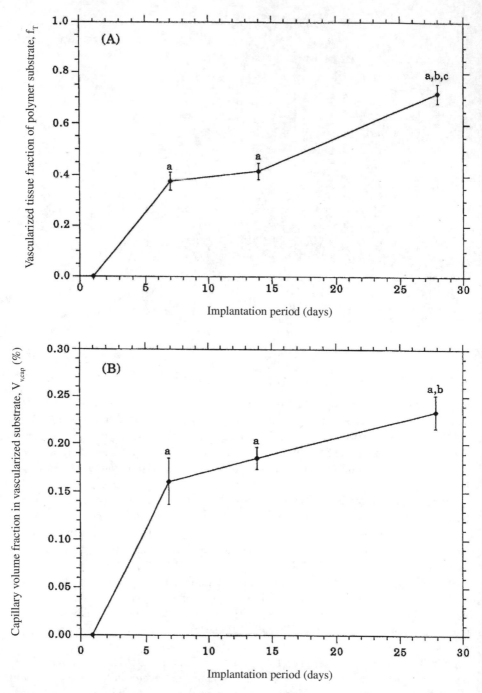

Figure 9.10 (A) Vascularized tissue fraction of polymer matrix for each period of implantation. (B) Capillary volume fraction within vascularized region of polymer matrix for each period of implantation. (from Sarver *et al.* 1995, with permission).

9.8 CAPILLARY VOLUME FRACTIONS

The capillary volume fractions determined in the studies above can be compared with what are found in other tissues throughout the body. A rough estimate for the average capillary volume fraction in the body can be found by the following argument. The total cross-sectional area of all the capillaries in the body is estimated to be 2500 cm^2 (Guyton 1991). Assuming an average capillary length is 0.05 cm, then in a 70 kg human (with density of about 1 g/cm^3) the volume fraction of capillaries would be: (2500 cm^2 × 0.05 cm/70,000 cm^3) × 100 = 0.18%. Therefore, in a sense, the above polymeric support structures have capillary volume fractions comparable to the "average" value for the body. Certainly, there will be quite a bit of variation around this average considering the many different functions the various tissues in the body perform. For example, the brain has capillary volume fractions in the range of 1–4%, cardiac muscle is on the order of 7–10%, and skeletal muscle is around 1%. Tumors typically have capillary volume fractions in the range of 0.1–1%. The higher capillary volume fractions of the brain and cardiac muscle reflect their much higher demand for oxygen. Skeletal muscle has a reserve supply of capillaries to meet the increased oxygen demands during strenuous exercise.

9.9 MEASURING THE BLOOD FLOW WITHIN POLYMERIC SUPPORT STRUCTURES

Another way to monitor the ingrowth of tissue and the development of a vascular supply within the support structure, is through the use of radioactive or fluorescent microsphere techniques (Selman et al. 1984; Sarver et al. 1995). This technique is illustrated for the rat in Figure 9.11.

In this technique, a syringe pump, which we will see later serves as a reference flowrate, is connected to the left femoral artery and blood is removed at a constant rate. After about 1 minute, 15 micron diameter radioactive or fluorescent microspheres are injected into the left ventricle of the heart via a cannula in the right common carotid artery. The microspheres are then allowed to distribute throughout the animal's body for 1 minute. Because of their small size, the microspheres become trapped within the capillaries in the body. They are also being removed from the body by the syringe pump at a known flowrate. The animal is then euthenized, and the major organs and support matrix are removed from the animal. The radioactivity or fluorescent levels in samples from the organs, the support matrix, and the syringe pump are determined. The blood flowrate to each of these samples is then proportional to their level of radioactivity or fluorescence, recognizing that the amount of radioactivity or fluorescence in the syringe pump sample is used to calibrate the flowrate.

Sarver et al. (1995) used this technique to also determine the blood flowrate within the polyester support matrix for various times after implantation. The results are shown in Figure 9.12. Recall that two layers of the support material were used in the device shown in Figure 9.9. The layer adjacent to the mesentery clearly shows the progressive development of the capillary bed as reflected by the increase in support matrix blood flow. These results are also in general agreement with the capillary volume fractions

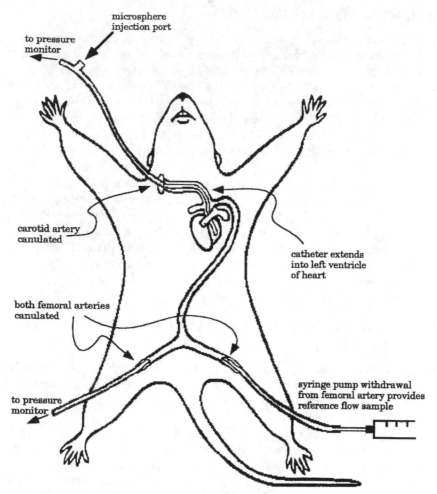

Figure 9.11 Experimental setup for measuring blood flow using the radioactive microsphere technique (Sarver 1994).

determined from the histological sections of the support material that were shown earlier in Figure 9.10. After about 14 days, a plateau is reached with little further improvement in the blood flowrate, indicating that perhaps 14 days after implantation is an appropriate time for cell transplantation into a support matrix of this thickness. The innermost layer shows a delay in blood flowrate development that is expected recognizing the time needed for the tissue front to reach this layer.

Focusing on the first layer, we see that the maximum blood flowrate is about 0.042 ml/min after 28 days of implantation. The support material has a volume of: $\pi/4 \times 1 \text{ cm}^2 \times 0.15 \text{ cm} = 0.118 \text{ cm}^3$, of which the tissue could occupy 84% of this volume giving a potential tissue volume of 0.1 cm^3. The blood perfusion rate for the tissue in this layer is then 0.042 ml min^{-1} $\times$ 1 0.1^{-1} cm^{-3} = 0.42 ml cm^{-3} min^{-1} or 42 ml min^{-1} 100^{-1} g^{-1} of

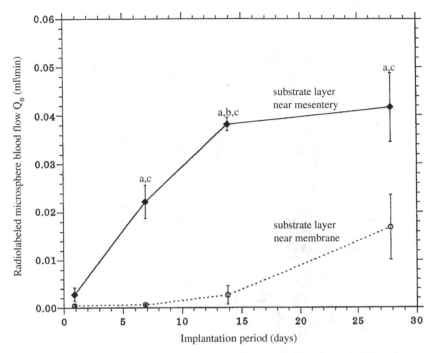

Figure 9.12   Blood flow measured in polymer matrix for each implantation period (from Sarver *et al.* 1995, with permission).

tissue. This value of the tissue blood perfusion rate within the support structure compares favorably with values obtained for other tissues within the body as summarized in Table 6.2.

9.10 MEASURING MASS TRANSFER RATES

The implant device shown in Figure 9.9 can also be used to determine the mass transfer characteristics of the capillary bed (Sarver *et al.* 1995). Note the presence of a small chamber in the region above the vascularized support structure. This chamber can also be immunoisolated from the vascularized support structure by a semipermeable membrane that typically has a nominal molecular weight cutoff of around 50,000 Daltons. This nominal pore size will stop components of the host's immune system (immune cells and antibodies, which have molecular weights > 150,000) from entering the chamber region of the device.

If the chamber region is loaded with a suitable tracer compound as shown in Figure 9.13, then the tracer's uptake and elimination from the body can be followed and analyzed using the pharmacokinetic techniques developed in Chapter 7. In this way, the transport properties of the capillary bed can be determined.

An ideal tracer compound for this type of study is radioactive inulin. Inulin is a

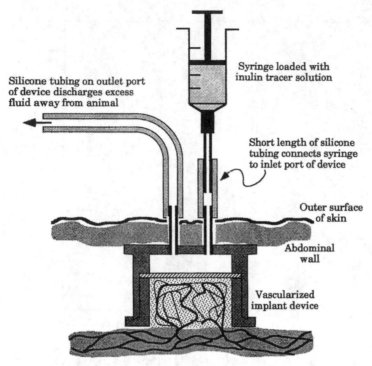

Syringe loaded with inulin tracer solution

Silicone tubing on outlet port of device discharges excess fluid away from animal

Short length of silicone tubing connects syringe to inlet port of device

Outer surface of skin

Abdominal wall

Vascularized implant device

Figure 9.13   The radiolabeled inulin tracer technique (Sarver *et al.* 1995, with permission).

polysaccharide molecule with a molecular weight of around 5500 Daltons. As mentioned in Chapter 7, inulin is only removed from the blood through the glomerulus and is not secreted or reabsorbed within the renal tubules. Furthermore, it is not metabolized and is non-toxic. Therefore, if a dilute solution of inulin is placed within the chamber of the device as shown in Figure 9.13, the only elimination pathway for inulin from the body is through the kidneys, and is equivalent, at any time, to the product of the GFR and the inulin plasma concentration. If the GFR and volume of distribution are known for the test subject, for example, from an inulin infusion study as in Example 7.1, then a pharmacokinetic model can be readily developed to correlate the transient plasma inulin levels (Sarver 1994; Sarver *et al.* 1995). Correlation of the pharmacokinetic model to the plasma inulin levels then provides a variety of transport information such as the matrix blood flow and the capillary wall and immunoisolation membrane permeability to inulin. With this information, one can then predict the permeabilities of other compounds of interest using the solute permeability relationships developed in Chapter 5. This type of information can then be used to perform scaleup studies for bioartificial organs as discussed in Chapter 10.

Figure 9.14 shows the dimensionless plasma inulin levels for the period of time after injection of inulin into the device chamber. The time since device implantation is

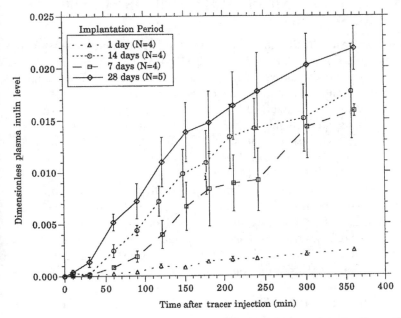

Figure 9.14   Plasma inulin level as a function of the time after tracer injection (Sarver 1994).

shown as a parameter. The devices, once again, contained two layers of the polyester surgical support felt and were implanted in the abdominal cavity of laboratory rats. The dimensionless plasma inulin concentration (X) is defined by the following equation.

$$X = \frac{C}{\left(\dfrac{D}{V_B}\right)} \tag{9.2}$$

In the above equation, C is the measured plasma inulin concentration, and the quantity (D/V_B) represents the ratio of the amount of inulin injected into the device chamber to the inulin distribution volume. This ratio represents the maximum possible plasma inulin concentration that would occur if the inulin dose were instantly mixed throughout its distribution volume after injection.

For a given time period since device implantation, we see from Figure 9.14 that the plasma inulin concentration increases steadily with time following its introduction into the chamber of the device. We also observe that at any time following addition of the inulin tracer to the device, the plasma inulin levels are higher for longer implantation periods. The rate of increase of the plasma inulin levels also increases for longer times since device implantation. This reflects the increase in capillary surface area and blood flow with time since implantation and serves as another measure of tissue ingrowth and vascularization in the support matrix material.

9.11 PHARMACOKINETIC MODELING OF INULIN TRANSPORT IN A POLYMERIC SUPPORT STRUCTURE

We can develop a pharmacokinetic model to analyze the plasma inulin levels for the period of time after injection into the chamber of the device shown in Figure 9.13. A compartmental model approach can be used as outlined in Figure 9.15. Four separate well-mixed compartments are used to define the distribution of inulin within the device and the body. The device itself is assumed to consist of three compartments. The first compartment is the implantation chamber where the dilute solution of inulin is initially injected. Inulin is then assumed to diffuse out of this chamber and across the semipermeable membrane and an adjacent layer of yet to be vascularized matrix. The inulin tracer then enters the interstitial fluid space of the tissue region that occupies the void space within the remaining portion of the support structure. This matrix tissue region is then the second compartment. The third compartment is the capillary bed itself that lies within the matrix tissue. These capillaries have blood flowing through them at the rate Q_{CB}. Inulin then diffuses from the tissue region into these capillaries where it is carried to the body represented by a single compartment having an inulin concentration of C_B. Inulin is then eliminated from the fourth or body compartment (V_B) by simple filtration through the kidneys.

Unsteady mass balances for the dimensionless inulin concentration can then be written as follows for each of the compartments.

$$V_I \frac{dX_I}{dt} = -P_{IT} S_{IT} \left(X_I - X_T \right) \tag{9.3}$$

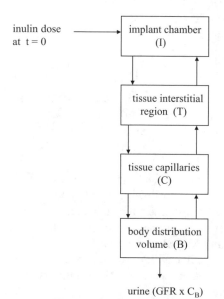

inulin dose
at t = 0

implant chamber
(I)

tissue interstitial
region (T)

tissue capillaries
(C)

body distribution
volume (B)

urine (GFR x C_B)

Figure 9.15 Compartmental model for analysis of plasma inulin concentrations.

$$V_T \frac{dX_T}{dt} = P_{IT} S_{IT} \left(X_I - X_T \right) - P_{TC} S_{TC} \left(X_T - X_C \right)$$

(9.4)

$$V_C \frac{dX_C}{dt} = Q_{CB} \left(X_B - X_C \right) + P_{TC} S_{TC} \left(X_T - X_C \right)$$

(9.5)

$$V_B \frac{dX_B}{dt} = -Q_{CB} \left(X_B - X_C \right) - \text{GFR} \times X_B$$

(9.6)

Four initial conditions are needed to solve these differential equations. Therefore, at $t = 0$, we have that $X_I = V_B/V_I$ and $X_T = X_C = X_B = 0$.

In Equations 9.3 through 9.6, the P's represent the solute permeability between the various compartments. P_{IT} is the overall permeability between the implantation chamber and the tissue region. This value times the bulk concentration difference between the implantation chamber and the tissue region, $\left(\dfrac{D}{V_B} \right) \left(X_I - X_T \right)$, provides the solute transport flux.

P_{IT} can be related to the membrane permeability (P_m). The membrane permeability (P_m) can be estimated using the equations outlined in Chapter 5 or measured by in vitro experiments (Baker *et al.* 1997; Sarver *et al.* 1995; Dionne *et al.* 1996; Iwata *et al.* 1996). However, in vitro estimates of the membrane permeability may be significantly higher than the actual in vivo permeability of the membrane because of protein adsorption on the surface of the membrane.

The additional resistance presented by the avascular matrix (P_{AV}) adjacent to the membrane also must be considered and can be estimated as discussed in Chapter 5 by Equation 5.83. The membrane resistance, and the resistance of the adjacent avascular matrix, may then be combined to give the apparent membrane permeability, i.e. $\dfrac{1}{\bar{P}_m} = \dfrac{1}{P_m} + \dfrac{1}{P_{AV}}$. As we will soon show in Example 9.1, experimental data such as that shown in Figure 9.14 can be used to obtain the apparent membrane permeability and an estimate of the in vivo membrane permeability. Therefore, P_{IT} is given by the following equation.

$$P_{IT} = \frac{1}{\dfrac{1}{P_m} + \dfrac{1}{P_{AV}}} = \bar{P}_m$$

(9.7)

P_{TC} represents the solute permeability of the capillary wall (P_{CW}) which is assumed to be the controlling resistance for transport from the tissue region and into the capillary. S_{IT} is the surface area available for transport between the implantation chamber and the tissue region. This is equal to the immunoisolation membrane surface area S_M. S_{TC} represents the surface area available for solute transport between the tissue region and the capillaries. This is equivalent to the total capillary surface area S_{CW} and is given by this relationship, $\varepsilon_m S_M \delta_T f_T s$, where s is the capillary surface area per volume of tissue and is given by Equation 5.119. Recall that f_T is the fraction of the support structure volume that contains tissue.

Q_{CB} is the blood flowrate within the tissue region corrected for the extracellular volume fraction of blood, ζ_C, which is about 0.63 for inulin. This accounts for the fact that the inulin concentration measurements were based on plasma volume. If the inulin concentration was based on the total blood volume, then ζ_C would be equal to one. V_I is the volume of the implantation chamber and is based on the given device dimensions. V_T is the volume of the tissue region extracellular fluid and is equal to $f_T \zeta_T S_M \delta_T \varepsilon_M$, where ζ_T is the total extracellular volume fraction of the tissue within the support structure. Inulin only distributes within the extracellular fluid space in the tissue region, and the extracellular volume fraction (ζ_T) of most tissues is about 0.15. This fluid surrounding the cells is also assumed to be well-mixed and the inulin concentration homogeneous. The capillary volume in the tissue region (V_C) is given by $S_M \delta_T \varepsilon_M f_T v$, where v is the capillary volume fraction given by Equation 5.120.

Equations 9.3 through 9.6 can be simplified somewhat by recognizing that the capillary volume, V_C, is much smaller than the other volumes. Since V_C is so small, it has the effect of making the left-hand side of Equation 9.5 very nearly equal to zero. This creates a pseudo-steady state situation for the solute concentration in the capillaries and means that we can replace the differential equation (Equation 9.5) with the following algebraic equation.

$$Q_{CB}\left(X_B - X_C\right) + P_{TC}S_{TC}\left(X_T - X_C\right) = 0 \tag{9.8}$$

This equation can be simplified further by recognizing that the body compartment solute concentration (X_B) is much smaller than the solute concentration within the tissue capillaries of the implant device (X_C). This makes sense because the volume of the body compartment is several orders of magnitude larger than that of the capillaries within the device's support structure. Once the solute enters the body compartment, it becomes significantly diluted.

Equation 9.8 can then be rearranged to provide the solute capillary concentration within the tissue region in terms of the solute concentration in the tissue compartment only.

$$X_C = \left[\frac{P_{TC}S_{TC}}{Q_{CB} + P_{TC}S_{TC}}\right]X_T = K_{TC}X_T \tag{9.9}$$

This equation also shows that the capillary solute concentration is essentially at equilibrium with the tissue solute concentration as given by the constant distribution coefficient, K_{TC}. We can now rewrite Equations 9.3, 9.4, and 9.6, using Equation 9.9 for the capillary concentration, to obtain the following set of equations.

$$V_I \frac{dX_I}{dt} = -P_{IT}S_m\left(X_I - X_T\right) \tag{9.10}$$

$$V_T \frac{dX_T}{dt} = P_{IT}S_m\left(X_I - X_T\right) - P_{CW}S_{CW}\left(1 - K_{TC}\right)X_T \tag{9.11}$$

$$V_B \frac{dX_B}{dt} = Q_{CB} K_{TC} X_T - \mathrm{GFR} \times X_B \qquad (9.12)$$

These equations may be solved analytically using Laplace transforms (Sarver 1994; Gibaldi and Perrier 1982) to provide the following results for the dimensionless solute concentrations in the implantation chamber, the tissue region, and within the body compartment. Recall that Equation 9.9 provides the corresponding value of the concentration within the capillary compartment.

$$X_I = \left[\frac{V_B \left(\lambda_1 - k_{TI} - k_{CW} \right)}{V_I \left(\lambda_1 - \lambda_2 \right)} \right] e^{-\lambda_1 t} + \left[\frac{V_B \left(\lambda_2 - k_{TI} - k_{CW} \right)}{V_I \left(\lambda_2 - \lambda_1 \right)} \right] e^{-\lambda_2 t} \qquad (9.13)$$

$$X_T = \left[\frac{V_B k_{IT}}{V_T \left(\lambda_1 - \lambda_2 \right)} \right] \left(e^{-\lambda_2 t} - e^{-\lambda_1 t} \right) \qquad (9.14)$$

$$X_B = \left[\frac{k_{IT} k_{CW}}{\left(\lambda_1 - \lambda_2 \right)\left(\lambda_1 - k_{TE} \right)} \right] e^{-\lambda_1 t} + \left[\frac{k_{IT} k_{CW}}{\left(\lambda_2 - \lambda_1 \right)\left(\lambda_2 - k_{TE} \right)} \right] e^{-\lambda_2 t} + \left[\frac{k_{IT} k_{CW}}{\left(k_{TE} - \lambda_1 \right)\left(k_{TE} - \lambda_2 \right)} \right] e^{-k_{TE} t}$$
$$(9.15)$$

The rate constants (k's) are given by the following set of equations.

$$k_{IT} = \frac{P_{IT} S_m}{V_I} \qquad (9.16)$$

$$k_{TI} = \frac{P_{IT} S_m}{V_T} \qquad (9.17)$$

$$k_{CW} = -\frac{P_{CW} S_{CW} \left(K_{TC} - 1 \right)}{V_T} = \frac{Q_{CB} K_{TC}}{V_T} \qquad (9.18)$$

$$k_{TE} = \frac{\mathrm{GFR}}{V_B} \qquad (9.19)$$

The constants λ_1 and λ_2 are given by the equations below.

$$\lambda_1 = \frac{1}{2} \left[k_{IT} + k_{TI} + k_{CW} + \sqrt{\left(k_{IT} + k_{TI} + k_{CW} \right)^2 - 4 k_{IT} k_{CW}} \right] \qquad (9.20)$$

$$\lambda_2 = \frac{1}{2} \left[k_{IT} + k_{TI} + k_{CW} - \sqrt{\left(k_{IT} + k_{TI} + k_{CW} \right)^2 - 4 k_{IT} k_{CW}} \right] \qquad (9.21)$$

Solution of the above equations then provides a description of the solute concentrations in the various compartments following the injection of the solute in the implantation chamber. These equations can also be used to determine the membrane permeability, the capillary wall permeability, and the total elimination rate constant using data such as that shown in Figure 9.14. This is illustrated in the example below.

Example 9.1 Use the pharmacokinetic model for inulin distribution developed above and perform a non-linear regression analysis of the 28 day data shown in Figure 9.14. Obtain best estimates of the glomerular filtration rate (GFR), the in vivo membrane permeability, and the capillary wall permeability (PS_{cw}) using this data which was obtained from a total of five rats. The measured in vitro membrane permeability for inulin is 7.8×10^{-6} cm sec^{-1}. Also, recall that the blood perfusion rate for the vascularized matrix was 0.42 ml cm^{-3} min^{-1} (see Figure 9.12) and the vascularized tissue fraction of the matrix was 0.72 (see Figure 9.10).

SOLUTION Additional data needed to solve this problem are as follows. The depth of the implantation chamber (δ_l) is 0.07 cm and the volume of this chamber (V_l) is 0.052 ml. The inulin diffusivity in the bulk isotonic solution (D_l) at 37°C can be estimated from Figure 5.4 to be about 2.3×10^{-6} cm^2 sec^{-1}. D_T is the effective diffusivity in the tissue region and can be found using the data presented earlier in Figure 5.12. For a molecule the size of inulin, the diffusivity will be reduced by about 85% giving a value for D_T of 3.45×10^{-7} cm^2 sec^{-1}. The inulin distribution volume for a rat was found in Example 7.1. The same example also provides an initial estimate of the GFR. The GFR will then be determined from the inulin data since GFR has a significant effect on the inulin clearance. The rats used in these experiments also had an average body weight of 481 grams.

Figure 9.16 shows the results of a non-linear regression analysis of the data using the pharmacokinetic model developed above. We see that a non-linear regression on the unknown apparent membrane permeability ($\bar{P}_m$), the product of the capillary permeability and the capillary wall surface area (PS_{CW}), and the glomerular filtration rate (GFR) results in an excellent fit of the inulin data. The apparent membrane permeability ($\bar{P}_m$) is found to be 6.45×10^{-7} cm sec^{-1}. The in vitro inulin permeability for this membrane (P_m) was found to be 7.8×10^{-6} cm sec^{-1}. The lower in vivo value is reasonable considering the permeability reduction that would result

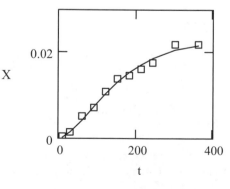

Figure 9.16 Comparison of plasma inulin concentrations with the PK model.

from protein adsorption within the 250 micron thick fibrous backing material of the membrane (Opong and Zydney 1991; Robertson and Zydney 1990a,b) and the resistance of the adjacent unvascularized layer of surgical support felt. The capillary wall inulin permeability and surface area product (PS_{cw}) was found to be 0.0016 cm^3 sec^{-1} 100^{-1} g^{-1} of tissue. Based on the data shown in Figure 5.13, a molecule the size of inulin (1.5 nm) should be expected to have a PS_{CW} value on the order of 0.01 cm^3 sec^{-1} 100^{-1} g^{-1} of tissue. The observed order of magnitude reduction in the value of PS_{CW} could be a result of reduced capillary surface area and/or capillary wall permeability. The Krogh tissue cylinder radius was obtained from analysis of histological sections and found to equal 68 microns. This would give a capillary surface area per volume of tissue equal to 15.6 1 cm^{-1}. Renkin (1977) presents estimates of the capillary surface area in tissues like that developing in this example that are on the order of 70 1 cm^{-1} with inulin permeabilities on the order of 10^{-6} cm/sec, thus giving a PS_{CW} on the order of 0.007 cm^3 sec^{-1} 100^{-1} g^{-1} of tissue. Therefore, it appears that the less than expected value of PS_{CW} obtained from the data used in this example is a result of reduced capillary surface area.

9.12 CELL TRANSPLANTATION INTO POLYMERIC SUPPORT STRUCTURES

The previous discussion shows that porous polymeric support structures are capable of developing within their confines a neovascularized tissue region. Furthermore, we find that this new tissue region has blood flow rates and capillary wall permeabilities comparable to those found in other tissues. We may now ask the following questions related to cell transplantation into these vascularized support structures. When is the best time to transplant cells? Is it possible to maintain the viability of transplanted cells, and can they continue to perform their normal function? To answer these questions consider the results reported by Thompson *et al.* (1988, 1989). They induced the formation of *organoid* neovascular structures in rats using polytetrafluoroethylene (PTFE) fibers coated with collagen and the angiogenesis initiator acidic fibroblast growth factor (aFGF). The fibers are about 20 microns in diameter and a bundle, or cotton ball-like mass, of these fibers was implanted within the abdominal cavity adjacent to the liver.

To demonstrate the efficacy of a cellular transplant using the vascularized PTFE support structure, they used the Gunn rat as a host for the implant. *Homozygous* Gunn rats lack the liver enzyme needed for the conjugation of bilirubin. Recall that bilirubin is a product obtained from hemoglobin at the end of a red blood cell's lifespan. The conjugated form of bilirubin is readily passed from the liver via the bile and then removed from the body in the feces. Since the Gunn rat cannot conjugate bilirubin, its plasma bilirubin levels are increased significantly from the normal level of < 1 mg dl^{-1}. The Wistar rat is genetically identical to the Gunn rat except that it has the ability to conjugate bilirubin. In their initial set of experiments, they obtained hepatocytes from the livers of the syngeneic Wistar rats. These hepatocytes were then seeded onto collagen coated PTFE fibers that did not contain the angiogenesis promoter aFGF. This structure was then implanted adjacent to the liver of the Gunn rat. Because the Gunn and Wistar rats are genetically the same, there is no need for immunosuppressive agents or immunoisolation of the transplanted cells. For the first 10 days after implantation, the plasma bilirubin levels remained unchanged as shown in Figure 9.17A. By the 20th day, the plasma bilirubin levels had decreased by about 50%. Shortly thereafter, the plasma bilirubin levels returned to their original high levels and remained there for the remaining days of the experiment.

The lack of long-term function of the hepatocytes in these preliminary studies can be attributed to several effects. First, the angiogenesis factor, aFGF, was not used in these first experiments. This would result in less vascularization of the support structure. Furthermore, the cells were seeded into the support structure at the time of implant. Lack of a good vascular supply at the time of cell seeding is clearly not optimal for cell growth and the transport of nutrients and bilirubin. This would account for the delayed response with regard to any effect on the plasma bilirubin levels. For the first 10 days, the implant was still in the process of tissue ingrowth and the development of a blood supply that is needed for efficient mass transport. It also allowed for the accumulation of bile acids that led to the death of the cells after about 20 days. Clearly,

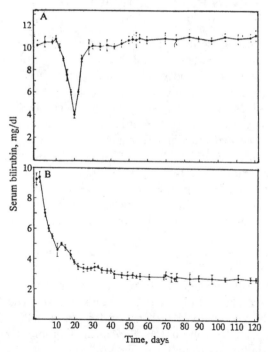

Figure 9.17 Plasma bilirubin levels following transplantation of hepatocytes into Gunn rats (A) collagen-coated PTFE support structure with hepatocytes added at time of implantation; (B) collagen-coated PTFE support structure with aFGF. Hepatocytes added 28 days after implantation. (from Thompson *et al.* 1989, with permission).

these results show that seeding the cells at the time the support structure is implanted is not the best approach.

In their second set of experiments, the collagen coated PTFE fibers containing adsorbed aFGF was first implanted. After 28 days, a suspension of Wistar rat hepatocytes was seeded into the network of the now vascularized fibers. As shown in Figure 9.17B, we see that the plasma bilirubin levels began to decrease within 1 day. After about 10 days, the plasma bilirubin levels had decreased by about 50% and at the end of the 120 day experiment this reduction was > 60%.

These results strongly suggest that polymeric support structures, through their ability to form their own vascular supply, can be used to sustain the long-term function of transplanted cells. Additionally, it appears better first to vascularize the support structure for several weeks before seeding the implant with the transplanted cells. Furthermore, these results demonstrate that transplanted cells carrying a normal gene are able to restore the function lost by the host's own genetically compromised cells. If the host's genetically deficient cells can be removed, the appropriate gene inserted, and their numbers expanded in tissue culture prior to their transplantation, then this approach becomes a powerful treatment method for a variety of diseases. The only limitation that needs to be addressed is the problem of immunorejection of the transplanted cells if they are not genetically identical to those of the host. The next chapter, therefore, addresses how immunoisolation of the transplanted cells can be incorporated to create bioartificial organs.

PROBLEMS

1. Consider the steady state diffusion of glucose from a well-stirred bulk liquid solution across a thin hydrogel layer and then reaction and diffusion of the glucose within a multilayer of epithelium which lies on top of the hydrogel layer (Perez *et al.* 1995). The epithelium consists of N layers of cells, each cell h_c in thickness. The cells consume glucose at a first order volumetric rate given by $k_1 C$, where C is the local concentration of glucose and k_1 is the first order rate constant. The surface of the cells exposed to the gas phase above is a no flux boundary for glucose, and C will, therefore, attain its minimum value at that interface (C_{min}). Calculate the fractional drop (C_b / C_{min}) in the glucose concentration from the bulk liquid medium (C_b) to the outermost epithelial surface (C_{min}). Assume there is no mass transfer resistance between the bulk solution and the surface of the hydrogel. Some additional information needed to solve this problem is shown in the table below.

Parameter	Value	Description
h_c	15 microns	thickness of single cell layer
k_1	1.5×10^{-3} 1 sec^{-1}	glucose rate constant
C_b	3.2 mg ml^{-1}	bulk glucose concentration
D_G	3×10^{-6} cm^2 sec^{-1}	diffusivity of glucose in the hydrogel
K_p	2	glucose hydrogel partition coefficient, bulk-gel or gel-epithelium
L	0.05 cm	hydrogel thickness
N	5	layers of epithelial cells
D_e	6×10^{-6} cm^2 sec^{-1}	glucose diffusivity in epithelium

2. The following dimensionless inulin concentrations were obtained for a device like that shown in Figure 9.9. In this case, the immunoisolation membrane was that developed by Baker *et al.* (1997). This membrane consisted of a polyvinyl alcohol hydrogel imbedded within the pores of a 0.2 micron polyethersulfone microporous membrane. Using this data, determine the apparent membrane permeability, the value of PS_{CW} for the capillaries, and the GFR. The device used in these experiments had a membrane radius of 0.65 cm. The vascularized matrix was 0.15 cm thick. The chamber containing the inulin was 0.07 cm thick. The rat body weight was 345 grams. All other properties are as in Example 9.1. How does the value of PS_{CW} for inulin compare with the data shown in Figure 5.13?

Time, minutes	Dimensionless inulin concentration, X
0	0.0
10	0.0188
30	0.0714
60	0.1433
90	0.1210
120	0.0943
150	0.1148
180	0.1044
210	0.0773

3. Select from Table 9.1 a subject for tissue engineering. Prepare a presentation for your class that summarizes recent advancements in the area you selected. Topics covered in your presentation should include a description of the clinical need, the methodology used, results that have been obtained (in vitro, in animals, and the status of any human clinical trials), the potential clinical impact, potential market impact, safety issues, and additional development needs.

4. Design a tissue engineered system for the delivery of human growth hormone (hGH) to the systemic circulation. hGH has a molecular weight of 22 kDaltons. Assume that the hGH has an apparent distribution volume of 30 liters. Autologous cells were transfected with a recombinant gene for hGH. The production rate of hGH from these cells is about 2500 ng 10^{-6} cells^{-1} 24^{-1} hr^{-1}. The plasma concentration of hGH should be maintained at about 10 ng ml^{-1}. Assume hGH has a half-life in the body of about 2.3 hours. Describe your system for delivering hGH and carefully state all assumptions. How many cells will be required for the delivery of the hGH?

BIOARTIFICIAL ORGANS

10.1 BACKGROUND

Tissue engineering is a very promising approach for the treatment of a variety of medical conditions. Treatment of a disease or medical problem using this approach requires the availability of the appropriate cells and the creation of an artificial support structure to contain them. The cells may be obtained from a variety of sources, for example, from an expanded population of the host's own (perhaps genetically modified) cells, from other compatible human donors, from animal sources, or even genetically engineered cell lines. The cells to be transplanted are then seeded at the appropriate time into the prevascularized polymeric support structure as discussed in the previous chapter. However, with the exception of *autologous cells*, one of the major obstacles that must be overcome is rejection of the transplanted cells by the host's immune system. The transplanted cells will be destroyed quickly by the host's immune system unless they are immunologically similar to the host's own cells (Benjamin and Leskowitz 1991).

Immunosuppressive drugs can be used to suppress the host's immune system and prolong the function of transplanted cells that are a relatively close match to the host. However, immunosuppressive drugs have potent side effects and, for example in the case of islet of Langerhans transplants to treat insulin-dependent diabetes, these drugs may provide a situation where the cure is worse than the disease. Genetically engineered cell lines, in addition to their possibly being rejected by the host's immune system, also pose additional risks that need to be considered. For example, direct implantation of an immortal (usually of tumor origin) cell line could lead to the unchecked growth of the implanted cells, they could move to and proliferate at a site other than the desired site, and they could change to a potentially hazardous form with a loss of their original therapeutic function. The full potential of tissue engineering will, therefore, be limited unless techniques can be developed either to restrict the host's immune response, or somehow modify the transplanted cells to make them more acceptable to the host's immune system, i.e. *immunomodification* (Lanza and Chick 1994b).

10.2 SOME IMMUNOLOGY

The immune response to foreign materials (*antigens*) such as transplanted cells consists primarily of a *cell-mediated component* and a *humoral component* (Benjamin and Leskowitz 1991). The major cellular components are the *B lymphocytes* and the *T lymphocytes*. The B lymphocytes, or B cells, form in the *bone marrow*, and, when properly activated by antigens, form proteins called *antibodies* that comprise the active agents of the humoral component of the immune system. The T lymphocytes originate in the *thymus* and come in two basic types, the *CD4⁺* (*helper*) *T cells* and the *CD8⁺* (*killer or cytotoxic*) *T cells*.

The immune system is activated by its intimate contact with foreign molecules called *antigens* (*anti*body *gen*erating). Antigens must possess "foreignness", meaning the antigen is unlike anything the immune system has seen before. Immature B and T lymphocytes that react against *self-antigens* (the body's own proteins for example) are eliminated during their maturation phase in the bone marrow or the thymus. This is the so-called *clonal deletion theory*. Thus, the mature B and T cells are *self-tolerant* and do not normally react against the body's own tissues.

Antigens generally have molecular weights > 6000 Daltons and possess some degree of molecular complexity. In fact, there can be many sites along the surface of such large molecules that are *immunogenic*[1]. These sites are referred to as *epitopes*. Fortunately, many synthetic polymers, although of high molecular weight, do not possess a sufficient amount of molecular complexity to provoke an immune response. Lower molecular weight materials, although not immunogenic by themselves, can associate for example with larger carrier molecules, like a protein, to give an immune response directed at the lower molecular weight compound. In this case, the lower molecular weight compound is referred to as a *hapten*.

10.2.1 B Lymphocytes

Each B cell has a unique antibody receptor on its surface that only recognizes a specific antigen. The B cell inventory within an individual is capable of producing as many as 100 million distinct antibodies. This allows the immune system to respond to almost any known or unknown antigen. If a specific B cell comes into contact with its corresponding antigen, it becomes activated through a process that we shall later see also involves the CD4⁺ or helper T cells. The activated B cell then begins to reproduce, rapidly increasing the number of B cells that are specific for a given antigen. This

---

[1]An immunogen is a molecule that can induce a specific response of the immune system. An antigen refers to the ability of a given molecule to react with the products of an immune response. For example, the binding of an antigen to an antibody. In this discussion, we will assume that antigen and immunogen are synonymous.

expanded collection of lymphocytes with a given specificity is referred to as a *clone of lymphocytes*. This selection by the antigen of a specific reactive clone of lymphocytes from a large pool of existing lymphocytes, each with their own unique antigen specificity, is called the *clonal selection theory*. These B cells then differentiate to form *plasma cells* that begin to secrete antibodies with the same antigen specificity.

10.2.2 Antibodies

Antibodies are proteins and are also called *gamma globulins* or *immunoglobulins*. They comprise about 20% of the total amount of protein found in plasma. There are five classes of antibodies that are called IgA, IgD, IgE, IgG, and IgM, where Ig stands for immunoglobulin and the letter designates the class. The basic structure of an antibody molecule consists of two light polypeptide chains and two heavy polypeptide chains held together by disulfide bonds as shown in Figure 10.1.

Conceptually, the antibody molecule is "Y-shaped", with each light chain paired in the upper branches of the "Y" with the heavy chains that form the "Y" structure. The molecule consists of three fragments. Two of these fragments are identical and reside at the top of the "Y" and each of these binds with antigen. These fragments are referred to as Fab, for fragment antigen binding. The base of the "Y" is called the Fc fragment, for fragment crystallizable. The Fc fragment does not bind with antigen but is responsible for the biological activity of the antibody after it binds with antigen. The Fc fragment is referred to as a *constant region* in terms of its amino acid sequence since it is the same for all antibodies within a given class. Since the Fab fragments bind to antigen, their structure is highly *variable* in order to provide the multitudinous shapes or conformations required for antigen specificity and recognition. The specific antigen binding characteristics of the Fab fragments is determined by their unique sequence of amino acids. Although an activated B cell (or plasma cell) makes antibodies with only a single antigen specificity, it can switch to make a different class of antibody, while still retaining the same antigen specificity.

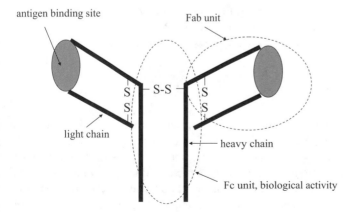

Figure 10.1   Structure of an antibody.

IgG is the major immunoglobulin in the body and is found in all the fluid spaces. It consists of a single "Y" shaped molecule with a molecular weight of about 150,000 Daltons. IgG has several important biological properties that are found in some of the other antibody classes as well. Because of its ability to bind two antigens per IgG molecule, IgG can cause the clumping or *agglutination* of particulate antigens such as those of invading microorganisms. These large antibody–antigen complexes are then readily phagocytized by phagocytic cells such as neutrophils and macrophages. In the case of smaller soluble antigens, this cross-linking of antigen and IgG leads to much larger complexes that become insoluble and *precipitate* out of solution and are then phagocytized.

IgG is also capable of binding to a variety of epitopes found on the surfaces of invading cells. The cell then becomes covered with IgG molecules that have their Fc fragments sticking out. Many phagocytic cells have surface receptors for the IgG Fc fragments allowing the phagocyte to bind to, engulf, and then destroy the invading organism. IgG is, therefore, an *opsonizing* antibody (opsonin is a Greek word that means to prepare for eating) since it facilitates the subsequent "eating" of the invading microorganism or cell by phagocytic cells.

The Fc receptor on an IgG molecule bound to a cellular antigen also binds to specific receptors on so-called *natural killer lymphocytes* (*NK cells*). The NK cells will then be attracted to and destroy the invading cell by the release of toxic substances called *perforins* (Ojcius *et al.* 1998). Perforins form channels in the cell membrane of the target cell making them leak and resulting in death of the cell. This process is called *antibody dependent cell-mediated cytotoxicity* or ADCC.

IgG is also capable of neutralizing many toxins by binding to their active site. Viruses are also neutralized in a similar manner. In this case, the IgG molecule binds and blocks sites on the virus's surface coat that are used by the virus for attachment to the cells it normally invades. IgG, when bound to antigen, is also capable of activating the *complement system*. The complement system consists of a group of protein enzyme precursors that, when activated, undergo an amplifying cascade of reactions that generate cell membrane attack complexes (MACs) that literally punch holes through the cell membrane. This results in lysis and death of the targeted cell.

IgA is found primarily in the fluid secretions of the body, for example in saliva, tears, mucus, and in the gastrointestinal fluids. It is found in a dimeric form, that is it consists of two "Y" shaped antibodies linked at the base of the "Y". The molecular weight of this antibody is about 400,000 Daltons. Its primary role is as a defense against local infections where it stops the invading microorganism from penetrating the body's epithelial surfaces that line the respiratory and gastrointestinal tract. It does not activate the complement system.

IgM forms a planar star-like pentamer and consists of five antibody molecules joined together at the base of the Fc regions. Its molecular weight is about 900,000 Daltons, and it is found mainly in the intravascular spaces. It has only five antigen binding sites because the restricted pentameric structure does not allow the Fab fragments to be open fully. Therefore, large antigens bound to one Fab will block the adjacent Fab. IgM is a very efficient agglutinating antibody since its large structure provides for binding between epitopes that are widely spaced on the antigen. IgM is also an excellent activator of the complement system. IgM is also found on the surface of mature B lymphocytes where it serves as the specific antigen receptor.

IgD is also present primarily on the cell surface of the B lymphocytes and is believed to be involved in the maturation of these cells. It exists as a single antibody molecule and, therefore, has a molecular weight of about 150,000 Daltons. Not much is really known about this antibody.

Finally, *IgE* also exists as a single antibody molecule and has a molecular weight of about 200,000 Daltons. The Fc region of the IgE molecule binds with high affinity to specific receptors found on *mast cells* and *basophils*. Mast cells are found outside of the capillaries within the connective tissue region (see Figure 9.3). They are involved in the process of inflammation and are responsible for the secretion of *heparin* into the blood as well as *histamine, bradykinin*, and *serotonin*. Basophils are a type of white blood cell or, *granulocyte*, that mediates the inflammation process through release of these same substances. When antigen binds to two adjacent cell surface bound IgE molecules, the cell (mast or basophil) becomes activated and releases a host of potent biologically active compounds such as histamine. These agents are responsible for the dilation and increased permeability of the blood vessels. In normal situations, these changes facilitate the movement of the immune system components, such as white blood cells, antibodies, and complement, into localized sites of inflammation and infection. However, in people with allergies, a particular antigen, in this case called an *allergen*, stimulates the IgE on the surface of mast cells and basophils leading to the unwanted effects of an allergic reaction.

10.2.3 T Lymphocytes

The T lymphocytes are the other major cell type that forms the basis of the cell-mediated immune response. The T cells are characterized by the presence of an antigen specific T cell receptor or TcR. The T cell receptor consists of an antigen recognizing molecule called Ti in close association with another polypeptide complex called CD3. The T cells, unlike the B cells, are not, however, activated by free antigen. T cell activation requires that the antigen be *presented* by other cells such as macrophages or other B cells. These cells that present antigen to the T cells are collectively called *accessory cells* or *antigen presenting cells* (APCs).

The APCs ingest and breakdown the polypeptide antigen into much smaller fragments. These fragments then become associated with special molecules called the *major histocompatibility complex* (MHC). There are two major types of MHC molecules, called *MHC class I* and *MHC class II*. Recall that these molecules belong to the class of cell adhesion molecules (CAMs). The MHC class I molecule is expressed by almost every nucleated cell found in the body. The MHC class II molecule is only found in specialized antigen presenting cells such as B cells and macrophages. The MHC molecule forms an antigen binding area in the shape of a cleft or pocket. This pocket can accept polypeptide antigens consisting of up to 20 amino acids. The complex of small antigen and MHC is then transported to the surface of the cell where the antigen is presented for recognition by the T cell receptor. The T cell, therefore, only recognizes antigen in combination with the MHC molecule.

T cell antigen recognition is shown in Figure 10.2. On the left side of this figure, we see a foreign material such as a dead virus, a cancer cell, a microorganism, or a large polypeptide, being ingested and broken down by an APC. The antigen, in combination with an MHC class II molecule, is then transported to the cell surface for antigen presentation. On the right side of this figure, we see a live virus infecting the cell. As the virus takes over the operation of the cell to increase the number of viruses, many of the viral peptides that are produced are transported to the cell surface by MHC class I molecules for antigen presentation. The presence of antigen and MHC will then be recognized by the T cell.

However, in either case, the T cell receptor by itself has a low affinity for antigen bound to MHC. To facilitate the recognition and binding of the T cell receptor with the antigen/MHC complex, there exist two important accessory molecules called CD4 and CD8 that also serve to distinguish between the two principal types of T cells.

T cells that have the CD4 molecule are known as $CD4^+$ or helper T cells, whereas those that have the CD8 molecule are known as $CD8^+$ or cytotoxic T cells. $CD4^+$ T cells only recognize antigen that is bound to the MHC class II molecule on the surface of an APC, see the left side of Figure 10.2. The $CD8^+$ T cell can only recognize antigen that is bound to the MHC class I molecule on the surface of all other nucleated cells found in the body as shown on the right side of Figure 10.2. This is called *MHC restriction*, that is the response of $CD4^+$ T cells is restricted to only that antigen bound to MHC class II molecules, whereas the response of $CD8^+$ T cells is restricted to only that antigen bound to MHC class I molecules.

10.2.4 Interaction Between APCs, B Cells, and T Cells

With this background, we can now examine the cooperative relationship that exists between the APCs, B cells, and the T cells during the immune response. First let us consider the interaction between a resting $CD4^+$ T cell, an APC, and a B cell. This is illustrated in Figure 10.3. The first step involves ingestion by an APC of an antigen that, in this example, contains both a B cell epitope and a T cell epitope. The APC could be

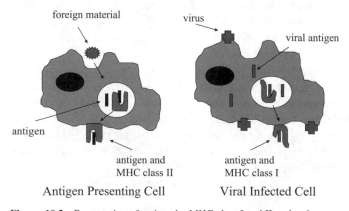

Figure 10.2 Presentation of antigen by MHC class I and II molecules.

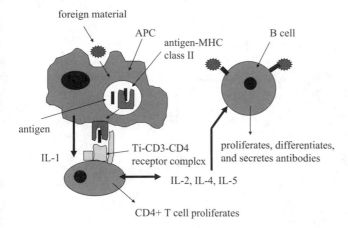

Figure 10.3 Interaction between APC, CD4⁺ T cell, and the B cell.

a macrophage that engulfs the antigen, or a B cell that internalizes the antibody–antigen complex. The T cell epitope is then expressed on the surface of the APC in combination with an MHC class II molecule. A CD4⁺ T cell that has a Ti-CD3 receptor that is specific for this particular antigen then binds with the antigen/MHC class II complex. Note that the CD4⁺ molecule stabilizes the binding of the antigen/MHC complex with the T cell receptor.

This binding of the CD4⁺ T cell and the antigen/MHC class II complex constitutes the first signal for activation of the CD4⁺ T cell. This is then followed by the release of a second signal by the APC of a soluble substance called interleukin-1 (IL-1) that is also essential for T cell activation. IL-1 is a small protein with a molecular weight of 15,000 Daltons that belongs to a class of substances called *lymphokines*. Lymphokines are cellular messengers that have an effect on other lymphocytes and are a subcategory of a broader class of intercellular messengers called *cytokines*. The CD4⁺ T cell then becomes activated and begins to secrete its own interleukins, specifically IL-2, IL-4, and IL-5. IL-2 induces the CD4⁺ T cell to proliferate, rapidly forming a clone of CD4⁺ T cells that are reactive to the specific antigen presented by the APC. IL-4 then activates the B cells and IL-5 induces the activated B cells to proliferate in number, forming a clone of B cells.

It is important to note that binding of antigen with the B cell receptor is not sufficient for the B cell to become activated and then differentiate to antibody secreting plasma cells. The B cell must become activated through the process shown in Figure 10.3 which also involves the CD4⁺ T cell. Additionally, the B cell need not bind with the same epitope that activated the T cell. Hence, the activated B cell may release antibodies with a different epitope specificity. Since the CD4⁺ T cell is responsible for the activation of the B cell, it is referred to as the "helper" T cell. The activated CD4⁺ T cell also secretes other lymphokines, such as *γ-interferon*, that serve to attract and activate macrophages and natural (NK) killer cells and inhibit viral replication.

Figure 10.4 illustrates how the CD8⁺ T cell becomes activated. Here, we see that the CD8⁺ T cell receptor binds with the antigen/MHC class I complex presented for

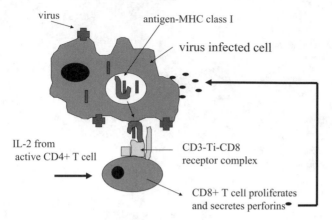

Figure 10.4   Activation of CD8$^+$ T cells.

example, by a virally infected cell. Once again the CD8 molecule stabilizes the interaction between the T cell receptor and the antigen/MHC complex. This binding of T cell receptor and antigen/MHC complex represents the first signal. In order for the CD8$^+$ T cells to proliferate (i.e. form a clone of CD8$^+$ T cells) and become activated, it also must receive a second signal that is provided by the IL-2 that is released as a result of the activation of the CD4$^+$ T cell shown in Figure 10.3. The activated CD8$^+$ T cell will then kill any cell that expresses the appropriate combination of antigen and MHC class I. The activated CD8$^+$ T cell accomplishes this cellular destruction through the release of special molecules called *perforins* that destroy the cell membranes of the target cell. Because of their role in cell death, the CD8$^+$ T cells are also known as cytotoxic T cells or killer T cells.

10.2.5 The Immune System and Transplanted Cells

The above discussion illustrates the complexity as well as the coordination that exists between the different components of the immune system. The immune system, through an elegant process, is capable of recognizing "self" from "non-self". Our interest here is to prevent or restrict the action of the immune system towards the transplanted cells. The immune system response to transplanted cells involves a combination of effects resulting from antibodies, complement, macrophages, B cells, and T cells.

Antibodies recognize the foreign antigens presented by the transplanted cells and induce the destruction of the transplanted cells through activation of the complement system and NK cells by the process of ADCC. Rejection of the transplanted cells also occurs by a T cell response against the MHC molecules that are expressed by the transplanted cells. The transplanted cells contain foreign MHC class I and class II molecules. The foreign MHC class II molecules are present on "passenger" leukocytes and macrophages that are present in the transplanted tissue. The foreign MHC class II molecules are sufficiently similar to the host's own combination of antigen/MHC class II complex to trigger the activation of the CD4$^+$ T cells. This is also true for the foreign

MHC class I molecules, leading to the activation of the host's CD8[+] T cells. Activation of these T cells and the antibody producing B cells, then leads to the destruction of the transplanted cells.

10.3 IMMUNOISOLATION

A promising method for restricting the host's immune response is to immunoisolate the transplanted cells (Lanza *et al.* 1995; Zielinski *et al.* 1997). This concept was shown earlier in Figure 9.2. Figure 10.5 presents a more detailed view of immunoisolation showing the possible pathways for rejection of the transplanted cells (Colton 1995). Immunoisolation can be accomplished through the use of a specially designed polymeric membrane that prevents passage of the major components of the immune system, that is the immune cells, antibodies, and complement.

The immune response occurs as a result of antigens shed by the transplanted cells. These antigens may be products secreted by the functioning cells or materials released by the death of the transplanted cells. These antigens will cross the immunoisolation membrane, and when recognized and presented by the host's immune system, lead to the cellular and humoral immune responses discussed earlier.

To restrict the passage of cells such as lymphocytes and macrophages, the pores or spaces within the immunoisolation membrane must be no larger than a micron or so. Recent evidence indicates that blocking the entry of the cellular component of the immune system is sufficient to prevent the rejection of allografts (Colton 1995). This is of special interest for applications involving allogeneic cells for gene therapy (Chang *et*

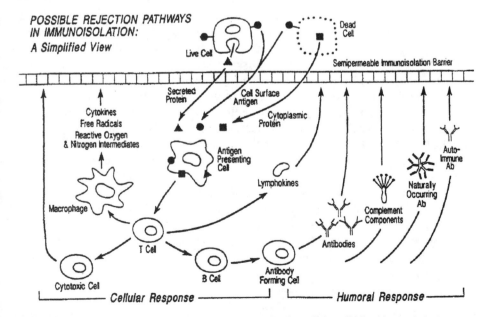

Figure 10.5   Pathways for rejection of immunoisolated cells (from Colton 1995, with permission).

al. 1993; Liu *et al.* 1993; Hughes *et al.* 1994; Al-Hendy *et al.* 1995; Chang 1997; Zalzman *et al.* 2003; Narushima *et al.* 2005).

However, openings that restrict the entry of immune cells will still allow the passage of antibodies and complement, which must be blocked for successful use in a bioartificial organ of xenogeneic cells. Therefore, the solute permeability characteristics of the immunoisolation membrane are critical for achieving successful immunoisolation of xenogeneic transplanted cells. This requires that the membrane have significant permeability to essential small molecular weight solutes such as oxygen, glucose, growth factors, and molecular carriers such as albumin and transferrin[2]. The membrane also must be permeable to waste products and to the therapeutic product, for example in the case of diabetes, this would be insulin. On the other hand, the membrane must have negligible permeability to the humoral components of the immune system, i.e antibodies, complement, and various cytokines and lymphokines.

As we discussed earlier, antibodies have molecular weights ranging from 150,000 to 900,000 Daltons. Naturally occurring antibodies of the IgM class exist in the host and are reactive against the MHC of xenografts. In addition, antibodies produced from autoimmune diseases such as diabetes would be expected to bind to antigens on xenograft islets of Langerhans.

In the absence of immune cells, antibody binding to antigens on the surface of the transplanted cells is usually not sufficient to damage the cells. However, if both antibodies and complement are present, then the transplanted cells can be destroyed by complement activation and formation of the membrane attack complex (Iwata *et al.* 1996).

The complement system is activated by two different routes called the *classical pathway* and the *alternative pathway*. The classical pathway requires the formation of antigen–antibody complexes for its initiation, whereas the alternative pathway does not. The alternative pathway becomes activated in response to recognition of large polysaccharide molecules present on the surface of the cell membranes of invading microorganisms. As far as the immunoisolation of transplanted cells is concerned, the classical pathway is of most interest for this discussion.

Activation of the classical pathway involves nine protein components known as C1 through C9. C1 is the first component to become activated and requires the binding with antigen of either two or more adjacent IgG antibodies or a single IgM antibody. C1 consists of three subunits called C1q, C1r, and C1s. Their molecular weights are respectively 400,000, 95,000, and 85,000 Daltons. The C1q subunit is the first subunit activated, and it binds with the Fc region of the bound antibodies. Its activation then activates the C1r subunit which then activates the C1s subunit which activates C4 and so on.

As we discussed earlier, in the absence of immune cells, binding of antibody with antigen is not sufficient to damage the transplanted cells. Destruction of the antibody laden cell will require the presence and activation of complement. Since the C1q subunit must bind with either IgM or several IgG to start the cascade of complement reactions, the immunoisolation membrane must be capable of preventing the passage of the C1q molecule.

---

[2]Albumin carries fatty acids and transferrin carries iron.

Figure 10.6   Complement C1q molecule.

C1q has an interesting molecular shape as shown in Figure 10.6. The critical dimension of C1q is the span of the six projections from the cylindrical base, which is about 30 nm. Because of the presence of proteins coating the walls of the pores in the membrane, the maximum pore size allowable for blocking C1q is on the order of 50 nm (Colton and Avgoustiniatos 1991; Colton 1995; Zielinski *et al.* 1997).

Several other issues with regard to immunoisolation also need to be considered. The immunoisolation membrane may be capable of preventing the passage of immune cells and complement C1q. However, activation of the immune cells by antigens released by the transplanted cells will also result in the release of such lymphokines as IL-1 and other cytotoxic agents such as nitric oxide, peroxides, and free radicals. These substances can be toxic to the transplanted cells. Lymphokines have small molecular weights, typically around 20,000 Daltons, and, because of their small size, may not be retained by the immunoisolation membrane. Recent studies however, have shown that immunoisolated cells were not affected by the presence of IL-1 (Zekorn *et al.* 1990). Substances such as IL-1 and other highly reactive species may be consumed by other reactions before they can penetrate the immunoisolation membrane to any significant distance (Colton 1995). In addition, many bioartificial organs operate at very low tissue densities in order to provide for effective oxygenation of the transplanted cells. The reduced tissue density results in a lower concentration of shed antigens and, hence, a reduced concentration of humoral agents.

The shed antigens could also cause the host to undergo a life-threatening *anaphylactic reaction*[3] or to suffer from *antibody-mediated hypersensitivity reactions*[4]. Lanza *et al.* (1994) examined these issues by transplanting immunoprotected canine and porcine islets into rats. The islets were encased within an acrylic hollow fiber membrane with a nominal molecular weight cutoff of about 80,000 Daltons. Their studies showed that the immunoprotected islet xenografts caused the host to generate antibodies to antigens given off by the transplanted islets. However, there was no evidence of any other pathological effects of these antibody–antigen immune

---

[3]A result of a significant release of inflammatory mediators, such as histamine, causing severe hypotension and bronchiolar constriction. Can cause death due to circulatory and respiratory failure.

[4]An inappropriate or exaggerated response of the immune system to an antigen. Binding of antibodies to the antigen starts the process.

complexes. Subsequent studies using immunoprotected human islets in human patients with diabetes (Scharp *et al.* 1994; Shiroki *et al.* 1995) have shown that the immunoisolation membrane can protect against not only the allogeneic immune response, but also against the autoimmune components responsible for the development of insulin dependent diabetes mellitus.

10.4 PERMEABILITY OF IMMUNOISOLATION MEMBRANES

Figure 10.7 presents permeability data obtained on a hollow fiber immunoisolation membrane for a variety of solutes of different molecular weight (Dionne *et al.* 1996). This particular membrane was made from a copolymer of acrylonitrile and vinyl chloride. The permselective membrane consists of a thin skin on either side of a much thicker spongy wall region that provides overall structural strength. The thickness of the spongy wall was about 100 microns, and these membranes were similar to those that were successfully used in preliminary tests to immunoisolate human islets in patients with diabetes (Scharp *et al.* 1994). These membranes had a reported nominal molecular weight cutoff of 65,000 Daltons.

Notice how the permeability of this membrane at first decreases gradually as the solute molecular weight increases. As the solute molecular weight continues to increase, the permeability decreases much more rapidly. This can be explained by recalling our equation for solute permeability given by Equation 5.83. The size of low molecular weight solutes is much smaller than the pores in the membrane. For these solutes, steric exclusion (K) and solvent drag (ω_r) effects caused by the pore wall are negligible. The decrease in permeability is, therefore, directly related to the decrease in solute diffusivity, which according to Equations 5.48 and 5.49, would be inversely proportional to the solute size, or the one-third power of the solute molecular weight. This is also shown as the dashed line in Figure 10.7. As the solute size reaches a critical fraction of the membrane average pore size, steric exclusion and solvent drag become more important, and the solute permeability rapidly decreases with increasing solute molecular weight.

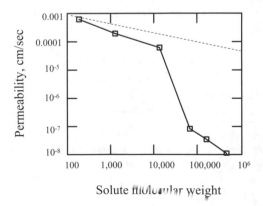

Solute molecular weight

Figure 10.7 Immunoisolation membrane permeability (data from Dionne *et al.* 1996).

These data also clearly show the problem of defining the "molecular weight cutoff" for the membrane. There does not exist a molecular weight above which the solute permeability drops to zero. Surprisingly, we see that even very large molecules, here at the extreme represented by a molecular weight of about 440,000 in Figure 10.7, have a measurable permeability. Therefore, a membrane's molecular weight cutoff is very subjective and dependent on the definition that one chooses to apply. For example, some membrane manufacturers define the molecular weight cutoff on the basis of 90% retention of a given solute after so many hours of dialysis. Others base it on ultrafiltration of a solution in which a particular solute is 90% retained.

Although the exact nature of the pores or openings within the membrane is not known, descriptions such as that based on the hydrodynamic pore model developed in Chapter 5 (Equation 5.83) provide a useful framework for understanding the transport of various solutes, and also provide a means for correlating data. The data shown in Figure 10.7 can, therefore, be used to obtain an estimate of the average pore size in the membrane. This is shown in the following example.

Example 10.1 Using the solute permeability data shown in Figure 10.7, determine the average size of the pores in the membrane. Recall that the membrane is 100 microns thick. The table below summarizes the physical properties of the solutes that were used. Use the hindered diffusion model developed by Bungay and Brenner (1973) to describe the effect of the pore size on the diffusion of the solute. This equation is used, rather than the Renkin model given in Equation 5.55, because the Bungay and Brenner equation is generally valid for the entire range of the ratio of solute to pore size, i.e. $0 \leq \dfrac{a}{r} \leq 1$. The Renkin equation is only valid for $\dfrac{a}{r} < 0.4$ and may not provide a correct description of the diffusion of larger molecules in comparably sized pores. In the Bungay and Brenner model, $\omega_r = \dfrac{6\pi}{K_t}$ in the permeability equation given by Equation 5.83. K_t is given by the following equation. In this equation (a) is the molecular radius and (r) is the pore radius.

$$K_t = \frac{9}{4}\pi^2 \sqrt{2}\left(1 - \frac{a}{r}\right)^{-5/2}\left[1 + \sum_{n=1}^{2} z_n\left(1 - \frac{a}{r}\right)^n\right] + \sum_{n=0}^{4} z_{n+3}\left(\frac{a}{r}\right)^n$$

with $z_1 = -1.2167$; $z_2 = 1.5336$; $z_3 = -22.5083$; $z_4 = -5.6117$; $z_5 = -0.3363$; $z_6 = 1.216$; $z_7 = 1.647$.

Solute	Molecular weight	Stokes radius, nm	$D_{water} \times 10^6$ cm^2 sec^{-1} 37°C	$P_m \times 10^8$ cm sec^{-1}
Glucose	180	0.35	9.24	63,200
Vitamin B$_{12}$	1300	0.77	5.00	20,000
Cytochrome C	13,400	1.65	1.78	6160
Bovine serum albumin (BSA)	67,000	3.61	0.964	7.95
IgG	155,000	5.13	0.629	3.59
Apoferritin	440,000	5.93	0.611	1.07

Source: Dionne et al. (1996).

SOLUTION Since the membrane consists of two permselective skins and a much thicker spongy wall region, the observed solute permeabilities must account for the resistances of all three regions. Therefore, the observed permeability is given by the following equation assuming the two skins are equivalent in their resistance to mass transfer.

$$\frac{1}{P_m} = \frac{2}{\dfrac{\varepsilon_{skin} D K \omega_r}{\tau_{skin} t_{skin}}} + \frac{1}{\dfrac{\varepsilon_{sponge} D}{\tau_{sponge} t_{sponge}}}$$

The porosity of the sponge is assumed to be about 0.8 and the tortuosity of the sponge is assumed to be unity. The non-linear regression analysis is based on two variables, the pore radius (r) and a paramter α. The parameter α accounts for the group $\dfrac{\varepsilon_{skin}}{\tau_{skin} t_{skin}}$ for which the individual values of the parameters are not known. Since the permeability values vary over a considerable range, it is appropriate to define the objective function for the regression in terms of the natural logarithm of the permeabilities. Hence, we write the sum of the square of the errors as follows.

$$SSE(r,\alpha) = \sum_i \left(\ln(P_i) - \ln(P_{calc}(i,r,\alpha)) \right)^2$$

The result of the regression analysis is shown in Figure 10.8 and provides an estimated pore diameter of 14.2 nm.

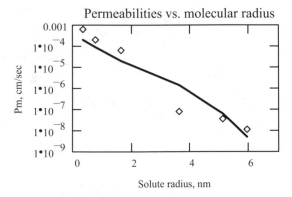

Figure 10.8 Measured and predicted permeabilities in an immunoisolation membrane.

10.5 MEMBRANE SHERWOOD NUMBER

The solute permeabilities given in Figure 10.7 can also be compared with the permeability the solute would have in an aqueous layer that has the same thickness as the membrane itself. Therefore, we may write the following equation that expresses the ratio of these respective quantities.

$$\frac{P_m}{P_{\text{aqueous layer}}} = \frac{P_m}{D / t_m} = \frac{P_m t_m}{D} = \frac{D_{\text{effective}}}{D} = Sh_m \tag{10.1}$$

We see that this ratio is equivalent to defining a Sherwood number for the membrane, i.e. Sh_m, that represents the ratio of the effective diffusivity of the solute through membrane to the aqueous solute diffusivity. This process has the effect of eliminating membrane thickness as a variable when correlating membrane permeability data. For a given solute and membrane structure, the Sh_m should be independent of the membrane thickness.

The data shown in Figure 10.7 are replotted as the membrane Sherwood number versus solute molecular weight in Figure 10.9. We see that for low molecular weight solutes, the membrane Sherwood number is relatively constant. In this example, it is about 0.68. As the solute molecular weight is increased, and the solute size becomes comparable to that of the pores, the membrane Sherwood number begins to decrease rapidly.

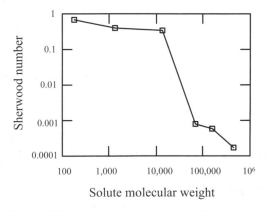

Figure 10.9   Immunoisolation membrane Sherwood number (data from Dionne *et al.* 1996).

10.6 EXAMPLES OF BIOARTIFICIAL ORGANS

Bioartificial organs contain living tissue or cells that are immunoisolated by polymeric membranes with the properties discussed above. These devices are also known as *hybrid artificial organs* since they consist of both artificial materials and living tissue or cells. Bioartificial organs are being considered for the treatment of a variety of diseases such as diabetes, liver failure and kidney failure, for neurological disorders such as Parkinson's or Alzheimer's disease, for control of pain, and for delivery of a variety of therapeutic products secreted by genetically engineered cell lines.

For diabetes, the bioartificial organ contains cells that secrete minute-by-minute the appropriate amount of insulin needed to maintain blood glucose levels within a very narrow range. Bioartificial livers and kidneys attempt to replace all of the physiological functions of these complex organs. Treatment of neurological disorders involves the use of specialized genetically engineered cells that secrete neuroprotective products such as dopamine, nerve growth factor, glial cell-line derived neurotrophic factor, and brain-derived neurotrophic factor. Immunoprotected cells that secrete pain-reducing neuroactive compounds such as β-endorphin are being considered for the treatment of pain resulting from cancer. Genetically engineered cell lines can also be used to treat a variety of other diseases. Examples include providing a specific hormone (for example growth hormone), replacing missing clotting factors in hemophilia (factor VIII or IX), and providing needed enzymes in such diseases as adenosine deaminase deficiency that results in severe combined immunodeficiency disease (SCID). More details about these specific applications of bioartificial organs may be found in the list of references. In the following discussion, we will look at bioartificial organ replacements for the endocrine pancreas, the liver, and the kidney as representative systems.

10.6.1 The Bioartificial Pancreas

The focus of much of the research on bioartificial organs has been on the treatment of insulin dependent diabetes mellitus (IDDM), or type I diabetes, with a *bioartificial pancreas*. This is because IDDM is a chronic disease with many serious complications such as blindness, kidney disease, gangrene, heart disease, and stroke that lead to a poorer quality of life for those afflicted with this disease. IDDM, therefore, represents a major health problem in the USA. Over 1 million people in the USA have IDDM and the cost of their healthcare exceeds several billion dollars each year. Additionally, methods have recently been developed that allow for the isolation of mass quantities of the cells needed to treat diabetes, i.e. the islets of Langerhans, from the pancreas of humans and from large mammals such as pigs, thus providing a potentially unlimited supply of donor islet tissue (Lanza and Chick 1994a; Ricordi *et al.* 1988, 1990a,b; Warnock *et al.* 1988, 1989; Inoue *et al.* 1992; Ricordi 1992; Lacy 1995; Maki *et al.* 1996; Lakey *et al.* 1996; Cheng *et al.* 2004). Use of these allogeneic or xenogeneic cells without immunosuppression requires that they be immunoisolated in a device such as a bioartificial organ. Research on the treatment of IDDM with a bioartificial pancreas also provides a well-defined disease state[5] that can be used to explore alternative approaches for the design of a bioartificial organ that may have broader application in the treatment of other endocrine diseases, liver disease, and renal failure.

IDDM is believed to be caused by an autoimmune process (Notkins 1979; Atkinson and MacLaren 1990) that destroys the insulin-secreting β cells found within the islets of Langerhans. Both environmental (viruses or chemicals) and hereditary factors are involved in its development. The islets of Langerhans are found scattered throughout the pancreas and represent about 1–2% of the pancreas mass.

Insulin is a small protein (6000 Daltons), and the key hormone involved in the

---

[5]In that the transplanted cells (islets) respond to an easily measured stimulus (glucose) and respond by secretion of a therapeutic agent (insulin) with a measurable outcome, i.e. improved blood glucose.

regulation of the body's blood glucose levels. In the absence of insulin, glucose levels in the blood exceed normal values by several times. This is a result of the body's cells not being able to metabolize glucose. Without the ability to use glucose as an energy source, the body responds by metabolizing fats and proteins with a corresponding increase in the body fluids of ketoacids such as acetoacetic acid. This excess acid results in a condition known as ketoacidosis which, if left untreated, can lead to death.

The conventional treatment for IDDM involves the daily administration of exogenous insulin to replace the insulin that is no longer produced by the patient's β cells found within the islets of Langerhans. This results in the almost normal metabolism of carbohydrates, fats, and proteins. However, even with exogenous insulin, the patient's blood glucose levels are not controlled as well as normal. The healthy islet β cells are able to regulate the release of insulin in such a manner so as to maintain blood glucose levels within a narrow range, about 80–120 mg/dl. Insulin injections are not capable of providing such close control because of the difficulty of properly timing the injections and the inherent insulin transport delays from the site of injection. The abnormally high blood glucose levels, or *hyperglycemia*, in patients treated with exogenous insulin is now believed to be the primary cause of the long-term complications of diabetes (American Diabetes Association 1993). These elevated blood glucose levels lead to protein *glycosylation* that damages the microvasculature and other tissues in the body.

The goal of the bioartificial pancreas is, therefore, to utilize the glucose regulating capability of healthy donor islets, β cells, or engineered β cells (Zalzman *et al.* 2003; Narushima *et al.* 2005) to provide improved blood glucose control in patients with IDDM. This approach should minimize the complications of the disease and improve the quality of the patient's life. Transplanting the donor islets without immuno-protection is only warranted in those cases where the patient is already receiving immunosuppressive therapy, for example as a result of a kidney transplant to treat end-stage renal failure. In an otherwise healthy patient with diabetes, immunosuppressive drugs and their attendant problems would not be needed if the islets were immunoisolated. Immunoisolation of the islets would not only protect them from the host's normal response to foreign tissue, but also from those agent's of the immune system that caused the original autoimmune destruction of the patient's own cells. These immune system *memory cells* are still capable of responding to the antigens that resulted in the original destruction of the patient's own islets and could possibly destroy the donor islets or donor β cells.

10.6.1.1 Bioartificial pancreas approaches A variety of approaches have been described over the years for the bioartificial pancreas, and most of these have been tested extensively in laboratory animals. An excellent review of these devices through 1990 may be found in the paper by Colton and Avgoustiniatos (1991), and more recent reviews may be found in Mikos *et al.* (1994), Lanza *et al.* (1995), and Colton (1995). A series of books related to islet transplantation and the bioartificial pancreas have also been recently published (Ricordi 1992; Lanza and Chick 1994c).

These devices can be broadly classified into the following four categories: *intravascular*, *microencapsulation*, *macroencapsulation*, and *organoid*. These

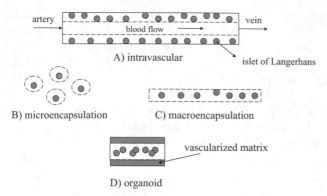

Figure 10.10   Approaches for a bioartificial pancreas.

approaches are illustrated in Figure 10.10. These devices, are for the most part, general and are being considered for the immunoisolation of a variety of other cell types.

Of concern in any of these approaches for a bioartificial pancreas is the ability of the device to be readily fabricated and seeded with islets or cells, their reliability over an extended period of time, and their cost. These issues also relate to the overall biocompatibility of the device. Specifically in regard to device toxicity, immune system response, potential for fibrotic encapsulation, and for intravascular devices, the potential for thrombosis. Furthermore, the devices must be capable of maintaining long-term viablity of the donor tissue and have the proper mass transfer characteristics to normalize blood glucose control. This requires proper understanding and integration during the design process of the islet glucose-insulin kinetics, the mass transfer properties of the immunoisolation membrane, and the islet or cell density within the device.

10.6.1.2 Intravascular devices The intravascular devices (Figure 10.10A) (Chick *et al.* 1977; Moussy *et al.* 1989; Lepeintre *et al.* 1990; Lanza *et al.* 1992a,c,d,e, 1993; Sarver and Fournier 1990; Maki *et al.* 1991, 1993; Petruzzo *et al.* 1991; Sullivan *et al.* 1991) involve a direct connection to the vascular system of the host via an arterio-venous shunt. Blood flows through the lumen of small polymeric hollow fibers or in the spaces between flat membrane sheets. Surrounding the blood flow path, and protected by the membrane, are the islets or cells. An advantage of this type of device is that the convective flow of the blood provides the potential for better mass transfer rates. In some cases, an ultrafiltration flow can cross the membrane at the arterial end and reenter at the venous end, much like the Starling flow found in capillaries. The close contact with a sizable flow of blood also offers the potential for good tissue oxygenation. A disadvantage of the intravascular approach is the surgical risk associated with device implantation, connection to the host's vascular system, potential to form blood clots, and removal at a later time.

One of the first experiments to demonstrate the feasibility of the intravascular approach, and for that manner the overall concept of a bioartificial pancreas, was performed by Chick *et al.* (1977). A rat with chemically-induced diabetes was connected ex vivo to an intravascular device containing cells obtained from neonatal rats. The device used in these experiments consisted of a bundle of 100 hollow fibers with diameters less than 1 mm and a length of 11 cm. The fibers had a nominal molecular weight cutoff of about 50,000 Daltons.

After connection of the animal to the ex vivo intravascular device, there followed a rapid decrease in blood glucose levels reaching normalized values of 110–130 mg/dl (considered normal for a rat) after about 6 hours. After the removal of the device, blood glucose levels rapidly increased and returned to their diabetic levels. These results indicated the feasibility and the potential of the bioartificial pancreas for the treatment of IDDM.

A particularly serious problem with devices that contained numerous hollow-fiber membranes was the formation of blood clots in the entrance and exit header regions of the device. These regions of the device, where the fluid stream is either expanding or contracting, tend to form secondary flows or swirls that induce the formation of blood clots.

In an effort to minimize the formation of blood clots without the use of anticoagulants such as heparin, a device with a single coiled hollow fiber membrane tube was developed by Maki *et al.* (1991). Their device is shown in Figure 10.11. This device consists of an annular-shaped acrylic chamber 9 cm in diameter and 2 cm thick. The chamber contains a 30–35 cm length of a coiled hollow fiber tubule with an inner diameter of 5–6 mm and a wall thickness of 120–140 microns. The hollow fiber tubule was connected at each end to polytetrafluoroethylene (PTFE) arterial grafts that provided connection of the device to the vascular system by an arteriovenous shunt. The

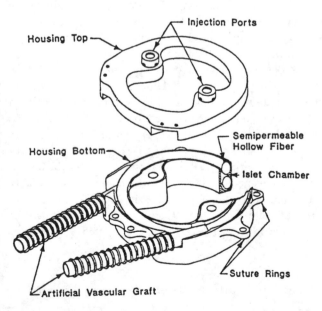

Figure 10.11   The biohybrid artificial pancreas (from Maki *et al.* 1991, with permission).

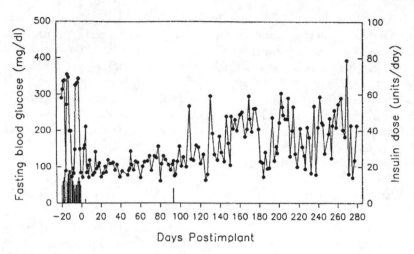

Figure 10.12 Fasting glucose and insulin requirements after implantation of device shown in Figure 10.11 (from Lanza *et al.* 1992d, with permission).

larger bore size of the single hollow fiber tubule eliminates the thrombosis observed in smaller bore hollow fiber tubes and in the header regions of the device described earlier. Low dose aspirin therapy was also found to contribute to long-term patency presumably through aspirin's ability to prevent platelet activation.

The hollow fiber membrane had a nominal molecular weight cutoff of about 80,000 Daltons. A cavity surrounded the coiled tubular membrane and provided a 5 to 6 ml volume for the placement of the islets of Langerhans. This device was evaluated in pancreatectomized dogs using allogeneic canine islets and xenogeneic bovine and porcine islets. The device was implanted within the abdominal cavity and *anastomosed* to the left common iliac artery and the right common iliac vein.

This device was able to demonstrate, in many cases, improved blood glucose control for varying periods of time using both allo- and xenogeneic islets (Lanza *et al.* 1992d). Figure 10.12 illustrates an example of long-term function in a dog implanted with a single device containing about 42,000 canine islet equivalents[6]. Prior to device implantation, this particular dog had a fasting blood glucose level of 230 mg dl^{-1} and required about 16 U day^{-1} of insulin[7]. After implantation, no exogenous insulin was provided for the 280 day period of the experiment. During the first 140 days, the fasting blood glucose was nearly normal and averaged 116 mg dl^{-1} which compares favorably with the value of 91 mg dl^{-1} observed in normal dogs. After 140 days, glucose control began to deteriorate over the rest of the experiment. A higher fraction of animals was insulin free for longer periods of time when two devices were implanted as a result of the greater mass of islets.

---

[6]The equivalent islet number (EIN) is defined as the number of islets 150 microns in diameter that are equivalent in volume to a given sample of islets.

[7]Insulin dose is based on U(nits) of insulin where 1 U of insulin = 40 micrograms of insulin.

Results obtained with xenogeneic implants containing porcine islets in this device were also very encouraging (Maki *et al.* 1996). Prior to receiving porcine islets, one particular dog had a fasting blood glucose level of 479 mg/dl and required 39 U/day of insulin. After this dog received 216,000 EIN of porcine islets, fasting blood glucose averaged 185 mg/dl and exogenous insulin averaged about 10 U/day. This device failed after 271 days of operation. Analysis of the porcine islets after device removal did not show any evidence of immune cell infiltration. These tests, therefore, presented clear evidence that xenogeneic porcine islets could be protected from the host's immune system using immunoisolation membranes. The only major complication in the use of these devices in dogs was vascular thrombosis. However, this clotting problem may be unique to dogs because of their known hypercoagulability.

10.6.1.3 Microencapsulation The microencapsulation approach (see Figure 10.10B) involves the placement of one or several islets within small polymeric capsules (Altman *et al.* 1986; Gharapetian *et al.* 1986; Lum *et al.* 1991; Soon-Shiong *et al.* 1992a,b, 1993; Sugamori and Sefton 1989; Sun *et al.* 1977; Lanza *et al.* 1995; Sun *et al.* 1996; Tun *et al.* 1996). The microcapsules are then injected within the peritoneal cavity. The diameter of the microcapsule is typically in the range of 300 to 600 microns. Immunoprotective surface coatings of the islets have also been described (Pathak *et al.* 1992).

Microencapsulation provides diffusion distances on the order of 100 to 200 microns and very high surface areas per volume of islet tissue. Accordingly, their small size provides good diffusion characteristics for nutrients and oxygen which improves islet or cell viability. This also provides for a good glucose/insulin response that offers the potential for normalization of blood glucose levels.

The microcapsules usually consist of the islet or cellular aggregates immersed within a hydrogel material with another eggshell-like layer that provides the immunoisolation characteristics and mechanical strength. A wide variety of polymer chemistries have been described for the hydrogel and the immunoprotective layer (Douglas and Sefton 1990; Gharapetian *et al.* 1986, 1987; Matthew *et al.* 1993; Sefton *et al.* 1987; Sun *et al.* 1996; Lanza *et al.* 1995). Of particular concern in selecting the microcapsule chemistry is the formation of a fibrotic capsule around the encapsulated islet. Fibrotic capsule formation can severely limit the diffusion of nutrients and oxygen resulting in loss of islet or cellular function. The success of the microencapsulation approach is, therefore, strongly dependent on choosing membrane materials that minimize this fibrotic reaction.

A particularly promising microencapsulation approach was described by Soon-Shiong (1994) and demonstrated in large animals and several human patients. They used an alginate-poly-L-lysine encapsulation system. Alginates are natural polymers composed of the polysaccharides mannuronic acid and guluronic acid. Soon-Shiong *et al.* (1994) showed that the high mannuronic acid residues in the alginate are responsible for the fibrotic response. Mannuronic acid was shown to induce the lymphokines IL-1 and TNF (tumor necrosis factor) which are known to promote the proliferation of fibroblasts and lead to fibrotic capsule formation. By reducing the alginate's mannuronic acid content, and increasing the guluronic content (> 64%), they were able to minimize the fibrotic response. The higher guluronic acid content also provided another benefit. It was found that alginates with higher guluronic contents were mechanically stronger.

These modified alginate microcapsules were evaluated in a series of nine spontaneously diabetic dogs. Three of the dogs received free unencapsulated islets and the other six dogs received encapsulated islets. The donor islets were obtained from other dogs and each recipient received their quantity of islets by an intraperitoneal injection. The islets were provided at an average dose of about 20,000 EIN/kg of body weight.

Figure 10.13 summarizes the plasma glucose levels prior to and following the injection of the islets. Exogenous insulin was stopped four days prior to islet injection, and the plasma glucose levels at that time averaged 312 mg/dl. The first day after receiving islets, the blood glucose levels were reduced to an average of 116 mg/dl in those animals that received encapsulated islets, and averaged 120 mg/dl in those receiving free islets. Rejection of the unprotected free islets occurred rapidly with hyperglycemia returning in about 6 days. The animals receiving encapsulated islets exhibited normoglycemia for periods of time ranging from 63 to 172 days, for a median period of 105 days. Failure of the encapsulated islets was attributed to membrane failure as a result of the water soluble nature of the alginate system.

Sun *et al.* (1996) developed a microencapsulation system for islets using alginate-polylysine-alginate microcapsules. Most microcapsules contained only a single islet and had diameters in the range of 250 to 350 microns. Preclinical studies were done

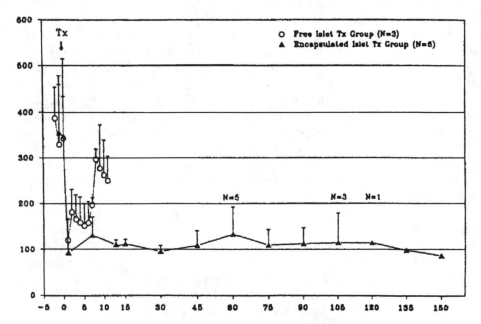

Figure 10.13   Fasting glucose levels (mg dl^{-1}) vs. time (days) following treatment with either free or microencapsulated islets (from Soon-Shiong *et al.* 1992a, with permission).

using microencapsulated porcine islets implanted into spontaneously diabetic cynomologus monkeys. Seven monkeys receiving an average of 16,100 EIN islets/kg body weight were insulin independent for an average of 289 days. The shortest duration of control was 120 days and the longest 803 days. Prior to receiving the islets, fasting blood glucose levels averaged 353 mg/dl. Following the transplant, fasting blood glucose levels averaged 112 mg/dl. During the experiments, the sera of the animals was tested repeatedly for the presence of anti-porcine islet antibodies. No antibodies were detected confirming the presence of effective immunoisolation of the porcine islets. Diabetic control animals also received unencapsulated islets. However, euglycemia was only maintained for about 9 days after which, the islets were destroyed by the host's immune system with a return to pretransplant diabetic levels of the blood glucose.

These results using the microencapsulation approach are very promising. However, there still are some technical difficulties with this approach that must be overcome. The strength and integrity of the microcapsules needs to be improved to provide long-term functioning and viability of the islets. Aggregation of the beads is also a problem as this affects their mass transfer characteristics and impacts glucose control and islet survival. The total volume for the smaller microcapsules of Sun et al. (1996) is modest. For example, using their monkey results, and assuming a 70 kg patient, a total of 1.1 million EIN would be required (16,100 EIN/kg body weight). With one islet per 300 micron diameter microcapsule, this amounts to a total bead volume of only 16 ml. Loss of islet function with time can be compensated for by injection of fresh microcapsules. Retrieval of the small microcapsules could be difficult if this should be necessary after their introduction.

10.6.1.4 Macroencapsulation Macroencapsulation (see Figure 10.10C) involves placing numerous islets within the immunoisolation membrane structure. Typical approaches involve the use of multiple hollow fibers that encase the islets (Altman et al. 1986; Lanza et al. 1991, 1992 a,b,c; Lacy et al. 1991) or a much larger flat bag-like structure (Gu et al. 1994; Inoue et al. 1992; Hayashi et al. 1996). Their larger size and smaller number also makes retrieval less of a problem in comparison to microcapsules. The membranes used for macroencapsulation must not only possess the needed permeability requirements, but because of their larger size, must also provide sufficient mechanical strength to maintain their integrity. This is a difficult objective since the membrane wall must be thin for diffusional purposes and this may compromise overall strength. In many cases, it has been found that thin-wall hollow fiber membranes were prone to breaking (Lanza et al. 1995). Rupture of the membrane wall will result in rejection of the islet tissue and loss of function. As is the case for microcapsules, membrane chemistry is an important factor in minimizing the formation of fibrous tissue around the implant.

Most of the attention on the macroencapsulation technique has been focused on the use of hollow fibers. A tradeoff exists on the selection of the fiber diameter. A large diameter fiber will result in a shorter overall length, but can lead to diffusional limitations. Lack of oxygen transport and the accumulation of waste products can, therefore, result in a central core of necrotic tissue. On the other hand, a small fiber diameter, while improving the transport characteristics, can result in an incredibly long fiber length. This longer length increases the probability of breakage and makes implantation of the fibers more difficult.

An example of the use of hollow fibers as a vehicle for the macroencapsulation of the islets of Langerhans has recently been reported (Lanza *et al.* 1991, 1992 a,b,c, 1995). This particular system has been studied in some detail in both rats and dogs using a variety of sources for the islets. The islets were enclosed within semipermeable hollow fiber membranes that were 2–3 cm in length and had an internal diameter of 1.8–4.8 mm. The membrane wall thickness ranged from 69–105 microns and the membrane had a nominal molecular weight cutoff of 50,000–80,000 Daltons. Depending on the fiber diameter, approximately 9–50 fibers were implanted within the peritoneal cavity of each rat.

Figure 10.14 (Lanza and Chick 1997) summarizes fasting plasma glucose levels after fibers containing bovine islets (20,000 EIN) were implanted in the peritoneal cavity of rats whose diabetes was induced chemically using the drug, *streptozotocin*. In each case, the diabetic state was reversed within 24 hours after implantation. In some cases normoglycemia was maintained for 1 year. The wider bore membranes (4.5–4.8 mm ID) that were retrieved after several months were found to contain a necrotic core. Viable islets were only found to exist within a 0.5–1 mm layer along the membrane wall. Some of the late failure of the implants was also attributable to membrane breakage as a result of the thin membrane wall.

Similar results were obtained using canine islets contained within the same hollow fibers and implanted within the peritoneal cavity of diabetic BB/Wor rats (Lanza *et al.* 1992b). This type of rat develops autoimmune diabetes with a pathology similar to that found in humans. The canine islet xenografts provided normoglycemia in these rats for more than 8 months. This is an important result since it shows that the immunoisolation membrane protected the islet xenografts from the host's normal graft rejection response, as well as from the autoimmune disease process that produced the original diabetes. Protection of the islets from the autoimmune disease process is important in order to treat successfully human patients with diabetes.

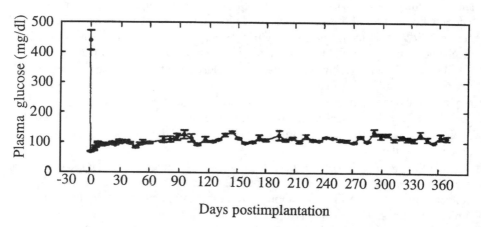

Figure 10.14 Fasting glucose levels in rats that received hollow fiber implants containing bovine islets (from Lanza and Chick 1997, with permission).

The hollow fiber macroencapsulation of islets was also tested in pancreatectomized dogs. Figure 10.15 (Lanza *et al.* 1992a,c,d,e) summarizes results obtained for two dogs that were insulin-free for at least 70 days after receiving islet allografts contained within these fibers. Each dog received between 155 and 248 fibers, containing a total of about 300,000 EIN. This equates to a total fiber length of about 500 cm. The dogs had an average weight of about 17 kg, providing an islet loading of nearly 20,000 EIN/kg of body weight. For a 70 kg human, the total number of islets would, therefore, be about 1.4 million EIN, or a total fiber length on the order of 2100 cm (70 feet).

10.6.1.5 Organoid The organoid approach (see Figure 10.10D) (Sarver *et al.* 1995; Hill *et al.* 1994) represents the union of the tissue engineering and macroencapsulation concepts. The bioartificial organoid consists of a thin circular or rectangular disk-like compartment for the tissue that is enclosed by an immunoisolation membrane. Adjacent to the membrane is a thin layer of a porous scaffold material that becomes vascularized by the host. The use of tissue engineering allows control over the environment adjacent to the immunoisolation membrane. Therefore, the fibrotic response can be minimized, and through the growth of a vascular bed, intimate contact with the host's vasculature can be obtained. This has the potential to improve the viability and long-term functioning of the transplanted tissue. Additionally, these devices, unlike the microcapsule or

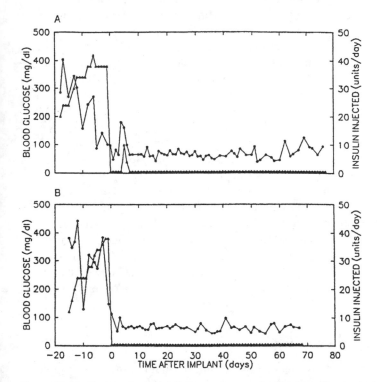

Figure 10.15 Fasting glucose levels in two dogs receiving hollow fiber implants containing canine islets (from Lanza *et al.* 1992a,c,d,e, with permission).

hollow fiber approaches, have ports that allow the device to be seeded with the islet tissue once the porous material has become sufficiently vascularized. The ports also allow for islet tissue to be replaced on a regular basis by a relatively minor procedure.

10.6.2 Number of Islets Needed

One issue confronting all of the devices discussed thus far concerns the mass of tissue required to achieve normoglycemia in a human patient with diabetes. The above studies in rats and dogs have used islet loadings on the order of 20,000 EIN/kg body weight. This is considerably higher than the benchmark study reported by Warnock *et al.* (1988). They found that the threshold islet loading for treating pancreatectomized dogs corresponded to ≈ 4.3 μL of islet tissue/kg. This equates to about 2500 EIN/kg. It is important to point out, however, that these were unencapsulated autologous islets that were implanted within the liver or the spleen. These islets had no artificial mass transfer resistance to overcome and no host immune response that would lead to rejection. This result, most likely, defines the lower limit on islet loading. Certainly, improvements in the mass transfer characteristics of the devices discussed so far could lead to islet loadings more on the order of 5000 to 10,000 EIN/kg. Overall device size is dependent on the amount of tissue needed, as well as on the device mass transfer characteristics for nutrients, oxygen, waste products and the therapeutic agent itself. The mass transfer characteristics, therefore, are a critical factor in defining the islet tissue loading within the device itself.

The results presented so far demonstrate the potential of the bioartificial pancreas as a method for treating diabetes. All of the approaches, however, suffer from the problem of poor long-term islet survival. Islets have the potential to function for many years as evidenced by the fact that, unless they become diseased, they exist for as long as the lifespan of the animal they are found in. Therefore, islets should be capable of surviving within a bioartificial pancreas for many years.

Islet survival is a function of many variables including the amount of trauma they experience during their isolation and purification; the device mass transfer characteristics for nutrients, oxygen, and waste products; the integrity of the membrane immunoisolation system; and, finally, the host's response to the implant. This latter factor determines the nature of the physiological environment in the vicinity of the implant. Certainly, the formation of a region of under-vascularized fibrotic tissue will not provide the conditions needed for the long-term functioning and survival of the islets. Because of questions concerning the long-term viability and functioning of the islet tissue, it may be necessary periodically to recharge the patients with islets. A bioartificial pancreas that has the ability to be reseeded would certainly offer an advantage.

10.6.3 Islet Insulin Release Model

The design of a bioartificial pancreas requires an understanding of the insulin release rate from an islet or insulin producing cells and its dependence on plasma glucose levels. Islets that have been isolated from a pancreas or genetically engineered cells are

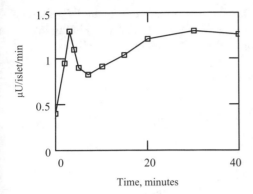

Figure 10.16   Rat islet insulin secretion rate (data from Nomura *et al*. 1984).

usually challenged in culture with a step change in the glucose concentration. The glucose challenge is used to assess islet or cell viability and glucose responsiveness. When challenged in this manner, islets or insulin-secreting cells release insulin in a biphasic fashion (Grodsky 1972).

Figure 10.16 (Nomura *et al*. 1984) illustrates the insulin response of rat islets to a rapid ramp increase in the glucose concentration from an initial value of 80 mg/dl to a final value of 200 mg dl^{-1}. Initially, the islets respond to the rate of change of the plasma glucose concentration, resulting in a maximum of the first phase insulin release rate within a few minutes of the glucose challenge. Following this, the second phase islet insulin release rate is proportional to the difference between the plasma glucose concentration and its fasting value of about 80 mg dl^{-1}. We see that this second phase release rate is relatively constant once the glucose level has achieved a steady state value.

A mathematical description of the islet insulin release rate during a glucose challenge can be obtained through application of control theory. Nomura *et al*. (1984) have shown that the dynamics of the glucose induced secretion of insulin can be expressed as the sum of the proportional response to the glucose concentration, and a derivative response to the rate of change in the glucose concentration, each with a first order lag time, respectively given by T_1 and T_2. The Laplace transform of the islet insulin release rate can, therefore, be expressed as follows:

$$r_{\text{islet}}(s) = \left[\frac{K_p}{1+T_1 s} + \frac{T_d s}{1+T_2 s} \right] C_G(s) \tag{10.2}$$

This equation may be inverted (in the Laplace transform sense) to give the following equation for the islet insulin release rate in terms of the glucose concentration and its rate of change.

$$r_{\text{islet}}(t) = \int_{-\infty}^{t} C_p C_G(z) e^{-(t-z)/T_1} dz + \int_{-\infty}^{t} C_d \frac{dC_G(z)}{dz} e^{-(t-z)/T_2} dz \tag{10.3}$$

The parameters in this equation; i.e. $C_p \ (= K_p/T_1)$, $C_d \ (= T_d/T_2)$, T_1, and T_2, may be obtained by performing a non-linear regression analysis of islet or cellular insulin release rate data such as that shown in Figure 10.16. This can be accomplished by first representing the ramp glucose profile as follows:

$$C_G(t) = C_G^0 \ , \ \text{for } t < 0$$

$$C_G(t) = \left(\frac{C_G^{SS} - C_G^0}{t_0} \right) t + C_G^0 \ , \ \text{for } 0 \le t \le t_0 \tag{10.4}$$

$$C_G(t) = C_G^{SS} \ , \ \text{for } t \ge t_0$$

These equations may be substituted into Equation 10.3 to obtain algebraic expressions for the islet insulin release rate for a ramp change in the glucose concentration.

$$r_{islet}(t) = C_p T_1 C_G^0 \, e^{-t/T_1} + C_p T_1^2 \left(\frac{C_G^{SS} - C_G^0}{t_0} \right) \left[\frac{t}{T_1} - \left(1 - e^{-t/T_1} \right) \right]$$

$$+ C_p C_G^0 T_1 \left(1 - e^{-t/T_1} \right) + C_d T_2 \left(\frac{C_G^{SS} - C_G^0}{t_0} \right) \left(1 - e^{-t/T_2} \right) \ , \ \textit{for } 0 \le t \le t_0 \tag{10.5}$$

$$r_{islet}(t) = C_p C_G^0 T_1 e^{-t/T_1} + C_p T_1^2 \left(\frac{C_G^{SS} - C_G^0}{t_0} \right) \left[e^{-(t-t_0)/T_1} \left(\frac{t_0}{T_1} - 1 \right) + e^{-t/T_1} \right]$$

$$+ C_p C_G^0 T_1 \left[e^{-(t-t_0)/T_1} - e^{-t/T_1} \right] + C_p C_G^{SS} T_1 \left(1 - e^{-(t-t_0)/T_1} \right)$$

$$+ C_d \left(\frac{C_G^{SS} - C_G^0}{t_0} \right) T_2 \left[e^{-(t-t_0)/T_2} - e^{-t/T_2} \right] \ , \ \text{for } t > t_0 \tag{10.6}$$

The example below illustrates how the parameters in the Nomura *et al.* islet insulin release model may be determined for rat islets through the use of the above equations.

Example 10.2 Determine the constants in the Nomura *et al.* insulin release model for the rat islet perifusion data shown in Figure 10.16.

SOLUTION A non-linear regression analysis finds the values of $T_1 = 10.11$ minutes, $T_2 = 1.88$ minutes, $C_p = 6.26 \times 10^{-4}$ µU dl mg^{-1} min^{-2} islet^{-1}, and $C_d = 0.011$ µU dl mg^{-1} min^{-1} islet^{-1}, best fits the data shown in Figure 10.16. Figure 10.17 shows that the Nomura model provides an excellent fit to the islet insulin release data.

Rat islets have a very potent response to a step change in glucose concentration. Islets from dogs, pigs, and human pancreas typically show significantly lower insulin secretion rates on a per islet basis.

The actual in vivo response of islets depends not only on glucose levels, but also depends, in a significant way, on the levels of other nutrients such as amino acids and

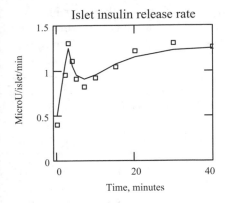

Figure 10.17 Comparison of measured and predicted insulin release rate.

gastrointestinal hormones[8]. These later substances, in the presence of rising glucose levels, can almost double the insulin secretion rate of an islet. However, no quantitative framework is available to describe the effect of these additional secretagogues on the insulin secretion rate.

Under physiological conditions, the variation in plasma glucose levels is less pronounced than a glucose challenge test on islets, and the biphasic insulin release pattern is not as evident. In fact, the insulin secretion rate, under these conditions, displays a strong sigmoidal dependence on the blood glucose levels (Guyton 1991; Sturis *et al.* 1991). The insulin secretion rate for the human pancreas saturates at a value of about 200,000 μU min^{-1} when the glucose concentration reaches 300 mg dl^{-1}. Since the human pancreas contains about 1 million islets, this saturation insulin release rate is 0.2 μU islet^{-1} min^{-1}.

Sturis *et al.* (1991) represented the insulin secretion rate under physiological conditions as a sigmoidal function of the plasma glucose concentration. The following equation proposed by Sturis *et al.* (1991) may then be used to describe the in vivo insulin response of an islet.

$$r_{islet} = \frac{0.209}{1+\exp\left(-3.33 G_B +6.6\right)} \tag{10.7}$$

In this equation, the islet insulin release rate is in units of μU islet^{-1} min^{-1}, and the glucose concentration is in mg ml^{-1}.

10.6.4 Pharmacokinetic Modeling of Glucose and Insulin Interactions

Although device testing in experimental diabetic animals provides conclusive proof of a device's efficacy and potential, mathematical models are also useful tools for investigating the myriad parameters that affect device performance. Mathematical

---

[8]Gastrointestinal hormones include gastrin, secretin, cholecystokinin, and gastric inhibitory peptide.

models also provide the opportunity to perform scaleup studies. For example, they allow for critical examination of the results obtained in small laboratory animals and extrapolation of these results to understand how the device may perform in humans.

Several physiological pharmacokinetic models of glucose and insulin metabolism have been described (Guyton *et al.* 1978; Sorensen *et al.* 1982; Berger and Rodbard 1989; Sturis *et al.* 1991, 1995). These models can be combined with an islet glucose–insulin response model, and a device model, to assess quantitatively the level of glucose control that may be achievable (Smith *et al.* 1991).

The pharmacokinetic model for glucose and insulin metabolism proposed by Sturis *et al.* (1991, 1995) is particularly attractive because of its relative simplicity and its ability to represent the experimentally observed temporal oscillations of insulin and glucose levels (Kraegen *et al.* 1972). Secretion of insulin in humans has been found to exhibit two distinct types of periodic oscillations. A rapid oscillation with a period of 10–15 minutes, and a longer or ultradian oscillation, with a period of 100–150 minutes. The rapid oscillations are of small amplitude (insulin < 1–2 μU/ml and glucose < 1 mg/dl) and may be the result of an intrinsic pacemaker in the islets. The ultradian oscillations exhibit a much larger amplitude and are self-sustained when the stimulus is continuously presented, for example by a constant infusion of glucose or by continuous enteral glucose feeding. On the other hand, the ultradian oscillations are damped when the glucose stimulus is presented as a discrete event, for example by either a meal or ingestion of a fixed quantity of glucose.

Figure 10.18 presents a block diagram of the Sturis *et al.* (1991) model for describing the interactions between glucose and insulin. This model is based on four negative feedback loops involving interactions between glucose and insulin; 1) elevated glucose levels stimulates the secretion of insulin and the resulting increase in insulin levels decreases the endogenous production of glucose, which then has the effect of lowering glucose levels; 2) elevated glucose levels stimulates the secretion of insulin, the resulting increase in insulin levels enhances glucose utilization, which also has the

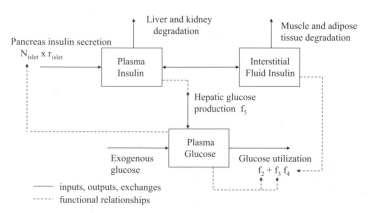

Figure 10.18   Model of glucose and insulin interactions in the body (after Sturis *et al.* 1991).

effect of reducing glucose levels; 3) rising levels of glucose inhibits production of glucose; and 4) increasing levels of glucose stimulates its utilization. These four loops control the amount of glucose and insulin in the body. However, because of the non-linear and dynamic interaction between glucose and insulin, the amounts of these substances in the body is never at a stable equilibrium.

The model also includes two time delays which are important for describing the observed oscillatory dynamics. The first delay affects the suppression of glucose production by insulin or the recovery of this process when insulin levels decrease. The second delay considers the fact that the biological action of insulin correlates better with the concentration of insulin in an intersititial compartment that equilibrates slowly with the plasma insulin concentration.

The three main variables in the model are the concentration of glucose in the plasma (G_B, mg ml^{-1}), the concentration of insulin in the plasma (I_B, μU ml^{-1}), and the concentration of insulin in the interstitial fluid (I_{IF}, μU ml^{-1}). There are also three additional variables that are introduced to account for the delay between the plasma insulin level and its effect on glucose production (x_1, x_2, x_3 with time lag τ_d).

A total of six differential equations are needed to describe the state of the system as a function of time (here in minutes). These equations are as follows:

$$V_{Bi}\frac{dI_B}{dt} = r_{islet}\left(G_B\right)\times 10^6 - E\left(I_B - I_{IF}\right) - \frac{I_B V_{Bi}}{\tau_p}$$

$$\frac{dI_{IF}}{dt} = E\left(\frac{I_B}{V_{IF}} - \frac{I_{IF}}{V_{IF}}\right) - \frac{I_{IF}}{\tau_i}$$

$$V_{Bg}\frac{dG_B}{dt} = r_{Gin}\left(t\right) - f_2\left(G_B\right) - f_3\left(G_B\right)f_4\left(I_{IF}\right) + f_5\left(x_3\right)$$

$$\frac{dx_1}{dt} = 3\frac{I_B V_{Bi} - x_1}{\tau_d}, \quad \frac{dx_2}{dt} = 3\frac{x_1 - x_2}{\tau_d}, \quad \frac{dx_3}{dt} = 3\frac{x_2 - x_3}{\tau_d}$$

(10.8)

The insulin secretion rate of an islet, r_{islet}, was given earlier by Equation 10.7 and is multiplied by one million to account for all of the islets in the pancreas.

The insulin that is released distributes in the plasma space of volume given by V_{Bi}, or it enters the intersitial fluid space represented by V_{IF}. The insulin transport rate into the interstitial fluid is proportional to the difference in the insulin concentration in the two compartments and is described by a rate constant E. Insulin is degraded within the plasma space by a first order rate process with time constant given by τ_p. Interstitial insulin is also degraded with a time constant of τ_i and enhances glucose utilization as described by the function $f_4(I_{IF})$.

Glucose is assumed to distribute throughout a single compartment (V_{Bg}) and affects its own utilization through the functions f_2 and f_3. Glucose utilization represented by f_2 is insulin independent, whereas f_3 is multiplied by an additional term f_4 that is dependent on the interstitial insulin concentration. The functions f_2 (mg/min) and $f_3 \times f_4$ (mg/min) are given by the following equations.

$$f_2 = 72\left[1 - \exp\left(-6.94\,G_B\right)\right]$$

$$f_3 \times f_4 = 10\,G_B \times \left[\frac{90}{1 + \exp\left(-1.772\ln\left(I_{IF}\left(1 + \dfrac{V_{IF}}{E\,\tau_i}\right)\right) + 7.76\right)} + 4\right] \tag{10.9}$$

Insulin also inhibits production of glucose by the liver (f_5) via a process that is dependent on a time delay (τ_d) between the appearance of insulin in the plasma and its inhibitory effect on glucose production. The function f_5 (mg/min) is given by the next equation and is dependent on a time delayed plasma insulin concentration represented by x_3.

$$f_5 = \frac{180}{1 + \exp\left(\dfrac{0.29\,x_3}{V_{Bi}} - 7.5\right)} \tag{10.10}$$

Glucose can enter the body in an arbitrary time varying fashion as given by the function $r_{G\,in}(t)$ (mg min^{-1}). For example, the glucose input function for a meal or an oral glucose tolerance test (OGTT) can be described by the following equation that describes the absorption of glucose from the gastrointestinal tract. In this equation, D (mg) represents the total amount of glucose ingested, k_a (1 min^{-1}) is the absorption rate constant, and k_e (1 min^{-1}) is the elimination rate constant.

$$r_{G\,in}(t) = D\,k_e\left(\frac{k_a}{k_a - k_e}\right)\left(e^{-k_e t} - e^{-k_a t}\right) \tag{10.11}$$

Table 10.1 summarizes the values of all the parameters used in the model of Sturis *et al.* (1991). For a given glucose input function, the above equations can be solved numerically to provide the time course of the plasma glucose and insulin levels.

Table 10.1 Parameter values for the Sturis *et al.* model of glucose and insulin interactions

Parameter	Value
E (rate constant for exchange of insulin between plasma and the interstitial fluid compartment)	200 ml min^{-1}
V_{Bi} (insulin plasma distribution volume)	3000 ml
V_{IF} (insulin interstitial fluid distribution volume)	11,000 ml
V_{Bg} (glucose plasma distribution volume)	10,000 ml
τ_p (time constant for plasma insulin degradation)	6 min
τ_i (time constant for interstitial fluid insulin degradation)	100 min
τ_d (time delay between plasma insulin and glucose production)	36 min

Source: Sturis *et al.* (1991 and 1995).

10.6.5 Using the Pharmacokinetic Model to Evaluate the Performance of a Bioartificial Pancreas

We can now use the Sturis *et al.* (1991) pharmacokinetic model outlined above, along with an insulin release model for an islet, to explore the level of glucose control that is possible with a bioartificial pancreas. There are two basic types of tests that can be used to measure the level of glucose control. One stringent measure of glucose control that is frequently used is called the *k-value*. The *k*-value represents the slope of the least squares fit of the logarithm of the plasma glucose concentration plotted as a function of time for the 50 minutes following the intravenous administration of a dose of glucose. This glucose dose is typically 0.5 grams/kg of body weight and is referred to as an intravenous glucose tolerance test or IVGTT.

Following the glucose dose for the IVGTT, plasma glucose levels rapidly rise within a few minutes to levels exceeding 300 mg/dl from the initial fasting level of around 80 mg/dl. The rate of glucose decay following an IVGTT is indicative of the glucose and insulin responsiveness of the patient's pancreas. The *k*-value is expressed in units of %/minute. If the *k*-value following an IVGTT is less than 1%/min, then the patient is assumed to have diabetes.

Normal *k*-values range from about 2–3%/min. A bioartificial pancreas would need to provide a *k*-value greater than about 1.5%/min to be considered effective at controlling blood glucose levels in a patient with diabetes. This somewhat lower than normal *k*-value is acceptable because the IVGTT provides a major glucose challenge that is not normally observed following a meal.

The oral glucose tolerance test (OGTT) involves the oral administration of a beverage containing 50 grams of glucose. Glucose levels greater than 140 mg/dl two hours after administration of the glucose may be indicative of diabetes.

The simulation of blood glucose and insulin levels following an IVGTT or an OGTT using the pharmacokinetic model outlined above is straightforward. Assume that the immunoisolation membrane and the layer of islets or insulin secreting cells is the controlling mass transfer resistance for glucose and insulin transport. As shown in Figure 10.19, two well-mixed compartments are used to define the distribution of glucose and insulin within the device and the rest of the body. The distribution of

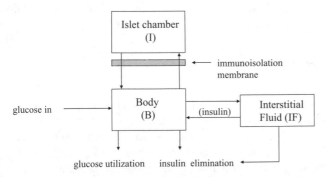

Figure 10.19 Compartmental model for evaluation of a bioartificial pancreas.

insulin also includes an additional interstitial fluid compartment to account for the delay in its action as discussed earlier. Unsteady mass balances for glucose and insulin can now be written for each of the compartments.

islet or cell chamber

$$V_I \frac{dI_I}{dt} = N_{\text{islets}} \, r_{\text{islet}} - P_{mI} S_m \left(I_I - I_B \right)$$

$$V_I \frac{dG_I}{dt} = P_{mG} S_m \left(G_B - G_I \right) \tag{10.12}$$

body distribution volume

$$V_{Bi} \frac{dI_B}{dt} = P_{mI} S_m \left(I_I - I_B \right) - E \left(I_B - I_{\text{IF}} \right) - \frac{I_B V_{Bi}}{\tau_p}$$

$$V_{Bg} \frac{dG_B}{dt} = r_{G\text{in}} (t) - P_{mG} S_m \left(G_B - G_I \right) - f_2 \left(G_B \right) - f_3 \left(G_B \right) f_4 \left(I_{\text{IF}} \right) + f_5 \left(x_3 \right) \tag{10.13}$$

interstitial fluid space for insulin

$$V_{\text{IF}} \frac{dI_{\text{IF}}}{dt} = E \left(I_B - I_{\text{IF}} \right) - \frac{V_{\text{IF}} I_{\text{IF}}}{\tau_i} \tag{10.14}$$

equations for delayed insulin action on glucose production

$$\frac{dx_1}{dt} = 3 \frac{V_{Bi} I_B - x_1}{\tau_d}, \quad \frac{dx_2}{dt} = 3 \frac{x_1 - x_2}{\tau_d}, \quad \frac{dx_3}{dt} = 3 \frac{x_2 - x_3}{\tau_d} \tag{10.15}$$

For the special case of an IVGTT, the value of $r_{G\text{in}}(t)$ would be set equal to zero.

The initial conditions for the solution of the above equations for either an IVGTT or an OGTT are based on fasting levels. Fasting plasma glucose levels are typically about 80 mg/dl. Assuming negligible consumption of glucose by the tissue in the organoid, one may reasonably assume that the glucose concentrations in the organoid regions are the same as that found in the rest of the body. However, since the IVGTT glucose dose is rapidly distributed throughout the body distribution volume, in comparison to the length of time required for its subsequent removal, we assume that at $t = 0$ (+), the plasma glucose level is given by the equation below.

$$G_B \left(t = 0+ \right) = G_B \left(\text{fasting} \right) + \frac{500 \left(\text{mg/kg bw} \right) \times \left(\text{body weight, kg} \right)}{V_{Bg}} \tag{10.16}$$

Fasting plasma insulin levels are about 10 μU/ml. During the fasting period, we assume that the organoid regions are at a steady state equilibrium with the rest of the body as far as insulin is concerned. Accordingly, the islet insulin secretion rate based on the fasting glucose concentration must match the rate of insulin removal from the body. The steady state solution of the set of compartmental insulin equations provides the following relationships for the initial insulin concentration in each of the compartments.

$$I_B^0 = \frac{N_{islets} r_{islet}^0 \left(G_B^0\right) \tau_i}{V_{IF}} \times \frac{1}{\dfrac{V_{Bi} \tau_i}{V_{IF} \tau_p} \dfrac{1}{\dfrac{V_{IF}}{\tau_i E} + 1}}$$

$$I_{IF}^0 = \frac{I_B^0}{\left(\dfrac{1}{\tau_i E} + \dfrac{1}{V_{IF}}\right) V_{IF}}$$

$$I_I^0 = I_B^0 + \frac{N_{islet} r_{islet}^0 \left(G_B^0\right)}{P_{mI} S_m}$$

(10.17)

Solution of the above set of equations provides the temporal change in the plasma glucose concentration following administration of either an OGTT or an IVGTT to a patient with a bioartificial pancreas. The above equations can be easily modified to evaluate the glucose control for a variety of bioartificial pancreas approaches. The following example illustrates a prediction of the plasma glucose levels using a bioartificial pancreas following an OGTT.

Example 10.3 Estimate the glucose concentrations following an OGTT using islets that are macroencapsulated between two immunoisolation membrane disks. A total of 50 grams of glucose is taken orally. The glucose absorption rate constant is 0.042 1 min^{-1} and the glucose elimination rate constant is 0.0083 1 min^{-1}. Base the calculation on a 70 kg patient receiving a total of 750,000 EIN. Assume the islet insulin secretion rate is given by Equation 10.7. The half-thickness of the islet chamber is 75 microns. The void volume within the islet chamber must be at least 65% for sufficient oxygen transport. The immunoisolation membrane permeabilities for glucose and insulin are based on the membrane developed by Baker *et al.* (1997), i.e. 4×10^{-4} cm sec^{-1} for glucose and 8×10^{-5} cm sec^{-1} for insulin.

SOLUTION Figure 10.20 provides the solution for the glucose and insulin concentrations as a function of time. Note the damped ultradian oscillations of the

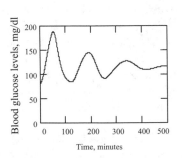
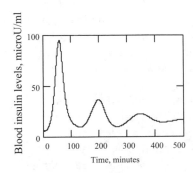

Figure 10.20   Predicted glucose and insulin levels in a bioartificial pancreas following an OGTT.

glucose and insulin concentrations with time. The bioartificial pancreas provides a glucose and insulin response that is very similar to that observed in normal patients (Kraegen *et al.* 1972). We therefore would expect that the bioartificial pancreas has the potential to restore normoglycemia in patients with diabetes. The size of the device is 18 cm (7 inches) in diameter and only several millimeters in thickness. Rather than one large unit, several smaller units could be implanted. For example, four units each 9 cm (3.5 inches) in diameter would be equivalent.

10.7 THE BIOARTIFICIAL LIVER

The liver is a very complex organ that performs a variety of life sustaining functions (Yarmush *et al.* 1992; Galletti and Jauregui 1995). The human liver weighs about 1500 grams and is highly vascularized. From Table 6.2 we see that the liver receives over 25% of the cardiac output. The liver blood supply comes from two major sources. These are the hepatic artery and the portal vein. The portal vein collects the venous drainage from the spleen, the pancreas, and the intestines. The portal vein input from these regions passes through the liver before entering the rest of the systemic circulation. The liver, therefore, has the ability to detoxify potentially harmful substances that are absorbed from the gastrointestinal tract.

The basic functional cellular units of the liver are called the *hepatocytes*. Each hepatocyte is about 25 microns in diameter, and there are close to 250 billion of them in the human liver, accounting for 75% of the liver volume. The hepatocytes are metabolically very active and provide an incredible variety of functions. Their role in carbohydrate metabolism includes the storage of excess glucose as *glycogen* and the release of this stored form of glucose (*glycogenolysis*) when blood glucose levels are low. In addition, the liver converts other sugars such as galactose and fructose to glucose. If blood glucose levels are low, and the glycogen stores are also depleted, then the liver performs a process known as *gluconeogenesis* wherein glucose is synthesized from amino acids.

The liver also plays a major role in fat metabolism. The liver is principally responsible for the body's ability to derive energy from fats. The liver can also convert excess carbohydrates and proteins into fat, synthesize cholesterol and phospholipids, and form the lipoprotein carrier molecules. The lipoproteins are responsible for transporting cholesterol, phospholipids, and fats to other tissues throughout the body. Cholesterol and phospholipids are important components of cellular membranes. The liver has an extremely important role in protein metabolism. These functions include the deamination of amino acids as a prelude to their use as an energy source, or their conversion to carbohydrates or fats. Deamination of amino acids produces large amounts of ammonia. The liver converts the ammonia into urea which is then excreted by the kidneys. Without this function, ammonia levels in the blood rapidly increase leading to hepatic coma and death.

With the exception of the gamma globulins (immunoglobulins or antibodies), practically all of the plasma proteins are made in the liver. The most important being albumin. The liver also makes most of the substances found in the blood that are responsible for the clotting of blood. These include fibrinogen, prothrombin, and most of the other clotting factors. The liver also provides storage for a variety of nutrients such as vitamins and iron, which is stored as a protein complex called *ferritin*. The hepatocytes, because of their high enzyme content, also provide a major role in the detoxification and conjugation of a variety of materials such as drugs, environmental toxins, and hormones that are produced elsewhere in the body. These byproduct materials are water soluble and are excreted in the bile or by the kidneys.

The liver has the unique capability of being able to regenerate itself following tissue damage. This is extremely beneficial considering the role it plays in detoxifying materials from the gastrointestinal tract. However, if the tissue damage rate exceeds the liver's ability to regenerate, then liver failure results, leading to a life threatening situation. Liver failure accounts for over 30,000 deaths each year in the USA.

There are principally two types of liver failure: *cirrhosis* and *fulminant hepatic failure*. Cirrhosis of the liver accounts for over half of the deaths due to liver disease. In cirrhosis, fibrotic tissue forms in place of the damaged liver tissue severely compromising the liver's ability to regenerate. Common causes of cirrhosis includes alcoholism and chronic hepatitis. Fulminant hepatic failure is a rapidly progressing failure of the liver that can lead to death within several weeks of onset. It can be caused by chemical and viral hepatitis.

Liver failure causes a variety of life-threatening abnormalities to include the accumulation in the plasma of ammonia, bilirubin, and decreased levels of albumin and clotting factors. There is also a buildup of toxins and overactivity of the hormonal systems that are believed to lead to a condition known as *hepatic encephalopathy* that can cause irreversible brain damage, coma, and death.

The only long-term and relatively successful treatment method for liver failure is transplantation. However, liver transplantation is severely limited by the shortage of donor organs. Many patients, therefore, die before a liver becomes available, or they succumb shortly after the transplant because of the complications and brain damage associated with liver failure.

10.7.1 Artificial Liver Systems

Treatment of liver failure by artificial means has been the focus of considerable research. Table 10.2 provides a summary of some of the artificial liver systems that have been evaluated (Yarmush *et al.* 1992). The primary goal of these artificial liver systems is to provide a means to maintain the patient in a stable state until a liver transplant is possible. In some cases, artificial liver support could provide the liver with the chance to regenerate, and the patient could return to a normal life without a transplant.

Table 10.2 Artificial liver systems

Hemodialysis	Cross-circulation
Hemoperfusion	Extracorporeal perfusion
Immobilized enzymes	Cross-hemodialysis
Plasma exchange	Hepatocyte hemoperfusion

The difficulty in developing an artificial liver is the multitude of functions the liver performs. Initially, it was thought that the best approach would be to minimize the buildup of toxins. Accordingly, efforts focused on such approaches as hemodialysis, hemoperfusion, and immobilized enzyme reactors for removal of these materials.

The *hemodialysis* systems were similar to those used in kidney dialysis, with the exception that they used membranes with a higher molecular weight cutoff to allow passage of the larger-sized toxic materials. This approach was still not effective at removing large protein bound toxins. *Hemoperfusion* employed beds of activated carbon that adsorbed the toxic molecules. A particular problem with this approach is its non-specificity, removing beneficial substances as well. *Immobilized enzyme reactors* used liver enzymes to provide for more specific removal of the toxic molecules. The major difficulty is providing a complete set of liver enzymes. These three approaches also suffered from the disadvantage that they do not restore any of the synthetic functions of the liver.

Other approaches that were used attempted to both reduce the level of toxins and restore substances normally synthesized by the liver. These methods included plasma exchange, cross-circulation, extracorporeal perfusion, and cross-hemodialysis. *Plasma exchange* involves replacing the patients plasma with donor plasma. Provided the exchange rate is sufficiently high, this approach can replace the lost liver function. However, the major limitation is the large amount of donor plasma needed and the increased risk to the patient of viral infections. *Cross-circulation* involved connecting the patient's circulation to that of another human. In this way, the healthy liver is shared. However, the risk to the human donor limits the wide spread use of this approach. *Extracorporeal perfusion* of the patient's blood through a xenogeneic liver has also been tried. Some success with this approach has been achieved using pig and baboon livers. However, the liver function degrades quickly as a result of the host's immune rejection response. *Cross-hemodialysis* attempts to minimize the host's immune response to the donor liver by perfusing the donor liver with a separate supply of blood. This blood is then sent through a hemodialyzer where it is contacted across a dialysis membrane with the patient's own blood. In this way, toxins are removed and synthetic substances are added to the patient's blood. *Hemoperfusion of liver tissue* has also been tried. Liver tissue pieces are directly contacted with the patient's blood. Frozen or freeze dried dead tissue was comparable in activity to fresh tissue indicating that only those few layers of cells near the surface of the fresh liver pieces are adequately oxygenated. The dead tissue still contained enzymes with some residual activity.

10.7.2 Bioartificial Livers

Hemoperfusion of liver tissue pieces is not capable of providing optimal conditions for the long-term function of the liver tissue. In the absence of an intact vasculature, the mass transfer resistances within the liver pieces are just too large for simple diffusion to overcome. Recently, attention has focused on the development of *bioartificial livers* that are based on the use of isolated hepatocytes (Jauregui *et al.* 1997; Legallais *et al.* 2001; Kobayashi *et al.* 2003). Isolated hepatocytes would not be as severely affected by mass transfer limitations and, furthermore, they can be immunoprotected. Isolated hepatocytes have been proposed for both implantable and extracorporeal systems. Implantable systems are based on either the tissue engineering concepts discussed in Chapter 9, or involve the encapsulation of the hepatocytes in a manner similar to the techniques already discussed for islets (Cai *et al.* 1988; Cima *et al.* 1991a; Johnson *et al.* 1994; Yang *et al.* 1994; Powers *et al.* 2002).

The mass of hepatocytes that is needed is estimated to be on the order of 10% of the original liver mass (Yarmush *et al.* 1992; Rozga *et al.* 1994). This amounts to 25 billion cells or a cellular volume of nearly 100 ml. Attachment to a support structure, or encapsulation of this quantity of cells, while still maintaining a reasonable device volume, is a major obstacle with implantation approaches. For example, for hepatocytes attached to microcarriers[9], the estimated total volume is 500 ml. For micro-encapsulation within a polymeric capsule, this volume is 2500 ml, whereas macro-encapsulation in hollow fibers yields a volume that is estimated to be 1300 ml.

Because of these size constraints, extracorporeal bioartificial livers are an attractive approach for providing temporary liver function until an organ is available for transplantation or the patient's liver can regenerate. The patient's blood or plasma flows through an external circuit that provides contact with the hepatocytes.

The hepatocytes can be presented to the patient's blood or plasma in a number of ways. The simplest is as a suspension of cells. However, this provides no immunoprotection, and like most mammalian cells, hepatocytes do better if they are attached to something. In some cases, the suspension of cells can be placed within a hemodialyzer and separated from the blood or plasma stream by a dialysis membrane. Although this approach provides immunoprotection, the low molecular weight cutoff of the dialysis membrane severely limits the molecular sizes that can be detoxified or synthesized. Other approaches involve perfusing the patient's blood or plasma through columns packed with microcarrier attached hepatocytes, or hepatocytes encapsulated within polymeric beads. Hepatocytes attached to microcarriers, however, provide no immunoprotection. Hollow fiber units have also been proposed. In these systems the hepatocytes are within the shell space and attach themselves to the outer surface of the hollow fibers. Blood or plasma flows through the lumen of the hollow fibers.

One major limitation of the membrane based approaches discussed so far involves the significant mass transfer resistance created by the presence of the membrane. Solute transport through the membrane and into the space containing the hepatocytes is by diffusion only. This not only limits the supply of essential nutrients and oxygen to the hepatocytes, but also limits the detoxification and biosynthesis rates of the hepatocytes as well. Three particularly promising approaches attempt to overcome these mass transfer limitations. All are based on the use of shell and tube hollow fiber modules.

10.7.3 Three Extracorporeal Bioartificial Livers

Figure 10.21 illustrates one approach that provides three distinct compartments (Nyberg *et al.* 1992, 1993 a,b). The first compartment consists of a gel-like collagen

---

[9]A microcarrier is a small particle about 200 microns in diameter, sometimes coated with ECM materials, that provides for cell attachment and formation of a confluent monolayer of cells.

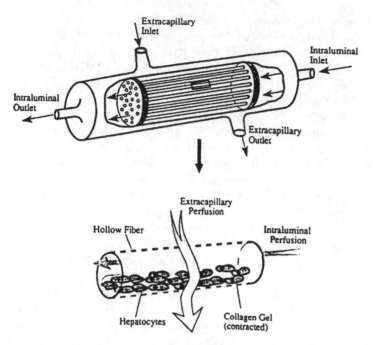

Figure 10.21   A 3 compartment hollow fiber bioartificial liver (from Nyberg *et al.* 1993b, with permission).

matrix that traps the hepatocytes within the hollow fiber lumen. The second compartment consists of the space within the lumen of the hollow fiber that results after contraction of the original collagen suspension containing the hepatocytes. The third compartment is the shell space that surrounds the hollow fibers. The hollow fibers have a nominal molecular weight cutoff of 100,000 Daltons and immunoprotect the hepatocytes from the patient's immune system.

In operation, the patient's blood flows through the shell space. Substances to be detoxified readily diffuse through the hollow fiber membrane and into the collagen gel that contains the hepatocytes. Similarly, substances produced by the hepatocytes, diffuse across the membrane and are carried away by the blood to the patient. Of particular note in this system is the presence of the space within the lumen that is formed after the gel contracts. This space provides an additional flow path that allows simultaneous perfusion of the hepatocytes with specific nutrients, growth factors, and hormones that are needed to maintain their differentiated function and viability. ECM type materials can also be incorporated into the gel that contains the hepatocytes. In this way, the hepatocytes can be presented with an optimal culture environment while performing their liver support functions.

This particular bioartificial liver was evaluated in vitro over a period of seven days. Approximately 5×10^7 cells were loaded into the device. Hepatocyte function was evaluated on the basis of albumin production, oxygen consumption, and lidocaine clearance. It is important to point out that, unlike the bioartificial pancreas, there is no single measure of differentiated hepatocyte function. However, albumin was steadily produced during the seven day period at rates comparable to that seen in static cultures. Furthermore, oxygen consumption remained relatively constant following an initial decline over the first two days of culture. This initial decrease in the oxygen consumption, although possibly being related to cell death, was also considered to be a result of the increased metabolic demands on the hepatocytes as a result of the trauma during their isolation and placement in a new environment.

The cytochrome P-450 system (Hantsen 1998) is a major pathway for the biotransformation in the liver of many substances and is thought to be an important function that must be provided by a bioartificial liver. Lidocaine clearance can be used as a measure of oxidative metabolism provided by the cytochrome P-450 system. Lidocaine clearance in this bioartificial liver was relatively constant, and the presence of lidocaine metabolites demonstrated that lidocaine biotransformation was occurring. Electron micrographs after seven days of operation showed the presence of differentiated viable hepatocytes.

Figure 10.22 illustrates an extracorporeal bioartificial liver system that consists of several integrated components (Rozga *et al.* 1993, 1994; Giorgio *et al.* 1993; Jauregui *et al.* 1997). A plasmapheresis unit is first used to form a plasma stream from arterial blood. This plasma stream feeds into a high flow plasma recirculation loop that forms the core of the bioartificial liver support system. Within the recirculation loop, plasma first enters a column loaded with activated cellulose-coated charcoal. The activated charcoal column is used to enhance the detoxification capability of the overall system and to protect the hepatocytes from any toxic materials found in the patient's plasma. The detoxified plasma then enters a membrane oxygenator. This ensures an adequate supply of oxygen to maintain the viability and function of the hepatocytes. The oxygenated plasma then flows through the hollow fiber module that contains the hepatocytes. The plasma flows through the hollow fiber lumens, and the hepatocytes are

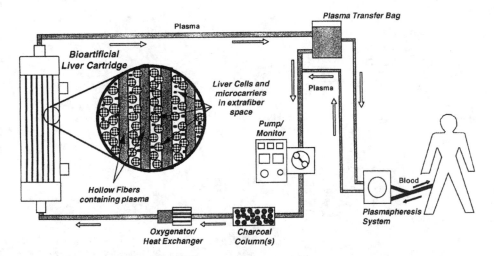

Figure 10.22 HepatAssist bioartificial liver (from Jauregui *et al.* 1997, with permission).

in the shell space that surrounds the hollow fibers. Plasma exiting the hollow fiber cartridge is then recombined with the cellular components of the blood and returned to the patient's body.

The hepatocytes within the bioartificial liver cartridge are attached to collagen coated microcarrier beads that occupy the shell space. About 6 billion porcine hepatocytes are used. The hollow fibers have large pore sizes (0.2 microns) that allow for significant fluid convection through the hollow fiber membrane as a result of the trans-fiber pressure drop. From the fiber entrance to its midpoint, there is a significant flow of plasma out of the fiber and into the shell space that contains the anchored hepatocytes. From the fiber midpoint to the fiber exit, this flow reverses itself and reenters the fiber lumen. This is much like the Starling flow phenomena discussed in Chapter 3 for capillaries. This convective flow into the shell space passes freely between the microcarrier beads that contain the hepatocytes on their surfaces. This convective flow through the shell space provides for very efficient contacting of the plasma solutes and the hepatocytes, and far surpasses what is achievable by diffusion alone.

One limitation of the approach shown in Figure 10.22 is the lack of complete immunoisolation of the xenogeneic hepatocytes. Plasma perfusion eliminates the cellular components of the immune system, however, antibodies and complement will readily flow across the membrane and into the shell space that contains the hepatocytes. The treatment time proposed for this device is relatively short and the patient is immunosuppressed in anticipation of a liver transplant. So the tradeoff between a high transmembrane flow resulting in good mass transport, and the resultant permeability to the humoral components of the immune system, seems appropriate. Certainly, longer term use of this device will require improved methods of immunoisolation that do not compromise the mass transport properties of the hepatocyte bioreactor.

Figure 10.23 illustrates the Extracorporeal Liver Assist Device or ELAD (Sussman *et al.* 1992; Kelly and Sussman 1994). This is a hollow fiber device that uses a cloned

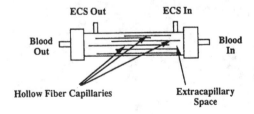

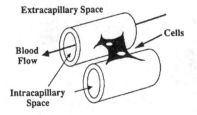

Figure 10.23 The Extracorporeal Liver Assist Device (ELAD) (from Sussman *et al.* 1992, with permission).

human cell line in place of hepatocytes. The cell line is derived from an hepatoblastoma and selected for liver specific functions. These cells exhibit the ability to synthesize protein, urea, and glucose, and are capable of detoxifying substances via the cytochrome P-450 system. The use of a human cell line provides an unlimited supply of cells and minimizes the immune problems that may result from the use of xenogeneic hepatocytes. The ELAD is seeded with 2 grams of cells and is placed in culture until 200 grams of cells are obtained. This quantity of cells is capable of producing therapeutic levels of plasma proteins. For example, 5 grams of albumin are produced each day, which is about one-half the normal daily production rate of a healthy adult human liver.

Blood flows from the patient's vein through a standard hemodialysis pump. The blood exiting the pump is mixed with heparin to prevent blood clots and then passes through the lumen of the hollow fibers of the ELAD device. A portion of the blood flowing through the hollow fibers forms a plasma ultrafiltrate that permeates the shell space where it comes into direct contact with the cells. Ultrafiltration of the blood allows delivery of higher molecular weight solutes at rates significantly higher than that possible by diffusion alone. The ultrafiltrate then passes through a 0.45 micron filter before it is returned to the patient.

Evaluation of this device in dogs with fulminant hepatic failure provided rapid improvement in blood chemistry. There was also a rapid rise in plasma human albumin and α-fetoprotein levels following connection of the device. Factor V dependent clotting times showed a corresponding rapid decrease. After 48 hours of treatment, dogs with fulminant hepatic failure have sufficiently recovered that they can be disconnected from the device. Compared with control animals that die, the treated animals regenerated their own livers.

These results are very promising and demonstrate that the bioartificial liver has the potential to provide a safe and effective means of stabilizing liver failure patients until a liver is available for transplantation.

10.8 THE BIOARTIFICIAL KIDNEY

Another potential area for the application of bioartificial organ technology is in the development of a bioartificial kidney (Ip and Aebischer 1989; Cieslinski and Humes 1994; Humes 1995; Humes 1997). Patients with end stage renal disease can be treated for many years either by hemodialysis or by the newer technique of *continuous ambulatory peritoneal dialysis* (CAPD). However, these are not permanent solutions and these patients ultimately require a kidney transplant in order to survive.

The kidney not only provides a filtration and waste removal function, but also provides several other important functions as well. For example, *erythropoietin* is released by specialized cells found in the kidney in response to hypoxia. Erythropoietin is a major stimulus for the production of red blood cells in the bone marrow. Uremic patients, therefore, suffer from *anemia*. Low blood pressure causes the release of *renin* from the juxtaglomerular cells that are found in the kidney. Renin initiates the formation of *angiotensin II*, a potent vasoconstrictor, that results in an increase in blood pressure. Angiotensin II also acts on the kidneys, decreasing the excretion of both salt and water. This expands the extracellular fluid volume with the result that the blood pressure is increased. The kidneys are also responsible for the conversion of vitamin D into a substance[10] that promotes the absorption of calcium from the intestine. Without this substance, the bones become severely weakened because of the loss of calcium. These other very important functions that are performed by the healthy kidney are, therefore, compromised as a result of kidney failure.

As shown earlier in Figure 7.3, the functional unit of the kidney is called the nephron. Recall that it consists of two major components, the glomerulus and the renal tubule. The glomerulus is responsible primarily for the selective ultrafiltration of waste products from the blood. It must perform this waste removal function and, at the same time, retain essential blood components such as albumin. The glomerular filtrate that is formed then passes through the various segments of the renal tubule. The specialized segments of the renal tubule regulate the amount of urine that is formed and its final solute composition. The renal tubule cells, therefore, have the ability to control both the fluid reabsorption rate and the transport rate of an individual solute. This is accomplished in such a manner so as to maintain homeostasis with regard to the body's fluid volume and overall composition. Because of the chemical sensing and selective transport ability of the renal tubule, it is unlikely that this sophisticated function could ever be reproduced artificially. Accordingly, artificial kidneys will primarily function at the level of simply removing waste products by dialysis.

A bioartificial kidney has the potential of reproducing many of the homeostatic, regulatory, and endocrine functions that are performed by the healthy kidney. This would far surpass the filtration function of existing dialysis systems and allow blood purification to occur in a more physiologic fashion. A bioartificial kidney would consist of two main components, an artificial glomerulus and a bioartificial renal tubule (Ip and Aebischer 1989). The artificial glomerulus can be fabricated from polymeric membranes that have a high hydraulic conductance. The transmembrane pressure gradient can be controlled to provide the desired bulk flow rate of plasma across the membrane. This bulk flow, or ultrafiltration of the plasma, will also provide for significant convective transport of solutes across the membrane. The convective transport of all solutes will be essentially the same up to the molecular weight cutoff of the membrane. This is especially important for the larger size solutes where much more can be removed by convection than by simple diffusion. The artificial glomerular membranes could be configured as hollow fibers or perhaps spiral wound sheets. The major requirement being that the membrane arrangement is conducive to the production of a significant ultrafiltration flow.

Several problems still need to be addressed with regard to the fabrication of an artificial glomerulus. These include uncontrolled bleeding as a result of anticoagulation, a decreased ultrafiltration rate over time as a result of protein deposition in the membrane, and the challenge of developing a non-thrombogenic membrane. One exciting possibility for overcoming the clotting tendency of most polymeric materials is to cover the blood contacting surfaces with a monolayer of autologous endothelial

[10] 1,25-dihydroxycholecalciferol

cells. Endothelial cells form the blood contacting lining of the body's blood vessels and, in the absence of injury, are responsible for preventing the blood from clotting.

The bioartificial renal tubule could consist of viable renal epithelial cells supported within a tubular structure such as a hollow fiber membrane. The hollow fiber membrane, in addition to its role as a structural support, would also provide immunoprotection of these cells. The feasibility of creating a bioartificial tubule was demonstrated by Ip and Aebischer (1989). They grew renal epithelial cell lines to confluence on a permeable membrane. The cells inhibited the passive diffusive transport of solutes and exhibited active transport of specific solutes. Therefore, it appears possible to construct a bioartificial renal tubule that can regulate the transport rates of a variety of solutes. This is an important first step towards the development of a bioartificial kidney.

Another possibility for creating a renal tubule is to use renal tubule stem cells (Cieslinski and Humes 1994; Humes and Cieslinski 1992). *Stem cells* are a class of highly proliferative cells that give rise to cells that have a differentiated function. The resulting differentiated cells provide the ultimate physiologic structure and function found in the body's tissues and organs. Perhaps the best known class of stem cells are the *pluripotential hemopoietic stem cells* (PHSC) (Koller and Palsson 1993) that are responsible for the formation of all the differentiated cells found in blood. The differentiated cells arising from the PHSC consist of the red blood cells, white blood cells, and platelets. In many cases, the terminally differentiated state of the cell is incapable of reproducing. Therefore, the stem cell has the responsibility of producing and replacing those terminally differentiated cells that have become damaged or injured for a variety of reasons. For example, renal tubule stem cells would regenerate renal tubules following their injury. In the presence of the proper growth factors, renal tubule stem cells have the capability to form tubule-like structures in culture. The isolation and culture of human renal tubule stem cells could, therefore, be an important source of tissue for the formation of bioartificial renal tubules.

10.8.1 Two Configurations for a Bioartificial Kidney

Figure 10.24 illustrates two possible arrangements of the artificial glomerulus and the bioartificial tubule. In the first system, blood flows through the luminal spaces of the hollow fibers that comprise the hemofilter and the two tubule sections. A flow control valve, after the hemofilter, regulates the filtration rate of the hemofilter. The hyperosmotic blood then leaves the hemofilter and continues through the proximal and distal tubule modules. The ultrafiltrate generated in the hemofilter flows through the shell side of the two tubule modules. Renal tubule cells are grown on the outer surfaces of the hollow fibers in the tubule modules. Note the cells do not contact the blood and are immunoprotected by the hollow fiber membrane. These cells selectively transport solutes from the shell side to the blood side or vice versa. The hyperosomotic blood that is formed in the hemofilter facilitates the reabsorption of water from the ultrafiltrate.

The second system has the advantage that blood only contacts an artificial membrane in the hemofiltration unit. The blood flows within the hollow fibers of the hemofilter, and the ultrafiltrate that is formed collects within the hemofilter shell space. The ultrafiltrate then enters both the lumens and the shell spaces of the proximal and

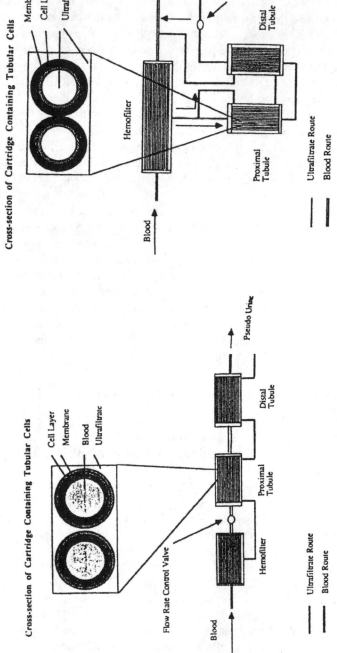

Figure 10.24 The bioartificial kidney (from Ip and Aebischer 1989, with permission).

distal tubule hollow fiber modules. In this case, the renal tubule cells are grown on the luminal side of the hollow fiber membranes. Selective solute transport then occurs between the luminal and shell spaces of the tubule modules. The shell side ultrafiltrate is then returned to the blood exiting the hemofilter. The ultrafiltrate that leaves the luminal side of the hollow fibers within the distal tubule module passes through a flow control valve that regulates the ultrafiltration rate of the hemofilter.

For an implantable device, the blood supply for both configurations could originate from an arteriovenous circuit that uses the common iliac artery and vein. The pseudo urine that is formed could be drained into a ureter for collection and final excretion via the bladder.

PROBLEMS

1. Prepare a paper and presentation that describes a specific application of a bioartificial organ. Include a description of the clinical need, the approach used, results that have been obtained (in vitro, in animals, and the status of any clinical trials), the potential clinical impact, the potential market impact, safety issues, and additional development needs.

2. In Example 10.1, the hindered diffusion model of Bungay and Brenner (1973) was used to estimate the effective pore size of the hollow fiber immunoisolation membrane based on the measured solute permeabilities. This hindered diffusion model provides, at best, a semi-quantitative description of the membrane's solute permeability. Discuss factors that may impact the ability of the hindered diffusion model to describe solute permeability data in immunoisolation membranes.

3. Baker *et al.* (1997) developed an immunoisolation membrane by incorporating a high water content polyvinyl alcohol (PVA) hydrogel into a thin disk-shaped microporous polyethersulfone (PES) filter. The PES filter had a thickness of 150 microns, a porosity of 81%, and a mean pore size of 0.2 microns. The hydrogel formed in the pores of the microporous membrane was found to contain 86% water by weight. The 0.2 micron pores of the microporous membrane should block the cellular components of the immune system, and the PVA hydrogel, through control of the PVA cross-linking, can be tailored to control the passage of high molecular weight solutes. The table below summarizes the physical properties of the solutes that were used and their measured permeability in this composite membrane.

Solute	Molecular weight	Solute radius, nm	$D \times 10^6$, cm^2 sec^{-1}, 23°C	$P_m \times 10^8$, cm sec^{-1}
Glucose	180	0.36	6.38	40,000
Vitamin B$_{12}$	1355	0.75	3.10	16,000
Inulin	5500	1.41	1.65	8000
Lysozyme	14,500	1.92	1.22	80
Myoglobin	16,890	1.9	1.21	70
Chymotrypsinogen	25,000	2.24	1.04	60
Ovalbumin	45,000	2.73	0.851	25
Albumin	69,000	3.55	0.655	4
IgG	160,000	5.35	0.434	4.5
Blue dextran	2,000,000	9.26	0.251	0.5

Source: Baker *et al.* (1997).

What is the effective pore size of the composite membrane based on the hindered diffusion model? An alternative model for solute diffusion through this composite membrane can be based on the methods outlined in Chapter 5 for gels. In this case, the solute moves through a random fibrous network consisting of the crosslinked polymer chains of the PVA hydrogel. Use Equations 5.79 and 5.80 or Equation 5.81 to describe the solute permeablity results.

4. It is estimated that perhaps as many as 500,000 patients with diabetes could benefit from a bioartificial pancreas. Assuming that each patient needs about 1 million islets, and that the islets need to be replaced each year, then about 1000 pancreatic islet isolations need to be performed each day to meet this demand for islets. Prepare a report that describes a process for the massive isolation of islets of Langerhans from the pancreas of large mammals such as the dog or the pig. Be sure to describe any safety issues. The references should first be consulted for an assessment of the current state of islet isolation methods for a single pancreas.

5. Derive an expression for the steady state islet insulin secretion rate based on the Nomura *et al.* (1984) model.

6. Derive Equations 10.5 and 10.6 for the islet insulin release rate following a step change in the glucose concentration.

7. The following perifusion data was obtained by Lakey *et al.* (*Transplantation* 59: 689–694, 1995) for human islets of Langerhans. The islets were challenged with glucose that was initially at 50 mg/100ml followed by a prompt increase of glucose to 500 mg/100ml for 60 minutes. Use the Nomura *et al.* (1984) islet insulin release model to describe this data.

Time, minutes	Islet insulin release rate, μU islet^{-1} min^{-1}
0	0.10
5	0.83
8	0.75
28	0.61
50	0.68
60	0.50

8. Use the Sturis *et al.* (1991, 1995) pharmacokinetic model for glucose and insulin to describe an OGTT for a normal person. How do the normal person glucose and insulin profiles compare to those obtained in Example 10.3 using a bioartificial pancreas?

9. Repeat Example 10.3 assuming the 70 kg patient receives an IVGTT. What is the k-value?

10. Repeat Example 10.3 assuming human islets are used. Use the Nomura *et al.* islet model and the parameters obtained in Problem 7 for human islets.

11. Repeat Example 10.3 assuming the islets are macroencapsulated in the hollow fibers described in Example 10.2. These hollow fiber membranes have a nominal external diameter of 950 microns. What is the entire length of fibers needed?

12. For the device described in Example 10.3, the oxygen permeability of the membrane is estimated to be 0.001 cm sec^{-1}. Based on the method outlined in Example 6.4, show that the maximum volume fraction of the islets is 0.35.

13. Repeat Example 10.3 assuming the half-thickness of the device is 350 microns. The membrane oxygen permeability is estimated to be 0.001 cm sec^{-1}. How are the glucose and insulin profiles affected by a change in device thickness? What volume fraction of islets is allowable based on oxygen transport limitations?

REFERENCES

Abbas, M., and V.P. Tyagi. 1987. Analysis of a hollow fiber artificial kidney performing simultaneous dialysis and ultrafiltration. *Chem. Eng. Sci.* 42:133–142.

Agrawal, C.M., G.G. Niederauer, and K.A. Athanasiou. 1995. Fabrication and characterization of PLA-PGA orthopedic implants. *Tissue Engineering.* 1:241–252.

Alberts, B., D. Bray, J. Lewis, M. Raff, K. Roberts, and J.D. Watson. 1989. *Molecular biology of the cell*, 2nd ed. New York: Garland Publishing, Inc.

Al-Hendy, A., G. Hortelano, G.S. Tannenbaum, and P.L. Chang. 1995. Correction of the growth defect in dwarf mice with nonautologous microencapsulated myoblasts – an alternate approach to somatic gene therapy. *Hum. Gene Ther.* 6:165–175.

Allen, J.W., and S.N. Bhatia. 2003. Formation of steady state oxygen gradients in vitro. *Biotechnol. Bioeng.* 82:253–262.

Altman, J.J., D. Houlbert, P. Callard, P. McMillan, B.A. Solomon, J. Rosen, and P.M. Galletti. 1986. Longterm plasma glucose normalization in experimental diabetic rats with microencapsulated implants of benign human insulinomas. *Diabetes* 35:625–633.

Ameer, G.A., W. Harmon, R. Sasisekharan, and R. Langer. 1999a. Investigation of a whole blood fluidized bed Taylor-Couette flow device for enzymatic heparin neutralization. *Biotech. Bioeng.* 62:602–608.

Ameer, G.A.S., S. Raghavan, R. Sasisekharan, W. Harmon, C.L. Cooney, and R. Langer. 1999b. Regional heparinization via simultaneous separation and reaction in a novel Taylor-Couette flow device. *Biotech. Bioeng.* 63:618–624.

American Diabetes Association. 1993. Position statement. Implications of the diabetes control and complications trial. *Diabetes* 42:1555–1558.

Anderson, J.L., and J.A. Quinn. 1974. Restricted transport in small pores. *Biophys. J.* 14:130–149.

Anderson, J.M. 1994. The extracellular matrix and biomaterials. In *Implantation biology*, ed. R.S. Greco. 113–130. Boca Raton, FL: CRC Press, Inc.

Armer, T.A., and T.R. Hanley. 1986. Characterization of mass transfer in the hollow fiber artificial kidney. *Chem. Eng. Commun.* 47:49–71.

Arnold F., and D.C. West. 1991. Angiogenesis in wound healing. *Pharmacol. Therapy* 52:407–422.

Arpaci V.S. 1966. *Conduction heat transfer*, Addison-Wesley, Reading, Mass.

Atkinson, M.A., and N.K. Maclaren. 1990. What causes diabetes? *Sci. Am.* July:62–71.

Baker, A.R., R.L. Fournier, J.G. Sarver, J.L. Long, P.J. Goldblatt, J.M. Horner, and S.H. Selman. 1997. Evaluation of an immunoisolation membrane formed by incorporating a polyvinyl alcohol hydrogel within a microporous filter support. *Cell Transplant* 6:585–595.

Barbee, J.H., and G.R. Cokelet. 1971a. Prediction of blood flow in tubes with diameters as small as 29μ. *Microvas. Res.* 3:17–21.

Barbee, J.H., and G.R. Cokelet. 1971b. The Fahraeus effect. *Microvascular Res.* 3:6–16.

Bass, L., and S. Keiding. 1988. Physiologically based models and strategic experiments in hepatic pharmacology. *Biochem. Pharmacol.* 37:1425–1431.

Bawa, R., R.A. Siegel, B. Marasca, M. Karel, and R. Langer. 1985. An explanation for the controlled release of macromolecules from polymers. *J. Controlled Release* 1:259–267.

Beck, R.E., and J.S. Schultz. 1970. Hindered diffusion in microporous membranes with known pore geometry. *Science* 170:1302–1305.

Benjamin, E., and S. Leskowitz. 1991. *Immunology: a short course.* New York: Wiley-Liss.

Berger, M., and D. Rodbard. 1989. Computer simulation of plasma and glucose dynamics after subcutaneous insulin injection. *Diabetes Care* 12:725–736.

Bernstein, H., V.C. Yang, and R. Langer. 1987a. Distribution of heparinase covalently immobilized to agarose: experimental and theoretical studies. *Biotechnol. Bioeng.* 30:196–207.

Bernstein, H., V.C. Yang, and R. Langer. 1987b. Immobilized heparinase: in vitro reactor model. *Biotechnol. Bioeng.* 30:239–250.

Bird R.B., W.E. Stewart, and E.N. Lightfoot. 2002. *Transport phenomena.*, 2nd ed. New York: John Wiley & Sons.

Brinkman, H.C. 1947. A calculation of the viscous force exerted by a flowing fluid on a dense swarm of particles. *Appl. Sci. Res.* A1:27–34.

Brown, A.N., B.-S. Kim, E. Alsberg, and D.J. Mooney. 2000. Combining chondrocytes and smooth muscle cells to engineer hybrid soft tissue constructs. *Tissue Eng.* 6: 297–305.

Bungay, P.M. and H. Brenner. 1973. The motion of a closely fitting sphere in a fluid filled tube. *Int. J. Multiph. Flow.* 1:25.

Burton, A.C. 1972. *Physiology and biophysics of the circulation*, 2nd ed. Chicago: Year Book Medical Publishers, Inc.

Cai, Z., Z. Shi, G.M. O'Shea, and A.M. Sun. 1988. Microencapsulated hepatocytes for bioartificial liver support. *Artif. Orgs.* 12:388–393.

Cappello, A., F. Grandi, C. Lamberti, and A. Santoro. 1994. Comparative evaluation of different methods to estimate urea distribution volume and generation rate. *Int. J. Artif. Orgs.* 17:322–330.

Catapano, G., A. Wodetzki, and U. Baurmeister. 1992. Blood flow outside regularly spaced hollowfibers: the future concept of membrane devices? *Int. J. Artif. Orgs.* 15:327–330.

Cavallaro, J.F., P.D. Kemp, and K.H. Kraus. 1994. Collagen fabrics as biomaterials. *Biotechnol. Bioeng.* 43:781–791.

Chang, P.L. 1997. Nonautologous gene therapy with implantable devices. *IEEE Eng. Med. Biol.* September/October:145–150.

Chang, P.L., N. Shen, and A.J. Wescott. 1993. Delivery of recombinant gene products with microencapsulated cells in vivo. *Hum. Gene Ther.* 4:433–440.

Charm, S.E., and G.S. Kurland. 1974. *Blood flow and microcirculation.* New York: John Wiley & Sons.

Cheng, S.-Y., J. Gross, and A. Sambanis. 2004. Hybrid pancreatic tissue substitute consisting of recombinant insulin-secreting cells and glucose responsive material. *Biotechnol. Bioeng.* 87:863–873.

Chick, W.L., J.J. Perna, V. Lauris, D. Low, P.M. Galletti, G. Panol, A.D. Whittemore, A.A. Like, C.K. Colton, and M.J. Lysaght. 1977. Artificial pancreas using living beta cells: effects on glucose homeostasis in diabetic rats. *Science* 197:780–782.

Cieslinski, D.A., and H.D. Humes. 1994. Tissue engineering of a bioartificial kidney. *Biotechnol. Bioeng.* 43:678–681.

Cima, L.G., D.E. Ingber, J.P. Vacanti, and R. Langer. 1991a. Hepatocyte culture on biodegradable polymeric substrates. *Biotechnol. Bioeng.* 38:145–158.

Cima, L.G., J.P. Vacanti, C. Vacanti, D. Ingber, D. Mooney, and R. Langer. 1991b. Tissue engineering by cell transplantation using degradable polymer substrates. *J. Biomech. Eng.* 113:143–151.

Colton, C.K. 1995. Implantable biohybrid artificial organs. *Cell Transplant* 4:415–436.

Colton, C.K., and E.S. Avgoustiniatos. 1991. Bioengineering in development of the hybrid artificial pancreas. *J. Biomech. Eng.* 113:152–170.

Colton, C.K., and E.G. Lowrie. 1981. Hemodialysis: physical principles and technical considerations. In *The kidney*, 2nd ed., Volume II, ed. by B.M. Brenner, and F.C. Rector, Jr., Chapter 47. Philadelphia: WB Saunders Co.

Cooney, D.O. 1976. *Biomedical engineering principles.* New York: Marcel Dekker, Inc.

Curry, F.E., and C.C. Michel. 1980. A fiber matrix model of capillary permeability. *Microvas. Res.* 20:96–99.

Cussler, E.L. 1984. *Diffusion: mass transfer in fluid systems.* Cambridge, UK: Cambridge University Press.

Dee, K.C., D.A. Puleo, and R. Bizios. 2002. *Tissue–biomaterial interactions.* Horsoken, NJ: John Wiley & Sons.

Deen, W.M. 1987. Hindered transport of large molecules in liquid-filled pores. *AIChE J* 33:1409–1425.

Dionne, K.E., C.K. Colton, and M.L. Yarmush. 1989. Effect of oxygen on isolated pancreatic tissue. *ASAIO Trans.* 35:739–741.

Dionne, K.E., C.K. Colton, and M.L. Yarmush. 1991. A microperifusion system with environmental control for studying insulin secretion by pancreatic tissue. *Biotechnol. Prog.* 7:359–368.

Dionne, K.E., B.M. Cain, R.H. Li, W.J. Bell, E.J. Doherty, D.H. Rein, M.J. Lysaght, and F.T. Gentile. 1996. Transport characterization of membranes for immunoisolation. *Biomaterials.* 17:257–266.

Douglas, J.A., and M.V. Sefton. 1990. The permeability of EUDRAGIT RL and HEMA-MMA microcapsules to glucose and inulin. *Biotechnol. Bioeng.* 36:653–664.

Durbin, R.P. 1960. Osmotic flow of water across permeable cellulose membranes. *J. General Physiol.* 44:315–326.

Fogler, H.S. 1992. *Elements of chemical reaction engineering*, 2nd ed. New York: Prentice-Hall, Inc.

Folkman, J. 1985. Tumor angiogenesis. *Can. Res.* 43:175–203.

Folkman, J., and M. Klagsbrun. 1987. Angiogenic factors. *Science* 235:442–447.

Freeman, B. 1995. Osmosis. In *Encyclopedia of applied physics*, Volume 13, ed. G.L. Trigg. Berlin: VCH Publishers.

Friedman, D.W., P.J. Orland, and R.S. Greco. 1994. Biomaterials: an historical perspective. In *Implantation Biology*, Chapter 1, edited by R.S. Greco. Boca Raton, FL: CRC Press, Inc.

Friedman, E.A. 1989. Toward a hybrid bioartificial pancreas. *Diabetes Care* 12:415–419.

Fukui, Y., A. Funakubo, and T. Kawamura. 1994. Development of an intra blood circuit membrane oxygenator. *ASAIO J.* M732–M734.

Fung, Y.C. 1993. *Biomechanics: Mechanical properties of living tissues.* 2nd ed. New York: Springer-Verlag.

Gabriel, J.L., T.F. Miller, M.R. Wolfson, and T.H. Shaffer. 1996. Quantitative structure–activity relationships of perfluorinated hetero-hydrocarbons as a potential respiratory media. *ASAIO J.* 42:968–973.

Gaehtgens, P. 1980. Flow of blood through narrow capillaries: rheological mechanisms determining capillary hematocrit and apparent viscosity. *Biorheology* 17:183–189.

Galban, C.J., and B.R. Locke. 1997. Analysis of cell growth in a polymer scaffold using a moving boundary approach. *Biotechnol. Bioeng.* 56:422–432.

Galletti, P.M., and C.K. Colton. 1995. Artificial lungs and blood–gas exchange devices. In *The biomedical engineering handbook.* ed. J.D. Bronzino, 1879–1997. Boca Raton, FL: CRC Press, Inc.

Galletti, P.M., and H.O. Jauregui. 1995. Liver support systems. In *The Biomedical engineering handbook*, ed. J.D. Bronzino, 1952–1966. Boca Raton, FL: CRC Press, Inc.

Galletti, P.M., C.K. Colton, and M.J. Lysaght. 1995. Artificial kidney. In *The biomedical engineering handbook.* ed. J.D. Bronzino. 1898–1922. Boca Raton, FL: CRC Press, Inc.

Garred, L.J., B. Canaud, and P.C. Farrell. 1983. A simple kinetic model for assessing peritoneal mass transfer in chronic ambulatory peritoneal dialysis. *ASAIO J.* 6:131–137.

Gharapetian, H., N.A. Davies, and A.M. Sun. 1986. Encapsulation of viable cells within polyacrylate membranes. *Biotechnol. Bioeng.* 28:1595–1600.

Gharapetian, H., M. Maleki, G.M. O'Shea, R.C. Carpenter, and A.M. Sun. 1987. Polyacrylate microcapsules for cell encapsulation: effects of copolymer structure on membrane properties. *Biotechnol. Bioeng.* 29:775–779.

Gibaldi, M., and D. Perrier. 1982. *Pharmacokinetics*, 2nd ed. New York: Marcel Dekker, Inc.

Gibbon, J.H. 1954. Application of a mechanical heart and lung apparatus to cardiac surgery. *Minnesota Med.* 37:71.

Giorgio, T.D., A.D. Moscioni, J. Rozga, and A.A. Demetriou. 1993. Mass transfer in a hollow fiber device used as a bioartificial liver. *ASAIO J.* 39:886–892.

Goldstein, A.S., G. Zhu, G.E. Morris, R.K. Meszlenyi, and A.G. Mikos. 1999. Effect of osteoblastic culture conditions on the structure of poly(DL-lactic-co-glycolic acid) foam scaffolds. *Tissue Eng.* 5:421–432.

Gray, D.N. 1981. The status of olefin-SO2 copolymers as biomaterials. In *Biomedical and dental applications of polymers*, ed C.G. Gebelein, and F.F. Koblitz, 21–27. New York: Plenum Publishers.

Gray, D.N. 1984. Polymeric Membranes for Artificial Lungs, ACS Symposium Series No. 256, *Polymer materials and artificial organs*, ed. C.G. Gebelein, 151–161. Washington, DC: American Chemical Society.

Grodsky, G.M. 1972. A threshold distribution hypothesis for packet storage of insulin and its mathematical modeling. *J. Clin. Investig.* 51:2047–2059.

Gu, Y.J., K. Inoue, S. Shinohara, R. Doi, M. Kogire, T. Aung, S. Sumi, M. Imamura, T. Fujisato, S. Maetani, and Y. Ikada. 1994. Xenotransplantation of bioartificial pancreas using a mesh-reinforced polyvinyl alcohol bag. *Cell Transplant.* 3:S19–S21.

Guyton, A.C. 1991. *Textbook of medical physiology*, 8th ed. Philadelphia: W.B. Saunders Co.

Guyton, J.R., R.O. Foster, J.S. Soeldner, M.H. Tan, C.B. Kahn, L. Koncz, and R.E. Gleason. 1978. A model of glucose insulin homeostasis in man that incorporates the heterogeneous fast pool theory of pancreatic insulin release. *Diabetes* 27:1027–1042.

Hachiya, H.L., P.A. Halban, and G.L. King. 1988. Intracellular pathways of insulin transport across vascular endothelial cells. *Am. J. Physiol.* 255 (*Cell Physiol.* 24), C459–C464.

Hammar, H. 1993. Wound healing. *Int. J. Dermatol.* 32:6–15.

Hantsen, P.D. 1998. Understanding drug–drug interactions. *Science Med.* 5:16–25.

Hayashi, H., K. Inoue, T. Aung, T. Tun, G. Yuanjun, W. Wenjing, S. Shinohara, H. Kaji, R. Doi, H. Setoyama, M. Kato, M. Imamura, S. Maetani, N. Morikawa, H. Iwata, Y. Ikada, and J.-I. Miyazaki. 1996. Application of a novel B cell line MIN6 to a mesh reinforced polyvinyl alcohol hydrogel tube and three layer agarose microcapsules: an in vitro study. *Cell Transplant.* 5:S65–S69.

Haynes, R.F. 1960. Physical basis of the dependence of blood viscosity on tube radius. *Am. J. Physiol.* 198:1193.

Hill, R.S., L.A. Martinson, S.K. Young, and R.W. Dudek. 1994. Chapter 12: Macroporous devices for diabetes correction. In *Immunoisolation of pancreatic islets*, edited by R.P. Lanza and W.L. Chick, Chapter 12. Boulder, CO: R.G. Landes Co.

Holder W.D., H.E. Gruber, W.D. Roland, A.L. Moore, C.R. Culberson, A.B. Loebsack, K.J.L. Burg, and D.J. Mooney. 1997. Increased vascularization and heterogeneity of vascular structures occurring in polyglycolide matrices containing aortic endothelial cells implanted in the rat. *Tissue Eng.* 3:149–160.

Horwitz, A.E. 1997. Integrins and health. *Sci. Am.* 276:68–75.

Hubbell, J.A. 1997. Matrix effects. In *Principles of tissue engineering*, ed. R.P. Lanza, R. Langer, and W.L. Chick, 247–262. Boulder, CO: R.G. Landes Co.

Hughes, M., A. Vassilakos, D.W. Andrews, G. Hortelano, J. Belmont, and P.L. Chang. 1994. Construction of a secretable adenosine deaminase for a novel approach to somatic gene therapy. *Hum. Gene Ther.* 5:1445–1454.

Humes, H.D. 1995. Tissue engineering of the kidney. In *The biomedical engineering handbook*, ed. J.D. Bronzino, 1807–1824. Boca Raton, FL: CRC Press, Inc.

Humes, H.D. 1997. Application of cell and gene therapies in the tissue engineering of renal replacement devices. In *Principles of tissue engineering*. ed. R.P. Lanza, R. Langer, and W.L. Chick, 577–589. Boulder, CO: R.G. Landes Co.

Humes, H.D., and D.A Cieslinski. 1992. Interaction between growth factors and retinoic acid in the induction of kidney tubulogenesis. *Exp. Cell Res.* 201:8–15.

Inoue, K., T. Fujisato, Y.J. Gu, K. Burczak, S. Sumi, M. Kogire, T. Tobe, K. Uchida, I. Nakai, S. Maetani, and Y. Ikada. 1992. Experimental hybrid islet transplantation: Application of polyvinyl alcohol membrane for entrapment of islets. *Pancreas* 7:562–568.

Intaglietta, M. 1997. Whitaker Lecture 1996. Microcirculation, biomedical engineering, and artificial blood. *Ann. Biomed. Eng.* 25:593–603.

Intaglietta, M., and R.M. Winslow. 1995. Artificial blood. In *The Biomedical Engineering Handbook*, ed. J.D. Bronzino, 2011–2024. Boca Raton, FL: CRC Press, Inc.

Ip, T.K., and P. Aebischer. 1989. Renal epithelial-cell-controlled solute transport across permeable membranes as the foundation for a bioartificial kidney. *Artificial Organs* 13:58–65.

Iwata, H., N. Morikawa, and Y. Ikada. 1996. Permeability of filters used for immunoisolation. *Tissue Eng.* 2:289–298.

James, K., and J. Kohn. 1996. New biomaterials for tissue engineering. *MRS Bulletin.* 21:22–26.

Jarvik, R.K. 1981. The total artificial heart. *Sci. Am.* 244:74–80.

Jauregui, H.O., C.J.P. Mullon, and B.A. Solomon. 1997. Extracorporeal artificial liver support. In *Principles of tissue engineering.* ed. R.P. Lanza, R. Langer, and W.L. Chick, 463–479. Boulder, CO: R.G. Landes Co.

Johnson, L.B., J. Aiken, D. Mooney, B.L. Schloo, L. Griffith-Cima, R. Langer, and J.P. Vacanti. 1994. The mesentary as a laminated vascular bed for hepatocyte transplantation. *Cell Transplant.* 3:273–281.

Joshi, A., and J. Raje. 2002. Sonicated transdermal drug transport. *J. Controlled Release* 83:13–22.

Jurmann, M.J., S. Demertzis, H.-J. Schaeffers, T. Wahlers, and A. Haverich. 1992. Intravascular oxygenation for advanced respiratory therapy. *ASAIO J.* 38:120–124.

Karoor, S., J. Molina, C.R. Buchmann, C. Colton, J.S. Logan, and L.W. Henderson. 2003. Immunoaffinity removal of xenoreactive antibodies using modified dialysis or microfiltration membranes. *Biotechnol. Bioeng.* 81:134–148.

Kaufmann, P.M., S. Heimrath, B.S. Kim, and D.J. Mooney. 1997. Highly porous polymer matrices as a three-dimensional culture system for hepatocytes. *Cell Transplant.* 6:463–468.

Kedem, O., and A. Katchalsky. 1958. Thermodynamic analysis of the permeability of biological membranes to nonelectrolytes. *Biochem. Biophys. Acta* 27:229.

Kelly, J.H., and N.L. Sussman. 1994. The Hepatix extracorporeal liver assist device in the treatment of fulminant hepatic failure. *ASAIO J.* 40:83–85.

Kobayashi, N., T. Okitsu, S. Nakaji, and N. Tanaka. 2003. Hybrid bioartificial liver: establishing a reversibly immortalized human hepatocyte line and developing a bioartificial liver for practical use. *J. Artif. Organs* 6:236–244.

Kolff, W.J. 1947. *New ways of treating uremia: The artificial kidney, peritoneal lavage, intestinal lavage.* London: J.A. Churchill.

Koller, M.R., and B.O. Palsson. 1993. Tissue engineering: reconstitution of human hematopoiesis ex vivo. *Biotechnol. Bioeng.* 42:909–930.

Koushanpour, E. 1976. *Renal physiology.* Philadelphia: W.B. Saunders Co.

Kraegen, E.W., J.D. Young, E.P. George, and L. Lazarus. 1972. Oscillations in blood glucose and insulin after oral glucose. *Horm. Met. Res.* 4:409–413.

Krogh, A. 1919. The number and distribution of capillaries in muscles with calculations of the oxygen pressure head necessary for supplying the tissue. *J. Physiol.* 52:409–415.

Lacy, P.E. 1995. Treating diabetes with transplanted cells. *Sci. Am.* 273:501–548.

Lacy, P.E., O.D. Hegre, A. Gerasimidi-Vazeou, F.T. Gentile, and K.E. Dionne. 1991. Maintenance of normoglycemia in diabetic mice by subcutaneous xenografts of encapsulated islets. *Science* 254:1782–1784.

Lagerlund, T.D., and P.A. Low. 1993. Mathematical modeling of time dependent oxygen transport in rat peripheral nerve. *Comput. Biol. Med.* 23:29–47.

Lakey, J.R.T., G.L. Warnock, Z. Ao, and R.V. Rajotte. 1996. Bulk cryopreservation of isolated islets of Langerhans. *Cell Transplant.* 5:395–404.

Langer, R., and J.P. Vacanti. 1993. Tissue engineering. *Science* 260:920–926.

Langer, R., J.P. Vacanti, C.A. Vacanti, A. Atala, L.E. Freed, and G. Vunjak-Novakovic. 1995. Tissue engineering: biomedical applications. *Tissue Eng.* 1:151–161.

Lanza, R.P., and W.L. Chick. 1994a. (eds.) *Procurement of pancreatic islets,* volume 1. Boulder CO: R.G. Landes Co.

Lanza, R.P., and W.L. Chick. 1994b. (eds.) *Immunomodulation of pancreatic islets,* volume 2. Boulder CO: R.G. Landes Co.

Lanza, R.P., and W.L. Chick. 1994c. (eds.) *Immunoisolation of pancreatic islets,* volume 3. Boulder CO: R.G. Landes Co.

Lanza, R.P., and W.L. Chick. 1997. Endocrinology: pancreas. In *Principles of tissue engineering,* ed. R.P. Lanza, R. Langer, and W.L. Chick. 405–425. Boulder, CO: R.G. Landes Co.

Lanza, R.P., D.H. Butler, K.M. Borland, J.E. Staruk, D.L. Faustman, B.A. Solomon, T.E. Muller, R.G. Rupp, T. Maki, A.P. Monaco, and W.L. Chick. 1991. Xenotransplantation of canine, bovine, and porcine islets in diabetic rats without immunosuppression. *Proc. Natl. Acad. Sci. USA* 88:11100–11104.

Lanza, R.P., D.H. Butler, K.M. Borland, J.M. Harvey, D.L. Faustman, B.A. Solomon, T.E. Muller, R.G. Rupp,

T. Maki, A.P. Monaco, and W.L. Chick. 1992a. Successful xenotransplantation of a diffusion based bio-hybrid artificial pancreas: a study using canine, bovine, and porcine islets. *Transplant. Proc.* 24:669–671.

Lanza, R.P., K.M. Borland, J.E. Staruk, M.C. Appel, B.A. Solomon, and W.L. Chick. 1992b. Transplantation of encapsulated canine islets into spontaneously diabetic BB/Wor rats without immunosuppression. *Endocrinology* 131:637–642.

Lanza, R.P., K.M. Borland, P. Lodge, M. Carretta, S.J. Sullivan, T.E. Muller, B.A. Solomon, T. Maki, A.P. Monaco, and W.L. Chick. 1992c. Treatment of severely diabetic pancreatectomized dogs using a diffusion-based hybrid pancreas. *Diabetes*, 41:886–889.

Lanza, R.P., A.P. Monaco, S.J. Sullivan, and W.L. Chick. 1992d. *Pancreatic Islet Cell Transplantation*, ed, C Ricordi, Chapter 22. Boulder, CO: R.G. Landes Co.

Lanza, R.P., S.J. Sullivan, and W.L. Chick. 1992e. Islet transplantation with immunoisolation. *Diabetes* 41:1503–1510.

Lanza, R.P., A.M. Beyer, J.E. Staruk, and W.L. Chick. 1993. Biohybrid artificial pancreas. *Transplantation* 56:1067–1072.

Lanza, R.P., A.M. Beyer, and W.L. Chick. 1994. Xenogeneic humoral responses to islets transplanted in biohybrid diffusion chambers. *Transplantation* 57:1371–1375.

Lanza, R.P., W.M. Kuhtreiber, and W.L. Chick. 1995. Encapsulation technologies. *Tissue Eng.* 1:181–196.

Lauffenburger, D.A., and J.J. Linderman. 1993. *Receptors*. New York: Oxford University Press.

Lavin, A., C. Sung, A.L. Klibanov, and R. Langer. 1985. A potential treatment for neonatal jaundice. *Science* 230:543–545.

Leach, J.B., K.A. Bivens, C.W. Patrick, Jr., and C.E. Schmidt. 2003. Photocrosslinked hyaluronic acid hydrogels: natural, biodegradable tissue engineering scaffolds. *Biotech. Bioeng.* 82:579–589.

Lee, C.-J., and Y.-L. Chang. 1988. Theoretical evaluation of the performance characteristics of a combined hemodialysis/hemoperfusion system – I. Single pass flow. *Can. J. Chem. Eng.* 66:263–270.

Lee, C.-J., S-T. Hsu, and S.-C. Hu. 1989. A one compartment single pore model for extracorporeal hemoperfusion. *Comput. Biol. Med.* 19:83–94.

Legallais, C., B. David, and E. Dore. 2001. Bioartificial livers (BAL): current technological aspects and future developments. *J. Membr. Sci.* 181:81–95.

Lepeintre, J., H. Briandet, F. Moussy, D. Chicheportiche, S. Darquy, J. Rouchette, P. Imbaud, J.J. Duron, and G. Reach. 1990. Ex vivo evaluation in normal dogs of insulin released by a bioartificial pancreas containing isolated rat islets of Langerhans. *Artificial Organs*, 14:20–27.

Levick, J.R. 1991. *An introduction to cardiovascular physiology*. Boston, MA: Butterworths & Co.

Lewis, R. 1997. New directions in research on blood substitutes. *Genetic Eng. News*. 17:1.

Li, Z., T. Yipintsoi, and J.B. Bassingthwaighte. 1997. Nonlinear model for capillary-tissue oxygen transport and metabolism. *Ann. Biomed. Eng.* 25:604–619.

Lifson, N., K.G. Kramlinger, R.R. Mayrand, and E.J. Lender. 1980. Blood flow to the rabbit pancreas with special reference to the islets of Langerhans. *Gastroenterology* 79:466–473.

Lightfoot, E.N. 1995. The roles of mass transfer in tissue function. In *The biomedical engineering handbook*, ed. J.D. Bronzino, 1656–1670. Boca Raton, FL: CRC Press, Inc.

Liu, H., F.A. Ofosu, and P.L. Chang. 1993. Expression of human factor IX by microencapsulated recombinant fibroblasts. *Hum. Gene Ther.* 4:291–301.

Long, M.W. 1995. Tissue microenvironments. In *The biomedical engineering handbook*, ed. J.D. Bronzino, 1692–1709. Boca Raton, FL: CRC Press, Inc.

Lu, L., and A.G. Mikos. 1996. The importance of new processing techniques in tissue engineering. *MRS Bulletin.* 21:28–32.

Lum, Z.-P., I.T. Tai, M. Krestow, J. Norton, I. Vacek, and A.M. Sun. 1991. Prolonged reversal of diabetic state in NOD mice by xenografts of microencapsulated rat islets. *Diabetes* 40:1511–1516.

Lysaght, M.J., and P.C. Farrell. 1989. Membrane phenomena and mass transfer kinetics in peritoneal dialysis. *J. Mem. Sci.* 44:5–33.

Lysaght, M.J., and J. Moran. 1995. Peritoneal dialysis equipment. In *The biomedical engineering handbook*. ed. J.D. Bronzino. 1923–1935. Boca Raton, FL: CRC Press, Inc.

Ma, P.X., and J.-W. Choi. 2001. Biodegradable polymer scaffolds with well-defined interconnected spherical pore network. *Tissue Eng.* 7:23–33.

Maki, T., C.S. Ubhi, H. Sanchez-Farpon, S.J. Sullivan, K. Borland, T.E. Muller, B.A. Solomon, W.L. Chick, and A.P. Monaco, 1991. Successful treatment of diabetes with the biohybrid artificial pancreas in dogs. *Transplantation* 51:43–51.

Maki, T., J.P.A. Lodge, M. Carretta, H. Ohzato, K.M. Borland, S.J. Sullivan, J. Staruk, T.E. Muller, B.A. Solomon, W.L. Chick, and A.P. Monaco. 1993. Treatment of severe diabetes mellitus for more than one year using a vascularized hybrid artificial pancreas. *Transplantation* 55:713–718.

Maki, T., I. Otsu, J.J. O'Neil, K. Dunleavy, C.J.P. Mullon, B.A. Solomon, and A.P. Monaco. 1996. Treatment of diabetes by xenogeneic islets without immunosuppression. *Diabetes.* 45:342–347.

Makarewicz, A.J., L.F. Mockros, and R.W. Anderson, 1993. A pumping intravascular lung with active mixing. *ASAIO J.* 39:M466–M469.

Malda, J., J. Rouwkema, D.E. Martens, E.P. le Comte, F.K. Kooy, J. Tramper, C.A. van Blitterswijk, and J. Riesle. 2004. Oxygen gradients in tissue engineered PEGT/PBT cartilaginous constructs: measurement and modeling. *Biotechnol. Bioeng.* 86:9–18.

Mann, G.E., L.H. Smaje, and D.L. Yudilevich. 1979. Permeability of the fenestrated capillaries in the cat submandibular gland to lipid-insoluble molecules. *J. Physiol.* 297:335–354.

Marchant, R.E., and I.-W. Wang. 1994. Physical and chemical aspects of biomaterials used in humans. *Implantation biology*, ed. R.S. Greco, 13–38. Boca Raton, FL: CRC Press, Inc.

Margiotta, M.S., L.D. Benton, R.S. Greco. 1994. Integrins, adhesion molecules, and biomaterials. In *Implantation biology*, ed. R.S. Greco, 149–163. Boca Raton, FL: CRC Press, Inc.

Matthew, H.W., S.O. Salley, W.D. Peterson, and M.D. Klein. 1993. Complex coacervate microcapsules for mammalian cell culture and artificial organ development. *Biotechnol. Prog.* 9:510–519.

Maxwell, J.C. 1873. *A treatise on electricity and magnetism*, volume 1. Oxford, UK: Clarendon Press.

McCabe, W.L., J.C. Smith, and P. Harriott. 1985. *Unit operations of chemical engineering*, 4th ed. New York: McGraw-Hill.

McNabb, M.E., R.V. Ebert, and K. McCusker. 1982. Plasma nicotine levels produced by chewing nicotine gum. *JAMA* 248:865–868.

Merrill, E.W., A.M. Benis, E.R. Gilliland, T.K. Sherwood, and E.W. Salzman. 1965. Pressure flow relations of human blood in hollow fibers at low flow rates. *J. Appl. Physiol.* 20:954–967.

Mikos, A.G., G. Sarakinos, M.D. Lyman, D.E. Ingber, J.P. Vacanti, and R. Langer. 1993. Revascularization of porous biodegradable polymers. *Biotechnol. Bioeng.* 42:716–723.

Mikos, A.G., M.G. Papadaki, S. Kouvroukoglou, S.L. Ishaug, and R.C. Thomson. 1994. Mini-review: islet transplantation to create a bioartificial pancreas. *Biotechnol. Bioeng.* 43:673–677.

Mirhashemi, S., K. Messmer, and M. Intaglietta. 1987. Tissue perfusion during normovolemic hemodilution investigated by a hydraulic model of the cardiovascular system. *Int. J. Microcirc. Clin. Exp.* 6:123–136.

Mitragotri, S. 2003. Modeling skin permeability to hydrophilic and hydrophobic solutes based on four permeation pathways. *J. Control. Release* 86:69–92.

Mooney, D.J., G. Organ, J.P. Vacanti, and R. Langer. 1994. Design and fabrication of biodegradable polymer devices to engineer tubular tissues. *Cell Transplant.* 3:203–210.

Mooney, D.J., S. Park, P.M. Kaufmann, K. Sano, K. McNamara, J.P. Vacanti, and R. Langer. 1995a. Biodegradable sponges for hepatocyte transplantation. *J. Biomed. Mat. Res.* 29:959–965.

Mooney, D.J., C. Breuer, K. McNamara, J.P. Vacanti, and R. Langer. 1995b. Fabricating tubular devices from polymers of lactic and glycolic acid for tissue engineering. *Tissue Eng.* 1:107–117.

Mooney, D.J., P.M. Kaufmann, K. Sano, S.P. Schwendeman, K. Majahod, B. Schloo, J.P. Vacanti, and R. Langer. 1996. Localized delivery of epidermal growth factor improves the survival of transplanted hepatocytes. *Biotech. Bioeng.* 50:422–429.

Mooney, D.J., B.-S. Kim, J.P. Vacanti, R. Langer, and A. Atala. 1997. Tissue engineering: genitourinary system. In *Principles of tissue engineering.* ed. R.P. Lanza, R. Langer, and W.L. Chick, 591–600. Boulder, CO: R.G. Landes Co.

Morgan, J.R., and M.L. Yarmush. 1997. Bioengineered skin substitutes. *Science Med.* 4:6–15.

Morgan, J.R., R.G. Tompkins, and M.L. Yarmush. 1994. Genetic engineering and therapeutics. In *Implantation biology*, ed. R.S. Greco, 387–400. Boca Raton, FL: CRC Press, Inc.

Moussy, F, J. Rouchette, G. Reach, R. Cannon, M.Y. Jaffrin. 1989. In vitro evaluation of a bioartificial pancreas under various hemodynamic conditions. *Artificial Organs* 13:109–115.

Narushima, M., N. Kobayashi, T. Okitsu *et al.* 2005. A human β-cell line for transplantation therapy to control Type 1 diabetes. *Nature Biotech.* 23:1274–1282.

Naughton, G.K., W.R. Tolbert, and T.M. Grillot. 1995. Emerging developments in tissue engineering and cell technology. *Tissue Engineering* 1:211–219.

Niro, R., J.P. Byers, R.L. Fournier, and K. Bachmann. 2003. Application of a convective–dispersion model to predict in vivo hepatic clearance from in vitro measurements using cryopreserved human hepatocytes. *Curr. Drug Metab.* 4:357–369.

Nomura, N., M. Schihiri, R. Kawamori, Y. Yamasaki, N. Iwama, and H. Abe. 1984. A mathematical insulin secretion model and its validation in isolated rat pancreatic islets perifusion. *Comput. Biomed. Res.* 17:570–579.

Notari, R.E. 1987. *Biopharmaceutics and clinical pharmacokinetics*, 4th ed. New York: Marcel Dekker, Inc.

Notkins, A.L. 1979. The causes of diabetes. *Sci. Am.* 241:62–73.

Nugent, L.J., and R.K. Jain. 1984a. Extravascular diffusion in normal and neoplastic tissues. *Canc. Res.* 44:238–244.

Nugent, L.J., and R.K. Jain. 1984b. Pore and fiber-matrix models for diffusive transport in normal and neoplastic tissues. *Microvas. Res.* 28:270–274.

Nyberg, S.L., J.L. Platt, K. Shirabe, W.D. Payne, W.-S. Hu, and F.B. Cerra. 1992. Immunoprotection of xenocytes in a hollow fiber bioartificial liver. *ASAIO J.* 38:M463–M467.

Nyberg, S.L., H.J. Mann, R.P. Remmel, W.-S. Hu, and F.B. Cerra. 1993a. Pharmacokinetic analysis verifies P450 function during in vitro and in vivo application of a bioartificial liver. *ASAIO J.* 39:M252–M256.

Nyberg, S.L., R.A. Shatford, M.V. Peshwa, J.G. White, F.B. Cerra, and W.-S. Hu. 1993b. Evaluation of a hepatocyte entrapment hollow fiber bioreactor: a potential bioartificial liver. *Biotechnol. Bioeng.* 41:194–203.

Oerther, S., H. LeGall, E. Payan, F. Lapicque, N. Presle, P. Hubert, J. Dexheimer, P. Netter, and F. Lapicque. 1999. Hyaluronate-alginate gel as a novel biomaterial: mechanical properties and formation mechanism. *Biotechnol. Bioeng.* 63:206–215.

Ogston, A.G. 1958. The spaces in a uniform random suspension of fibers. *Trans. Faraday Soc.* 54:1754–1757.

Oie, S., and T.N. Tozer. 1979. Effect of altered plasma protein binding on apparent volume of distribution. *J. Pharm. Sci.* 68:1203–1205.

Ojcius, D.M., C.-C. Liu, and J.D. Young. 1998. Pore-forming proteins. *Science Med.* 5:44–53.

Opong, W.S., and A.L. Zydney. 1991. Diffusive and convective protein transport through asymmetric membranes. *AIChE J.* 37:1497–1510.

Owen, D.H., J.J. Peters, and D.F. Katz. 2000. Rheological properties of contraceptive gels. *Contraception* 62:321–326.

Pachence, J.M., and J. Kohn. 1997. Biodegradable polymers for tissue engineering. In *Principles of tissue engineering*, ed. R.P. Lanza, R. Langer, and W.L. Chick, 273–293. Boulder, CO: R.G. Landes Co.

Palsson, B.O., and S. Bhatia. 2004. *Tissue engineering*, Upper Saddle River, NJ: Pearson Prentice Hall.

Panchagnula, R. 1997. Transdermal delivery of drugs. *Ind. J. Pharmacol.* 29:140–156.

Parsons-Wingerter, P., and E.H. Sage. 1997. Regulation of cell behavior by extracellular proteins. In *Principles of tissue engineering*, ed. R.P. Lanza, R. Langer, and W.L. Chick, 111–131. Boulder, CO: R.G. Landes Co.

Pathak, C.P., A.S. Sawhney, and J.A. Hubbell. 1992. Rapid photopolymerization of immunoprotective gels in contact with cells and tissues. *J. Am. Chem. Soc.* 114:8311–8312.

Perez, E.P., E.W. Merrill, D. Miller, and L.G. Cima. 1995. Corneal epithelial wound healing on bilayer composite hydrogels. *Tissue Eng.* 1:263–277.

Petersen, E.F., R.G.S. Spencer, and E.W. McFarland. 2002. Microengineering neocartilage scaffolds. *Biotechnol. Bioeng.* 78:802–805.

Petruzzo, P., L. Pibiri, M.A. DeGiudici, G. Basta, R. Calafiore, A. Falorni, P. Brunetti, and G. Brotzu. 1991. Xenotransplantation of microencapsulated pancreatic islets contained in a vascular prosthesis: preliminary results. *Transplant Int.* 4:200–204.

Poling, B.E., J.M. Prausnitz, and J.P. O'Connell. 2001. *The properties of gases and liquids*, 5th ed, New York, McGraw-Hill.

Potts, R.O., and R.H. Guy. 1992. Predicting skin permeability. *Pharm. Res.* 9:663–669.

Powers, M.J., K. Domansky, M.R. Kaazempur-Mofrad, A. Kalezi, A. Capitano, A. Upadhyaya, P. Kurzawski, K.E. Wack, D.B. Stolz, R. Kamm, and L.G. Griffith. 2002. A microfabricated array bioreactor for perfused 3D liver culture. *Biotechnol. Bioeng.* 78:257–269.

Pratt, A.B, F.E. Weber, H.G. Schmoekel, R. Muller, and J.A. Hubbell. 2004. Synthetic extracellular matrices for in situ tissue engineering. *Biotech. Bioeng.* 86:27–36.

Prausnitz, J.M., and F.H. Shair. 1961. A thermodynamic correlation of gas solubilities. *AIChE J* 7:682.

Prausnitz, J.M, R.N. Lichtenthaler, and E.G. de Azevedo. 1986. *Molecular thermodynamics of fluid-phase equilibria*, 2nd ed, Englewood Ciffs, NJ: Prentice-Hall.

Radisic, M., M. Euloth, L. Yang, R. Langer, L. Freed, and G. Vunjak-Novakovic. 2003. High density seeding of myocyte cells in cardiac tissue engineering. *Biotech. Bioeng.* 82:403–414.

Ramachandran, PA., and RA. Mashelkar. 1980. Lumped parameter model for hemodialyzer with application to simulation of patient-artificial kidney system. *Med. Biol. Eng. Comput.* 18:179–188.

Reid, R.C., J.M. Praunsnitz, and T.K. Sherwood. 1977. *The properties of gases and liquids*, 3rd ed, New York: McGraw-Hill.

Renkin, E.M. 1954. Filtration, diffusion, and molecular sieving through porous cellulose membranes. *J. Gen. Physiol.* 38:225.

Renkin, E.M. 1977. Multiple pathways of capillary permeability. *Circ. Res.* 41:735–743.

Renkin, E.M, and F.E. Curry. 1979. Transport of water and solutes across capillary endothelium. In *Membrane transport in biology*, Volume 4, Chapter 1, ed G. Giebisch, and D.C. Tosteson. New York: Springer-Verlag.

Replogle, R.L., H.J. Meiselman, and E.W. Merrill. 1967. Clinical implications of blood rheology studies. *Circulation* 36:148–160.

Richardson, P.D. 1987. Artificial lungs and oxygenation devices. In *Handbook of bioengineering*, ed. Skalak, R., and S. Chien, Chapter 27. New York: McGraw-Hill Co.

Ricordi, C. 1992. (ed.) *Pancreatic islet cell transplantation.* Boulder, CO: R.G. Landes Co.

Ricordi, C., P.E. Lacy, E.H. Finke, B.J. Olack, and D.W. Scharp. 1988. Automated method for isolation of human pancreatic islets. *Diabetes* 37:413–420.

Ricordi, C., C. Socci, A. Davalli, A. Vertova, P. Baro, I. Sassi, S. Braghi, N. Giuzzi, G. Pozza, and V. DiCarlo. 1990a. Application of the automated method to islet isolation in swine. *Transplant. Proc.* 22:784–785.

Ricordi, C., C. Socci, A. Davalli, C. Staudacher, A. Vertova, P. Baro, M. Freschi, F. Gavazzi, F. Bertuzzi, G. Pozza, and V. DiCarlo. 1990b. Swine islet isolation and transplantation. *Horm. Metabol. Res.* 25:26–30.

Riley, M.R., F.J. Muzzio, H.M. Buettner, and S.C. Reyes. 1994. Monte Carlo calculation of effective diffusivities in two- and three-dimensional heterogeneous materials of variable structure. *Physical Rev. E.* 49:3500–3503.

Riley, M.R., F.J. Muzzio, H.M. Buettner, and S.C. Reyes. 1995a. Diffusion in heterogeneous media: application to immobilized cell systems. *AIChE J.* 41:691–700.

Riley, M.R., H.M. Buettner, F.J. Muzzio, and S.C. Reyes. 1995b. Monte Carlo simulation of diffusion and reaction in two dimensional cell structures. *Biophysical J.* 68:1716–1726.

Riley, M.R., F.J. Muzzio, and H.M. Buettner. 1995c. The effect of structure on diffusion and reaction in immobilized cell systems. *Chem. Eng. Sci.* 50:3357–3367.

Riley, M.R., F.J. Muzzio, H.M. Buettner, and S.C. Reyes. 1996. A simple correlation for predicting effective diffusivities in immobilized cell systems. *Biotech. Bioeng.* 49:223–227.

Robertson, B.C., and A.L. Zydney. 1990a. Hindered protein diffusion in asymmetric ultrafiltration membranes with highly constricted pores. *J. Membr. Sci.* 49:287–303.

Robertson, B.C., and Zydney, A.L. 1990b. Protein adsorption in asymmetric ultrafiltration membranes with highly constricted pores. *J. Col. Int. Sci.* 134:563–575.

Rosen, S.L. 1993. *Fundamental principles of polymeric materials*, 2nd ed. New York: John Wiley & Sons.

Rosenberg, G. 1995. Artificial heart and circulatory assist devices. In *The biomedical engineering handbook*, ed. J.D. Bronzino, 1839–1846. Boac Raton, FL: CRC Press, Inc.

Rozga, J., F. Williams, M.-S. Ro, D.F. Neuzil, T.D. Giorgio, G. Backfisch, A.D. Moscioni, R. Hakim, and A.A. Demetriou. 1993. Development of a bioartificial liver: properties and function of a hollow fiber module inoculated with liver cells. *Hepatology* 17:258–265.

Rozga, J., E. Morsiani, E. LePage, A.D. Moscioni, T. Giorgio, and A.A. Demetriou. 1994. Isolated hepatocytes in a bioartificial liver: a single group view and experience. *Biotechnol. Bioeng.* 43:645–653.

Rubin, E., and J.L. Farber. 1994. *Pathology*, 2nd ed. Philadelphia: J.B. Lippincott Co.

Saltzman, W.M. 1997. Cell interactions with polymers. In *Principles of tissue engineering*, ed. R.P. Lanza, R. Langer, and W.L. Chick, 225–246. Boulder, CO: R.G. Landes Co.

Sandler, S.I. 1989. *Chemical engineering thermodynamics*, 2nd ed. Wiley Series in Chemical Engineering. New York: John Wiley & Sons.

Sarver, J.G. 1994. Development of a tracer technique that utilizes a physiological pharmacokinetic model to quantitatively assess the change in mass transfer rates associated with vascular growth and the application of this technique to the evaluation of a novel bioartificial organoid. PhD dissertation. The University of Toledo.

Sarver, J.G., and R.L. Fournier. 1990. Numerical investigation of a novel spiral wound membrane sandwich design for an implantable bioartificial pancreas. *Comput. Biol. Med.* 20:105–119.

Sarver, J.G., R.L. Fournier, P.J. Goldblatt, T.L. Phares, S.E. Mertz, A.R. Baker, R.J. Mellon, J.M. Horner, and S.H. Selman. 1995. Tracer technique to measure in vivo chemical transport rates within an implantable cell transplantation device. *Cell Transplant.* 4: 201 217.

Scharp, D.W., C.J. Swanson, B.J. Olack, P.P. Latta, O.D. Hegre, E.J. Doherty, F.T. Gentile, K.S. Flavin, M.F. Ansara, and P.E. Lacy. 1994. Protection of encapsulated human islets implanted without immunosuppression in patients with type I or type II diabetes and in nondiabetic control subjects. *Diabetes* 43:1167–1170.

Schlichting, H. 1979. *Boundary Layer Theory*. New York: McGraw-Hill.

Schrezenmeir, J., J. Kirchgessner, L. Gero, L.A. Kunz, J. Beyer, and W. Mueller-Klieser. 1994. Effect of microencapsulation on oxygen distribution in islets organs. *Transplantation* 57:1308–1314.

Scott, M.D., K.L. Murad, F. Koumpouras, M. Talbot, and J.W. Eaton. 1997. Chemical camouflage of antigenic determinants: stealth erythrocytes. *Proc. Natl. Acad. Sci. USA*. 94:7566–7571.

Secomb, T.W., R. Hsu, M.W. Dewhirst, B. Klitzman, and J.F. Gross. 1993. Analysis of oxygen transport to tumor tissue by microvascular networks. *Int. J. Radiation Oncology Biol. Phys.* 25:481–489.

Sefton, M.V., R.M. Dawson, R.L. Broughton, J. Blysniuk, and M.E. Sugamori. 1987. Microencapsulation of mammalian cells in a water insoluble polyacrylate by coextrusion and interfacial precipitation. *Biotechnol. Bioeng.* 29:135–1143.

Selman, S.L., M. Kreimer-Birnbaum, J.E. Klaunig, P.J. Goldblatt, R.W. Keck, and S.L. Britton. 1984. Blood flow in transplantable bladder tumors treated with hematoporphyrin derivative and light. *Cancer Res.* 44:1924–1927.

Shah, N., and A. Mehra. 1996. Modeling of oxygen uptake in perfluorocarbon emulsions. *ASAIO J.* 42:181–189.

Shaoting, W., Z. Fengbao, M. Tengxiang, and G. Hanqing. 1990. Investigation on patient-artificial kidney system using compartment models. *Chem. Eng. Sci.* 45:2943–2948.

Shiroki, R., T. Mohanakumar, and D.W. Scharp. 1995. Analysis of the serological and cellular sensitization induced by encapsulated human islets transplantation in type I and type II diabetes patients. *Cell Tranplant.* 4:535–538.

Shuler, M., and F. Kargi. 2001. *Bioprocess engineering: basic concepts*, 2nd ed. Upper Saddle River, NJ: Prentice Hall.

Smith, B., J.G. Sarver, and R.L. Fournier. 1991. A comparison of islet transplantation and subcutaneous insulin injections for the treatment of diabetes mellitus. *Comput. Biol. Med.* 21:417–427.

Soon-Shiong, P. 1994. Encapsulated islet transplantation: pathway to human clinical trials. In *Immunoisolation of Pancreatic Islets*, ed. R.P. Lanza, and W.L. Chick, Chapter 6, Volume 3. Boulder, CO: R.G. Landes Co.

Soon-Shiong, P., E. Feldman, R. Nelson, R. Heintz, N. Merideth, P. Sandford, T. Zheng, and J. Komtebedde. 1992a. *Trans. Proc.* 24:2946–2947.

Soon-Shiong, P., E. Feldman, R. Nelson, J. Komtebedde, O. Smidsrod, G. Skjak-Braek, T. Espevik, R. Heintz, and M. Lee. 1992b. Successful reversal of spontaneous diabetes in dogs by intraperitoneal microencapsulated islets. *Transplantation* 54:769–774.

Soon-Shiong, P., E. Feldman, R. Nelson, R. Heintz, Q. Yao, O. Smidsrod, and P. Sandford. 1993. Longterm reversal of diabetes by the injection of immunoprotected islets. *Proc. Nat. Acad. Sci. USA* 90:5843–5847.

Sorenson, J.T., C.K. Colton, R.S. Hillman, and J.S. Soeldner. 1982. Use of a pharmacokinetic model of glucose homeostasis for assessment of performance requirements for improved insulin therapies. *Diabetes Care* 5:148–157.

Spotnitz, H.M. 1987. Circulatory assist devices. In *Handbook of Bioengineering*. ed. R. Skalak, and S. Chien.38.1–38.18. McGraw-Hill, Inc.

Staverman, A.J. 1948. Non-Equilibrium thermodynamics of membrane processes. 48:176–185. *Trans. Faraday Soc.* 48:176.

Stryer, L. 1988. *Biochemistry*, New York: W.H. Freeman and Company.

Sturis, J., K.S. Polonsky, E. Mosekilde, and E. Van Cauter. 1991. Computer model for mechanisms underlying ultradian oscillations of insulin and glucose. *Am. J. Physiol.* 260:E801–E809.

Sturis, J., C. Knudsen, N.M. O'Meara, J.S. Thomsen, E. Mosekilde, E. Van Cauter, and K.S. Polonsky. 1995. Phase-locking regions in a forced model of slow insulin and glucose oscillations. In *Dynamical disease: mathematical analysis of human illness*, ed. J. Belair, L. Glass, U. An der Heiden, and J. Milton. Woodbury, NY: AIP Press.

Sugamori, M.E., and M.V. Sefton. 1989. Microencapsulation of pancreatic islets in a water insoluble polyacrylate. *Trans. Am. Soc. Artif. Organs* 35:791–799.

Sullivan, S.J., T. Maki, K.M. Borland, M.D. Mahoney, B.A. Solomon, T.E. Muller, A.P Monaco, and W.L. Chick. 1991. Biohybrid artificial pancreas: longterm implantation studies in diabetic, pancreatectomized dogs. *Science* 252:718–721.

Sun, A.M., W. Parisius, G.M. Healy, I. Vacek, and H.G. Macmorine. 1977. The use, in diabetic rats and monkeys, of artificial capillary units containing cultured islets of Langerhans. *Diabetes* 26:1136–1139.

Sun, Y., X. Ma, D. Zhou, I. Vacek, and A.M. Sun. 1996. Normalization of diabetes in spontaneously diabetic cynomologus monkeys by xenografts of microencapsulated porcine islets without immunosuppression. *J. Clin. Invest.* 98:1417–1422.

Sung, C, A. Lavin, A.M. Klibanov, and R. Langer. 1986. An immobilized enzyme reactor for the detoxification of bilirubin. *Biotechnol. Bioeng.* 28:1531–1539.

Sussman, N.L., M.G. Chong, T. Koussayer, D. He, T.A. Shang, H.H. Whisennand, and J.H. Kelly. 1992. Reversal of fulminant hepatic failure using an extracorporeal liver assist device. *Hepatology* 16:60–65.

Takeda, T., S. Murphy, S. Uyama, G.M. Organ, B.L. Schloo, and J.P. Vacanti. 1995. Hepatocyte transplantation in swine using prevascularized polyvinyl alcohol sponges. *Tissue Eng.* 1:253–262.

Thomas, H.W. 1962. The wall effect in capillary instruments: an improved analysis suitable for application to blood and other particulate suspensions. *Biorheology* 1:41–56.

Thomas, L.C. 1992. *Heat transfer*. New York: Prentice-Hall.

Thompson, J.A., K.D. Anderson, J.M. DiPietro, J.A. Zwiebel, M. Zametta, W.F. Anderson, and T. Maciag. 1988. Site-directed neovessel formation in vivo. *Science* 241:1349–1352.

Thompson, J.A., C.C. Haudenschild, K.D. Anderson, J.M. DiPietro, W.F. Anderson, and T. Maciag. 1989. Heparin-binding growth factor 1 induces the formation of organoid neovascular structures in vivo. *Proc. Nat. Acad. Sci. USA* 86:7928–7932.

Thomson, R.C., M.J. Yaszemski, and A.G. Mikos. 1997. Polymer scaffold processing. In *Principles of tissue engineering*. ed. R.P. Lanza, R. Langer, and W.L. Chick, 263–272. Boulder, CO: R.G. Landes Co.

Tilles, A.W., H. Baskaaran, P. Roy, M.L. Yarmush, and M. Toner. 2001. Effects of oxygenation and flow on the viability and function of rat hepatocytes cocultured in a microchannel flat-plate bioreactor. *Biotechnol. Bioeng.* 74:379–389.

Tong, J., and J.L. Anderson. 1996. Partitioning and diffusion of proteins and linear polymers in polyacrylamide gels. *Biophysical J.* 70:1505–1513.

Tsai, A.G., K.-E. Arfors, and M. Intaglietta. 1990. Analysis of oxygen transport to tissue during extreme hemodilution. *Adv. Exp. Med. Biol.* 277:881–887.

Tun, T., H. Hayashi, T. Aung, Y.-J. Gu, R. Doi, H. Kaji, Y. Echigo, W.-J. Wang, H. Setoyama, M. Imamura, S. Maetani, N. Morikawa, I. Iwata, and Y. Ikada. 1996. A newly developed three layer agarose microcapsule for a promising biohybrid artificial pancreas: rat to mouse xenotransplantation. *Cell Transplant.* 5:S59–S63.

Tyn, M.T., and T.W. Gusek. 1990. Prediction of diffusion coefficients of proteins, *Biotech. Bioeng.* 35:327–338.

Vaslef, S.N., L.F. Mockros, and R.W. Anderson. 1989. Development of an intravascular lung assist device. *ASAIO Trans.* 35:660–664.

Vaslef, S.N., L.F. Mockros, R.W. Anderson, and R.J. Leonard. 1994. Use of a mathematical model to predict oxygen transfer rates in hollow fiber membrane oxygenators. *ASAIO J.* 40:990–996.

Verma, R.K., D. Murali Krishna, and S. Garg. 2002. Formulation aspects in the development of osmotically controlled oral drug delivery systems. *J. Controlled Release* 79:7–27.

Wake, M.C., C.W. Patrick, Jr., and A.G. Mikos. 1994. Pore morphology effects on the fibrovascular tissue growth in porous polymer substrates. *Cell Transplant.* 3:339–343.

Warnock, W.L., and R.V. Rajotte. 1988. Critical mass of purified islets that induce normoglycemia after implantation into dogs. *Diabetes* 37:467–470.

Warnock, W.L., D. Ellis, R.V. Rajotte, I. Dawidson, S. Baekkeskov, and J. Egebjerg. 1988. Studies of the isolation and viability of human islets of Langerhans. *Transplantation* 45:957–963.

Warnock, W.L., D.K. Ellis, M. Cattral, D. Untch, N.M. Kneteman, and R.V. Rajotte. 1989. Viable purified islets of Langerhans from collagenase perfused human pancreas. *Diabetes*, 38:136–139.

Welling, P.G. 1986. *Pharmacokinetics: processes and mathematics*, ACS monograph 185. Washington, DC: American Chemical Society.

Wickramasinghe, S.R., J.D. Garcia, and B. Han. 2002a Mass and momentum transfer in hollow fiber blood oxygenators. *J. Membr. Sci.* 208:247.

Wickramasinghe, S.R., C.M. Kahr, and B. Han. 2002b. Mass transfer in blood oxygenators using blood analogue fluids. *Biotechnol. Prog.* 18:867.

Wickramasinghe, S.R., B. Han, J.D. Garcia, and P. Specht. 2005. Microporous membrane blood oxygenators. *AIChE J.* 51:656–670.

Winslow, R.M. 1997. Blood substitutes. *Science Med.* 4:54–63.

Wright, J.C., S.T. Leonard, C.L. Stevenson, J.C. Beck, G. Chen, R.M. Jao, P.A. Johnson, J. Leonard, and R.J. Skowronski. 2001. An in vivo/in vitro comparison with a leuprolide somotic implant for the treatment of prostate cancer. *J. Control. Release* 75:1–10.

Yang, M.B., J.P. Vacanti, and D.E. Ingber. 1994. Hollow fibers for hepatocyte encapsulation and transplantation: studies of survival and function in rats. *Cell Transplant.* 3:373–385.

Yang, M.-C., and E.L. Cussler. 1986. Designing hollow-fiber contactors. *AIChE J.* 32:1910–1916.

Yarmush, M.L., J.C.Y. Dunn, and R.G. Tompkins. 1992. Assessment of artificial liver support technology. *Cell Transplant.* 1:323–341.

Zalzman, M., S. Gupta, R.K. Giri, I. Berkovich, B.S. Sappal, O. Karnieli, M.A. Zern, N. Fleischer, and S. Efrat. 2003. Reversal of hyperglycemia in mice by using human expandable insulin-producing cells differentiated from fetal liver progenitor cells. *Proc. Natl. Acad. Sci.* 100:7253–7258.

Zekorn, T., U. Siebers, R.G. Bretzel, M. Renardy, H. Planck, P. Zschocke, and K. Federlin. 1990. Protection of islets of Langerhans from interleukin-1 toxicity by artificial membranes. *Transplantation* 50:391–394.

Zelman, A. 1987. The artificial kidney. In *Handbook of bioengineering*, ed. Skalak, R., and S. Chien, Chapter 39. New York: McGraw-Hill Co.

Zielinski, B.A., M.B. Goddard, and M.J. Lysaght. 1997. Immunoisolation. In *Principles of tissue engineering*. ed. R.P. Lanza, R. Langer, and W.L. Chick, 323–332. Boulder, CO: R.G. Landes Co.

Zydney, A.L. 1995. Therapeutic apheresis and blood fractionation. In *The biomedical engineering handbook*, ed. J.D. Bronzino, 1936–1951. Boca Raton: CRC Press, Inc.

Zydney, A.L., and C.K. Colton. 1986. A concentration polarization model for the filtrate flux in cross-flow microfiltration of particulate suspensions. *Chem. Eng. Comm.* 47:1–21.

INDEX